Springer Series in Statistics

Advisors:
P. Bickel, P. Diggle, S. Fienberg, K. Krickeberg,
I. Olkin, N. Wermuth, S. Zeger

Springer
New York
Berlin
Heidelberg
Hong Kong
London
Milan
Paris
Tokyo

Springer Series in Statistics

Andersen/Borgan/Gill/Keiding: Statistical Models Based on Counting Processes.
Atkinson/Riani: Robust Diagnostic Regression Analysis.
Atkinson/Riani/Cerioli: Exploring Multivariate Data with the Forward Search.
Berger: Statistical Decision Theory and Bayesian Analysis, 2nd edition.
Borg/Groenen: Modern Multidimensional Scaling: Theory and Applications.
Brockwell/Davis: Time Series: Theory and Methods, 2nd edition.
Chan/Tong: Chaos: A Statistical Perspective.
Chen/Shao/Ibrahim: Monte Carlo Methods in Bayesian Computation.
David/Edwards: Annotated Readings in the History of Statistics.
Devroye/Lugosi: Combinatorial Methods in Density Estimation.
Efromovich: Nonparametric Curve Estimation: Methods, Theory, and Applications.
Eggermont/LaRiccia: Maximum Penalized Likelihood Estimation, Volume I:
    Density Estimation.
Fahrmeir/Tutz: Multivariate Statistical Modelling Based on Generalized Linear
    Models, 2nd edition.
Fan/Yao: Nonlinear Time Series: Nonparametric and Parametric Methods.
Farebrother: Fitting Linear Relationships: A History of the Calculus of Observations
    1750-1900.
Federer: Statistical Design and Analysis for Intercropping Experiments, Volume I:
    Two Crops.
Federer: Statistical Design and Analysis for Intercropping Experiments, Volume II:
    Three or More Crops.
Ghosh/Ramamoorthi: Bayesian Nonparametrics.
Glaz/Naus/Wallenstein: Scan Statistics.
Good: Permutation Tests: A Practical Guide to Resampling Methods for Testing
    Hypotheses, 2nd edition.
Gouriéroux: ARCH Models and Financial Applications.
Gu: Smoothing Spline ANOVA Models.
Györfi/Kohler/Krzyżak/ Walk: A Distribution-Free Theory of Nonparametric
    Regression.
Haberman: Advanced Statistics, Volume I: Description of Populations.
Hall: The Bootstrap and Edgeworth Expansion.
Härdle: Smoothing Techniques: With Implementation in S.
Harrell: Regression Modeling Strategies: With Applications to Linear Models,
    Logistic Regression, and Survival Analysis.
Hart: Nonparametric Smoothing and Lack-of-Fit Tests.
Hastie/Tibshirani/Friedman: The Elements of Statistical Learning: Data Mining,
    Inference, and Prediction.
Hedayat/Sloane/Stufken: Orthogonal Arrays: Theory and Applications.
Heyde: Quasi-Likelihood and its Application: A General Approach to Optimal
    Parameter Estimation.
Huet/Bouvier/Poursat/Jolivet: Statistical Tools for Nonlinear Regression: A Practical
    Guide with S-PLUS and R Examples, 2nd edition.
Ibrahim/Chen/Sinha: Bayesian Survival Analysis.

(continued after index)

I.T. Jolliffe

Principal Component Analysis

Second Edition

With 28 Illustrations

 Springer

I.T. Jolliffe
Department of Mathematical Sciences
King's College
University of Aberdeen
Aberdeen AB24 3UE
UK
i.jolliffe@maths.abdn.ac.uk

Library of Congress Cataloging-in-Publication Data
Jolliffe, I.T.
    Principal component analysis / I.T. Jolliffe.—2nd ed.
      p. cm. — (Springer series in statistics)
    Includes bibliographical references and index.
    ISBN 0-387-95442-2 (alk. paper)
     1. Principal components analysis. I. Title. II. Series.
    QA278.5 .J65 2002
    519.5′35—dc21   2002019560

ISBN 0-387-95442-2   Printed on acid-free paper.

Printed in the United States of America.

9 8 7 6 5 4 3

www.springer-ny.com

Springer-Verlag New York Berlin Heidelberg
A member of BertelsmannSpringer Science+Business Media GmbH

Preface to the Second Edition

Since the first edition of the book was published, a great deal of new material on principal component analysis (PCA) and related topics has been published, and the time is now ripe for a new edition. Although the size of the book has nearly doubled, there are only two additional chapters. All the chapters in the first edition have been preserved, although two have been renumbered. All have been updated, some extensively. In this updating process I have endeavoured to be as comprehensive as possible. This is reflected in the number of new references, which substantially exceeds those in the first edition. Given the range of areas in which PCA is used, it is certain that I have missed some topics, and my coverage of others will be too brief for the taste of some readers. The choice of which new topics to emphasize is inevitably a personal one, reflecting my own interests and biases. In particular, atmospheric science is a rich source of both applications and methodological developments, but its large contribution to the new material is partly due to my long-standing links with the area, and not because of a lack of interesting developments and examples in other fields. For example, there are large literatures in psychometrics, chemometrics and computer science that are only partially represented. Due to considerations of space, not everything could be included. The main changes are now described.

Chapters 1 to 4 describing the basic theory and providing a set of examples are the least changed. It would have been possible to substitute more recent examples for those of Chapter 4, but as the present ones give nice illustrations of the various aspects of PCA, there was no good reason to do so. One of these examples has been moved to Chapter 1. One extra property (A6) has been added to Chapter 2, with Property A6 in Chapter 3 becoming A7.

Chapter 5 has been extended by further discussion of a number of ordination and scaling methods linked to PCA, in particular varieties of the biplot. Chapter 6 has seen a major expansion. There are two parts of Chapter 6 concerned with deciding how many principal components (PCs) to retain and with using PCA to choose a subset of variables. Both of these topics have been the subject of considerable research in recent years, although a regrettably high proportion of this research confuses PCA with factor analysis, the subject of Chapter 7. Neither Chapter 7 nor 8 have been expanded as much as Chapter 6 or Chapters 9 and 10.

Chapter 9 in the first edition contained three sections describing the use of PCA in conjunction with discriminant analysis, cluster analysis and canonical correlation analysis (CCA). All three sections have been updated, but the greatest expansion is in the third section, where a number of other techniques have been included, which, like CCA, deal with relationships between two groups of variables. As elsewhere in the book, Chapter 9 includes yet other interesting related methods not discussed in detail. In general, the line is drawn between inclusion and exclusion once the link with PCA becomes too tenuous.

Chapter 10 also included three sections in first edition on outlier detection, influence and robustness. All have been the subject of substantial research interest since the first edition; this is reflected in expanded coverage. A fourth section, on other types of stability and sensitivity, has been added. Some of this material has been moved from Section 12.4 of the first edition; other material is new.

The next two chapters are also new and reflect my own research interests more closely than other parts of the book. An important aspect of PCA is interpretation of the components once they have been obtained. This may not be easy, and a number of approaches have been suggested for simplifying PCs to aid interpretation. Chapter 11 discusses these, covering the well-established idea of rotation as well recently developed techniques. These techniques either replace PCA by alternative procedures that give simpler results, or approximate the PCs once they have been obtained. A small amount of this material comes from Section 12.4 of the first edition, but the great majority is new. The chapter also includes a section on physical interpretation of components.

My involvement in the developments described in Chapter 12 is less direct than in Chapter 11, but a substantial part of the chapter describes methodology and applications in atmospheric science and reflects my long-standing interest in that field. In the first edition, Section 11.2 was concerned with 'non-independent and time series data.' This section has been expanded to a full chapter (Chapter 12). There have been major developments in this area, including functional PCA for time series, and various techniques appropriate for data involving spatial and temporal variation, such as (multichannel) singular spectrum analysis, complex PCA, principal oscillation pattern analysis, and extended empirical orthogonal functions (EOFs). Many of these techniques were developed by atmospheric scientists and are little known in many other disciplines.

The last two chapters of the first edition are greatly expanded and become Chapters 13 and 14 in the new edition. There is some transfer of material elsewhere, but also new sections. In Chapter 13 there are three new sections, on size/shape data, on quality control and a final 'odds-and-ends' section, which includes vector, directional and complex data, interval data, species abundance data and large data sets. All other sections have been expanded, that on common principal component analysis and related topics especially so.

The first section of Chapter 14 deals with varieties of non-linear PCA. This section has grown substantially compared to its counterpart (Section 12.2) in the first edition. It includes material on the Gifi system of multivariate analysis, principal curves, and neural networks. Section 14.2 on weights, metrics and centerings combines, and considerably expands, the material of the first and third sections of the old Chapter 12. The content of the old Section 12.4 has been transferred to an earlier part in the book (Chapter 10), but the remaining old sections survive and are updated. The section on non-normal data includes independent component analysis (ICA), and the section on three-mode analysis also discusses techniques for three or more groups of variables. The penultimate section is new and contains material on sweep-out components, extended components, subjective components, goodness-of-fit, and further discussion of neural nets.

The appendix on numerical computation of PCs has been retained and updated, but, the appendix on PCA in computer packages has been dropped from this edition mainly because such material becomes out-of-date very rapidly.

The preface to the first edition noted three general texts on multivariate analysis. Since 1986 a number of excellent multivariate texts have appeared, including Everitt and Dunn (2001), Krzanowski (2000), Krzanowski and Marriott (1994) and Rencher (1995, 1998), to name just a few. Two large specialist texts on principal component analysis have also been published. Jackson (1991) gives a good, comprehensive, coverage of principal component analysis from a somewhat different perspective than the present book, although it, too, is aimed at a general audience of statisticians and users of PCA. The other text, by Preisendorfer and Mobley (1988), concentrates on meteorology and oceanography. Because of this, the notation in Preisendorfer and Mobley differs considerably from that used in mainstream statistical sources. Nevertheless, as we shall see in later chapters, especially Chapter 12, atmospheric science is a field where much development of PCA and related topics has occurred, and Preisendorfer and Mobley's book brings together a great deal of relevant material.

A much shorter book on PCA (Dunteman, 1989), which is targeted at social scientists, has also appeared since 1986. Like the slim volume by Daultrey (1976), written mainly for geographers, it contains little technical material.

The preface to the first edition noted some variations in terminology. Likewise, the notation used in the literature on PCA varies quite widely. Appendix D of Jackson (1991) provides a useful table of notation for some of the main quantities in PCA collected from 34 references (mainly textbooks on multivariate analysis). Where possible, the current book uses notation adopted by a majority of authors where a consensus exists.

To end this Preface, I include a slightly frivolous, but nevertheless interesting, aside on both the increasing popularity of PCA and on its terminology. It was noted in the preface to the first edition that both terms 'principal component analysis' and 'principal components analysis' are widely used. I have always preferred the singular form as it is compatible with 'factor analysis,' 'cluster analysis,' 'canonical correlation analysis' and so on, but had no clear idea whether the singular or plural form was more frequently used. A search for references to the two forms in key words or titles of articles using the *Web of Science* for the six years 1995–2000, revealed that the number of singular to plural occurrences were, respectively, 1017 to 527 in 1995–1996; 1330 to 620 in 1997–1998; and 1634 to 635 in 1999–2000. Thus, there has been nearly a 50 percent increase in citations of PCA in one form or another in that period, but most of that increase has been in the singular form, which now accounts for 72% of occurrences. Happily, it is not necessary to change the title of this book.

I. T. Jolliffe
April, 2002
Aberdeen, U. K.

Preface to the First Edition

Principal component analysis is probably the oldest and best known of the techniques of multivariate analysis. It was first introduced by Pearson (1901), and developed independently by Hotelling (1933). Like many multivariate methods, it was not widely used until the advent of electronic computers, but it is now well entrenched in virtually every statistical computer package.

The central idea of principal component analysis is to reduce the dimensionality of a data set in which there are a large number of interrelated variables, while retaining as much as possible of the variation present in the data set. This reduction is achieved by transforming to a new set of variables, the principal components, which are uncorrelated, and which are ordered so that the first *few* retain most of the variation present in *all* of the original variables. Computation of the principal components reduces to the solution of an eigenvalue-eigenvector problem for a positive-semidefinite symmetric matrix. Thus, the definition and computation of principal components are straightforward but, as will be seen, this apparently simple technique has a wide variety of different applications, as well as a number of different derivations. Any feelings that principal component analysis is a narrow subject should soon be dispelled by the present book; indeed some quite broad topics which are related to principal component analysis receive no more than a brief mention in the final two chapters.

Although the term 'principal component analysis' is in common usage, and is adopted in this book, other terminology may be encountered for the same technique, particularly outside of the statistical literature. For example, the phrase 'empirical orthogonal functions' is common in meteorology, and in other fields the term 'factor analysis' may be used when 'principal component analysis' is meant. References to 'eigenvector analysis ' or 'latent vector analysis' may also camouflage principal component analysis. Finally, some authors refer to principal components analysis rather than principal component analysis. To save space, the abbreviations PCA and PC will be used frequently in the present text.

The book should be useful to readers with a wide variety of backgrounds. Some knowledge of probability and statistics, and of matrix algebra, is necessary, but this knowledge need not be extensive for much of the book. It is expected, however, that most readers will have had some exposure to multivariate analysis in general before specializing to PCA. Many textbooks on multivariate analysis have a chapter or appendix on matrix algebra, e.g. Mardia et al. (1979, Appendix A), Morrison (1976, Chapter 2), Press (1972, Chapter 2), and knowledge of a similar amount of matrix algebra will be useful in the present book.

After an introductory chapter which gives a definition and derivation of PCA, together with a brief historical review, there are three main parts to the book. The first part, comprising Chapters 2 and 3, is mainly theoretical and some small parts of it require rather more knowledge of matrix algebra and vector spaces than is typically given in standard texts on multivariate analysis. However, it is not necessary to read all of these chapters in order to understand the second, and largest, part of the book. Readers who are mainly interested in applications could omit the more theoretical sections, although Sections 2.3, 2.4, 3.3, 3.4 and 3.8 are likely to be valuable to most readers; some knowledge of the singular value decomposition which is discussed in Section 3.5 will also be useful in some of the subsequent chapters.

This second part of the book is concerned with the various applications of PCA, and consists of Chapters 4 to 10 inclusive. Several chapters in this part refer to other statistical techniques, in particular from multivariate analysis. Familiarity with at least the basic ideas of multivariate analysis will therefore be useful, although each technique is explained briefly when it is introduced.

The third part, comprising Chapters 11 and 12, is a mixture of theory and potential applications. A number of extensions, generalizations and uses of PCA in special circumstances are outlined. Many of the topics covered in these chapters are relatively new, or outside the mainstream of statistics and, for several, their practical usefulness has yet to be fully explored. For these reasons they are covered much more briefly than the topics in earlier chapters.

The book is completed by an Appendix which contains two sections. The first section describes some numerical algorithms for finding PCs, and the second section describes the current availability of routines for performing PCA and related analyses in five well-known computer packages.

The coverage of individual chapters is now described in a little more detail. A standard definition and derivation of PCs is given in Chapter 1, but there are a number of alternative definitions and derivations, both geometric and algebraic, which also lead to PCs. In particular the PCs are 'optimal' linear functions of $\mathbf{x}$ with respect to several different criteria, and these various optimality criteria are described in Chapter 2. Also included in Chapter 2 are some other mathematical properties of PCs and a discussion of the use of correlation matrices, as opposed to covariance matrices, to derive PCs.

The derivation in Chapter 1, and all of the material of Chapter 2, is in terms of the *population* properties of a random vector $\mathbf{x}$. In practice, a *sample* of data is available, from which to estimate PCs, and Chapter 3 discusses the properties of PCs derived from a sample. Many of these properties correspond to population properties but some, for example those based on the singular value decomposition, are defined only for samples. A certain amount of distribution theory for sample PCs has been derived, almost exclusively asymptotic, and a summary of some of these results, together with related inference procedures, is also included in Chapter 3. Most of the technical details are, however, omitted. In PCA, only the first few PCs are conventionally deemed to be useful. However, some of the properties in Chapters 2 and 3, and an example in Chapter 3, show the potential usefulness of the last few, as well as the first few, PCs. Further uses of the last few PCs will be encountered in Chapters 6, 8 and 10. A final section of Chapter 3 discusses how PCs can sometimes be (approximately) deduced, without calculation, from the patterns of the covariance or correlation matrix.

Although the purpose of PCA, namely to reduce the number of variables from p to $m(\ll p)$, is simple, the ways in which the PCs can actually be used are quite varied. At the simplest level, if a few uncorrelated variables (the first few PCs) reproduce most of the variation in all of the original variables, and if, further, these variables are interpretable, then the PCs give an alternative, much simpler, description of the data than the original variables. Examples of this use are given in Chapter 4, while subsequent chapters took at more specialized uses of the PCs.

Chapter 5 describes how PCs may be used to look at data graphically, Other graphical representations based on principal coordinate analysis, biplots and correspondence analysis, each of which have connections with PCA, are also discussed.

A common question in PCA is how many PCs are needed to account for 'most' of the variation in the original variables. A large number of rules has been proposed to answer this question, and Chapter 6 describes many of them. When PCA replaces a large set of variables by a much smaller set, the smaller set are new variables (the PCs) rather than a subset of the original variables. However, if a subset of the original variables is preferred, then the PCs can also be used to suggest suitable subsets. How this can be done is also discussed in Chapter 6.

In many texts on multivariate analysis, especially those written by non-statisticians, PCA is treated as though it is part of the factor analysis. Similarly, many computer packages give PCA as one of the options in a factor analysis subroutine. Chapter 7 explains that, although factor analysis and PCA have similar aims, they are, in fact, quite different techniques. There are, however, some ways in which PCA can be used in factor analysis and these are briefly described.

The use of PCA to 'orthogonalize' a regression problem, by replacing a set of highly correlated regressor variables by their PCs, is fairly well known. This technique, and several other related ways of using PCs in regression are discussed in Chapter 8.

Principal component analysis is sometimes used as a preliminary to, or in conjunction with, other statistical techniques, the obvious example being in regression, as described in Chapter 8. Chapter 9 discusses the possible uses of PCA in conjunction with three well-known multivariate techniques, namely discriminant analysis, cluster analysis and canonical correlation analysis.

It has been suggested that PCs, especially the last few, can be useful in the detection of outliers in a data set. This idea is discussed in Chapter 10, together with two different, but related, topics. One of these topics is the robust estimation of PCs when it is suspected that outliers may be present in the data, and the other is the evaluation, using influence functions, of which individual observations have the greatest effect on the PCs.

The last two chapters, 11 and 12, are mostly concerned with modifications or generalizations of PCA. The implications for PCA of special types of data are discussed in Chapter 11, with sections on discrete data, non-independent and time series data, compositional data, data from designed experiments, data with group structure, missing data and goodness-offit statistics. Most of these topics are covered rather briefly, as are a number of possible generalizations and adaptations of PCA which are described in Chapter 12.

Throughout the monograph various other multivariate techniques are introduced. For example, principal coordinate analysis and correspondence analysis appear in Chapter 5, factor analysis in Chapter 7, cluster analysis, discriminant analysis and canonical correlation analysis in Chapter 9, and multivariate analysis of variance in Chapter 11. However, it has not been the intention to give full coverage of multivariate methods or even to cover all those methods which reduce to eigenvalue problems. The various techniques have been introduced only where they are relevant to PCA and its application, and the relatively large number of techniques which have been mentioned is a direct result of the widely varied ways in which PCA can be used.

Throughout the book, a substantial number of examples are given, using data from a wide variety of areas of applications. However, no exercises have been included, since most potential exercises would fall into two narrow categories. One type would ask for proofs or extensions of the theory given, in particular, in Chapters 2, 3 and 12, and would be exercises mainly in algebra rather than statistics. The second type would require PCAs to be performed and interpreted for various data sets. This is certainly a useful type of exercise, but many readers will find it most fruitful to analyse their own data sets. Furthermore, although the numerous examples given in the book should provide some guidance, there may not be a single 'correct' interpretation of a PCA.

I. T. Jolliffe
June, 1986
Kent, U. K.

Acknowledgments

My interest in principal component analysis was initiated, more than 30 years ago, by John Scott, so he is, in one way, responsible for this book being written.

A number of friends and colleagues have commented on earlier drafts of parts of the book, or helped in other ways. I am grateful to Patricia Calder, Chris Folland, Nick Garnham, Tim Hopkins, Byron Jones, Wojtek Krzanowski, Philip North and Barry Vowden for their assistance and encouragement. Particular thanks are due to John Jeffers and Byron Morgan, who each read the entire text of an earlier version of the book, and made many constructive comments which substantially improved the final product. Any remaining errors and omissions are, of course, my responsibility, and I shall be glad to have them brought to my attention.

I have never ceased to be amazed by the patience and efficiency of Mavis Swain, who expertly typed virtually all of the first edition, in its various drafts. I am extremely grateful to her, and also to my wife, Jean, who took over my rôle in the household during the last few hectic weeks of preparation of that edition. Finally, thanks to Anna, Jean and Nils for help with indexing and proof-reading.

Much of the second edition was written during a period of research leave. I am grateful to the University of Aberdeen for granting me this leave and to the host institutions where I spent time during my leave, namely the Bureau of Meteorology Research Centre, Melbourne, the Laboratoire de Statistique et Probabilités, Université Paul Sabatier, Toulouse, and the Departamento de Matemática, Instituto Superior Agronomia, Lisbon, for the use of their facilities. Special thanks are due to my principal hosts at these institutions, Neville Nicholls, Philippe Besse and Jorge Cadima. Discussions with Wasyl Drosdowsky, Antoine de Falguerolles, Henri Caussinus and David Stephenson were helpful in clarifying some of my ideas. Wasyl Drosdowsky, Irene Oliveira and Peter Baines kindly supplied figures, and John Sheehan and John Pulham gave useful advice. Numerous authors sent me copies of their (sometimes unpublished) work, enabling the book to have a broader perspective than it would otherwise have had.

I am grateful to John Kimmel of Springer for encouragement and to four anonymous reviewers for helpful comments.

The last word must again go to my wife Jean, who, as well as demonstrating great patience as the project took unsociable amounts of time, has helped with some the chores associated with indexing and proofreading.

I. T. Jolliffe
April, 2002
Aberdeen, U. K.

Contents

List of Figures

List of Tables

1
Introduction

The central idea of principal component analysis (PCA) is to reduce the dimensionality of a data set consisting of a large number of interrelated variables, while retaining as much as possible of the variation present in the data set. This is achieved by transforming to a new set of variables, the principal components (PCs), which are uncorrelated, and which are ordered so that the first *few* retain most of the variation present in *all* of the original variables.

The present introductory chapter is in two parts. In the first, PCA is defined, and what has become the standard derivation of PCs, in terms of eigenvectors of a covariance matrix, is presented. The second part gives a brief historical review of the development of PCA.

1.1 Definition and Derivation of Principal Components

Suppose that $\mathbf{x}$ is a vector of p random variables, and that the variances of the p random variables and the structure of the covariances or correlations between the p variables are of interest. Unless p is small, or the structure is very simple, it will often not be very helpful to simply look at the p variances and all of the $\frac{1}{2}p(p-1)$ correlations or covariances. An alternative approach is to look for a few ($\ll p$) derived variables that preserve most of the information given by these variances and correlations or covariances.

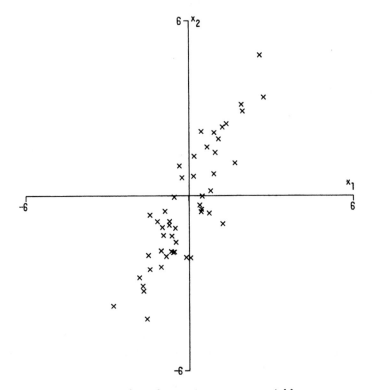

Figure 1.1. Plot of 50 observations on two variables x_1, x_2.

Although PCA does not ignore covariances and correlations, it concentrates on variances. The first step is to look for a linear function $\alpha_1'\mathbf{x}$ of the elements of $\mathbf{x}$ having maximum variance, where α_1 is a vector of p constants $\alpha_{11}, \alpha_{12}, \ldots, \alpha_{1p}$, and $'$ denotes transpose, so that

$$\alpha_1'\mathbf{x} = \alpha_{11}x_1 + \alpha_{12}x_2 + \cdots + \alpha_{1p}x_p = \sum_{j=1}^{p} \alpha_{1j}x_j.$$

Next, look for a linear function $\alpha_2'\mathbf{x}$, uncorrelated with $\alpha_1'\mathbf{x}$ having maximum variance, and so on, so that at the kth stage a linear function $\alpha_k'\mathbf{x}$ is found that has maximum variance subject to being uncorrelated with $\alpha_1'\mathbf{x}, \alpha_2'\mathbf{x}, \ldots, \alpha_{k-1}'\mathbf{x}$. The kth derived variable, $\alpha_k'\mathbf{x}$ is the kth PC. Up to p PCs could be found, but it is hoped, in general, that most of the variation in $\mathbf{x}$ will be accounted for by m PCs, where $m \ll p$. The reduction in complexity achieved by transforming the original variables to PCs will be demonstrated in many examples later in the book, but it will be useful here to consider first the unrealistic, but simple, case where $p = 2$. The advantage of $p = 2$ is, of course, that the data can be plotted exactly in two dimensions.

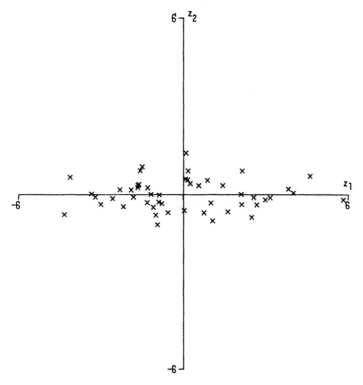

Figure 1.2. Plot of the 50 observations from Figure 1.1 with respect to their PCs z_1, z_2.

Figure 1.1 gives a plot of 50 observations on two highly correlated variables x_1, x_2 . There is considerable variation in both variables, though rather more in the direction of x_2 than x_1. If we transform to PCs z_1, z_2, we obtain the plot given in Figure 1.2.

It is clear that there is greater variation in the direction of z_1 than in either of the original variables, but very little variation in the direction of z_2. More generally, if a set of p (> 2) variables has substantial correlations among them, then the first few PCs will account for most of the variation in the original variables. Conversely, the last few PCs identify directions in which there is very little variation; that is, they identify near-constant linear relationships among the original variables.

As a taster of the many examples to come later in the book, Figure 1.3 provides a plot of the values of the first two principal components in a 7-variable example. The data presented here consist of seven anatomical measurements on 28 students, 11 women and 17 men. This data set and similar ones for other groups of students are discussed in more detail in Sections 4.1 and 5.1. The important thing to note here is that the first two PCs account for 80 percent of the total variation in the data set, so that the

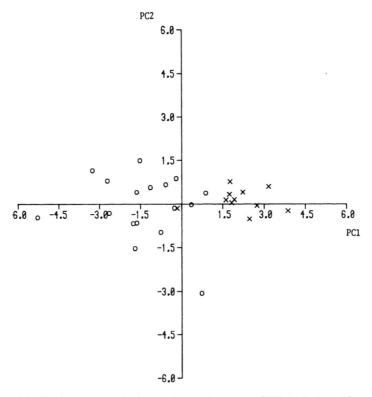

Figure 1.3. Student anatomical measurements: plots of 28 students with respect to their first two PCs. × denotes women; o denotes men.

2-dimensional picture of the data given in Figure 1.3 is a reasonably faithful representation of the positions of the 28 observations in 7-dimensional space. It is also clear from the figure that the first PC, which, as we shall see later, can be interpreted as a measure of the overall size of each student, does a good job of separating the women and men in the sample.

Having defined PCs, we need to know how to find them. Consider, for the moment, the case where the vector of random variables $\mathbf{x}$ has a known covariance matrix Σ. This is the matrix whose (i,j)th element is the (known) covariance between the ith and jth elements of $\mathbf{x}$ when $i \neq j$, and the variance of the jth element of $\mathbf{x}$ when $i = j$. The more realistic case, where Σ is unknown, follows by replacing Σ by a sample covariance matrix $\mathbf{S}$ (see Chapter 3). It turns out that for $k = 1, 2, \cdots, p$, the kth PC is given by $z_k = \boldsymbol{\alpha}_k' \mathbf{x}$ where $\boldsymbol{\alpha}_k$ is an eigenvector of Σ corresponding to its kth largest eigenvalue λ_k. Furthermore, if $\boldsymbol{\alpha}_k$ is chosen to have unit length ($\boldsymbol{\alpha}_k' \boldsymbol{\alpha}_k = 1$), then $\text{var}(z_k) = \lambda_k$, where $\text{var}(z_k)$ denotes the variance of z_k.

The following derivation of these results is the standard one given in many multivariate textbooks; it may be skipped by readers who mainly are interested in the applications of PCA. Such readers could also skip much of Chapters 2 and 3 and concentrate their attention on later chapters, although Sections 2.3, 2.4, 3.3, 3.4, 3.8, and to a lesser extent 3.5, are likely to be of interest to most readers.

To derive the form of the PCs, consider first $\alpha_1'\mathbf{x}$; the vector α_1 maximizes $\text{var}[\alpha_1'\mathbf{x}] = \alpha_1'\Sigma\alpha_1$. It is clear that, as it stands, the maximum will not be achieved for finite α_1 so a normalization constraint must be imposed. The constraint used in the derivation is $\alpha_1'\alpha_1 = 1$, that is, the sum of squares of elements of α_1 equals 1. Other constraints, for example $\text{Max}_j|\alpha_{1j}| = 1$, may more useful in other circumstances, and can easily be substituted later on. However, the use of constraints other than $\alpha_1'\alpha_1 = constant$ in the derivation leads to a more difficult optimization problem, and it will produce a set of derived variables different from the PCs.

To maximize $\alpha_1'\Sigma\alpha_1$ subject to $\alpha_1'\alpha_1 = 1$, the standard approach is to use the technique of Lagrange multipliers. Maximize

$$\alpha_1'\Sigma\alpha_1 - \lambda(\alpha_1'\alpha_1 - 1),$$

where λ is a Lagrange multiplier. Differentiation with respect to α_1 gives

$$\Sigma\alpha_1 - \lambda\alpha_1 = \mathbf{0},$$

or

$$(\Sigma - \lambda\mathbf{I}_p)\alpha_1 = \mathbf{0},$$

where $\mathbf{I}_p$ is the $(p \times p)$ identity matrix. Thus, λ is an eigenvalue of Σ and α_1 is the corresponding eigenvector. To decide which of the p eigenvectors gives $\alpha_1'\mathbf{x}$ with maximum variance, note that the quantity to be maximized is

$$\alpha_1'\Sigma\alpha_1 = \alpha_1'\lambda\alpha_1 = \lambda\alpha_1'\alpha_1 = \lambda,$$

so λ must be as large as possible. Thus, α_1 is the eigenvector corresponding to the largest eigenvalue of Σ, and $\text{var}(\alpha_1'\mathbf{x}) = \alpha_1'\Sigma\alpha_1 = \lambda_1$, the largest eigenvalue.

In general, the kth PC of $\mathbf{x}$ is $\alpha_k'\mathbf{x}$ and $\text{var}(\alpha_k'\mathbf{x}) = \lambda_k$, where λ_k is the kth largest eigenvalue of Σ, and α_k is the corresponding eigenvector. This will now be proved for $k = 2$; the proof for $k \geq 3$ is slightly more complicated, but very similar.

The second PC, $\alpha_2'\mathbf{x}$, maximizes $\alpha_2'\Sigma\alpha_2$ subject to being uncorrelated with $\alpha_1'\mathbf{x}$, or equivalently subject to $\text{cov}[\alpha_1'\mathbf{x}, \alpha_2'\mathbf{x}] = 0$, where $\text{cov}(x, y)$ denotes the covariance between the random variables x and y . But

$$\text{cov}\,[\alpha_1'\mathbf{x}, \alpha_2'\mathbf{x}] = \alpha_1'\Sigma\alpha_2 = \alpha_2'\Sigma\alpha_1 = \alpha_2'\lambda_1\alpha_1' = \lambda_1\alpha_2'\alpha_1 = \lambda_1\alpha_1'\alpha_2.$$

Thus, any of the equations

$$\alpha_1'\Sigma\alpha_2 = 0, \quad \alpha_2'\Sigma\alpha_1 = 0,$$
$$\alpha_1'\alpha_2 = 0, \qquad \alpha_2'\alpha_1 = 0$$

could be used to specify zero correlation between $\alpha_1'\mathbf{x}$ and $\alpha_2'\mathbf{x}$. Choosing the last of these (an arbitrary choice), and noting that a normalization constraint is again necessary, the quantity to be maximized is

$$\alpha_2'\Sigma\alpha_2 - \lambda(\alpha_2'\alpha_2 - 1) - \phi\alpha_2'\alpha_1,$$

where λ, ϕ are Lagrange multipliers. Differentiation with respect to α_2 gives

$$\Sigma\alpha_2 - \lambda\alpha_2 - \phi\alpha_1 = \mathbf{0}$$

and multiplication of this equation on the left by α_1' gives

$$\alpha_1'\Sigma\alpha_2 - \lambda\alpha_1'\alpha_2 - \phi\alpha_1'\alpha_1 = 0,$$

which, since the first two terms are zero and $\alpha_1'\alpha_1 = 1$, reduces to $\phi = 0$. Therefore, $\Sigma\alpha_2 - \lambda\alpha_2 = \mathbf{0}$, or equivalently $(\Sigma - \lambda\mathbf{I}_p)\alpha_2 = \mathbf{0}$, so λ is once more an eigenvalue of Σ, and α_2 the corresponding eigenvector.

Again, $\lambda = \alpha_2'\Sigma\alpha_2$, so λ is to be as large as possible. Assuming that Σ does not have repeated eigenvalues, a complication that is discussed in Section 2.4, λ cannot equal λ_1. If it did, it follows that $\alpha_2 = \alpha_1$, violating the constraint $\alpha_1'\alpha_2 = 0$. Hence λ is the second largest eigenvalue of Σ, and α_2 is the corresponding eigenvector.

As stated above, it can be shown that for the third, fourth, ..., pth PCs, the vectors of coefficients $\alpha_3, \alpha_4, \ldots, \alpha_p$ are the eigenvectors of Σ corresponding to $\lambda_3, \lambda_4, \ldots, \lambda_p$, the third and fourth largest, ..., and the smallest eigenvalue, respectively. Furthermore,

$$\mathrm{var}[\alpha_k'\mathbf{x}] = \lambda_k \qquad \text{for } k = 1, 2, \ldots, p.$$

This derivation of the PC coefficients and variances as eigenvectors and eigenvalues of a covariance matrix is standard, but Flury (1988, Section 2.2) and Diamantaras and Kung (1996, Chapter 3) give alternative derivations that do not involve differentiation.

It should be noted that sometimes the vectors α_k are referred to as 'principal components.' This usage, though sometimes defended (see Dawkins (1990), Kuhfeld (1990) for some discussion), is confusing. It is preferable to reserve the term 'principal components' for the derived variables $\alpha_k'\mathbf{x}$, and refer to α_k as the vector of coefficients or loadings for the kth PC. Some authors distinguish between the terms 'loadings' and 'coefficients,' depending on the normalization constraint used, but they will be used interchangeably in this book.

1.2 A Brief History of Principal Component Analysis

The origins of statistical techniques are often difficult to trace. Preisendorfer and Mobley (1988) note that Beltrami (1873) and Jordan (1874) independently derived the singular value decomposition (SVD) (see Section 3.5) in a form that underlies PCA. Fisher and Mackenzie (1923) used the SVD in the context of a two-way analysis of an agricultural trial. However, it is generally accepted that the earliest descriptions of the technique now known as PCA were given by Pearson (1901) and Hotelling (1933). Hotelling's paper is in two parts. The first, most important, part, together with Pearson's paper, is among the collection of papers edited by Bryant and Atchley (1975).

The two papers adopted different approaches, with the standard algebraic derivation given above being close to that introduced by Hotelling (1933). Pearson (1901), on the other hand, was concerned with finding lines and planes that best fit a set of points in p-dimensional space, and the geometric optimization problems he considered also lead to PCs, as will be explained in Section 3.2

Pearson's comments regarding computation, given over 50 years before the widespread availability of computers, are interesting. He states that his methods 'can be easily applied to numerical problems,' and although he says that the calculations become 'cumbersome' for four or more variables, he suggests that they are still quite feasible.

In the 32 years between Pearson's and Hotelling's papers, very little relevant material seems to have been published, although Rao (1964) indicates that Frisch (1929) adopted a similar approach to that of Pearson. Also, a footnote in Hotelling (1933) suggests that Thurstone (1931) was working along similar lines to Hotelling, but the cited paper, which is also in Bryant and Atchley (1975), is concerned with factor analysis (see Chapter 7), rather than PCA.

Hotelling's approach, too, starts from the ideas of factor analysis but, as will be seen in Chapter 7, PCA, which Hotelling defines, is really rather different in character from factor analysis.

Hotelling's motivation is that there may be a smaller 'fundamental set of independent variables ... which determine the values' of the original p variables. He notes that such variables have been called 'factors' in the psychological literature, but introduces the alternative term 'components' to avoid confusion with other uses of the word 'factor' in mathematics. Hotelling chooses his 'components' so as to maximize their successive contributions to the total of the variances of the original variables, and calls the components that are derived in this way the 'principal components.' The analysis that finds such components is then christened the 'method of principal components.'

Hotelling's derivation of PCs is similar to that given above, using Lagrange multipliers and ending up with an eigenvalue/eigenvector problem, but it differs in three respects. First, he works with a correlation, rather than covariance, matrix (see Section 2.3); second, he looks at the original variables expressed as linear functions of the components rather than components expressed in terms of the original variables; and third, he does not use matrix notation.

After giving the derivation, Hotelling goes on to show how to find the components using the power method (see Appendix A1). He also discusses a different geometric interpretation from that given by Pearson, in terms of ellipsoids of constant probability for multivariate normal distributions (see Section 2.2). A fairly large proportion of his paper, especially the second part, is, however, taken up with material that is not concerned with PCA in its usual form, but rather with factor analysis (see Chapter 7).

A further paper by Hotelling (1936) gave an accelerated version of the power method for finding PCs; in the same year, Girshick (1936) provided some alternative derivations of PCs, and introduced the idea that sample PCs were maximum likelihood estimates of underlying population PCs.

Girshick (1939) investigated the asymptotic sampling distributions of the coefficients and variances of PCs, but there appears to have been only a small amount of work on the development of different applications of PCA during the 25 years immediately following publication of Hotelling's paper. Since then, however, an explosion of new applications and further theoretical developments has occurred. This expansion reflects the general growth of the statistical literature, but as PCA requires considerable computing power, the expansion of its use coincided with the widespread introduction of electronic computers. Despite Pearson's optimistic comments, it is not really feasible to do PCA by hand, unless p is about four or less. But it is precisely for larger values of p that PCA is most useful, so that the full potential of the technique could not be exploited until after the advent of computers.

Before ending this section, four papers will be mentioned; these appeared towards the beginning of the expansion of interest in PCA and have become important references within the subject. The first of these, by Anderson (1963), is the most theoretical of the four. It discussed the asymptotic sampling distributions of the coefficients and variances of the sample PCs, building on the earlier work by Girshick (1939), and has been frequently cited in subsequent theoretical developments.

Rao's (1964) paper is remarkable for the large number of new ideas concerning uses, interpretations and extensions of PCA that it introduced, and which will be cited at numerous points in the book.

Gower (1966) discussed links between PCA and various other statistical techniques, and also provided a number of important geometric insights.

Finally, Jeffers (1967) gave an impetus to the really practical side of the subject by discussing two case studies in which the uses of PCA go beyond that of a simple dimension-reducing tool.

To this list of important papers the book by Preisendorfer and Mobley (1988) should be added. Although it is relatively unknown outside the disciplines of meteorology and oceanography and is not an easy read, it rivals Rao (1964) in its range of novel ideas relating to PCA, some of which have yet to be fully explored. The bulk of the book was written by Preisendorfer over a number of years, but following his untimely death the manuscript was edited and brought to publication by Mobley.

Despite the apparent simplicity of the technique, much research is still being done in the general area of PCA, and it is very widely used. This is clearly illustrated by the fact that the *Web of Science* identifies over 2000 articles published in the two years 1999–2000 that include the phrases 'principal component analysis' or 'principal components analysis' in their titles, abstracts or keywords. The references in this book also demonstrate the wide variety of areas in which PCA has been applied. Books or articles are cited that include applications in agriculture, biology, chemistry, climatology, demography, ecology, economics, food research, genetics, geology, meteorology, oceanography, psychology and quality control, and it would be easy to add further to this list.

2
Mathematical and Statistical Properties of Population Principal Components

In this chapter many of the mathematical and statistical properties of PCs are discussed, based on a known population covariance (or correlation) matrix Σ. Further properties are included in Chapter 3 but in the context of sample, rather than population, PCs. As well as being derived from a statistical viewpoint, PCs can be found using purely mathematical arguments; they are given by an orthogonal linear transformation of a set of variables optimizing a certain algebraic criterion. In fact, the PCs optimize several different algebraic criteria and these optimization properties, together with their statistical implications, are described in the first section of the chapter.

In addition to the algebraic derivation given in Chapter 1, PCs can also be looked at from a geometric viewpoint. The derivation given in the original paper on PCA by Pearson (1901) is geometric but it is relevant to samples, rather than populations, and will therefore be deferred until Section 3.2. However, a number of other properties of population PCs are also geometric in nature and these are discussed in the second section of this chapter.

The first two sections of the chapter concentrate on PCA based on a covariance matrix but the third section describes how a correlation, rather than a covariance, matrix may be used in the derivation of PCs. It also discusses the problems associated with the choice between PCAs based on covariance versus correlation matrices.

In most of this text it is assumed that none of the variances of the PCs are equal; nor are they equal to zero. The final section of this chapter explains briefly what happens in the case where there is equality between some of the variances, or when some of the variances are zero.

Most of the properties described in this chapter have sample counterparts. Some have greater relevance in the sample context, but it is more convenient to introduce them here, rather than in Chapter 3.

2.1 Optimal Algebraic Properties of Population Principal Components and Their Statistical Implications

Consider again the derivation of PCs given in Chapter 1, and denote by **z** the vector whose kth element is z_k, the kth PC, $k = 1, 2, \ldots, p$. (Unless stated otherwise, the kth PC will be taken to mean the PC with the kth largest variance, with corresponding interpretations for the 'kth eigenvalue' and 'kth eigenvector.') Then

$$\mathbf{z} = \mathbf{A}'\mathbf{x}, \tag{2.1.1}$$

where **A** is the orthogonal matrix whose kth column, $\boldsymbol{\alpha}_k$, is the kth eigenvector of $\boldsymbol{\Sigma}$. Thus, the PCs are defined by an orthonormal linear transformation of **x**. Furthermore, we have directly from the derivation in Chapter 1 that

$$\boldsymbol{\Sigma}\mathbf{A} = \mathbf{A}\boldsymbol{\Lambda}, \tag{2.1.2}$$

where $\boldsymbol{\Lambda}$ is the diagonal matrix whose kth diagonal element is λ_k, the kth eigenvalue of $\boldsymbol{\Sigma}$, and $\lambda_k = \mathrm{var}(\boldsymbol{\alpha}'_k\mathbf{x}) = \mathrm{var}(z_k)$. Two alternative ways of expressing (2.1.2) that follow because **A** is orthogonal will be useful later, namely

$$\mathbf{A}'\boldsymbol{\Sigma}\mathbf{A} = \boldsymbol{\Lambda} \tag{2.1.3}$$

and

$$\boldsymbol{\Sigma} = \mathbf{A}\boldsymbol{\Lambda}\mathbf{A}'. \tag{2.1.4}$$

The orthonormal linear transformation of **x**, (2.1.1), which defines **z**, has a number of optimal properties, which are now discussed in turn.

Property A1. *For any integer q, $1 \le q \le p$, consider the orthonormal linear transformation*

$$\mathbf{y} = \mathbf{B}'\mathbf{x}, \tag{2.1.5}$$

*where **y** is a q-element vector and $\mathbf{B}'$ is a $(q \times p)$ matrix, and let $\boldsymbol{\Sigma}_y = \mathbf{B}'\boldsymbol{\Sigma}\mathbf{B}$ be the variance-covariance matrix for **y**. Then the trace of $\boldsymbol{\Sigma}_y$, denoted $\mathrm{tr}\,(\boldsymbol{\Sigma}_y)$, is maximized by taking $\mathbf{B} = \mathbf{A}_q$, where $\mathbf{A}_q$ consists of the first q columns of **A**.*

PROOF.   Let β_k be the kth column of $\mathbf{B}$; as the columns of $\mathbf{A}$ form a basis for p-dimensional space, we have

$$\beta_k = \sum_{j=1}^{p} c_{jk}\alpha_j, \quad k = 1, 2, \ldots, q,$$

where c_{jk}, $j = 1, 2, \ldots, p$, $k = 1, 2, \ldots, q$, are appropriately defined constants. Thus $\mathbf{B} = \mathbf{AC}$, where $\mathbf{C}$ is the $(p \times q)$ matrix with (j, k)th element c_{jk}, and

$$\mathbf{B'\Sigma B} = \mathbf{C'A'\Sigma AC} = \mathbf{C'\Lambda C}, \quad \text{using (2.1.3)}$$

$$= \sum_{j=1}^{p} \lambda_j \mathbf{c}_j \mathbf{c}_j'$$

where $\mathbf{c}_j'$ is the jth row of $\mathbf{C}$. Therefore

$$\text{tr}(\mathbf{B'\Sigma B}) = \sum_{j=1}^{p} \lambda_j \, \text{tr}(\mathbf{c}_j \mathbf{c}_j')$$

$$= \sum_{j=1}^{p} \lambda_j \, \text{tr}(\mathbf{c}_j' \mathbf{c}_j)$$

$$= \sum_{j=1}^{p} \lambda_j \mathbf{c}_j' \mathbf{c}_j$$

$$= \sum_{j=1}^{p}\sum_{k=1}^{q} \lambda_j c_{jk}^2. \qquad (2.1.6)$$

Now

$$\mathbf{C} = \mathbf{A'B}, \quad \text{so}$$
$$\mathbf{C'C} = \mathbf{B'AA'B} = \mathbf{B'B} = \mathbf{I}_q,$$

because $\mathbf{A}$ is orthogonal, and the columns of $\mathbf{B}$ are orthonormal. Hence

$$\sum_{j=1}^{p}\sum_{k=1}^{q} c_{jk}^2 = q, \qquad (2.1.7)$$

and the columns of $\mathbf{C}$ are also orthonormal. The matrix $\mathbf{C}$ can be thought of as the first q columns of a $(p \times p)$ orthogonal matrix, $\mathbf{D}$, say. But the rows of $\mathbf{D}$ are orthonormal and so satisfy $\mathbf{d}_j' \mathbf{d}_j = 1$, $j = 1, \ldots, p$. As the rows of $\mathbf{C}$ consist of the first q elements of the rows of $\mathbf{D}$, it follows that $\mathbf{c}_j' \mathbf{c}_j \leq 1$, $j = 1, \ldots, p$, that is

$$\sum_{k=1}^{q} c_{jk}^2 \leq 1. \qquad (2.1.8)$$

Now $\sum_{k=1}^{q} c_{jk}^2$ is the coefficient of λ_j in (2.1.6), the sum of these coefficients is q from (2.1.7), and none of the coefficients can exceed 1, from (2.1.8). Because $\lambda_1 > \lambda_2 > \cdots > \lambda_p$, it is fairly clear that $\sum_{j=1}^{p} (\sum_{k=1}^{q} c_{jk}^2) \lambda_j$ will be maximized if we can find a set of c_{jk} for which

$$\sum_{k=1}^{q} c_{jk}^2 = \left\{ \begin{array}{ll} 1, & j = 1, \ldots, q, \\ 0, & j = q + 1, \ldots, p. \end{array} \right. \tag{2.1.9}$$

But if $\mathbf{B}' = \mathbf{A}_q'$, then

$$c_{jk} = \left\{ \begin{array}{ll} 1, & 1 \le j = k \le q, \\ 0, & \text{elsewhere}, \end{array} \right.$$

which satisfies (2.1.9). Thus $\mathrm{tr}(\mathbf{\Sigma}_y)$ achieves its maximum value when $\mathbf{B}' = \mathbf{A}_q'$. $\quad\square$

Property A2.   *Consider again the orthonormal transformation*

$$\mathbf{y} = \mathbf{B}'\mathbf{x},$$

with $\mathbf{x}$, $\mathbf{B}$, $\mathbf{A}$ *and* $\mathbf{\Sigma}_y$ *defined as before. Then* $\mathrm{tr}(\mathbf{\Sigma}_y)$ *is minimized by taking* $\mathbf{B} = \mathbf{A}_q^*$ *where* $\mathbf{A}_q^*$ *consists of the last* q *columns of* $\mathbf{A}$.

PROOF.   The derivation of PCs given in Chapter 1 can easily be turned around for the purpose of looking for, successively, linear functions of $\mathbf{x}$ whose variances are as *small* as possible, subject to being uncorrelated with previous linear functions. The solution is again obtained by finding eigenvectors of $\mathbf{\Sigma}$, but this time in reverse order, starting with the smallest. The argument that proved Property A1 can be similarly adapted to prove Property A2. $\quad\square$

The statistical implication of Property A2 is that the last few PCs are not simply unstructured left-overs after removing the important PCs. Because these last PCs have variances as small as possible they are useful in their own right. They can help to detect unsuspected near-constant linear relationships between the elements of $\mathbf{x}$ (see Section 3.4), and they may also be useful in regression (Chapter 8), in selecting a subset of variables from $\mathbf{x}$ (Section 6.3), and in outlier detection (Section 10.1).

Property A3. (the Spectral Decomposition of $\mathbf{\Sigma}$)

$$\mathbf{\Sigma} = \lambda_1 \boldsymbol{\alpha}_1 \boldsymbol{\alpha}_1' + \lambda_2 \boldsymbol{\alpha}_2 \boldsymbol{\alpha}_2' + \cdots + \lambda_p \boldsymbol{\alpha}_p \boldsymbol{\alpha}_p'. \tag{2.1.10}$$

PROOF.

$$\mathbf{\Sigma} = \mathbf{A}\mathbf{\Lambda}\mathbf{A}' \quad \text{from (2.1.4)},$$

and expanding the right-hand side matrix product shows that $\mathbf{\Sigma}$ equals

$$\sum_{k=1}^{p} \lambda_k \boldsymbol{\alpha}_k \boldsymbol{\alpha}_k',$$

as required (see the derivation of (2.1.6)). $\quad\square$

This result will prove to be useful later. Looking at diagonal elements, we see that

$$\text{var}(x_j) = \sum_{k=1}^{p} \lambda_k \alpha_{kj}^2.$$

However, perhaps the main statistical implication of the result is that not only can we decompose the combined variances of all the elements of $\mathbf{x}$ into decreasing contributions due to each PC, but we can also decompose the whole covariance matrix into contributions $\lambda_k \boldsymbol{\alpha}_k \boldsymbol{\alpha}_k'$ from each PC. Although not strictly decreasing, the elements of $\lambda_k \boldsymbol{\alpha}_k \boldsymbol{\alpha}_k'$ will tend to become smaller as k increases, as λ_k decreases for increasing k, whereas the elements of $\boldsymbol{\alpha}_k$ tend to stay 'about the same size' because of the normalization constraints

$$\boldsymbol{\alpha}_k' \boldsymbol{\alpha}_k = 1, \quad k = 1, 2, \ldots, p.$$

Property A1 emphasizes that the PCs explain, successively, as much as possible of $\text{tr}(\boldsymbol{\Sigma})$, but the current property shows, intuitively, that they also do a good job of explaining the off-diagonal elements of $\boldsymbol{\Sigma}$. This is particularly true when the PCs are derived from a correlation matrix, and is less valid when the covariance matrix is used and the variances of the elements of $\mathbf{x}$ are widely different (see Section 2.3).

It is clear from (2.1.10) that the covariance (or correlation) matrix can be constructed exactly, given the coefficients and variances of the first r PCs, where r is the rank of the covariance matrix. Ten Berge and Kiers (1999) discuss conditions under which the correlation matrix can be exactly reconstructed from the coefficients and variances of the first q $(< r)$ PCs.

A corollary of the spectral decomposition of $\boldsymbol{\Sigma}$ concerns the conditional distribution of $\mathbf{x}$, given the first q PCs, $\mathbf{z}_q$, $q = 1, 2, \ldots, (p-1)$. It can be shown that the linear combination of $\mathbf{x}$ that has maximum variance, conditional on $\mathbf{z}_q$, is precisely the $(q+1)$th PC. To see this, we use the result that the conditional covariance matrix of $\mathbf{x}$, given $\mathbf{z}_q$, is

$$\boldsymbol{\Sigma} - \boldsymbol{\Sigma}_{xz} \boldsymbol{\Sigma}_{zz}^{-1} \boldsymbol{\Sigma}_{zx},$$

where $\boldsymbol{\Sigma}_{zz}$ is the covariance matrix for $\mathbf{z}_q$, $\boldsymbol{\Sigma}_{xz}$ is the $(p \times q)$ matrix whose (j, k)th element is the covariance between x_j and z_k, and $\boldsymbol{\Sigma}_{zx}$ is the transpose of $\boldsymbol{\Sigma}_{xz}$ (Mardia et al., 1979, Theorem 3.2.4).

It is seen in Section 2.3 that the kth column of $\boldsymbol{\Sigma}_{xz}$ is $\lambda_k \boldsymbol{\alpha}_k$. The matrix $\boldsymbol{\Sigma}_{zz}^{-1}$ is diagonal, with kth diagonal element λ_k^{-1}, so it follows that

$$\boldsymbol{\Sigma}_{xz} \boldsymbol{\Sigma}_{zz}^{-1} \boldsymbol{\Sigma}_{zx} = \sum_{k=1}^{q} \lambda_k \boldsymbol{\alpha}_k \lambda_k^{-1} \lambda_k \boldsymbol{\alpha}_k'$$

$$= \sum_{k=1}^{q} \lambda_k \boldsymbol{\alpha}_k \boldsymbol{\alpha}_k',$$

and, from (2.1.10),

$$\Sigma - \Sigma_{xz}\Sigma_{zz}^{-1}\Sigma_{zx} = \sum_{k=(q+1)}^{p} \lambda_k \alpha_k \alpha_k'.$$

Finding a linear function of $\mathbf{x}$ having maximum conditional variance reduces to finding the eigenvalues and eigenvectors of the conditional covariance matrix, and it easy to verify that these are simply $(\lambda_{(q+1)}, \alpha_{(q+1)})$, $(\lambda_{(q+2)}, \alpha_{(q+2)}), \ldots, (\lambda_p, \alpha_p)$. The eigenvector associated with the largest of these eigenvalues is $\alpha_{(q+1)}$, so the required linear function is $\alpha_{(q+1)}'\mathbf{x}$, namely the $(q+1)$th PC.

Property A4.   *As in Properties A1, A2, consider the transformation* $\mathbf{y} = \mathbf{B}'\mathbf{x}$. *If* $\det(\Sigma_y)$ *denotes the determinant of the covariance matrix* $\mathbf{y}$, *then* $\det(\Sigma_y)$ *is maximized when* $\mathbf{B} = \mathbf{A}_q$.

PROOF.   Consider any integer, k, between 1 and q, and let $S_k = $ the subspace of p-dimensional vectors orthogonal to $\alpha_1, \ldots, \alpha_{k-1}$. Then $\dim(S_k) = p - k + 1$, where $\dim(S_k)$ denotes the dimension of S_k. The kth eigenvalue, λ_k, of Σ satisfies

$$\lambda_k = \underset{\substack{\alpha \in S_k \\ \alpha \neq 0}}{\text{Sup}} \left\{ \frac{\alpha'\Sigma\alpha}{\alpha'\alpha} \right\}.$$

Suppose that $\mu_1 > \mu_2 > \cdots > \mu_q$, are the eigenvalues of $\mathbf{B}'\Sigma\mathbf{B}$ and that $\gamma_1, \gamma_2, \cdots, \gamma_q$, are the corresponding eigenvectors. Let $T_k = $ the subspace of q-dimensional vectors orthogonal to $\gamma_{k+1}, \cdots, \gamma_q$, with $\dim(T_k) = k$. Then, for any non-zero vector γ in T_k,

$$\frac{\gamma'\mathbf{B}'\Sigma\mathbf{B}\gamma}{\gamma'\gamma} \geq \mu_k.$$

Consider the subspace $\tilde{S}_k$ of p-dimensional vectors of the form $\mathbf{B}\gamma$ for γ in T_k.

$$\dim(\tilde{S}_k) = \dim(T_k) = k \qquad \text{(because } \mathbf{B} \text{ is one-to-one; in fact,}$$
$$\mathbf{B} \text{ preserves lengths of vectors).}$$

From a general result concerning dimensions of two vector spaces, we have

$$\dim(S_k \cap \tilde{S}_k) + \dim(S_k + \tilde{S}_k) = \dim S_k + \dim \tilde{S}_k.$$

But

$$\dim(S_k + \tilde{S}_k) \leq p, \qquad \dim(S_k) = p - k + 1 \quad \text{and} \quad \dim(\tilde{S}_k) = k,$$

so

$$\dim(S_k \cap \tilde{S}_k) \geq 1.$$

There is therefore a non-zero vector $\boldsymbol{\alpha}$ in S_k of the form $\boldsymbol{\alpha} = \mathbf{B}\boldsymbol{\gamma}$ for a $\boldsymbol{\gamma}$ in T_k, and it follows that

$$\mu_k \leq \frac{\boldsymbol{\gamma}'\mathbf{B}'\boldsymbol{\Sigma}\mathbf{B}\boldsymbol{\gamma}}{\boldsymbol{\gamma}'\boldsymbol{\gamma}} = \frac{\boldsymbol{\gamma}'\mathbf{B}'\boldsymbol{\Sigma}\mathbf{B}\boldsymbol{\gamma}}{\boldsymbol{\gamma}\mathbf{B}'\mathbf{B}\boldsymbol{\gamma}} = \frac{\boldsymbol{\alpha}'\boldsymbol{\Sigma}\boldsymbol{\alpha}}{\boldsymbol{\alpha}'\boldsymbol{\alpha}} \leq \lambda_k.$$

Thus the kth eigenvalue of $\mathbf{B}'\boldsymbol{\Sigma}\mathbf{B} \leq k$th eigenvalue of $\boldsymbol{\Sigma}$ for $k = 1, \cdots, q$. This means that

$$\det(\boldsymbol{\Sigma}_y) = \prod_{k=1}^{q} (k\text{th eigenvalue of } \mathbf{B}'\boldsymbol{\Sigma}\mathbf{B}) \leq \prod_{k=1}^{q} \lambda_k.$$

But if $\mathbf{B} = \mathbf{A}_q$, then the eigenvalues of $\mathbf{B}'\boldsymbol{\Sigma}\mathbf{B}$ are

$$\lambda_1, \lambda_2, \cdots, \lambda_q, \quad \text{so that} \quad \det(\boldsymbol{\Sigma}_y) = \prod_{k=1}^{q} \lambda_k$$

in this case, and therefore $\det(\boldsymbol{\Sigma}_y)$ is maximized when $\mathbf{B} = \mathbf{A}_q$.   □

The result can be extended to the case where the columns of $\mathbf{B}$ are not necessarily orthonormal, but the diagonal elements of $\mathbf{B}'\mathbf{B}$ are unity (see Okamoto (1969)). A stronger, stepwise version of Property A4 is discussed by O'Hagan (1984), who argues that it provides an alternative derivation of PCs, and that this derivation can be helpful in motivating the use of PCA. O'Hagan's derivation is, in fact, equivalent to (though a stepwise version of) Property A5, which is discussed next.

Note that Property A1 could also have been proved using similar reasoning to that just employed for Property A4, but some of the intermediate results derived during the earlier proof of Al are useful elsewhere in the chapter.

The statistical importance of the present result follows because the determinant of a covariance matrix, which is called the *generalized variance*, can be used as a single measure of spread for a multivariate random variable (Press, 1972, p. 108). The square root of the generalized variance, for a multivariate normal distribution is proportional to the 'volume' in p-dimensional space that encloses a fixed proportion of the probability distribution of $\mathbf{x}$. For multivariate normal $\mathbf{x}$, the first q PCs are, therefore, as a consequence of Property A4, q linear functions of $\mathbf{x}$ whose joint probability distribution has contours of fixed probability enclosing the maximum volume.

Property A5.   *Suppose that we wish to predict each random variable, x_j in $\mathbf{x}$ by a linear function of $\mathbf{y}$, where $\mathbf{y} = \mathbf{B}'\mathbf{x}$, as before. If σ_j^2 is the residual variance in predicting x_j from $\mathbf{y}$, then $\Sigma_{j=1}^{p}\sigma_j^2$ is minimized if $\mathbf{B} = \mathbf{A}_q$.*

The statistical implication of this result is that if we wish to get the best linear predictor of $\mathbf{x}$ in a q-dimensional subspace, in the sense of minimizing the sum over elements of $\mathbf{x}$ of the residual variances, then this optimal subspace is defined by the first q PCs.

It follows that although Property A5 is stated as an algebraic property, it can equally well be viewed geometrically. In fact, it is essentially the population equivalent of sample Property G3, which is stated and proved in Section 3.2. No proof of the population result A5 will be given here; Rao (1973, p. 591) outlines a proof in which $\mathbf{y}$ is replaced by an equivalent set of uncorrelated linear functions of $\mathbf{x}$, and it is interesting to note that the PCs are the *only* set of p linear functions of $\mathbf{x}$ that are uncorrelated *and* have orthogonal vectors of coefficients. This last result is prominent in the discussion of Chapter 11.

A special case of Property A5 was pointed out in Hotelling's (1933) original paper. He notes that the first PC derived from a correlation matrix is the linear function of $\mathbf{x}$ that has greater mean square correlation with the elements of $\mathbf{x}$ than does any other linear function. We return to this interpretation of the property, and extend it, in Section 2.3.

A modification of Property A5 can be introduced by noting that if $\mathbf{x}$ is predicted by a linear function of $\mathbf{y} = \mathbf{B}'\mathbf{x}$, then it follows from standard results from multivariate regression (see, for example, Mardia et al., 1979, p. 160), that the residual covariance matrix for the best such predictor is

$$\boldsymbol{\Sigma}_x - \boldsymbol{\Sigma}_{xy}\boldsymbol{\Sigma}_y^{-1}\boldsymbol{\Sigma}_{yx}, \tag{2.1.11}$$

where $\boldsymbol{\Sigma}_x = \boldsymbol{\Sigma}, \boldsymbol{\Sigma}_y = \mathbf{B}'\boldsymbol{\Sigma}\mathbf{B}$, as defined before, $\boldsymbol{\Sigma}_{xy}$ is the matrix whose (j,k)th element is the covariance between the jth element of $\mathbf{x}$ and the kth element of $\mathbf{y}$, and $\boldsymbol{\Sigma}_{yx}$ is the transpose of $\boldsymbol{\Sigma}_{xy}$. Now $\boldsymbol{\Sigma}_{yx} = \mathbf{B}'\boldsymbol{\Sigma}$, and $\boldsymbol{\Sigma}_{xy} = \boldsymbol{\Sigma}\mathbf{B}$, so (2.1.11) becomes

$$\boldsymbol{\Sigma} - \boldsymbol{\Sigma}\mathbf{B}(\mathbf{B}'\boldsymbol{\Sigma}\mathbf{B})^{-1}\mathbf{B}'\boldsymbol{\Sigma}. \tag{2.1.12}$$

The diagonal elements of (2.1.12) are σ_j^2, $j = 1, 2, \ldots, p$, so, from Property A5, $\mathbf{B} = \mathbf{A}_q$ minimizes

$$\sum_{j=1}^{p} \sigma_j^2 = \text{tr}[\boldsymbol{\Sigma} - \boldsymbol{\Sigma}\mathbf{B}(\mathbf{B}'\boldsymbol{\Sigma}\mathbf{B})^{-1}\mathbf{B}'\boldsymbol{\Sigma}].$$

A derivation of this result in the sample case, and further discussion of it, is provided by Jong and Kotz (1999).

An alternative criterion is $\|\boldsymbol{\Sigma} - \boldsymbol{\Sigma}\mathbf{B}(\mathbf{B}'\boldsymbol{\Sigma}\mathbf{B})^{-1}\mathbf{B}'\boldsymbol{\Sigma}\|$, where $\|\cdot\|$ denotes the Euclidean norm of a matrix and equals the square root of the sum of squares of *all* the elements in the matrix. It can be shown (Rao, 1964) that this alternative criterion is also minimized when $\mathbf{B} = \mathbf{A}_q$.

This section has dealt with PCs derived from covariance matrices. Many of their properties are also relevant, in modified form, for PCs based on correlation matrices, as discussed in Section 2.3. That section also contains a further algebraic property which is specific to correlation matrix-based PCA.

2.2 Geometric Properties of Population Principal Components

It was noted above that Property A5 can be interpreted geometrically, as well as algebraically, and the discussion following Property A4 shows that A4, too, has a geometric interpretation. We now look at two further, purely geometric, properties.

Property G1.   *Consider the family of p-dimensional ellipsoids*

$$\mathbf{x}'\mathbf{\Sigma}^{-1}\mathbf{x} = \text{const.} \qquad (2.2.1)$$

The PCs define the principal axes of these ellipsoids.

PROOF.   The PCs are defined by the transformation (2.1.1) $\mathbf{z} = \mathbf{A}'\mathbf{x}$, and since $\mathbf{A}$ is orthogonal, the inverse transformation is $\mathbf{x} = \mathbf{A}\mathbf{z}$. Substituting into (2.2.1) gives

$$(\mathbf{A}\mathbf{z})'\mathbf{\Sigma}^{-1}(\mathbf{A}\mathbf{z}) = \text{const} = \mathbf{z}'\mathbf{A}'\mathbf{\Sigma}^{-1}\mathbf{A}\mathbf{z}.$$

It is well known that the eigenvectors of $\mathbf{\Sigma}^{-1}$ are the same as those of $\mathbf{\Sigma}$, and that the eigenvalues of $\mathbf{\Sigma}^{-1}$ are the reciprocals of those of $\mathbf{\Sigma}$, assuming that they are all strictly positive. It therefore follows, from a corresponding result to (2.1.3), that $\mathbf{A}\mathbf{\Sigma}^{-1}\mathbf{A} = \mathbf{\Lambda}^{-1}$ and hence

$$\mathbf{z}'\mathbf{\Lambda}^{-1}\mathbf{z} = \text{const.}$$

This last equation can be rewritten

$$\sum_{k=1}^{p} \frac{z_k^2}{\lambda_k} = \text{const} \qquad (2.2.2)$$

and (2.2.2) is the equation for an ellipsoid referred to its principal axes. Equation (2.2.2) also implies that the half-lengths of the principal axes are proportional to $\lambda_1^{1/2}, \lambda_2^{1/2}, \ldots, \lambda_p^{1/2}$. ☐

This result is statistically important if the random vector $\mathbf{x}$ has a multivariate normal distribution. In this case, the ellipsoids given by (2.2.1) define contours of constant probability for the distribution of $\mathbf{x}$. The first (largest) principal axis of such ellipsoids will then define the direction in which statistical variation is greatest, which is another way of expressing the algebraic definition of the first PC given in Section 1.1. The direction of the first PC, defining the first principal axis of constant probability ellipsoids, is illustrated in Figures 2.1 and 2.2 in Section 2.3. The second principal axis maximizes statistical variation, subject to being orthogonal to the first, and so on, again corresponding to the algebraic definition. This interpretation of PCs, as defining the principal axes of ellipsoids of constant density, was mentioned by Hotelling (1933) in his original paper.

It would appear that this particular geometric property is only of direct statistical relevance if the distribution of $\mathbf{x}$ is multivariate normal, whereas for most other properties of PCs no distributional assumptions are required. However, the property will be discussed further in connection with Property G5 in Section 3.2, where we see that it has some relevance even without the assumption of multivariate normality. Property G5 looks at the sample version of the ellipsoids $\mathbf{x}'\boldsymbol{\Sigma}\mathbf{x} = $ const. Because $\boldsymbol{\Sigma}$ and $\boldsymbol{\Sigma}^{-1}$ share the same eigenvectors, it follows that the principal axes of the ellipsoids $\mathbf{x}'\boldsymbol{\Sigma}\mathbf{x} = $ const are the same as those of $\mathbf{x}'\boldsymbol{\Sigma}^{-1}\mathbf{x} = $ const, except that that their order is reversed.

We digress slightly here to note that some authors imply, or even state explicitly, as do Qian et al. (1994), that PCA needs multivariate normality. This text takes a very different view and considers PCA as a mainly descriptive technique. It will become apparent that many of the properties and applications of PCA and related techniques described in later chapters, as well as the properties discussed in the present chapter, have no need for explicit distributional assumptions. It cannot be disputed that linearity and covariances/correlations, both of which play a central rôle in PCA, have especial relevance when distributions are multivariate normal, but this does not detract from the usefulness of PCA when data have other forms. Qian et al. (1994) describe what might be considered an additional property of PCA, based on minimum description length or stochastic complexity (Rissanen and Yu, 2000), but as they use it to define a somewhat different technique, we defer discussion to Section 14.4.

Property G2.   *Suppose that $\mathbf{x}_1$, $\mathbf{x}_2$ are independent random vectors, both having the same probability distribution, and that $\mathbf{x}_1$, $\mathbf{x}_2$, are both subjected to the same linear transformation*

$$\mathbf{y}_i = \mathbf{B}'\mathbf{x}_i, \quad i = 1, 2.$$

If $\mathbf{B}$ is a $(p \times q)$ matrix with orthonormal columns chosen to maximize $E[(\mathbf{y}_1 - \mathbf{y}_2)'(\mathbf{y}_1 - \mathbf{y}_2)]$, then $\mathbf{B} = \mathbf{A}_q$, using the same notation as before.

PROOF.   This result could be viewed as a purely algebraic property, and, indeed, the proof below is algebraic. The property is, however, included in the present section because it has a geometric interpretation. This is that the expected squared Euclidean distance, in a q-dimensional subspace, between two vectors of p random variables with the same distribution, is made as large as possible if the subspace is defined by the first q PCs.

To prove Property G2, first note that $\mathbf{x}_1$, $\mathbf{x}_2$ have the same mean $\boldsymbol{\mu}$ and covariance matrix $\boldsymbol{\Sigma}$. Hence $\mathbf{y}_1$, $\mathbf{y}_2$ also have the same mean and covariance matrix, $\mathbf{B}'\boldsymbol{\mu}$, $\mathbf{B}'\boldsymbol{\Sigma}\mathbf{B}$ respectively.

$$\begin{aligned}
E[(\mathbf{y}_1 - \mathbf{y}_2)'(\mathbf{y}_1 - \mathbf{y}_2)] &= E\{[(\mathbf{y}_1 - \mathbf{B}'\boldsymbol{\mu}) - (\mathbf{y}_2 - (\mathbf{B}'\boldsymbol{\mu})]'[(\mathbf{y}_1 - \mathbf{B}'\boldsymbol{\mu}) \\
&\quad - (\mathbf{y}_2 - \mathbf{B}'\boldsymbol{\mu})]\} \\
&= E[(\mathbf{y}_1 - \mathbf{B}'\boldsymbol{\mu})'(\mathbf{y}_1 - \mathbf{B}'\boldsymbol{\mu})] \\
&\quad + E[(\mathbf{y}_2 - \mathbf{B}'\boldsymbol{\mu})'(\mathbf{y}_2 - \mathbf{B}'\boldsymbol{\mu})].
\end{aligned}$$

The cross-product terms disappear because of the independence of $\mathbf{x}_1$, $\mathbf{x}_2$, and hence of $\mathbf{y}_1$, $\mathbf{y}_2$.

Now, for $i = 1, 2$, we have

$$
\begin{aligned}
E[(\mathbf{y}_i - \mathbf{B}'\boldsymbol{\mu})'(\mathbf{y}_i - \mathbf{B}'\boldsymbol{\mu})] &= E\{\mathrm{tr}[(\mathbf{y}_i - \mathbf{B}'\boldsymbol{\mu})'(\mathbf{y}_i - \mathbf{B}'\boldsymbol{\mu})]\} \\
&= E\{\mathrm{tr}[(\mathbf{y}_i - \mathbf{B}'\boldsymbol{\mu})(\mathbf{y}_i - \mathbf{B}'\boldsymbol{\mu})']\} \\
&= \mathrm{tr}\{E[(\mathbf{y}_i - \mathbf{B}'\boldsymbol{\mu})(\mathbf{y}_i - \mathbf{B}'\boldsymbol{\mu})']\} \\
&= \mathrm{tr}(\mathbf{B}'\boldsymbol{\Sigma}\mathbf{B}).
\end{aligned}
$$

But $\mathrm{tr}(\mathbf{B}'\boldsymbol{\Sigma}\mathbf{B})$ is maximized when $\mathbf{B} = \mathbf{A}_q$, from Property A1, and the present criterion has been shown above to be $2\,\mathrm{tr}(\mathbf{B}'\boldsymbol{\Sigma}\mathbf{B})$. Hence Property G2 is proved. □

There is a closely related property whose geometric interpretation is more tenuous, namely that with the same definitions as in Property G2,

$$
\det\{E[(\mathbf{y}_1 - \mathbf{y}_2)(\mathbf{y}_1 - \mathbf{y}_2)']\}
$$

is maximized when $\mathbf{B} = \mathbf{A}_q$ (see McCabe (1984)). This property says that $\mathbf{B} = \mathbf{A}_q$ makes the generalized variance of $(\mathbf{y}_1 - \mathbf{y}_2)$ as large as possible. Generalized variance may be viewed as an alternative measure of distance apart of $\mathbf{y}_1$ and $\mathbf{y}_2$ in q-dimensional space, though a less intuitively obvious measure than expected squared Euclidean distance.

Finally, Property G2 can be reversed in the sense that if $E[(\mathbf{y}_1 - \mathbf{y}_2)'(\mathbf{y}_1 - \mathbf{y}_2)]$ or $\det\{E[(\mathbf{y}_1 - \mathbf{y}_2)(\mathbf{y}_1 - \mathbf{y}_2)']\}$ is to be *minimized*, then this can be achieved by taking $\mathbf{B} = \mathbf{A}_q^*$.

The properties given in this section and in the previous one show that covariance matrix PCs satisfy several different optimality criteria, but the list of criteria covered is by no means exhaustive; for example, Devijver and Kittler (1982, Chapter 9) show that the first few PCs minimize *representation entropy* and the last few PCs minimize *population entropy*. Diamantaras and Kung (1996, Section 3.4) discuss PCA in terms of maximizing *mutual information* between $\mathbf{x}$ and $\mathbf{y}$. Further optimality criteria are given by Hudlet and Johnson (1982), McCabe (1984) and Okamoto (1969). The geometry of PCs is discussed at length by Treasure (1986).

The property of *self-consistency* is useful in a non-linear extension of PCA (see Section 14.1.2). For two p-variate random vectors $\mathbf{x}$, $\mathbf{y}$, the vector $\mathbf{y}$ is self-consistent for $\mathbf{x}$ if $E(\mathbf{x}|\mathbf{y}) = \mathbf{y}$. Flury (1997, Section 8.4) shows that if $\mathbf{x}$ is a p-variate random vector with a multivariate normal or elliptical distribution, and $\mathbf{y}$ is the orthogonal projection of $\mathbf{x}$ onto the q-dimensional subspace spanned by the first q PCs for $\mathbf{x}$, then $\mathbf{y}$ is self-consistent for $\mathbf{x}$. Tarpey (1999) uses self-consistency of principal components after linear transformation of the variables to characterize elliptical distributions.

2.3 Principal Components Using a Correlation Matrix

The derivation and properties of PCs considered above are based on the eigenvectors and eigenvalues of the covariance matrix. In practice, as will be seen in much of the remainder of this text, it is more common to define principal components as

$$\mathbf{z} = \mathbf{A}'\mathbf{x}^*, \tag{2.3.1}$$

where $\mathbf{A}$ now has columns consisting of the eigenvectors of the correlation matrix, and $\mathbf{x}^*$ consists of standardized variables. The goal in adopting such an approach is to find the principal components of a standardized version $\mathbf{x}^*$ of $\mathbf{x}$, where $\mathbf{x}^*$ has jth element $x_j/\sigma_{jj}^{1/2}$, $j = 1, 2, \ldots, p$, x_j is the jth element of $\mathbf{x}$, and σ_{jj} is the variance of x_j. Then the covariance matrix for $\mathbf{x}^*$ is the correlation matrix of $\mathbf{x}$, and the PCs of $\mathbf{x}^*$ are given by (2.3.1).

A third possibility, instead of using covariance or correlation matrices, is to use covariances of x_j/w_j, where the weights w_j are chosen to reflect some a priori idea of the relative importance of the variables. The special case $w_j = \sigma_{jj}^{1/2}$ leads to $\mathbf{x}^*$, and to PCs based on the correlation matrix, but various authors have argued that the choice of $w_j = \sigma_{jj}^{1/2}$ is somewhat arbitrary, and that different values of w_j might be better in some applications (see Section 14.2.1). In practice, however, it is relatively unusual that a uniquely appropriate set of w_j suggests itself.

All the properties of the previous two sections are still valid for correlation matrices, or indeed for covariances based on other sets of weights, except that we are now considering PCs of $\mathbf{x}^*$ (or some other transformation of $\mathbf{x}$), instead of $\mathbf{x}$.

It might seem that the PCs for a correlation matrix could be obtained fairly easily from those for the corresponding covariance matrix, since $\mathbf{x}^*$ is related to $\mathbf{x}$ by a very simple transformation. However, this is not the case; the eigenvalues and eigenvectors of the correlation matrix have no simple relationship with those of the corresponding covariance matrix. In particular, if the PCs found from the correlation matrix are expressed in terms of $\mathbf{x}$ by transforming back from $\mathbf{x}^*$ to $\mathbf{x}$, then these PCs are not the same as the PCs found from $\mathbf{\Sigma}$, except in very special circumstances (Chatfield and Collins, 1989, Section 4.4). One way of explaining this is that PCs are invariant under *orthogonal* transformations of $\mathbf{x}$ but not, in general, under other transformations (von Storch and Zwiers, 1999, Section 13.1.10). The transformation from $\mathbf{x}$ to $\mathbf{x}^*$ is not orthogonal. The PCs for correlation and covariance matrices do not, therefore, give equivalent information, nor can they be derived directly from each other. We now discuss the relative merits of the two types of PC.

A major argument for using correlation—rather than covariance—matrices to define PCs is that the results of analyses for different sets of random variables are more directly comparable than for analyses based on covariance matrices. The big drawback of PCA based on covariance matrices is the sensitivity of the PCs to the units of measurement used for each element of $\mathbf{x}$. If there are large differences between the variances of the elements of $\mathbf{x}$, then those variables whose variances are largest will tend to dominate the first few PCs (see, for example, Section 3.3). This may be entirely appropriate if all the elements of $\mathbf{x}$ are measured in the same units, for example, if all elements of $\mathbf{x}$ are anatomical measurements on a particular species of animal, all recorded in centimetres, say. Even in such examples, arguments can be presented for the use of correlation matrices (see Section 4.1). In practice, it often occurs that different elements of $\mathbf{x}$ are completely different types of measurement. Some might be lengths, some weights, some temperatures, some arbitrary scores on a five-point scale, and so on. In such a case, the structure of the PCs will depend on the choice of units of measurement, as is illustrated by the following artificial example.

Suppose that we have just two variables, x_1, x_2, and that x_1 is a length variable which can equally well be measured in centimetres or in millimetres. The variable x_2 is not a length measurement—it might be a weight, in grams, for example. The covariance matrices in the two cases are, respectively,

$$\mathbf{\Sigma}_1 = \begin{pmatrix} 80 & 44 \\ 44 & 80 \end{pmatrix} \quad \text{and} \quad \mathbf{\Sigma}_2 = \begin{pmatrix} 8000 & 440 \\ 440 & 80 \end{pmatrix}.$$

The first PC is $0.707x_1 + 0.707x_2$ for $\mathbf{\Sigma}_1$ and $0.998x_1 + 0.055x_2$ for $\mathbf{\Sigma}_2$, so a relatively minor change in one variable has the effect of changing a PC that gives *equal* weight to x_1 and x_2 to a PC that is almost entirely *dominated* by x_1. Furthermore, the first PC accounts for 77.5 percent of the total variation for $\mathbf{\Sigma}_1$, but 99.3 percent for $\mathbf{\Sigma}_2$.

Figures 2.1 and 2.2 provide another way of looking at the differences between PCs for the two scales of measurement in x_1. The plots give contours of constant probability, assuming multivariate normality for $\mathbf{x}$ for $\mathbf{\Sigma}_1$ and $\mathbf{\Sigma}_2$, respectively. It is clear from these figures that, whereas with $\mathbf{\Sigma}_1$ both variables have the same degree of variation, for $\mathbf{\Sigma}_2$ most of the variation is in the direction of x_1. This is reflected in the first PC, which, from Property G1, is defined by the major axis of the ellipses of constant probability.

This example demonstrates the general behaviour of PCs for a covariance matrix when the variances of the individual variables are widely different; the same type of behaviour is illustrated again for samples in Section 3.3. The first PC is dominated by the variable with the largest variance, the second PC is dominated by the variable with the second largest variance, and so on, with a substantial proportion of the total variation accounted for by just two or three PCs. In other words, the PCs differ little from

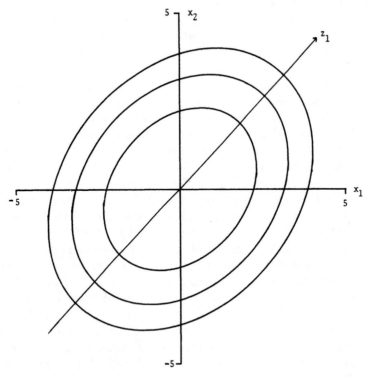

Figure 2.1. Contours of constant probability based on $\mathbf{\Sigma}_1 = \begin{pmatrix} 80 & 44 \\ 44 & 80 \end{pmatrix}$.

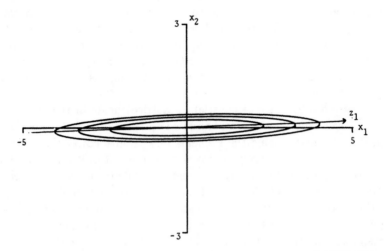

Figure 2.2. Contours of constant probability based on $\mathbf{\Sigma}_2 = \begin{pmatrix} 8000 & 440 \\ 440 & 80 \end{pmatrix}$.

the original variables rearranged in decreasing order of the size of their variances. Also, the first few PCs account for little of the off-diagonal elements of Σ in this case (see Property A3) above. In most circumstances, such a transformation to PCs is of little value, and it will not occur if the correlation, rather than covariance, matrix is used.

The example has shown that it is unwise to use PCs on a covariance matrix when $\mathbf{x}$ consists of measurements of different types, unless there is a strong conviction that the units of measurements chosen for each element of $\mathbf{x}$ are the only ones that make sense. Even when this condition holds, using the covariance matrix will not provide very informative PCs if the variables have widely differing variances. Furthermore, with covariance matrices and non-commensurable variables the PC scores are difficult to interpret—what does it mean to add a temperature to a weight? For correlation matrices, the standardized variates are all dimensionless and can be happily combined to give PC scores (Legendre and Legendre, 1983, p. 129).

Another problem with the use of covariance matrices is that it is more difficult than with correlation matrices to compare informally the results from different analyses. Sizes of variances of PCs have the same implications for different correlation matrices of the same dimension, but not for different covariance matrices. Also, patterns of coefficients in PCs can be readily compared for different correlation matrices to see if the two correlation matrices are giving similar PCs, whereas informal comparisons are often much trickier for covariance matrices. Formal methods for comparing PCs from different covariance matrices are, however, available (see Section 13.5).

The use of covariance matrices does have one general advantage over correlation matrices, and a particular advantage seen in a special case. The general advantage is that statistical inference regarding population PCs based on sample PCs is easier for covariance matrices than for correlation matrices, as will be discussed in Section 3.7. This is relevant when PCA is used in a context where statistical inference is important. However, in practice, it is more common to use PCA as a descriptive, rather than an inferential, tool, and then the potential advantage of covariance matrix PCA is irrelevant.

The second advantage of covariance matrices holds in the special case when all elements of $\mathbf{x}$ are measured in the same units. It can then be argued that standardizing the elements of $\mathbf{x}$ to give correlations is equivalent to making an arbitrary choice of measurement units. This argument of arbitrariness can also be applied more generally to the use of correlation matrices, but when the elements of $\mathbf{x}$ are measurements of different types, the choice of measurement units leading to a covariance matrix is even more arbitrary, so that the correlation matrix is again preferred.

Standardizing the variables may be thought of as an attempt to remove the problem of scale dependence from PCA. Another way of doing this is to compute PCs of the logarithms of the original data (Flury, 1997, Section 8.4), though this is only feasible and sensible for restricted types of data,

as in allometry (Section 13.2) and for compositional data (Section 13.3).

We conclude this section by looking at three interesting properties which hold for PCs derived from the correlation matrix. The first is that the PCs depend not on the absolute values of correlations, but only on their ratios. This follows because multiplication of all off-diagonal elements of a correlation matrix by the same constant leaves the eigenvectors of the matrix unchanged (Chatfield and Collins, 1989, p. 67).

The second property, which was noted by Hotelling (1933) in his original paper, is that if, instead of the normalization $\alpha'_k \alpha_k = 1$, we use

$$\tilde{\alpha}'_k \tilde{\alpha}_k = \lambda_k, \quad k = 1, 2, \ldots, p, \tag{2.3.2}$$

then, $\tilde{\alpha}_{kj}$ the jth element of $\tilde{\alpha}_k$, is the correlation between the jth standardized variable x^*_j and the kth PC. To see this note that for $k = 1, 2, \ldots, p$,

$$\tilde{\alpha}_k = \lambda_k^{1/2} \alpha_k, \quad \mathrm{var}(z_k) = \lambda_k,$$

and the p-element vector $\Sigma \alpha_k$ has as its jth element the covariance between x^*_j and z_k. But $\Sigma \alpha_k = \lambda_k \alpha_k$, so the covariance between x^*_j and z_k is $\lambda_k \alpha_{kj}$. Also $\mathrm{var}(x^*_j) = 1$, and the correlation between x^*_j and z_k is therefore

$$\frac{\lambda_k \alpha_{jk}}{[\mathrm{var}(x^*_j) \, \mathrm{var}(z_k)]^{1/2}} = \lambda_k^{1/2} \alpha_{kj}$$

$$= \tilde{\alpha}_{kj},$$

as required.

Because of this property the normalization (2.3.2) is quite often used, in particular in computer packages, but it has the disadvantage that it is less easy to informally interpret and compare a set of PCs when each PC has a different normalization on its coefficients. This remark is, of course, relevant to sample, rather than population, PCs, but, as with some other parts of the chapter, it is included here to avoid a possibly disjointed presentation.

Both of these properties that hold for correlation matrices can be modified for covariance matrices, but the results are, in each case, less straightforward.

The third property is sufficiently substantial to deserve a label. It is included in this section because, at first sight, it is specific to correlation matrix PCA although, as we will see, its implications are much wider. Proofs of the result are available in the references cited below and will not be reproduced here.

Property A6.   *For any integer q, $1 \leq q \leq p$, consider the orthonormal linear transformation*

$$\mathbf{y} = \mathbf{B}'\mathbf{x}, \tag{2.3.3}$$

as defined in Property A1. Let $R^2_{j:q}$ be the squared multiple correlation between x_j and the q variables $y_1, y_2, \ldots, y_q$, defined by the elements of $\mathbf{y}$.

The criterion

$$\sum_{j=1}^{p} R_{j:q}^2$$

is maximized when $y_1, y_2, \ldots, y_q$ are the first q correlation matrix PCs. The maximized value of the criterion is equal to the sum of the q largest eigenvalues of the correlation matrix.

Because the principal components are uncorrelated, the criterion in Property A6 reduces to

$$\sum_{j=1}^{p} \sum_{k=1}^{q} r_{jk}^2$$

where r_{jk}^2 is the squared correlation between the jth variable and the kth PC. The criterion will be maximized by any matrix **B** that gives **y** spanning the same q-dimensional space as the first q PCs. However, the correlation matrix PCs are special, in that they *successively* maximize the criterion for $q = 1, 2, \ldots, p$. As noted following Property A5, this result was given by Hotelling (1933) alongside his original derivation of PCA, but it has subsequently been largely ignored. It is closely related to Property A5. Meredith and Millsap (1985) derived Property A6 independently and noted that optimizing the multiple correlation criterion gives a *scale invariant* method (as does Property A5; Cadima, 2000). One implication of this scale invariance is that it gives added importance to correlation matrix PCA. The latter is not simply a variance-maximizing technique for standardized variables; its derived variables are also the result of optimizing a criterion which is scale invariant, and hence is relevant whether or not the variables are standardized. Cadima (2000) discusses Property A6 in greater detail and argues that optimization of its multiple correlation criterion is actually a new technique, which happens to give the same results as correlation matrix PCA, but is broader in its scope. He suggests that the derived variables be called *Most Correlated Components*. Looked at from another viewpoint, this broader relevance of correlation matrix PCA gives another reason to prefer it over covariance matrix PCA in most circumstances.

To conclude this discussion, we note that Property A6 can be easily modified to give a new property for covariance matrix PCA. The first q covariance marix PCs maximize, amongst all orthonormal linear transformations of **x**, the sum of squared *covariances* between $x_1, x_2, \ldots, x_p$ and the derived variables $y_1, y_2, \ldots, y_q$. Covariances, unlike correlations, are not scale invariant, and hence neither is covariance matrix PCA.

2.4 Principal Components with Equal and/or Zero Variances

The final, short, section of this chapter discusses two problems that may arise in theory, but are relatively uncommon in practice. In most of this chapter it has been assumed, implicitly or explicitly, that the eigenvalues of the covariance or correlation matrix are all different, and that none of them is zero.

Equality of eigenvalues, and hence equality of variances of PCs, will occur for certain patterned matrices. The effect of this occurrence is that for a group of q equal eigenvalues, the corresponding q eigenvectors span a certain unique q-dimensional space, but, within this space, they are, apart from being orthogonal to one another, arbitrary. Geometrically (see Property G1), what happens for $q = 2$ or 3 is that the principal axes of a circle or sphere cannot be uniquely defined; a similar problem arises for hyperspheres when $q > 3$. Thus individual PCs corresponding to eigenvalues in a group of equal eigenvalues are not uniquely defined. A further problem with equal-variance PCs is that statistical inference becomes more complicated (see Section 3.7).

The other complication, variances equal to zero, occurs rather more frequently, but is still fairly unusual. If q eigenvalues are zero, then the rank of Σ is $(p - q)$ rather than p, and this outcome necessitates modifications to the proofs of some properties given in Section 2.1 above. Any PC with zero variance defines an exactly constant linear relationship between the elements of $\mathbf{x}$. If such relationships exist, then they imply that one variable is redundant for each relationship, as its value can be determined exactly from the values of the other variables appearing in the relationship. We could therefore reduce the number of variables from p to $(p - q)$ without losing any information. Ideally, exact linear relationships should be spotted before doing a PCA, and the number of variables reduced accordingly. Alternatively, any exact or near-exact linear relationships uncovered by the last few PCs can be used to select a subset of variables that contain most of the information available in all of the original variables. This and related ideas are more relevant to samples than to populations and are discussed further in Sections 3.4 and 6.3.

There will always be the same number of zero eigenvalues for a correlation matrix as for the corresponding covariance matrix, since an exact linear relationship between the elements of $\mathbf{x}$ clearly implies an exact linear relationship between the standardized variables, and vice versa. There is not the same equivalence, however, when it comes to considering equal variance PCs. Equality of some of the eigenvalues in a covariance (correlation) matrix need not imply that any of the eigenvalues of the corresponding correlation (covariance) matrix are equal. A simple example is when the p variables all have equal correlations but unequal variances. If $p > 2$, then the last $(p-1)$ eigenvalues of the correlation matrix are equal (see Morrison, 1976, Section 8.6), but this relationship will not hold, in general, for the covariance matrix. Further discussion of patterns in covariance or correlation matrices, and their implications for the structure of the corresponding PCs, is given in Section 3.8.

3

Mathematical and Statistical Properties of Sample Principal Components

The first part of this chapter is similar in structure to Chapter 2, except that it deals with properties of PCs obtained from a sample covariance (or correlation) matrix, rather than from a population covariance (or correlation) matrix. The first two sections of the chapter, as in Chapter 2, describe, respectively, many of the algebraic and geometric properties of PCs. Most of the properties discussed in Chapter 2 are almost the same for samples as for populations. They will be mentioned again, but only briefly. There are, in addition, some properties that are relevant only to sample PCs, and these will be discussed more fully.

The third and fourth sections of the chapter again mirror those of Chapter 2. The third section discusses, with an example, the choice between correlation and covariance matrices, while the fourth section looks at the implications of equal and/or zero variances among the PCs, and illustrates the potential usefulness of the last few PCs in detecting near-constant relationships between the variables.

The last five sections of the chapter cover material having no counterpart in Chapter 2. Section 3.5 discusses the singular value decomposition, which could have been included in Section 3.1 as an additional algebraic property. However, the topic is sufficiently important to warrant its own section, as it provides a useful alternative approach to some of the theory surrounding PCs, and also gives an efficient practical method for actually computing PCs.

The sixth section looks at the probability distributions of the coefficients and variances of a set of sample PCs, in other words, the probability distributions of the eigenvectors and eigenvalues of a sample covariance matrix.

The seventh section then goes on to show how these distributions may be used to make statistical inferences about the population PCs, based on sample PCs.

Section 3.8 demonstrates how the approximate structure and variances of PCs can sometimes be deduced from patterns in the covariance or correlation matrix. Finally, in Section 3.9 we discuss models that have been proposed for PCA. The material could equally well have been included in Chapter 2, but because the idea of maximum likelihood estimation arises in some of the models we include it in the present chapter.

3.1   Optimal Algebraic Properties of Sample Principal Components

Before looking at the properties themselves, we need to establish some notation. Suppose that we have n independent observations on the p-element random vector $\mathbf{x}$; denote these n observations by $\mathbf{x}_1, \mathbf{x}_2, \ldots, \mathbf{x}_n$. Let $\tilde{z}_{i1} = \mathbf{a}_1' \mathbf{x}_i$, $i = 1, 2, \ldots, n$, and choose the vector of coefficients $\mathbf{a}_1'$ to maximize the sample variance

$$\frac{1}{n-1} \sum_{i=1}^{n} (\tilde{z}_{i1} - \bar{z}_1)^2$$

subject to the normalization constraint $\mathbf{a}_1' \mathbf{a}_1 = 1$. Next let $\tilde{z}_{i2} = \mathbf{a}_2' \mathbf{x}_i$, $i = 1, 2, \ldots, n$, and choose $\mathbf{a}_2'$ to maximize the sample variance of the $\tilde{z}_{i2}$ subject to the normalization constraint $\mathbf{a}_2' \mathbf{a}_2 = 1$, and subject also to the $\tilde{z}_{i2}$ being uncorrelated with the $\tilde{z}_{i1}$ in the sample. Continuing this process in an obvious manner, we have a sample version of the definition of PCs given in Section 1.1. Thus $\mathbf{a}_k' \mathbf{x}$ is defined as the kth sample PC, $k = 1, 2, \ldots, p$, and $\tilde{z}_{ik}$ is the *score* for the ith observation on the kth PC. If the derivation in Section 1.1 is followed through, but with sample variances and covariances replacing population quantities, then it turns out that the sample variance of the PC scores for the kth sample PC is l_k, the kth largest eigenvalue of the sample covariance matrix $\mathbf{S}$ for $\mathbf{x}_1, \mathbf{x}_2, \ldots, \mathbf{x}_n$, and $\mathbf{a}_k$ is the corresponding eigenvector for $k = 1, 2, \ldots, p$.

Define the $(n \times p)$ matrices $\tilde{\mathbf{X}}$ and $\tilde{\mathbf{Z}}$ to have (i, k)th elements equal to the value of the kth element $\tilde{x}_{ik}$ of $\mathbf{x}_i$, and to $\tilde{z}_{ik}$, respectively. Then $\tilde{\mathbf{Z}}$ and $\tilde{\mathbf{X}}$ are related by $\tilde{\mathbf{Z}} = \tilde{\mathbf{X}} \mathbf{A}$, where $\mathbf{A}$ is the $(p \times p)$ orthogonal matrix whose kth column is $\mathbf{a}_k$.

If the mean of each element of $\mathbf{x}$ is known to be zero, then $\mathbf{S} = \frac{1}{n} \tilde{\mathbf{X}}' \tilde{\mathbf{X}}$. It is far more usual for the mean of $\mathbf{x}$ to be unknown, and in this case the (j, k)th element of $\mathbf{S}$ is

$$\frac{1}{n-1} \sum_{i=1}^{n} (\tilde{x}_{ij} - \bar{x}_j)(\tilde{x}_{ik} - \bar{x}_k),$$

where

$$\bar{x}_j = \frac{1}{n} \sum_{i=1}^{n} \tilde{x}_{ij}, \qquad j = 1, 2, \ldots, p.$$

The matrix $\mathbf{S}$ can therefore be written as

$$\mathbf{S} = \frac{1}{n-1} \mathbf{X}'\mathbf{X}, \qquad (3.1.1)$$

where $\mathbf{X}$ is an $(n \times p)$ matrix with (i,j)th element $(\tilde{x}_{ij} - \bar{x}_j)$; the representation (3.1.1) will be very useful in this and subsequent chapters. The notation x_{ij} will be used to denote the (i,j)th element of $\mathbf{X}$, so that x_{ij} is the value of the jth variable *measured about its mean* $\bar{x}_j$ for the ith observation. A final notational point is that it will be convenient to define the matrix of PC scores as

$$\mathbf{Z} = \mathbf{X}\mathbf{A}, \qquad (3.1.2)$$

rather than as it was in the earlier definition. These PC scores will have exactly the same variances and covariances as those given by $\tilde{\mathbf{Z}}$, but will have zero means, rather than means $\bar{z}_k$, $k = 1, 2, \ldots, p$.

Another point to note is that the eigenvectors of $\frac{1}{n-1}\mathbf{X}'\mathbf{X}$ and $\mathbf{X}'\mathbf{X}$ are identical, and the eigenvalues of $\frac{1}{n-1}\mathbf{X}'\mathbf{X}$ are simply $\frac{1}{n-1}$ (the eigenvalues of $\mathbf{X}'\mathbf{X}$). Because of these relationships it will be convenient in some places below to work in terms of eigenvalues and eigenvectors of $\mathbf{X}'\mathbf{X}$, rather than directly with those of $\mathbf{S}$.

Turning to the algebraic properties A1–A5 listed in Section 2.1, define

$$\mathbf{y}_i = \mathbf{B}'\mathbf{x}_i \qquad \text{for } i = 1, 2, \ldots, n, \qquad (3.1.3)$$

where $\mathbf{B}$, as in Properties A1, A2, A4, A5, is a $(p \times q)$ matrix whose columns are orthonormal. Then Properties A1, A2, A4, A5, still hold, but with the sample covariance matrix of the observations $\mathbf{y}_i$, $i = 1, 2, \ldots, n$, replacing $\mathbf{\Sigma}_y$, and with the matrix $\mathbf{A}$ now defined as having kth column $\mathbf{a}_k$, with $\mathbf{A}_q$, $\mathbf{A}_q^*$, respectively, representing its first and last q columns. Proofs in all cases are similar to those for populations, after making appropriate substitutions of sample quantities in place of population quantities, and will not be repeated. Property A5 reappears as Property G3 in the next section and a proof will be given there.

The spectral decomposition, Property A3, also holds for samples in the form

$$\mathbf{S} = l_1 \mathbf{a}_1 \mathbf{a}_1' + l_2 \mathbf{a}_2 \mathbf{a}_2' + \cdots + l_p \mathbf{a}_p \mathbf{a}_p'. \qquad (3.1.4)$$

The statistical implications of this expression, and the other algebraic properties, A1, A2, A4, A5, are virtually the same as for the corresponding population properties in Section 2.1, except that they must now be viewed in a sample context.

In the case of sample correlation matrices, one further reason can be put forward for interest in the last few PCs, as found by Property A2. Raveh (1985) argues that the inverse $\mathbf{R}^{-1}$ of a correlation matrix is of greater interest in some situations than $\mathbf{R}$. It may then be more important to approximate $\mathbf{R}^{-1}$ than $\mathbf{R}$ in a few dimensions. If this is done using the spectral decomposition (Property A3) of $\mathbf{R}^{-1}$, then the first few terms will correspond to the last few PCs, since eigenvectors of $\mathbf{R}$ and $\mathbf{R}^{-1}$ are the same, except that their order is reversed. The rôle of the last few PCs will be discussed further in Sections 3.4 and 3.7, and again in Sections 6.3, 8.4, 8.6 and 10.1.

One further property, which is concerned with the use of principal components in regression, will now be discussed. Standard terminology from regression is used and will not be explained in detail (see, for example, Draper and Smith (1998)). An extensive discussion of the use of principal components in regression is given in Chapter 8.

Property A7.   *Suppose now that* $\mathbf{X}$, *defined as above, consists of n observations on p predictor variables* $\mathbf{x}$ *measured about their sample means, and that the corresponding regression equation is*

$$\mathbf{y} = \mathbf{X}\boldsymbol{\beta} + \boldsymbol{\epsilon}, \tag{3.1.5}$$

where $\mathbf{y}$ *is the vector of n observations on the dependent variable, again measured about the sample mean. (The notation* $\mathbf{y}$ *for the dependent variable has no connection with the usage of* $\mathbf{y}$ *elsewhere in the chapter, but is standard in regression.) Suppose that* $\mathbf{X}$ *is transformed by the equation* $\mathbf{Z} = \mathbf{XB}$, *where* $\mathbf{B}$ *is a $(p \times p)$ orthogonal matrix. The regression equation can then be rewritten as*

$$\mathbf{y} = \mathbf{Z}\boldsymbol{\gamma} + \boldsymbol{\epsilon},$$

where $\boldsymbol{\gamma} = \mathbf{B}^{-1}\boldsymbol{\beta}$. *The usual least squares estimator for* $\boldsymbol{\gamma}$ *is* $\hat{\boldsymbol{\gamma}} = (\mathbf{Z}'\mathbf{Z})^{-1}\mathbf{Z}'\mathbf{y}$. *Then the elements of* $\hat{\boldsymbol{\gamma}}$ *have, successively, the smallest possible variances if* $\mathbf{B} = \mathbf{A}$, *the matrix whose kth column is the kth eigenvector of* $\mathbf{X}'\mathbf{X}$, *and hence the kth eigenvector of* $\mathbf{S}$. *Thus* $\mathbf{Z}$ *consists of values of the sample principal components for* $\mathbf{x}$.

PROOF.   From standard results in regression (Draper and Smith, 1998, Section 5.2) the covariance matrix of the least squares estimator $\hat{\boldsymbol{\gamma}}$ is proportional to

$$\begin{aligned} (\mathbf{Z}'\mathbf{Z})^{-1} &= (\mathbf{B}'\mathbf{X}'\mathbf{XB})^{-1} \\ &= \mathbf{B}^{-1}(\mathbf{X}'\mathbf{X})^{-1}(\mathbf{B}')^{-1} \\ &= \mathbf{B}'(\mathbf{X}'\mathbf{X})^{-1}\mathbf{B}, \end{aligned}$$

as $\mathbf{B}$ is orthogonal. We require $\mathrm{tr}(\mathbf{B}_q'(\mathbf{X}'\mathbf{X})^{-1}\mathbf{B}_q)$, $q = 1, 2, \ldots, p$ be minimized, where $\mathbf{B}_q$ consists of the first q columns of $\mathbf{B}$. But, replacing $\boldsymbol{\Sigma}_y$ by $(\mathbf{X}'\mathbf{X})^{-1}$ in Property A2 of Section 2.1 shows that $\mathbf{B}_q$ must consist of the last q columns of a matrix whose kth column is the kth eigenvector of $(\mathbf{X}'\mathbf{X})^{-1}$. Furthermore, $(\mathbf{X}'\mathbf{X})^{-1}$ has the same eigenvectors as $\mathbf{X}'\mathbf{X}$, except that their order is reversed, so that $\mathbf{B}_q$ must have columns equal to the first q eigenvectors of $\mathbf{X}'\mathbf{X}$. As this holds for $q = 1, 2, \ldots, p$, Property A7 is proved. □

This property seems to imply that replacing the predictor variables in a regression analysis by their first few PCs is an attractive idea, as those PCs omitted have coefficients that are estimated with little precision. The flaw in this argument is that nothing in Property A7 takes account of the strength of the relationship between the dependent variable y and the elements of $\mathbf{x}$, or between y and the PCs. A large variance for $\hat{\gamma}_k$, the kth element of $\boldsymbol{\gamma}$, and hence an imprecise estimate of the degree of relationship between y and the kth PC, z_k, does not preclude a strong relationship between y and z_k (see Section 8.2). Further discussion of Property A7 is given by Fomby et al. (1978).

There are a number of other properties of PCs specific to the sample situation; most have geometric interpretations and are therefore dealt with in the next section.

3.2 Geometric Properties of Sample Principal Components

As with the algebraic properties, the geometric properties of Chapter 2 are also relevant for sample PCs, although with slight modifications to the statistical implications. In addition to these properties, the present section includes a proof of a sample version of Property A5, viewed geometrically, and introduces two extra properties which are relevant to sample, but not population, PCs.

Property G1 is still valid for samples if $\boldsymbol{\Sigma}$ is replaced by $\mathbf{S}$. The ellipsoids $\mathbf{x}'\mathbf{S}^{-1}\mathbf{x} = $ const no longer have the interpretation of being contours of constant probability, though they will provide estimates of such contours if $\mathbf{x}_1, \mathbf{x}_2, \ldots, \mathbf{x}_n$ are drawn from a multivariate normal distribution. Re-introducing a non-zero mean, the ellipsoids

$$(\mathbf{x} - \bar{\mathbf{x}})'\mathbf{S}^{-1}(\mathbf{x} - \bar{\mathbf{x}}) = \text{const}$$

give contours of equal Mahalanobis distance from the sample mean $\bar{\mathbf{x}}$. Flury and Riedwyl (1988, Section 10.6) interpret PCA as successively finding orthogonal directions for which the Mahalanobis distance from the data set to a hypersphere enclosing all the data is minimized (see Sections 5.3, 9.1 and 10.1 for discussion of Mahalanobis distance in a variety of forms).

Property G2 may also be carried over from populations to samples as follows. Suppose that the observations $\mathbf{x}_1, \mathbf{x}_2, \ldots \mathbf{x}_n$ are transformed by

$$\mathbf{y}_i = \mathbf{B}'\mathbf{x}_i, \qquad i = 1, 2, \ldots, n,$$

where $\mathbf{B}$ is a $(p \times q)$ matrix with orthonormal columns, so that $\mathbf{y}_1, \mathbf{y}_2, \ldots, \mathbf{y}_n$, are projections of $\mathbf{x}_1, \mathbf{x}_2, \ldots, \mathbf{x}_n$ onto a q-dimensional subspace. Then

$$\sum_{h=1}^{n} \sum_{i=1}^{n} (\mathbf{y}_h - \mathbf{y}_i)'(\mathbf{y}_h - \mathbf{y}_i)$$

is maximized when $\mathbf{B} = \mathbf{A}_q$. Conversely, the same criterion is minimized when $\mathbf{B} = \mathbf{A}_q^*$.

This property means that if the n observations are projected onto a q-dimensional subspace, then the sum of squared Euclidean distances between all pairs of observations in the subspace is maximized when the subspace is defined by the first q PCs, and minimized when it is defined by the last q PCs. The proof that this property holds is again rather similar to that for the corresponding population property and will not be repeated.

The next property to be considered is equivalent to Property A5. Both are concerned, one algebraically and one geometrically, with least squares linear regression of each variable x_j on the q variables contained in $\mathbf{y}$.

Property G3.   *As before, suppose that the observations $\mathbf{x}_1, \mathbf{x}_2, \ldots, \mathbf{x}_n$ are transformed by $\mathbf{y}_i = \mathbf{B}'\mathbf{x}_i$, $i = 1, 2, \ldots, n$, where $\mathbf{B}$ is a $(p \times q)$ matrix with orthonormal columns, so that $\mathbf{y}_1, \mathbf{y}_2, \ldots, \mathbf{y}_n$ are projections of $\mathbf{x}_1, \mathbf{x}_2, \ldots, \mathbf{x}_n$ onto a q-dimensional subspace. A measure of 'goodness-of-fit' of this q-dimensional subspace to $\mathbf{x}_1, \mathbf{x}_2, \ldots, \mathbf{x}_n$ can be defined as the sum of squared perpendicular distances of $\mathbf{x}_1, \mathbf{x}_2, \ldots, \mathbf{x}_n$ from the subspace. This measure is minimized when $\mathbf{B} = \mathbf{A}_q$.*

PROOF.   The vector $\mathbf{y}_i$ is an orthogonal projection of $\mathbf{x}_i$ onto a q-dimensional subspace defined by the matrix $\mathbf{B}$. Let $\mathbf{m}_i$ denote the position of $\mathbf{y}_i$ in terms of the original coordinates, and $\mathbf{r}_i = \mathbf{x}_i - \mathbf{m}_i$. (See Figure 3.1 for the special case where $p = 2$, $q = 1$; in this case $\mathbf{y}_i$ is a scalar, whose value is the length of $\mathbf{m}_i$.) Because $\mathbf{m}_i$ is an orthogonal projection of $\mathbf{x}_i$ onto a q-dimensional subspace, $\mathbf{r}_i$ is orthogonal to the subspace, so $\mathbf{r}_i'\mathbf{m}_i = 0$. Furthermore, $\mathbf{r}_i'\mathbf{r}_i$ is the squared perpendicular distance of $\mathbf{x}_i$ from the subspace so that the sum of squared perpendicular distances of $\mathbf{x}_1, \mathbf{x}_2, \ldots, \mathbf{x}_n$ from the subspace is

$$\sum_{i=1}^{n} \mathbf{r}_i'\mathbf{r}_i.$$

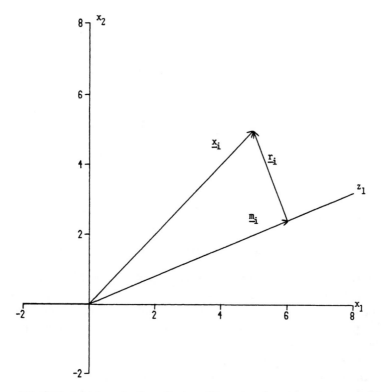

Figure 3.1. Orthogonal projection of a two-dimensional vector onto a one-dimensional subspace.

Now

$$\mathbf{x}_i'\mathbf{x}_i = (\mathbf{m}_i + \mathbf{r}_i)'(\mathbf{m}_i + \mathbf{r}_i)$$
$$= \mathbf{m}_i'\mathbf{m}_i + \mathbf{r}_i'\mathbf{r}_i + 2\mathbf{r}_i'\mathbf{m}_i$$
$$= \mathbf{m}_i'\mathbf{m}_i + \mathbf{r}_i'\mathbf{r}_i.$$

Thus

$$\sum_{i=1}^{n}\mathbf{r}_i'\mathbf{r}_i = \sum_{i=1}^{n}\mathbf{x}_i'\mathbf{x}_i - \sum_{i=1}^{n}\mathbf{m}_i'\mathbf{m}_i,$$

so that, for a given set of observations, minimization of the sum of squared perpendicular distances is equivalent to maximization of $\sum_{i=1}^{n}\mathbf{m}_i'\mathbf{m}_i$. Distances are preserved under orthogonal transformations, so the squared distance $\mathbf{m}_i'\mathbf{m}_i$ of $\mathbf{y}_i$ from the origin is the same in y coordinates as in x coordinates. Therefore, the quantity to be maximized is $\sum_{i=1}^{n}\mathbf{y}_i'\mathbf{y}_i$. But

$$\sum_{i=1}^{n}\mathbf{y}_i'\mathbf{y}_i = \sum_{i=1}^{n}\mathbf{x}_i'\mathbf{BB}'\mathbf{x}_i$$

$$= \operatorname{tr} \sum_{i=1}^{n} (\mathbf{x}_i' \mathbf{B} \mathbf{B}' \mathbf{x}_i)$$

$$= \sum_{i=1}^{n} \operatorname{tr}(\mathbf{x}_i' \mathbf{B} \mathbf{B}' \mathbf{x}_i)$$

$$= \sum_{i=1}^{n} \operatorname{tr}(\mathbf{B}' \mathbf{x}_i \mathbf{x}_i' \mathbf{B})$$

$$= \operatorname{tr}\left[\mathbf{B}' \left(\sum_{i=1}^{n} \mathbf{x}_i \mathbf{x}_i' \right) \mathbf{B} \right]$$

$$= \operatorname{tr}[\mathbf{B}' \mathbf{X}' \mathbf{X} \mathbf{B}]$$

$$= (n-1) \operatorname{tr}(\mathbf{B}' \mathbf{S} \mathbf{B}).$$

Finally, from Property A1, $\operatorname{tr}(\mathbf{B}'\mathbf{SB})$ is maximized when $\mathbf{B} = \mathbf{A}_q$.   □

Instead of treating this property (G3) as just another property of sample PCs, it can also be viewed as an alternative derivation of the PCs. Rather than adapting for samples the algebraic definition of population PCs given in Chapter 1, there is an alternative geometric definition of sample PCs. They are defined as the linear functions (projections) of $\mathbf{x}_1, \mathbf{x}_2, \ldots, \mathbf{x}_n$ that successively define subspaces of dimension $1, 2, \ldots, q, \ldots, (p-1)$ for which the sum of squared perpendicular distances of $\mathbf{x}_1, \mathbf{x}_2, \ldots, \mathbf{x}_n$ from the subspace is minimized. This definition provides another way in which PCs can be interpreted as accounting for as much as possible of the total variation in the data, within a lower-dimensional space. In fact, this is essentially the approach adopted by Pearson (1901), although he concentrated on the two special cases, where $q = 1$ and $q = (p-1)$. Given a set of points in p-dimensional space, Pearson found the 'best-fitting line,' and the 'best-fitting hyperplane,' in the sense of minimizing the sum of squared deviations of the points from the line or hyperplane. The best-fitting line determines the first principal component, although Pearson did not use this terminology, and the direction of the last PC is orthogonal to the best-fitting hyperplane. The scores for the last PC are simply the perpendicular distances of the observations from this best-fitting hyperplane.

Property G4.   *Let* $\mathbf{X}$ *be the* $(n \times p)$ *matrix whose* (i,j)th *element is* $\tilde{x}_{ij} - \bar{x}_j$, *and consider the matrix* $\mathbf{XX}'$. *The* ith *diagonal element of* $\mathbf{XX}'$ *is* $\sum_{j=1}^{p} (\tilde{x}_{ij} - \bar{x}_j)^2$, *which is the squared Euclidean distance of* $\mathbf{x}_i$ *from the centre of gravity* $\bar{\mathbf{x}}$ *of the points* $\mathbf{x}_1, \mathbf{x}_2, \ldots, \mathbf{x}_n$, *where* $\bar{\mathbf{x}} = \frac{1}{n} \sum_{i=1}^{n} \mathbf{x}_i$. *Also, the* (h,i)th *element of* $\mathbf{XX}'$ *is* $\sum_{j=1}^{p} (\tilde{x}_{hj} - \bar{x}_j)(\tilde{x}_{ij} - \bar{x}_j)$, *which measures the cosine of the angle between the lines joining* $\mathbf{x}_h$ *and* $\mathbf{x}_i$ *to* $\bar{\mathbf{x}}$, *multiplied by the distances of* $\mathbf{x}_h$ *and* $\mathbf{x}_i$ *from* $\bar{\mathbf{x}}$. *Thus* $\mathbf{XX}'$ *contains information about the configuration of* $\mathbf{x}_1, \mathbf{x}_2, \ldots, \mathbf{x}_n$ *relative to* $\bar{\mathbf{x}}$. *Now suppose that* $\mathbf{x}_1, \mathbf{x}_2, \ldots, \mathbf{x}_n$ *are projected onto a* q-*dimensional subspace with the usual orthogonal transformation* $\mathbf{y}_i = \mathbf{B}' \mathbf{x}_i$, $i = 1, 2, \ldots, n$. *Then the transfor-*

mation for which $\mathbf{B} = \mathbf{A}_q$ *minimizes the distortion in the configuration as measured by* $\|\mathbf{YY'} - \mathbf{XX'}\|$, *where* $\| \cdot \|$ *denotes Euclidean norm and* $\mathbf{Y}$ *is a matrix with* (i, j)*th element* $\tilde{y}_{ij} - \bar{y}_j$.

PROOF. $\mathbf{Y} = \mathbf{XB}$, so

$$\mathbf{YY'} = \mathbf{XBB'X} \quad \text{and} \quad \|\mathbf{YY'} - \mathbf{XX'}\| = \|\mathbf{XBB'X'} - \mathbf{XX'}\|.$$

A matrix result given by Rao (1973, p. 63) states that if $\mathbf{F}$ is a symmetric matrix of rank p with spectral decomposition

$$\mathbf{F} = f_1\boldsymbol{\phi}_1\boldsymbol{\phi}_1' + f_2\boldsymbol{\phi}_2\boldsymbol{\phi}_2' + \cdots + f_p\boldsymbol{\phi}_p\boldsymbol{\phi}_p',$$

and $\mathbf{G}$ is a matrix of rank $q < p$ chosen to minimize $\|\mathbf{F} - \mathbf{G}\|$, then

$$\mathbf{G} = f_1\boldsymbol{\phi}_1\boldsymbol{\phi}_1' + f_2\boldsymbol{\phi}_2\boldsymbol{\phi}_2' + \cdots + f_q\boldsymbol{\phi}_q\boldsymbol{\phi}_q'.$$

Assume that $\mathbf{X}$ has rank p, so that $\mathbf{x}_i - \bar{\mathbf{x}}$, $i = 1, 2, \ldots, n$, span p-dimensional space, and are not contained in any proper subspace. Then $\mathbf{XX'}$ also has rank p, and Rao's result can be used with $\mathbf{F} = \mathbf{XX'}$, and $\mathbf{G} = \mathbf{YY'}$.

Now, if l_k, $\mathbf{a}_k$ denote the kth eigenvalue and eigenvector, respectively, of $\mathbf{X'X}$, then the kth eigenvalue and eigenvector of $\mathbf{XX'}$ are l_k and $l_k^{-1/2}\mathbf{Xa}_k$, respectively, $k = 1, 2, \ldots, p$. The remaining $(n - p)$ eigenvalues of $\mathbf{XX'}$ are zero.

Using Rao's result above, $\|\mathbf{YY'} - \mathbf{XX'}\|$ is minimized when

$$\mathbf{G} = \mathbf{XBB'X'} = l_1^{-1}l_1\mathbf{Xa}_1\mathbf{a}_1'\mathbf{X'} + l_2^{-1}l_2\mathbf{Xa}_2\mathbf{a}_2'\mathbf{X'} + \cdots + l_q^{-1}l_q\mathbf{Xa}_q\mathbf{a}_q'\mathbf{X'},$$

or

$$\mathbf{XBB'X'} = \mathbf{Xa}_1\mathbf{a}_1'\mathbf{X'} + \mathbf{Xa}_2\mathbf{a}_2'\mathbf{X'} + \cdots + \mathbf{Xa}_q\mathbf{a}_q'\mathbf{X'}.$$

Multiplying both sides of this equation on the left by $(\mathbf{X'X})^{-1}\mathbf{X'}$ and on the right by $\mathbf{X}(\mathbf{X'X})^{-1}$, gives

$$\mathbf{BB'} = \mathbf{a}_1\mathbf{a}_1' + \mathbf{a}_2\mathbf{a}_2' + \cdots + \mathbf{a}_q\mathbf{a}_q',$$

from which it follows that the columns of $\mathbf{B}$ and the first q eigenvectors of $\mathbf{X'X}$, or equivalently of $\mathbf{S}$, span the same q-dimensional subspace. In other words, the transformation $\mathbf{B} = \mathbf{A}_q$ provides the required optimal subspace. □

Note that the result given by Rao (1973, p. 63) which was used in the above proof implies that the sum of the first q terms in the spectral decomposition of the sample covariance (or correlation) matrix $\mathbf{S}$ provides the rank q matrix $_q\mathbf{S}$ that minimizes $\|_q\mathbf{S} - \mathbf{S}\|$. Furthermore, $\|_q\mathbf{S} - \mathbf{S}\| = \sum_{k=q+1}^{p} l_k$, where l_k now denotes the kth eigenvalue of $\mathbf{S}$, rather than that of $\mathbf{X'X}$. The result follows because

$$\|_q\mathbf{S} - \mathbf{S}\| = \left\| \sum_{k=q+1}^{p} l_k\mathbf{a}_k\mathbf{a}_k' \right\|$$

$$= \sum_{k=q+1}^{p} l_k \left\| \mathbf{a}_k \mathbf{a}_k' \right\|$$

$$= \sum_{k=q+1}^{p} l_k \left[\sum_{i=1}^{p} \sum_{j=1}^{p} (a_{ki} a_{kj})^2 \right]^{1/2}$$

$$= \sum_{k=q+1}^{p} l_k \left[\sum_{i=1}^{p} a_{ki}^2 \sum_{j=1}^{p} a_{kj}^2 \right]^{1/2}$$

$$= \sum_{k=q+1}^{p} l_k,$$

as $\mathbf{a}_k' \mathbf{a}_k = 1$, $k = 1, 2, \ldots, p$.

Property G4 is very similar to another optimality property of PCs, discussed in terms of the so-called RV-coefficient by Robert and Escoufier (1976). The RV-coefficient was introduced as a measure of the similarity between two configurations of n data points, as described by $\mathbf{XX}'$ and $\mathbf{YY}'$. The distance between the two configurations is defined by Robert and Escoufier (1976) as

$$\left\| \frac{\mathbf{XX}'}{\{\text{tr}(\mathbf{XX}')^2\}^{1/2}} - \frac{\mathbf{YY}'}{\{\text{tr}(\mathbf{YY}')^2\}^{1/2}} \right\|, \tag{3.2.1}$$

where the divisors of $\mathbf{XX}'$, $\mathbf{YY}'$ are introduced simply to standardize the representation of each configuration in the sense that

$$\left\| \frac{\mathbf{XX}'}{\{\text{tr}(\mathbf{XX}')^2\}^{1/2}} \right\| = \left\| \frac{\mathbf{YY}'}{\{\text{tr}(\mathbf{YY}')^2\}^{1/2}} \right\| = 1.$$

It can then be shown that (3.2.1) equals $[2(1 - \text{RV}(\mathbf{X}, \mathbf{Y}))]^{1/2}$, where the RV-coefficient is defined as

$$\text{RV}(\mathbf{X}, \mathbf{Y}) = \frac{\text{tr}(\mathbf{XY}'\mathbf{YX}')}{\{\text{tr}(\mathbf{XX}')^2 \, \text{tr}(\mathbf{YY}')^2\}^{1/2}}. \tag{3.2.2}$$

Thus, minimizing the distance measure (3.2.1) which, apart from standardizations, is the same as the criterion of Property G4, is equivalent to maximization of $\text{RV}(\mathbf{X}, \mathbf{Y})$. Robert and Escoufier (1976) show that several multivariate techniques can be expressed in terms of maximizing $\text{RV}(\mathbf{X}, \mathbf{Y})$ for some definition of $\mathbf{X}$ and $\mathbf{Y}$. In particular, if $\mathbf{Y}$ is restricted to be of the form $\mathbf{Y} = \mathbf{XB}$, where $\mathbf{B}$ is a $(p \times q)$ matrix such that the columns of $\mathbf{Y}$ are uncorrelated, then maximization of $\text{RV}(\mathbf{X}, \mathbf{Y})$ leads to $\mathbf{B} = \mathbf{A}_q$, that is $\mathbf{Y}$ consists of scores on the first q PCs. We will meet the RV-coefficient again in Chapter 6 in the context of variable selection.

Property G5.   *The algebraic derivation of sample PCs reduces to finding, successively, vectors $\mathbf{a}_k$, $k = 1, 2, \ldots, p$, that maximize $\mathbf{a}_k' \mathbf{S} \mathbf{a}_k$ subject to $\mathbf{a}_k' \mathbf{a}_k = 1$, and subject to $\mathbf{a}_k' \mathbf{a}_l = 0$ for $l < k$. This statement of the problem can be viewed geometrically as follows (Stuart, 1982).*

Consider the first PC; this maximizes $\mathbf{a'Sa}$ *subject to* $\mathbf{a'a} = 1$. *But* $\mathbf{a'Sa}$ = const *defines a family of ellipsoids and* $\mathbf{a'a} = 1$ *defines a hypersphere in p-dimensional space, both centred at the origin. The hypersphere* $\mathbf{a'a} = 1$ *will intersect more than one of the ellipsoids in the family* $\mathbf{a'Sa}$ *(unless* $\mathbf{S}$ *is the identity matrix), and the points at which the hypersphere intersects the 'biggest' such ellipsoid (so that* $\mathbf{a'Sa}$ *is maximized) lie on the shortest principal axis of the ellipsoid. A simple diagram, as given by Stuart (1982), readily verifies this result when p = 2. The argument can be extended to show that the first q sample PCs are defined by the q shortest principal axes of the family of ellipsoids* $\mathbf{a'Sa}$ = const. *Although Stuart (1982) introduced this interpretation in terms of sample PCs, it is equally valid for population PCs.*

The earlier geometric property G1 was also concerned with ellipsoids but in the context of multivariate normality, where the ellipsoids $\mathbf{x'\Sigma^{-1}x}$ = const define contours of constant probability and where the first (longest) q principal axes of such ellipsoids define the first q population PCs. In light of Property G5, it is clear that the validity of the Property G1 does not really depend on the assumption of multivariate normality. Maximization of $\mathbf{a'Sa}$ is equivalent to minimization of $\mathbf{a'S^{-1}a}$, and looking for the 'smallest' ellipsoids in the family $\mathbf{a'S^{-1}a}$ = const that intersect the hypersphere $\mathbf{a'a} = 1$ will lead to the largest principal axis of the family $\mathbf{a'S^{-1}a}$. Thus the PCs define, successively, the principal axes of the ellipsoids $\mathbf{a'S^{-1}a}$ = const. Similar considerations hold for the population ellipsoids $\mathbf{a'\Sigma^{-1}a}$ = const, regardless of any assumption of multivariate normality. However, without multivariate normality the ellipsoids lose their interpretation as contours of equal probability, or as estimates of such contours in the sample case.

Further discussion of the geometry of sample PCs, together with connections with other techniques such as principal coordinate analysis (see Section 5.2) and special cases such as compositional data (Section 13.3), is given by Gower (1967).

As with population properties, our discussion of sample properties of PCA is not exhaustive. For example, Qian et al. (1994) consider the concept of stochastic complexity or minimum description length, as described by Rissanen and Yu (2000). They minimize the expected difference in complexity between a p-dimensional data set and the projection of the data onto a q-dimensional subspace. Qian et al. show that, if multivariate normality is assumed, the subset spanned by the first q PCs is obtained.

3.3   Covariance and Correlation Matrices: An Example

The arguments for and against using sample correlation matrices as opposed to covariance matrices are virtually identical to those given for populations in Section 2.3. Furthermore, it is still the case that there is no

Table 3.1. Correlations and standard deviations for eight blood chemistry variables.

	RBLOOD	PLATE	WBLOOD	NEUT	LYMPH	BILIR	SODIUM	POTASS
	Correlation matrix ($n = 72$)							
RBLOOD	1.000							
PLATE	0.290	1.000						
WBLOOD	0.202	0.415	1.000					
NEUT	−0.055	0.285	0.419	1.000				
LYMPH	−0.105	−0.376	−0.521	−0.877	1.000			
BILIR	−0.252	−0.349	−0.441	−0.076	0.206	1.000		
SODIUM	−0.229	−0.164	−0.145	0.023	0.034	0.192	1.000	
POTASS	0.058	−0.129	−0.076	−0.131	0.151	0.077	0.423	1.000
Standard deviations	0.371	41.253	1.935	0.077	0.071	4.037	2.732	0.297

straightforward relationship between the PCs obtained from a correlation matrix and those based on the corresponding covariance matrix. The main purpose of the present section is to give an example illustrating some of the properties of PCs based on sample covariance and correlation matrices.

The data for this example consist of measurements on 8 blood chemistry variables for 72 patients in a clinical trial. The correlation matrix for these data, together with the standard deviations of each of the eight variables, is given in Table 3.1. Two main points emerge from Table 3.1. First, there are considerable differences in the standard deviations, caused mainly by differences in scale for the eight variables, and, second, none of the correlations is particularly large in absolute value, apart from the value of −0.877 for NEUT and LYMPH.

The large differences in standard deviations give a warning that there may be considerable differences between the PCs for the correlation and covariance matrices. That this is indeed true can be seen in Tables 3.2 and 3.3, which give coefficients for the first four components, based on the correlation and covariance matrices respectively. For ease of comparison, the coefficients are rounded to the nearest 0.2. The effect of such severe rounding is investigated for this example in Section 10.3.

Each of the first four PCs for the correlation matrix has moderate-sized coefficients for several of the variables, whereas the first four PCs for the covariance matrix are each dominated by a single variable. The first component is a slight perturbation of the single variable PLATE, which has the largest variance; the second component is almost the same as the variable BILIR with the second highest variance; and so on. In fact, this pattern continues for the fifth and sixth components, which are not shown in Table 3.3. Also, the relative percentages of total variation accounted for by each component closely mirror the variances of the corresponding variables.

Table 3.2. Principal components based on the correlation matrix for eight blood chemistry variables.

Component number	1	2	3	4
	Coefficients			
RBLOOD	0.2	−0.4	0.4	0.6
PLATE	0.4	−0.2	0.2	0.0
WBLOOD	0.4	0.0	0.2	−0.2
NEUT	0.4	0.4	−0.2	0.2
LYMPH	−0.4	−0.4	0.0	−0.2
BILIR	−0.4	0.4	−0.2	0.6
SODIUM	−0.2	0.6	0.4	−0.2
POTASS	−0.2	0.2	0.8	0.0
Percentage of total variation explained	34.9	19.1	15.6	9.7

Table 3.3. Principal components based on the covariance matrix for eight blood chemistry variables.

Component number	1	2	3	4
	Coefficients			
RBLOOD	0.0	0.0	0.0	0.0
PLATE	1.0	0.0	0.0	0.0
WBLOOD	0.0	−0.2	0.0	1.0
NEUT	0.0	0.0	0.0	0.0
LYMPH	0.0	0.0	0.0	0.0
BILIR	0.0	1.0	−0.2	0.2
SODIUM	0.0	0.2	1.0	0.0
POTASS	0.0	0.0	0.0	0.0
Percentage of total variation explained	98.6	0.9	0.4	0.2

Because variable PLATE has a variance 100 times larger than any other variable, the first PC accounts for over 98 percent of the total variation. Thus the first six components for the covariance matrix tell us almost nothing apart from the order of sizes of variances of the original variables. By contrast, the first few PCs for the correlation matrix show that certain non-trivial linear functions of the (standardized) original variables account for substantial, though not enormous, proportions of the total variation in the standardized variables. In particular, a weighted contrast between the first four and the last four variables is the linear function with the largest variance.

This example illustrates the dangers in using a covariance matrix to find PCs when the variables have widely differing variances; the first few PCs will usually contain little information apart from the relative sizes of variances, information which is available without a PCA. There are, however, circumstances in which it has been argued that using the covariance matrix has some advantages; see, for example, Naik and Khattree (1996), although these authors transform their data (track record times for Olympic events are transformed to speeds) in order to avoid highly disparate variances.

Apart from the fact already mentioned in Section 2.3, namely, that it is more difficult to base statistical inference regarding PCs on correlation matrices, one other disadvantage of correlation matrix PCs is that they give coefficients for *standardized* variables and are therefore less easy to interpret directly. To interpret the PCs in terms of the original variables each coefficient must be divided by the standard deviation of the corresponding variable. An example which illustrates this is given in the next section. It must not be forgotten, however, that correlation matrix PCs, when re-expressed in terms of the original variables, are still linear functions of $\mathbf{x}$ that maximize variance with respect to the standardized variables and not with respect to the original variables.

An alternative to finding PCs for either covariance or correlation matrices is to calculate the eigenvectors of $\tilde{\mathbf{X}}'\tilde{\mathbf{X}}$ rather than $\mathbf{X}'\mathbf{X}$, that is, measure variables about zero, rather than about their sample means, when computing 'covariances' and 'correlations.' This idea was noted by Reyment and Jöreskog (1993, Section 5.4) and will be discussed further in Section 14.2.3. 'Principal component analysis' based on measures of association of this form, but for *observations* rather than variables, has been found useful for certain types of geological data (Reyment and Jöreskog, 1993). Another variant, in a way the opposite of that just mentioned, has been used by Buckland and Anderson (1985), among others. Their idea, which is also discussed further in Section 14.2.3, and which again is appropriate for a particular type of data, is to 'correct for the mean' in *both* the rows and columns of $\tilde{\mathbf{X}}$. Further possibilities, such as the use of weights or metrics, are described in Section 14.2.

3.4 Principal Components with Equal and/or Zero Variances

The problems that arise when some of the eigenvalues of a population covariance matrix are zero and/or equal were discussed in Section 2.4; similar considerations hold when dealing with a sample.

In practice, exactly equal non-zero eigenvalues are extremely rare. Even if the underlying population covariance or correlation matrix has a pattern that gives equal eigenvalues, sampling variation will almost always ensure that the sample eigenvalues are unequal. It should be noted, however, that nearly equal eigenvalues need careful attention. The subspace spanned by a set of nearly equal eigenvalues that are well-separated from all other eigenvalues is well-defined and stable, but individual PC directions within that subspace are unstable (see Section 10.3). This has implications for deciding how many components to retain (Section 6.1), for assessing which observations are influential (Section 10.2) and for deciding which components to rotate (Section 11.1).

With carefully selected variables, PCs with zero variances are a relatively rare occurrence. When q zero eigenvalues do occur for a sample covariance or correlation matrix, the implication is that the points $x_1, x_2, \ldots, x_n$ lie in a $(p - q)$-dimensional subspace of p-dimensional space. This means that there are q separate linear functions of the p original variables having constant values for each of the observations $x_1, x_2, \ldots, x_n$. Ideally, constant relationships between the variables should be detected before doing a PCA, and the number of variables reduced so as to avoid them. However, prior detection will not always be possible, and the zero-variance PCs will enable any unsuspected constant relationships to be detected. Similarly, PCs with very small, but non-zero, variances will define near-constant linear relationships. Finding such near-constant relationships may be of considerable interest. In addition, low-variance PCs have a number of more specific potential uses, as will be discussed at the end of Section 3.7 and in Sections 6.3, 8.4, 8.6 and 10.1.

3.4.1 Example

Here we consider a second set of blood chemistry data, this time consisting of 16 variables measured on 36 patients. In fact, these observations and those discussed in the previous section are both subsets of the same larger data set. In the present subset, four of the variables, x_1, x_2, x_3, x_4, sum to 1.00 for 35 patients and to 0.99 for the remaining patient, so that $x_1 + x_2 + x_3 + x_4$ is nearly constant. The last (sixteenth) PC for the correlation matrix has variance less than 0.001, much smaller than the fifteenth, and is (rounding coefficients to the nearest 0.1) $0.7x_1^* + 0.3x_2^* + 0.7x_3^* + 0.1x_4^*$, with all of the other 12 variables having negligible coefficients. Thus, the near-constant relationship has certainly been identified by the last PC, but not in an easily interpretable form. However, a simple interpretation can be restored if the standardized variables are replaced by the original variables by setting $x_j^* = x_j/s_{jj}^{1/2}$, where $s_{jj}^{1/2}$ is the sample standard deviation of x_j. When this is done, the last PC becomes (rounding coefficients to the nearest integer) $11x_1 + 11x_2 + 11x_3 + 11x_4$. The correct near-constant relationship has therefore been discovered exactly, to the degree of rounding used, by the last PC.

3.5 The Singular Value Decomposition

This section describes a result from matrix theory, namely the singular value decomposition (SVD), which is relevant to PCA in several respects. Given an arbitrary matrix $\mathbf{X}$ of dimension $(n \times p)$, which for our purposes will invariably be a matrix of n observations on p variables measured about their means, $\mathbf{X}$ can be written

$$\mathbf{X} = \mathbf{ULA'}, \tag{3.5.1}$$

where

(i) $\mathbf{U}$, $\mathbf{A}$ are $(n \times r)$, $(p \times r)$ matrices, respectively, each of which has orthonormal columns so that $\mathbf{U'U} = \mathbf{I}_r, \mathbf{A'A} = \mathbf{I}_r$;

(ii) $\mathbf{L}$ is an $(r \times r)$ diagonal matrix;

(iii) r is the rank of $\mathbf{X}$.

To prove this result, consider the spectral decomposition of $\mathbf{X'X}$. The last $(p - r)$ terms in (3.1.4) and in the corresponding expression for $\mathbf{X'X}$ are zero, since the last $(p - r)$ eigenvalues are zero if $\mathbf{X}$, and hence $\mathbf{X'X}$, has rank r. Thus

$$(n - 1)\,\mathbf{S} = \mathbf{X'X} = l_1\mathbf{a}_1\mathbf{a}_1' + l_2\mathbf{a}_2\mathbf{a}_2' + \cdots + l_r\mathbf{a}_r\mathbf{a}_r'.$$

[Note that in this section it is convenient to denote the eigenvalues of $\mathbf{X'X}$, rather than those of $\mathbf{S}$, as l_k, $k = 1, 2, \ldots, p$.] Define $\mathbf{A}$ to be the $(p \times r)$ matrix with kth column $\mathbf{a}_k$, define $\mathbf{U}$ as the $(n \times r)$ matrix whose kth column is

$$\mathbf{u}_k = l_k^{-1/2}\mathbf{X}\mathbf{a}_k, \qquad k = 1, 2, \ldots, r,$$

and define $\mathbf{L}$ to be the $(r \times r)$ diagonal matrix with kth diagonal element $l_k^{1/2}$. Then $\mathbf{U}$, $\mathbf{L}$, $\mathbf{A}$ satisfy conditions (i) and (ii) above, and we shall now show that $\mathbf{X} = \mathbf{ULA}'$.

$$\mathbf{ULA}' = \mathbf{U} \begin{bmatrix} l_1^{1/2}\mathbf{a}_1' \\ l_2^{1/2}\mathbf{a}_2' \\ \vdots \\ l_r^{1/2}\mathbf{a}_r' \end{bmatrix}$$

$$= \sum_{k=1}^{r} l_k^{-1/2}\mathbf{X}\mathbf{a}_k l_k^{1/2}\mathbf{a}_k' = \sum_{k=1}^{r} \mathbf{X}\mathbf{a}_k\mathbf{a}_k'$$

$$= \sum_{k=1}^{p} \mathbf{X}\mathbf{a}_k\mathbf{a}_k'.$$

This last step follows because $\mathbf{a}_k$, $k = (r+1), (r+2), \ldots, p$, are eigenvectors of $\mathbf{X}'\mathbf{X}$ corresponding to zero eigenvalues. The vector $\mathbf{X}\mathbf{a}_k$ is a vector of scores on the kth PC; the column-centering of $\mathbf{X}$ and the zero variance of the last $(p-r)$ PCs together imply that $\mathbf{X}\mathbf{a}_k = \mathbf{0}$, $k = (r+1), (r+2), \ldots, p$. Thus

$$\mathbf{ULA}' = \mathbf{X} \sum_{k=1}^{p} \mathbf{a}_k\mathbf{a}_k' = \mathbf{X},$$

as required, because the $(p \times p)$ matrix whose kth column is $\mathbf{a}_k$, is orthogonal, and so has orthonormal rows.

The importance of the SVD for PCA is twofold. First, it provides a computationally efficient method of actually finding PCs (see Appendix A1). It is clear that if we can find $\mathbf{U}, \mathbf{L}, \mathbf{A}$ satisfying (3.5.1), then $\mathbf{A}$ and $\mathbf{L}$ will give us the eigenvectors and the square roots of the eigenvalues of $\mathbf{X}'\mathbf{X}$, and hence the coefficients and standard deviations of the principal components for the sample covariance matrix $\mathbf{S}$. As a bonus we also get in $\mathbf{U}$ scaled versions of PC scores. To see this multiply (3.5.1) on the right by $\mathbf{A}$ to give $\mathbf{XA} = \mathbf{ULA}'\mathbf{A} = \mathbf{UL}$, as $\mathbf{A}'\mathbf{A} = \mathbf{I}_r$. But $\mathbf{XA}$ is an $(n \times r)$ matrix whose kth column consists of the PC scores for the kth PC (see (3.1.2) for the case where $r = p$). The PC scores z_{ik} are therefore given by

$$z_{ik} = u_{ik}l_k^{1/2}, \quad i = 1, 2, \ldots, n, \quad k = 1, 2, \ldots, r,$$

or, in matrix form, $\mathbf{Z} = \mathbf{UL}$, or $\mathbf{U} = \mathbf{ZL}^{-1}$. The variance of the scores for the kth PC is $\frac{l_k}{(n-1)}$, $k = 1, 2, \ldots, p$. [Recall that l_k here denotes the kth eigenvalue of $\mathbf{X}'\mathbf{X}$, so that the kth eigenvalue of $\mathbf{S}$ is $\frac{l_k}{(n-1)}$.] Therefore the scores given by $\mathbf{U}$ are simply those given by $\mathbf{Z}$, but scaled to have variance $\frac{1}{(n-1)}$. Note also that the columns of $\mathbf{U}$ are the eigenvectors of $\mathbf{XX}'$ corresponding to non-zero eigenvalues, and these eigenvectors are of potential interest if the rôles of 'variables' and 'observations' are reversed.

A second virtue of the SVD is that it provides additional insight into what a PCA actually does, and it gives useful means, both graphical and algebraic, of representing the results of a PCA. This has been recognized in different contexts by Mandel (1972), Gabriel (1978), Rasmusson et al. (1981) and Eastment and Krzanowski (1982), and will be discussed further in connection with relevant applications in Sections 5.3, 6.1.5, 9.3, 13.4, 13.5 and 13.6. Furthermore, the SVD is useful in terms of both computation and interpretation in PC regression (see Section 8.1 and Mandel (1982)) and in examining the links between PCA and correspondence analysis (Sections 13.1 and 14.2).

In the meantime, note that (3.5.1) can be written element by element as

$$x_{ij} = \sum_{k=1}^{r} u_{ik} l_k^{1/2} a_{jk}, \tag{3.5.2}$$

where u_{ik}, a_{jk} are the (i, k)th, (j, k)th elements of $\mathbf{U}$, $\mathbf{A}$, respectively, and $l_k^{1/2}$ is the kth diagonal element of $\mathbf{L}$. Thus x_{ij} can be split into parts

$$u_{ik} l_k^{1/2} a_{jk}, \quad k = 1, 2, \ldots, r,$$

corresponding to each of the first r PCs. If only the first m PCs are retained, then

$$_m \tilde{x}_{ij} = \sum_{k=1}^{m} u_{ik} l_k^{1/2} a_{jk} \tag{3.5.3}$$

provides an approximation to x_{ij}. In fact, it can be shown (Gabriel, 1978; Householder and Young, 1938) that $_m \tilde{x}_{ij}$ gives the best possible rank m approximation to x_{ij}, in the sense of minimizing

$$\sum_{i=1}^{n} \sum_{j=1}^{p} (_m x_{ij} - x_{ij})^2, \tag{3.5.4}$$

where $_m x_{ij}$ is any rank m approximation to x_{ij}. Another way of expressing this result is that the $(n \times p)$ matrix whose (i, j)th element is $_m \tilde{x}_{ij}$ minimizes $\|_m \mathbf{X} - \mathbf{X}\|$ over all $(n \times p)$ matrices $_m \mathbf{X}$ with rank m. Thus the SVD provides a sequence of approximations to $\mathbf{X}$ of rank $1, 2, \ldots, r$, which minimize the Euclidean norm of the difference between $\mathbf{X}$ and the approximation $_m \mathbf{X}$. This result provides an interesting parallel to the result given earlier (see the proof of Property G4 in Section 3.2): that the spectral decomposition of $\mathbf{X}'\mathbf{X}$ provides a similar optimal sequence of approximations of rank $1, 2, \ldots, r$ to the matrix $\mathbf{X}'\mathbf{X}$. Good (1969), in a paper extolling the virtues of the SVD, remarks that Whittle (1952) presented PCA in terms of minimizing (3.5.4).

Finally in this section we note that there is a useful generalization of the SVD, which will be discussed in Chapter 14.

3.6 Probability Distributions for Sample Principal Components

A considerable amount of mathematical effort has been expended on deriving probability distributions, mostly asymptotic, for the coefficients in the sample PCs and for the variances of sample PCs or, equivalently, finding distributions for the eigenvectors and eigenvalues of a sample covariance matrix. For example, the first issue of *Journal of Multivariate Analysis* in 1982 contained three papers, totalling 83 pages, on the topic. In recent years there has probably been less theoretical work on probability distributions directly connected to PCA; this may simply reflect the fact that there is little remaining still to do. The distributional results that have been derived suffer from three drawbacks:

 (i) they usually involve complicated mathematics;

 (ii) they are mostly asymptotic;

(iii) they are often based on the assumption that the original set of variables has a multivariate normal distribution.

Despite these drawbacks, the distributional results are useful in some circumstances, and a selection of the main circumstances is given in this section. Their use in inference about the population PCs, given sample PCs, is discussed in the next section.

Assume that $\mathbf{x} \sim N(\boldsymbol{\mu}, \boldsymbol{\Sigma})$, that is, $\mathbf{x}$ has a p-variate normal distribution with mean $\boldsymbol{\mu}$ and covariance matrix $\boldsymbol{\Sigma}$. Although $\boldsymbol{\mu}$ need not be given, $\boldsymbol{\Sigma}$ is assumed known. Then

$$(n-1)\mathbf{S} \sim W_p(\boldsymbol{\Sigma}, n-1),$$

that is $(n-1)\mathbf{S}$ has the so-called Wishart distribution with parameters $\boldsymbol{\Sigma}, (n-1)$ (see, for example, Mardia et al. (1979, Section 3.4)). Therefore, investigation of the sampling properties of the coefficients and variances of the sample PCs is equivalent to looking at sampling properties of eigenvectors and eigenvalues of Wishart random variables.

The density function of a matrix $\mathbf{V}$ that has the $W_p(\boldsymbol{\Sigma}, n-1)$ distribution is

$$c|\mathbf{V}|^{(n-p-2)/2} \exp\left\{-\frac{1}{2}\operatorname{tr}(\boldsymbol{\Sigma}^{-1}\mathbf{V})\right\},$$

where

$$c^{-1} = 2^{p(n-1)/2}\Pi^{p(1-p)/4}|\boldsymbol{\Sigma}|^{(n-1)/2}\prod_{j=1}^{p}\Gamma\left(\frac{n-j}{2}\right),$$

and various properties of Wishart random variables have been thoroughly investigated (see, for example, Srivastava and Khatri, 1979, Chapter 3).

Let l_k, $\mathbf{a}_k$, for $k = 1, 2, \ldots, p$ be the eigenvalues and eigenvectors of $\mathbf{S}$, respectively, and let λ_k, $\boldsymbol{\alpha}_k$, for $k = 1, 2, \ldots, p$, be the eigenvalues and eigenvectors of $\boldsymbol{\Sigma}$, respectively. Also, let $\mathbf{l}$, $\boldsymbol{\lambda}$ be the p-element vectors consisting of the l_k and λ_k, respectively and let the jth elements of $\mathbf{a}_k$, $\boldsymbol{\alpha}_k$ be a_{kj}, α_{kj}, respectively. [The notation a_{jk} was used for the jth element of $\mathbf{a}_k$ in the previous section, but it seems more natural to use a_{kj} in this and the next, section. We also revert to using l_k to denote the kth eigenvalue of $\mathbf{S}$ rather than that of $\mathbf{X}'\mathbf{X}$.] The best known and simplest results concerning the distribution of the l_k and the $\mathbf{a}_k$ assume, usually quite realistically, that $\lambda_1 > \lambda_2 > \cdots > \lambda_p > 0$; in other words all the population eigenvalues are positive and distinct. Then the following results hold *asymptotically*:

(i) all of the l_k are independent of all of the $\mathbf{a}_k$;

(ii) $\mathbf{l}$ and the $\mathbf{a}_k$ are jointly normally distributed;

(iii)

$$E(\mathbf{l}) = \boldsymbol{\lambda}, \quad E(\mathbf{a}_k) = \boldsymbol{\alpha}_k, \quad k = 1, 2, \ldots, p; \qquad (3.6.1)$$

(iv)

$$\mathrm{cov}(l_k, l_{k'}) = \begin{cases} \dfrac{2\lambda_k^2}{n-1} & k = k', \\ 0 & k \neq k', \end{cases} \qquad (3.6.2)$$

$$\mathrm{cov}(a_{kj}, a_{k'j'}) = \begin{cases} \dfrac{\lambda_k}{(n-1)} \displaystyle\sum_{\substack{l=1 \\ l \neq k}}^{p} \dfrac{\lambda_l \alpha_{lj} \alpha_{lj'}}{(\lambda_l - \lambda_k)^2} & k = k', \\ -\dfrac{\lambda_k \lambda_{k'} \alpha_{kj} \alpha_{k'j'}}{(n-1)(\lambda_k - \lambda_{k'})^2} & k \neq k'. \end{cases} \qquad (3.6.3)$$

An extension of the above results to the case where some of the λ_k may be equal to each other, though still positive, is given by Anderson (1963), and an alternative proof to that of Anderson can be found in Srivastava and Khatri (1979, Section 9.4.1).

It should be stressed that the above results are asymptotic and therefore only approximate for finite samples. Exact results are available, but only for a few special cases, such as when $\boldsymbol{\Sigma} = \mathbf{I}$ (Srivastava and Khatri, 1979, p. 86) and more generally for l_1, l_p, the largest and smallest eigenvalues (Srivastava and Khatri, 1979, p. 205). In addition, better but more complicated approximations can be found to the distributions of $\mathbf{l}$ and the $\mathbf{a}_k$ in the general case (see Srivastava and Khatri, 1979, Section 9.4; Jackson, 1991, Sections 4.2, 4.5; and the references cited in these sources). One specific point regarding the better approximations is that $E(l_1) > \lambda_1$ and $E(l_p) < \lambda_p$. In general the larger eigenvalues tend to be overestimated and the smaller ones underestimated. By expanding the bias of $\mathbf{l}$ as an estimator of $\boldsymbol{\lambda}$ in terms of powers of n^{-1}, 'corrected' estimates of λ_k can be constructed (Jackson, 1991, Section 4.2.2).

If a distribution other than the multivariate normal is assumed, distributional results for PCs will typically become less tractable. Jackson (1991, Section 4.8) gives a number of references that examine the non-normal case. In addition, for non-normal distributions a number of alternatives to PCs can reasonably be suggested (see Sections 13.1, 13.3 and 14.4).

Another deviation from the assumptions underlying most of the distributional results arises when the n observations are not independent. The classic examples of this are when the observations correspond to adjacent points in time (a time series) or in space. Another situation where non-independence occurs is found in sample surveys, where survey designs are often more complex than simple random sampling, and induce dependence ·between observations (see Skinner et al. (1986)). PCA for non-independent data, especially time series, is discussed in detail in Chapter 12.

As a complete contrast to the strict assumptions made in most work on the distributions of PCs, Efron and Tibshirani (1993, Section 7.2) look at the use of the 'bootstrap' in this context. The idea is, for a particular sample of n observations $\mathbf{x}_1, \mathbf{x}_2, \ldots, \mathbf{x}_n$, to take repeated random samples of size n from the distribution that has $P[\mathbf{x} = \mathbf{x}_i] = \frac{1}{n}$, $i = 1, 2, \ldots, n$, calculate the PCs for each sample, and build up empirical distributions for PC coefficients and variances. These distributions rely only on the structure of the sample, and not on any predetermined assumptions. Care needs to be taken in comparing PCs from different bootstrap samples because of possible reordering and/or sign switching in the PCs from different samples. Failure to account for these phenomena is likely to give misleadingly wide distributions for PC coefficients, and distributions for PC variances that may be too narrow.

3.7 Inference Based on Sample Principal Components

The distributional results outlined in the previous section may be used to make inferences about population PCs, given the sample PCs, provided that the necessary assumptions are valid. The major assumption that $\mathbf{x}$ has a multivariate normal distribution is often not satisfied and the practical value of the results is therefore limited. It can be argued that PCA should only ever be done for data that are, at least approximately, multivariate normal, for it is only then that 'proper' inferences can be made regarding the underlying population PCs. As already noted in Section 2.2, this is a rather narrow view of what PCA can do, as it is a much more widely applicable tool whose main use is descriptive rather than inferential. It can provide valuable descriptive information for a wide variety of data, whether the variables are continuous and normally distributed or not. The majority of applications of PCA successfully treat the technique as a purely descriptive tool, although Mandel (1972) argued that retaining m PCs in an analysis implicitly assumes a model for the data, based on (3.5.3). There has recently been an upsurge of interest in *models* related to PCA; this is discussed further in Section 3.9.

Although the purely inferential side of PCA is a very small part of the overall picture, the ideas of inference can sometimes be useful and are discussed briefly in the next three subsections.

3.7.1 Point Estimation

The maximum likelihood estimator (MLE) for Σ, the covariance matrix of a multivariate normal distribution, is not $\mathbf{S}$, but $\frac{(n-1)}{n}\mathbf{S}$ (see, for example, Press (1972, Section 7.1) for a derivation). This result is hardly surprising, given the corresponding result for the univariate normal. If $\boldsymbol{\lambda}$, $\mathbf{l}$, $\boldsymbol{\alpha}_k$, $\mathbf{a}_k$ and related quantities are defined as in the previous section, then the MLEs of $\boldsymbol{\lambda}$ and $\boldsymbol{\alpha}_k$, $k = 1, 2, \ldots, p$, can be derived from the MLE of Σ and are equal to $\hat{\boldsymbol{\lambda}} = \frac{(n-1)}{n}\mathbf{l}$, and $\hat{\boldsymbol{\alpha}}_k = \mathbf{a}_k$, $k = 1, 2, \ldots, p$, assuming that the elements of $\boldsymbol{\lambda}$ are all positive and distinct. The MLEs are the same in this case as the estimators derived by the method of moments. The MLE for λ_k is biased but asymptotically unbiased, as is the MLE for Σ. As noted in the previous section, $\mathbf{l}$ itself, as well as $\hat{\boldsymbol{\lambda}}$, is a biased estimator for $\boldsymbol{\lambda}$, but 'corrections' can be made to reduce the bias.

In the case where some of the λ_k are equal, the MLE for their common value is simply the average of the corresponding l_k, multiplied by $(n-1)/n$. The MLEs of the $\boldsymbol{\alpha}_k$ corresponding to equal λ_k are not unique; the $(p \times q)$ matrix whose columns are MLEs of $\boldsymbol{\alpha}_k$ corresponding to equal λ_k can be multiplied by any $(q \times q)$ orthogonal matrix, where q is the multiplicity of the eigenvalues, to get another set of MLEs.

Most often, point estimates of $\boldsymbol{\lambda}$, $\boldsymbol{\alpha}_k$ are simply given by $\mathbf{l}$, $\mathbf{a}_k$, and they are rarely accompanied by standard errors. An exception is Flury (1997, Section 8.6). Jackson (1991, Sections 5.3, 7.5) goes further and gives examples that not only include estimated standard errors, but also estimates of the correlations between elements of $\mathbf{l}$ and between elements of $\mathbf{a}_k$ and $\mathbf{a}_{k'}$. The practical implications of these (sometimes large) correlations are discussed in Jackson's examples. Flury (1988, Sections 2.5, 2.6) gives a thorough discussion of asymptotic inference for functions of the variances and coefficients of covariance-based PCs.

If multivariate normality cannot be assumed, and if there is no obvious alternative distributional assumption, then it may be desirable to use a 'robust' approach to the estimation of the PCs: this topic is discussed in Section 10.4.

3.7.2   Interval Estimation

The asymptotic marginal distributions of l_k and a_{kj} given in the previous section can be used to construct approximate confidence intervals for λ_k and α_{kj}, respectively. For l_k, the marginal distribution is, from (3.6.1) and (3.6.2), approximately

$$l_k \sim N(\lambda_k, \frac{2\lambda_k^2}{n-1}) \qquad (3.7.1)$$

so

$$\frac{l_k - \lambda_k}{\lambda_k[2/(n-1)]^{1/2}} \sim N(0,1),$$

which leads to a confidence interval, with confidence coefficient $(1-\alpha)$ for λ_k, of the form

$$\frac{l_k}{[1 + \tau z_{\alpha/2}]^{1/2}} < \lambda_k < \frac{l_k}{[1 - \tau z_{\alpha/2}]^{1/2}}, \qquad (3.7.2)$$

where $\tau^2 = 2/(n-1)$, and $z_{\alpha/2}$ is the upper $(100)\alpha/2$ percentile of the standard normal distribution $N(0,1)$. In deriving this confidence interval it is assumed that n is large enough so that $\tau z_{\alpha/2} < 1$. As the distributional result is asymptotic, this is a realistic assumption. An alternative approximate confidence interval is obtained by looking at the distribution of $\ln(l_k)$. Given (3.7.1) it follows that

$$\ln(l_k) \sim N(\ln(\lambda_k), \frac{2}{n-1}) \quad \text{approximately,}$$

thus removing the dependence of the variance on the unknown parameter λ_k. An approximate confidence interval for $\ln(\lambda_k)$, with confidence coefficient $(1-\alpha)$, is then $\ln(l_k) \pm \tau z_{\alpha/2}$, and transforming back to λ_k gives an approximate confidence interval of the form

$$l_k e^{-\tau z_{\alpha/2}} \le \lambda_k \le l_k e^{\tau z_{\alpha/2}}. \qquad (3.7.3)$$

The l_k are asymptotically independent, and joint confidence regions for several of the λ_k are therefore obtained by simply combining intervals of the form (3.7.2) or (3.7.3), choosing individual confidence coefficients so as to achieve an overall desired confidence level. Approximate confidence intervals for individual α_{kj} can be obtained from the marginal distributions of the a_{kj} whose means and variances are given in (3.6.1) and (3.6.3). The intervals are constructed in a similar manner to those for the λ_k, although the expressions involved are somewhat more complicated. Expressions become still more complicated when looking at joint confidence regions for several α_{kj}, partly because of the non-independence of separate a_{kj}. Consider $\mathbf{a}_k$: From (3.6.1), (3.6.3) it follows that, approximately,

$$\mathbf{a}_k \sim N(\boldsymbol{\alpha}_k, \mathbf{T}_k),$$

where

$$\mathbf{T}_k = \frac{\lambda_k}{(n-1)} \sum_{\substack{l=1 \\ l \neq k}}^{p} \frac{\lambda_l}{(\lambda_l - \lambda_k)^2} \boldsymbol{\alpha}_l \boldsymbol{\alpha}_l'.$$

The matrix $\mathbf{T}_k$ has rank $(p-1)$ as it has a single zero eigenvalue corresponding to the eigenvector $\boldsymbol{\alpha}_k$. This causes further complications, but it can be shown (Mardia et al., 1979, p. 233) that, approximately,

$$(n-1)(\mathbf{a}_k - \boldsymbol{\alpha}_k)'(l_k\mathbf{S}^{-1} + l_k^{-1}\mathbf{S} - 2\mathbf{I}_p)(\mathbf{a}_k - \boldsymbol{\alpha}_k) \sim \chi^2_{(p-1)}. \qquad (3.7.4)$$

Because $\mathbf{a}_k$ is an eigenvector of $\mathbf{S}$ with eigenvalue l_k, it follows that $l_k^{-1}\mathbf{S}\mathbf{a}_k = l_k^{-1}l_k\mathbf{a}_k = \mathbf{a}_k$, $l_k\mathbf{S}^{-1}\mathbf{a}_k = l_kl_k^{-1}\mathbf{a}_k = \mathbf{a}_k$, and

$$(l_k\mathbf{S}^{-1} + l_k^{-1}\mathbf{S} - 2\mathbf{I}_p)\mathbf{a}_k = \mathbf{a}_k + \mathbf{a}_k - 2\mathbf{a}_k = \mathbf{0},$$

so that the result (3.7.4) reduces to

$$(n-1)\boldsymbol{\alpha}_k'(l_k\mathbf{S}^{-1} + l_k^{-1}\mathbf{S} - 2\mathbf{I}_p)\boldsymbol{\alpha}_k \sim \chi^2_{(p-1)}. \qquad (3.7.5)$$

From (3.7.5) an approximate confidence region for $\boldsymbol{\alpha}_k$, with confidence coefficient $(1-\alpha)$, has the form $(n-1)\boldsymbol{\alpha}_k'(l_k\mathbf{S}^{-1} + l_k^{-1}\mathbf{S} - 2\mathbf{I}_p)\boldsymbol{\alpha}_k \leq \chi^2_{(p-1);\alpha}$ with fairly obvious notation.

Moving away from assumptions of multivariate normality, the non-parametric bootstrap of Efron and Tibshirani (1993), noted in Section 3.6, can be used to find confidence intervals for various parameters. In their Section 7.2, Efron and Tibshirani (1993) use bootstrap samples to estimate standard errors of estimates for α_{kj}, and for the proportion of total variance accounted for by an individual PC. Assuming approximate normality and unbiasedness of the estimates, the standard errors can then be used to find confidence intervals for the parameters of interest. Alternatively, the ideas of Chapter 13 of Efron and Tibshirani (1993) can be used to construct an interval for λ_k with confidence coefficient $(1-\alpha)$, for example, consisting of a proportion $(1-\alpha)$ of the values of l_k arising from the replicated bootstrap samples. Intervals for elements of $\boldsymbol{\alpha}_k$ can be found in a similar manner. Milan and Whittaker (1995) describe a related but different idea, the parametric bootstrap. Here, residuals from a model based on the SVD, rather than the observations themselves, are bootstrapped. An example of bivariate confidence intervals for $(\alpha_{1j}, \alpha_{2j})$ is given by Milan and Whittaker.

Some theory underlying non-parametric bootstrap confidence intervals for eigenvalues and eigenvectors of covariance matrices is given by Beran and Srivastava (1985), while Romanazzi (1993) discusses estimation and confidence intervals for eigenvalues of both covariance and correlation matrices using another computationally intensive distribution-free procedure, the jackknife. Romanazzi (1993) shows that standard errors of eigenvalue estimators based on the jackknife can have substantial bias and are sensitive to outlying observations. Bootstrapping and the jackknife have also been used to assess the stability of subspaces defined by a subset of the PCs, and hence to choose how many PCs to retain. In these circumstances there is more than one plausible way in which to conduct the bootstrap (see Section 6.1.5).

3.7.3 Hypothesis Testing

The same results, obtained from (3.6.1)–(3.6.3), which were used above to derive confidence intervals for individual l_k and a_{kj}, are also useful for constructing tests of hypotheses. For example, if it is required to test H_0 : $\lambda_k = \lambda_{k0}$ against $H_1 : \lambda_k \neq \lambda_{k0}$, then a suitable test statistic is

$$\frac{l_k - \lambda_{k0}}{\tau \lambda_{k0}},$$

which has, approximately, an $N(0,1)$ distribution under H_0, so that H_0 would be rejected at significance level α if

$$\left| \frac{l_k - \lambda_{k0}}{\tau \lambda_{k0}} \right| \geq z_{\alpha/2}.$$

Similarly, the result (3.7.5) can be used to test $H_0 : \boldsymbol{\alpha}_k = \boldsymbol{\alpha}_{k0}$ vs. $H_1 :$ $\boldsymbol{\alpha}_k \neq \boldsymbol{\alpha}_{k0}$. A test of H_0 against H_1 will reject H_0 at significance level α if

$$(n - 1)\boldsymbol{\alpha}'_{k0}(l_k \mathbf{S}^{-1} + l_k^{-1}\mathbf{S} - 2\mathbf{I}_p)\boldsymbol{\alpha}_{k0} \geq \chi^2_{(p-1);\alpha}.$$

This is, of course, an approximate test, although modifications can be made to the test statistic to improve the χ^2 approximation (Schott, 1987). Other tests, some exact, assuming multivariate normality of $\mathbf{x}$, are also available (Srivastava and Khatri, 1979, Section 9.7; Jackson, 1991, Section 4.6). Details will not be given here, partly because it is relatively unusual that a particular pattern can be postulated for the coefficients of an individual population PC, so that such tests are of limited practical use. An exception is the isometry hypothesis in the analysis of size and shape (Jolicoeur (1984)). Size and shape data are discussed briefly in Section 4.1, and in more detail in Section 13.2.

There are a number of tests concerning other types of patterns in $\boldsymbol{\Sigma}$ and its eigenvalues and eigenvectors. The best known of these is the test of $H_{0q} : \lambda_{q+1} = \lambda_{q+2} = \cdots = \lambda_p$, that is, the case where the last $(p-q)$ eigenvalues are equal, against the alternative H_{1q}, the case where at least two of the last $(p-q)$ eigenvalues are different. In his original paper, Hotelling (1933) looked at the problem of testing the equality of *two* consecutive eigenvalues, and tests of H_{0q} have since been considered by a number of authors, including Bartlett (1950), whose name is sometimes given to such tests. The justification for wishing to test H_{0q} is that the first q PCs may each be measuring some substantial component of variation in $\mathbf{x}$, but the last $(p-q)$ PCs are of equal variation and essentially just measure 'noise.' Geometrically, this means that the distribution of the last $(p-q)$ PCs has spherical contours of equal probability, assuming multivariate normality, and the last $(p - q)$ PCs are therefore not individually uniquely defined. By testing H_{0q} for various values of q it can be decided how many PCs are distinguishable from 'noise' and are therefore worth retaining. This idea for deciding how many components to retain will be discussed critically in Section 6.1.4. It is particularly relevant if a model similar to those described in Section 3.9 is assumed for the data.

A test statistic for H_{0q} against a general alternative H_{1q} can be found by assuming multivariate normality and constructing a likelihood ratio (LR) test. The test statistic takes the form

$$Q = \left\{ \prod_{k=q+1}^{p} l_k \bigg/ \left[\sum_{k=q+1}^{p} l_k/(p-q) \right]^{p-q} \right\}^{n/2}.$$

The exact distribution of Q is complicated, but we can use the well-known general result from statistical inference concerning LR tests, namely that $-2\ln(Q)$ has, approximately, a χ^2 distribution with degrees of freedom equal to the difference between the number of independently varying parameters under $H_{0q} \cup H_{1q}$ and under H_{0q}. Calculating the number of degrees of freedom is non-trivial (Mardia et al., 1979, p. 235), but it turns out to be $\nu = \frac{1}{2}(p - q + 2)(p - q - 1)$, so that approximately, under H_{0q},

$$n \left[(p-q)\ln(\bar{l}) - \sum_{k=q+1}^{p} \ln(l_k) \right] \sim \chi^2_\nu, \tag{3.7.6}$$

where

$$\bar{l} = \sum_{k=q+1}^{p} \frac{l_k}{p-q}.$$

In fact, the approximation can be improved if n is replaced by $n' = n - (2p + 11)/6$, so H_{0q} is rejected at significance level α if

$$n' \left[(p-q)\ln(\bar{l}) - \sum_{k=q+1}^{p} \ln(l_k) \right] \geq \chi^2_{\nu;\alpha}.$$

Another, more complicated, improvement to the approximation is given by Srivastava and Khatri (1979, p. 294). The test is easily adapted so that the null hypothesis defines equality of *any* subset of $(p - q)$ consecutive eigenvalues, not necessarily the smallest (Flury, 1997, Section 8.6). Another modification is to test whether the last $(p - q)$ eigenvalues follow a linear trend (Bentler and Yuan, 1998). The relevance of this null hypothesis will be discussed in Section 6.1.4.

A special case of the test of the null hypothesis H_{0q} occurs when $q = 0$, in which case H_{0q} is equivalent to all the variables being independent and having equal variances, a very restrictive assumption. The test with $q = 0$ reduces to a test that all variables are independent, with no requirement of equal variances, if we are dealing with a correlation matrix. However, it should be noted that all the results in this and the previous section are for covariance, not correlation, matrices, which restricts their usefulness still further.

In general, inference concerning PCs of correlation matrices is more complicated than for covariance matrices (Anderson, 1963; Jackson, 1991, Section 4.7), as the off-diagonal elements of a correlation matrix are non-trivial functions of the random variables which make up the elements of a covariance matrix. For example, the asymptotic distribution of the test statistic (3.7.6) is no longer χ^2 for the correlation matrix, although Lawley (1963) provides an alternative statistic, for a special case, which does have a limiting χ^2 distribution.

Another special case of the test based on (3.7.6) occurs when it is necessary to test

$$H_0 : \Sigma = \sigma^2 \begin{bmatrix} 1 & \rho & \cdots & \rho \\ \rho & 1 & \cdots & \rho \\ \vdots & \vdots & & \\ \rho & \rho & \cdots & 1 \end{bmatrix}$$

against a general alternative. The null hypothesis H_0 states that all variables have the same variance σ^2, and all pairs of variables have the same correlation ρ, in which case

$$\sigma^2[1 + (p-1)\rho] = \lambda_1 > \lambda_2 = \lambda_3 = \cdots = \lambda_p = \sigma^2(1 - \rho)$$

(Morrison, 1976, Section 8.6), so that the last $(p-1)$ eigenvalues are equal. If ρ, σ^2 are unknown, then the earlier test is appropriate with $q = 1$, but if ρ, σ^2 are specified then a different test can be constructed, again based on the LR criterion.

Further tests regarding λ and the a_k can be constructed, such as the test discussed by Mardia et al. (1979, Section 8.4.2) that the first q PCs account for a given proportion of the total variation. However, as stated at the beginning of this section, these tests are of relatively limited value in practice. Not only are most of the tests asymptotic and/or approximate, but they also rely on the assumption of multivariate normality. Furthermore, it is arguable whether it is often possible to formulate a particular hypothesis whose test is of interest. More usually, PCA is used to explore the data, rather than to verify predetermined hypotheses.

To conclude this section on inference, we note that little has been done with respect to PCA from a Bayesian viewpoint. Bishop (1999) is an exception. He introduces prior distributions for the parameters of a model for PCA (see Section 3.9). His main motivation appears to be to provide a means of deciding the dimensionality of the model (see Section 6.1.5).

Lanterman (2000) and Wang and Staib (2000) each use principal components in quantifying prior information in (different) image processing contexts.

Another possible use of PCA when a Bayesian approach to inference is adopted is as follows. Suppose that $\boldsymbol{\theta}$ is a vector of parameters, and that the posterior distribution for $\boldsymbol{\theta}$ has covariance matrix Σ. If we find PCs for $\boldsymbol{\theta}$, then the last few PCs provide information on which linear functions of the elements of $\boldsymbol{\theta}$ can be estimated with high precision (low variance). Conversely, the first few PCs are linear functions of the elements of $\boldsymbol{\theta}$ that can only be estimated with low precision. In this context, then, it would seem that the last few PCs may be more useful than the first few.

3.8   Principal Components for Patterned Correlation or Covariance Matrices

At the end of Chapter 2, and in Section 3.7.3, the structure of the PCs and their variances was discussed briefly in the case of a correlation matrix with equal correlations between all variables. Other theoretical patterns in correlation and covariance matrices can also be investigated; for example, Jolliffe (1970) considered correlation matrices with elements ρ_{ij} for which

$$\rho_{1j} = \rho, \quad j = 2, 3. \ldots, p,$$

and

$$\rho_{ij} = \rho^2, \quad 2 \leq i < j < p,$$

and Brillinger (1981, p. 108) discussed PCs for Töplitz matrices, which occur for time series data (see Chapter 12), and in which the ρ_{ij} depend only on $|i - j|$.

Such exact patterns will not, in general, occur in sample covariance or correlation matrices, but it is sometimes possible to deduce the approximate form of some of the PCs by recognizing a particular type of structure in a sample covariance or correlation matrix. One such pattern, which was discussed in Section 3.3, occurs when one or more of the variances in a covariance matrix are of very different sizes from all the rest. In this case, as illustrated in the example of Section 3.3, there will often be a PC associated with each such variable which is almost indistinguishable from that variable. Similar behaviour, that is, the existence of a PC very similar to one of the original variables, can occur for correlation matrices, but in rather different circumstances. Here the requirement for such a PC is that the corresponding variable is nearly uncorrelated with all of the other variables.

The other main type of pattern detected in many correlation matrices is one where there are one or more groups of variables within which all correlations are positive and not close to zero. Sometimes a variable in such a group will initially have entirely negative correlations with the other members of the group, but the sign of a variable is often arbitrary, and switching the sign will give a group of the required structure. If correlations between the q members of the group and variables outside the group are close to zero, then there will be q PCs 'associated with the group' whose coefficients for variables outside the group are small. One of these PCs will have a large variance, approximately $1 + (q - 1)\bar{r}$, where $\bar{r}$ is the average correlation within the group, and will have positive coefficients for all variables in the group. The remaining $(q - 1)$ PCs will have much smaller variances (of order $1 - \bar{r}$), and will have some positive and some negative coefficients. Thus the 'large variance PC' for the group measures, roughly, the average size of variables in the group, whereas the 'small variance PCs' give 'contrasts' between some or all of the variables in the group. There may be several such groups of variables in a data set, in which case each group will have one 'large variance PC' and several 'small variance PCs.' Conversely, as happens not infrequently, especially in biological applications when all variables are measurements on individuals of some species, we may find that all p variables are positively correlated. In such cases, the first PC is often interpreted as a measure of size of the individuals, whereas subsequent PCs measure aspects of shape (see Sections 4.1, 13.2 for further discussion).

The discussion above implies that the approximate structure and variances of the first few PCs can be deduced from a correlation matrix, provided that well-defined groups of variables are detected, including possibly single-variable groups, whose within-group correlations are high, and whose between-group correlations are low. The ideas can be taken further; upper and lower bounds on the variance of the first PC can be calculated, based on sums and averages of correlations (Friedman and Weisberg, 1981; Jackson, 1991, Section 4.2.3). However, it should be stressed that although data sets for which there is some group structure among variables are not uncommon, there are many others for which no such pattern is apparent. In such cases the structure of the PCs cannot usually be found without actually performing the PCA.

3.8.1 Example

In many of the examples discussed in later chapters, it will be seen that the structure of some of the PCs can be partially deduced from the correlation matrix, using the ideas just discussed. Here we describe an example in which *all* the PCs have a fairly clear pattern. The data consist of measurements of reflexes at 10 sites of the body, measured for 143 individuals. As with the examples discussed in Sections 3.3 and 3.4, the data were kindly supplied by Richard Hews of Pfizer Central Research.

Table 3.4. Correlation matrix for ten variables measuring reflexes.

	V1	V2	V3	V4	V5	V6	V7	V8	V9	V10
V1	1.00									
V2	0.98	1.00								
V3	0.60	0.62	1.00							
V4	0.71	0.73	0.88	1.00						
V5	0.55	0.57	0.61	0.68	1.00					
V6	0.55	0.57	0.56	0.68	0.97	1.00				
V7	0.38	0.40	0.48	0.53	0.33	0.33	1.00			
V8	0.25	0.28	0.42	0.47	0.27	0.27	0.90	1.00		
V9	0.22	0.21	0.19	0.23	0.16	0.19	0.40	0.41	1.00	
V10	0.20	0.19	0.18	0.21	0.13	0.16	0.39	0.40	0.94	1.00

The correlation matrix for these data is given in Table 3.4, and the coefficients of, and the variation accounted for by, the corresponding PCs are presented in Table 3.5. It should first be noted that the ten variables fall into five pairs. Thus, V1, V2, respectively, denote strength of reflexes for right and left triceps, with {V3, V4}, {V5, V6}, {V7, V8}, {V9, V10} similarly defined for right and left biceps, right and left wrists, right and left knees, and right and left ankles. The correlations between variables within each pair are large, so that the differences between variables in each pair have small variances. This is reflected in the last five PCs, which are mainly within-pair contrasts, with the more highly correlated pairs corresponding to the later components.

Turning to the first two PCs, there is a suggestion in the correlation matrix that, although all correlations are positive, the variables can be divided into two groups {V1–V6}, {V7–V10}. These correspond to sites in the arms and legs, respectively. Reflecting this group structure, the first and second PCs have their largest coefficients on the first and second groups of variables, respectively. Because the group structure is not clear-cut, these two PCs also have contributions from the less dominant group, and the first PC is a weighted average of variables from both groups, whereas the second PC is a weighted contrast between the groups.

The third, fourth and fifth PCs reinforce the idea of the two groups. The third PC is a contrast between the two pairs of variables in the second (smaller) group and the fourth and fifth PCs both give contrasts between the three pairs of variables in the first group.

It is relatively rare for examples with as many as ten variables to have such a nicely defined structure as in the present case for all their PCs. However, as will be seen in the examples of subsequent chapters, it is not unusual to be able to deduce the structure of at least a few PCs in this manner.

Table 3.5. Principal components based on the correlation matrix of Table 3.4

Component number	1	2	3	4	5	6	7	8	9	10
	Coefficients									
V1	0.3	−0.2	0.2	−0.5	0.3	0.1	−0.1	−0.0	−0.6	0.2
V2	0.4	−0.2	0.2	−0.5	0.3	0.0	−0.1	−0.0	0.7	−0.3
V3	0.4	−0.1	−0.1	−0.0	−0.7	0.5	−0.2	0.0	0.1	0.1
V4	0.4	−0.1	−0.1	−0.0	−0.4	−0.7	0.3	−0.0	−0.1	−0.1
V5	0.3	−0.2	0.1	0.5	0.2	0.2	−0.0	−0.1	−0.2	−0.6
V6	0.3	−0.2	0.2	0.5	0.2	−0.1	−0.0	0.1	0.2	0.6
V7	0.3	0.3	−0.5	−0.0	0.2	0.3	0.7	0.0	−0.0	0.0
V8	0.3	0.3	−0.5	0.1	0.2	−0.2	−0.7	−0.0	−0.0	−0.0
V9	0.2	0.5	0.4	0.0	−0.1	0.0	−0.0	0.7	−0.0	−0.1
V10	0.2	0.5	0.4	0.0	−0.1	0.0	0.0	−0.7	0.0	0.0
Percentage of total variation explained	52.3	20.4	11.0	8.5	5.0	1.0	0.9	0.6	0.2	0.2

3.9 Models for Principal Component Analysis

There is a variety of interpretations of what is meant by a *model* in the context of PCA. Mandel (1972) considers the retention of m PCs, based on the SVD (3.5.3), as implicitly using a model. Caussinus (1986) discusses three types of 'model.' The first is a 'descriptive algebraic model,' which in its simplest form reduces to the SVD. It can also be generalized to include a choice of metric, rather than simply using a least squares approach. Such generalizations are discussed further in Section 14.2.2. This model has no random element, so there is no idea of expectation or variance. Hence it corresponds to Pearson's geometric view of PCA, rather than to Hotelling's variance-based approach.

Caussinus's (1986) second type of model introduces probability distributions and corresponds to Hotelling's definition. Once again, the 'model' can be generalized by allowing a choice of metric.

The third type of model described by Caussinus is the so-called *fixed effects* model (see also Esposito (1998)). In this model we assume that the rows $x_1, x_2, \ldots, x_n$ of X are independent random variables, such that $E(x_i) = z_i$, where z_i lies in a q-dimensional subspace, F_q. Furthermore, if $e_i = x_i - z_i$, then $E(e_i) = 0$ and $var(e_i) = \frac{\sigma^2}{w_i}\Gamma$, where Γ is a positive definite symmetric matrix and the w_i are positive scalars whose sum is 1. Both Γ and the w_i are assumed to be known, but σ^2, the z_i and the subspace F_q all need to be estimated. This is done by minimizing

$$\sum_{i=1}^{n} w_i \|x_i - z_i\|_M^2, \tag{3.9.1}$$

where $\mathbf{M}$ denotes a metric (see Section 14.2.2) and may be related to $\boldsymbol{\Gamma}$. This statement of the model generalizes the usual form of PCA, for which $w_i = \frac{1}{n}, i = 1, 2, \ldots, n$ and $\mathbf{M} = \mathbf{I}_p$, to allow different weights on the observations and a choice of metric. When $\mathbf{M} = \boldsymbol{\Gamma}^{-1}$, and the distribution of the $\mathbf{x}_i$ is multivariate normal, the estimates obtained by minimizing (3.9.1) are maximum likelihood estimates (Besse, 1994b). An interesting aspect of the fixed effects model is that it moves away from the idea of a sample of identically distributed observations whose covariance or correlation structure is to be explored, to a formulation in which the variation among the *means* of the observations is the feature of interest.

Tipping and Bishop (1999a) describe a model in which column-centred observations $\mathbf{x}_i$ are independent normally distributed random variables with zero means and covariance matrix $\mathbf{BB'} + \sigma^2 \mathbf{I}_p$, where $\mathbf{B}$ is a $(p \times q)$ matrix. We shall see in Chapter 7 that this is a special case of a factor analysis model. The fixed effects model also has links to factor analysis and, indeed, de Leeuw (1986) suggests in discussion of Caussinus (1986) that the model is closer to factor analysis than to PCA. Similar models date back to Young (1941).

Tipping and Bishop (1999a) show that, apart from a renormalization of columns, and the possibility of rotation, the maximum likelihood estimate of $\mathbf{B}$ is the matrix $\mathbf{A}_q$ of PC coefficients defined earlier (see also de Leeuw (1986)). The MLE for σ^2 is the average of the smallest $(p - q)$ eigenvalues of the sample covariance matrix $\mathbf{S}$. Tipping and Bishop (1999a) fit their model using the EM algorithm (Dempster et al. (1977)), treating the unknown underlying components as 'missing values.' Clearly, the complication of the EM algorithm is not necessary once we realise that we are dealing with PCA, but it has advantages when the model is extended to cope with genuinely missing data or to mixtures of distributions (see Sections 13.6, 9.2.3). Bishop (1999) describes a Bayesian treatment of Tipping and Bishop's (1999a) model. The main objective in introducing a prior distribution for $\mathbf{B}$ appears to be as a means of deciding on its dimension q (see Section 6.1.5).

Roweis (1997) also uses the EM algorithm to fit a model for PCA. His model is more general than Tipping and Bishop's, with the error covariance matrix allowed to take any form, rather than being restricted to $\sigma^2 \mathbf{I}_p$. In this respect it is more similar to the fixed effects model with equal weights, but differs from it by not specifying different means for different observations. Roweis (1997) notes that a full PCA, with all p PCs, is obtained from his model in the special case where the covariance matrix is $\sigma^2 \mathbf{I}_p$ and $\sigma^2 \to 0$. He refers to the analysis based on Tipping and Bishop's (1999a) model with $\sigma^2 > 0$ as *sensible principal component analysis*.

Martin (1988) considers another type of probability-based PCA, in which each of the n observations has a probability distribution in p-dimensional space centred on it, rather than being represented by a single point. In the one non-trivial example considered by Martin (1988), the distributions are identical for each observation and spherical, so that the underlying covariance matrix has the form $\Sigma + \sigma^2 I_p$. Lynn and McCulloch (2000) use PCA to estimate latent fixed effects in a generalized linear model, and de Falguerolles (2000) notes that PCA can be viewed as a special case of the large family of generalized bilinear models.

Although PCA is a largely descriptive tool, it can be argued that building a model gives a better understanding of what the technique does, helps to define circumstances in which it would be inadvisable to use it, and suggests generalizations that explore the structure of a data set in a more sophisticated way. We will see how either the fixed effects model or Tipping and Bishop's (1999a) model can be used in deciding how many PCs to retain (Section 6.1.5); in examining mixtures of probability distributions (Section 9.2.3); in a robust version of PCA (Section 10.4); in analysing functional data (Section 12.3.4); in handling missing data (Section 13.6); and in generalizations of PCA (Section 14.1, 14.2). One application that belongs in the present chapter is described by Ferré (1995a). Here $\hat{\mu}_1, \hat{\mu}_2, \ldots, \hat{\mu}_k$ are estimates, derived from k samples of sizes $n_1, n_2, \ldots, n_k$ of vectors of p parameters $\mu_1, \mu_2, \ldots, \mu_k$. Ferré (1995a) proposes estimates that minimize an expression equivalent to (3.9.1) in which $w_i = \frac{n_i}{n}$ where $n = \sum_{i=1}^{k} n_i$; x_i, z_i are replaced by $\mu_i, \hat{\mu}_i$ where $\hat{\mu}_i$ is a projection onto an optimal q-dimensional space; and M is chosen to be S^{-1} where S is an estimate of the common covariance matrix for the data from which $\mu_1, \mu_2, \ldots, \mu_k$ are estimated. The properties of such estimators are investigated in detail by Ferré (1995a)

4
Principal Components as a Small Number of Interpretable Variables: Some Examples

The original purpose of PCA was to reduce a large number (p) of variables to a much smaller number (m) of PCs whilst retaining as much as possible of the variation in the p original variables. The technique is especially useful if $m \ll p$ and if the m PCs can be readily interpreted.

Although we shall see in subsequent chapters that there are many other ways of applying PCA, the original usage as a descriptive, dimension-reducing technique is probably still the most prevalent single application. This chapter simply introduces a number of examples from several different fields of application where PCA not only reduces the dimensionality of the problem substantially, but has PCs which are easily interpreted. Graphical representations of a set of observations with respect to the m retained PCs and discussion of how to choose an appropriate value of m are deferred until Chapters 5 and 6, respectively.

Of course, if m is very much smaller than p, then the reduction of dimensionality alone may justify the use of PCA, even if the PCs have no clear meaning, but the results of a PCA are much more satisfying if intuitively reasonable interpretations can be given to some or all of the m retained PCs.

Each section of this chapter describes one example in detail, but other examples in related areas are also mentioned in most sections. Some of the examples introduced in this chapter are discussed further in subsequent chapters; conversely, when new examples are introduced later in the book, an attempt will be made to interpret the first few PCs where appropriate. The examples are drawn from a variety of fields of application, demonstrating the fact that PCA has been found useful in a very large number of subject areas, of which those illustrated in this book form only a subset.

It must be emphasized that although in many examples the PCs can be readily interpreted, this is by no means universally true. There is no reason, a priori, why a mathematically derived linear function of the original variables (which is what the PCs are) should have a simple interpretation. It is remarkable how often it seems to be possible to interpret the first few PCs, though it is probable that some interpretations owe a lot to the analyst's ingenuity and imagination. Careful thought should go into any interpretation and, at an earlier stage, into the choice of variables and whether to transform them. In some circumstances, transformation of variables before analysis may improve the chances of a simple interpretation (see Sections 13.2, 13.3, 14.1 and 14.2). Conversely, the arbitrary inclusion of logarithms, powers, ratios, etc., of the original variables can make it unlikely that any simple interpretation will be found. Further discussion of the difficulties of interpretation, and of some alternative approaches, will be given in Chapter 11.

Many interesting applications have appeared since the first edition of this book, and some will be discussed in detail later in this edition. However, in the current chapter the original selection of examples, which illustrates a nice range of applications, has been kept. Extra references are given, but no new examples are discussed in detail. Texts such as Jackson (1991), Krzanowski (1988), Krzanowski and Marriott (1994) and Rencher (1995) are useful sources for additional examples. A non-exhaustive list of disciplines in which PCA has been applied was given at the end of Chapter 1.

4.1 Anatomical Measurements

One type of application where PCA has been found useful is identification of the most important sources of variation in anatomical measurements for various species. Typically, a large number of measurements are made on individuals of a species, and a PCA is done. The first PC almost always has positive coefficients for all variables and simply reflects overall 'size' of the individuals. Later PCs usually contrast some of the measurements with others, and can often be interpreted as defining certain aspects of 'shape' that are important for the species. Blackith and Reyment (1971, Chapter 12) mention applications to squirrels, turtles, ammonites, foraminifera (marine microorganisms) and various types of insects. The analysis of size and shape is a large topic in its own right, and will discussed in greater detail in Section 13.2. Here a small data set is examined in which seven measurements were taken for a class of 28 students (15 women, 13 men). The seven measurements are circumferences of chest, waist, wrist and head, lengths of hand and forearm, and overall height. A similar data set for a different group of students was introduced in Chapter 1.

Table 4.1. First three PCs: student anatomical measurements.

Component number		1	2	3
		Women		
Hand		0.33	0.56	0.03
Wrist		0.26	0.62	0.11
Height		0.40	−0.44	−0.00
Forearm	Coefficients	0.41	−0.05	−0.55
Head		0.27	−0.19	0.80
Chest		0.45	−0.26	−0.12
Waist		0.47	0.03	−0.03
Eigenvalue		3.72	1.37	0.97
Cumulative percentage of total variation		53.2	72.7	86.5
		Men		
Hand		0.23	0.62	0.64
Wrist		0.29	0.53	−0.42
Height		0.43	−0.20	0.04
Forearm	Coefficients	0.33	−0.53	0.38
Head		0.41	−0.09	−0.51
Chest		0.44	0.08	−0.01
Waist		0.46	−0.07	0.09
Eigenvalue		4.17	1.26	0.66
Cumulative percentage of total variation		59.6	77.6	87.0

The PCA was done on the correlation matrix, even though it could be argued that, since all measurements are made in the same units, the covariance matrix might be more appropriate (see Sections 2.3 and 3.3). The correlation matrix was preferred because it was desired to treat all variables on an equal footing: the covariance matrix gives greater weight to larger, and hence more variable, measurements, such as height and chest girth, and less weight to smaller measurements such as wrist girth and hand length.

Some of the results of the PC analyses, done separately for women and men, are given in Tables 4.1 and 4.2.

It can be seen that the form of the first two PCs is similar for the two sexes, with some similarity, too, for the third PC. Bearing in mind the small sample sizes, and the consequent large sampling variation in PC coefficients, it seems that the major sources of variation in the measurements, as given by the first three PCs, are similar for each sex. A combined PCA using all 28

Table 4.2. Simplified version of the coefficients in Table 4.1.

Component number	1	2	3
		Women	
Hand	+	+	
Wrist	+	+	
Height	+	−	
Forearm	+		−
Head	+	(−)	+
Chest	+	(−)	
Waist	+		
		Men	
Hand	+	+	+
Wrist	+	+	−
Height	+	(−)	
Forearm	+	−	+
Head	+		−
Chest	+		
Waist	+		

observations therefore seems appropriate, in order to get better estimates of the first three PCs. It is, of course, possible that later PCs are different for the two sexes, and that combining all 28 observations will obscure such differences. However, if we are interested solely in interpreting the first few, high variance, PCs, then this potential problem is likely to be relatively unimportant.

Before we attempt to interpret the PCs, some explanation of Table 4.2 is necessary. Typically, computer packages that produce PCs give the coefficients to several decimal places. When we interpret PCs, as with other types of tabular data, it is usually only the general *pattern* of the coefficients that is really of interest, not values to several decimal places, which may give a false impression of precision. Table 4.1 gives only two decimal places and Table 4.2 simplifies still further. A + or − in Table 4.2 indicates a coefficient whose absolute value is greater than half the maximum coefficient (again in absolute value) for the relevant PC; the sign of the coefficient is also indicated. Similarly, a (+) or (−) indicates a coefficient whose absolute value is between a quarter and a half of the largest absolute value for the PC of interest. There are, of course, many ways of constructing a simplified version of the PC coefficients in Table 4.1. For example, another possibility is to rescale the coefficients in each PC so that the maximum value is ±1, and tabulate only the values of the coefficients, rounded to one decimal place whose absolute values are above a certain cut-off, say 0.5 or 0.7. Values of coefficients below the cut-off are omitted, leaving blank spaces, as in Table 4.2. Some such simple representation is often helpful in interpreting PCs, particularly if a PCA is done on a large number of variables.

Sometimes a simplification such as that given in Table 4.2 may be rather too extreme, and it is therefore advisable to present the coefficients rounded to one or two decimal places as well. Principal components with rounded coefficients will no longer be optimal, so that the variances of the first few will tend to be reduced, and exact orthogonality will be lost. However, it has been shown (Bibby, 1980; Green, 1977) that fairly drastic rounding of coefficients makes little difference to the variances of the PCs (see Section 10.3). Thus, presentation of rounded coefficients will still give linear functions of x with variances very nearly as large as those of the PCs, while at the same time easing interpretations.

It must be stressed that interpretation of PCs is often more subtle than is generally realised. Simplistic interpretation can be misleading. As well as truncation or rounding of PC coefficients, a number of other ideas are available to aid interpretation. Some of these involve manipulation of the PC coefficients themselves, whilst others are based on alternative, but similar, techniques to PCA. In this chapter we concentrate on simple interpretation. Its dangers, and various alternative ways of tackling interpretation, are discussed in Chapter 11.

Turning now to the interpretation of the PCs in the present example, the first PC clearly measures overall 'size' for both sexes, as would be expected (see Section 3.8), as all the correlations between the seven variables are positive. It accounts for 53% (women) or 60% (men) of the total variation. The second PC for both sexes contrasts hand and wrist measurements with height, implying that, after overall size has been accounted for, the main source of variation is between individuals with large hand and wrist measurements relative to their heights, and individuals with the converse relationship. For women, head and chest measurements also have some contribution to this component, and for men the forearm measurement, which is closely related to height, partially replaces height in the component. This second PC accounts for slightly less than 20% of the total variation, for both sexes.

It should be noted that the sign of any PC is completely arbitrary. If every coefficient in a PC, $z_k = a_k'x$, has its sign reversed, the variance of z_k is unchanged, and so is the orthogonality of a_k with all other eigenvectors. For example, the second PC for men as recorded in Tables 4.1 and 4.2 has large positive values for students with large hand and wrist measurements relative to their height. If the sign of a_2, and hence z_2, is reversed, the large positive values now occur for students with small hand and wrist measurements relative to height. The interpretation of the PC remains the same, even though the roles of 'large' and 'small' are reversed.

The third PCs differ more between the sexes but nevertheless retain some similarity. For women it is almost entirely a contrast between head and forearm measurements; for men these two measurements are also important, but, in addition, hand and wrist measurements appear with the same signs as forearm and head, respectively. This component contributes 9%–14% of total variation.

Overall, the first three PCs account for a substantial proportion of total variation, 86.5% and 87.0% for women and men respectively. Although discussion of rules for deciding how many PCs to retain is deferred until Chapter 6, intuition strongly suggests that these percentages are large enough for three PCs to give an adequate representation of the data.

A similar but much larger study, using seven measurements on 3000 criminals, was reported by Macdonell (1902) and is quoted by Maxwell (1977). The first PC again measures overall size, the second contrasts head and limb measurements, and the third can be readily interpreted as measuring the shape (roundness versus thinness) of the head. The percentages of total variation accounted for by each of the first three PCs are 54.3%, 21.4% and 9.3%, respectively, very similar to the proportions given in Table 4.1.

The sample size (28) is rather small in our example compared to that of Macdonnell's (1902), especially when the sexes are analysed separately, so caution is needed in making any inference about the PCs in the population of students from which the sample is drawn. However, the same variables have been measured for other classes of students, and similar PCs have been found (see Sections 5.1 and 13.5). In any case, a description of the sample, rather than inference about the underlying population, is often what is required, and the PCs describe the major directions of variation within a sample, regardless of the sample size.

4.2   The Elderly at Home

Hunt (1978) described a survey of the 'Elderly at Home' in which values of a large number of variables were collected for a sample of 2622 elderly individuals living in private households in the UK in 1976. The variables collected included standard demographic information of the type found in the decennial censuses, as well as information on dependency, social contact, mobility and income. As part of a project carried out for the Departments of the Environment and Health and Social Security, a PCA was done on a subset of 20 variables from Hunt's (1978) data. These variables are listed briefly in Table 4.3. Full details of the variables, and also of the project as a whole, are given by Jolliffe et al. (1982a), while shorter accounts of the main aspects of the project are available in Jolliffe et al. (1980, 1982b). It should be noted that many of the variables listed in Table 4.3 are discrete, or even dichotomous.

Some authors suggest that PCA should only be done on continuous variables, preferably with normal distributions. However, provided that

Table 4.3. Variables used in the PCA for the elderly at home.

1. Age	11. Separate kitchen
2. Sex	12. Hot water
3. Marital status	13. Car or van ownership
4. Employed	14. Number of elderly in household
5. Birthplace	15. Owner occupier
6. Father's birthplace	16. Council tenant
7. Length of residence in present household	17. Private tenant
	18. Lives alone
8. Density: persons per room	19. Lives with spouse or sibling
9. Lavatory	20. Lives with younger generation
10. Bathroom	

inferential techniques that depend on assumptions such as multivariate normality (see Section 3.7) are not invoked, there is no real necessity for the variables to have any particular distribution. Admittedly, correlations or covariances, on which PCs are based, have particular relevance for normal random variables, but they are still valid for discrete variables provided that the possible values of the discrete variables have a genuine interpretation. Variables should not be defined with more than two possible values, unless the values have a valid meaning relative to each other. If 0, 1, 3 are possible values for a variable, then the values 1 and 3 must really be twice as far apart as the values 0 and 1. Further discussion of PCA and related techniques for discrete variables is given in Section 13.1.

It is widely accepted that old people who have only just passed retirement age are different from the 'very old,' so that it might be misleading to deal with all 2622 individuals together. Hunt (1978), too, recognized possible differences between age groups by taking a larger proportion of elderly whose age was 75 or over in her sample—compared to those between 65 and 74—than is present in the population as a whole. It was therefore decided to analyse the two age groups 65–74 and 75+ separately, and part of each analysis consisted of a PCA on the correlation matrices for the 20 variables listed in Table 4.3. It would certainly not be appropriate to use the covariance matrix here, where the variables are of several different types.

It turned out that for both age groups as many as 11 PCs could be reasonably well interpreted, in the sense that not too many coefficients were far from zero. Because there are relatively few strong correlations among the 20 variables, the effective dimensionality of the 20 variables is around 10 or 11, a much less substantial reduction than occurs when there are large correlations between most of the variables (see Sections 4.3 and 6.4, for example). Eleven PCs accounted for 85.0% and 86.6% of the total variation for the 65–74 and 75+ age groups, respectively.

Table 4.4. Interpretations for the first 11 PCs for the 'elderly at home.'

65–74	75+
Component 1 (16.0%; 17.8%)*	
Contrasts single elderly living alone with others.	Contrasts single elderly, particularly female, living alone with others.
Component 2 (13.0%; 12.9%)	
Contrasts those lacking basic amenities (lavatory, bathroom, hot water) in private rented accommodation with others.	Contrasts those lacking basic amenities (lavatory, bathroom, hot water), who also mainly lack a car and are in private, rented accommodation, not living with the next generation with others.
Component 3 (9.5%; 10.1%)	
Contrasts council tenants, living in crowded conditions with others.	Contrasts those who have a car, do not live in council housing (and tend to live in own accommodation) and tend to live with the next generation with others.
Component 4 (9.2%; 9.2%)	
Contrasts immigrants living with next generation with others. There are elements here of overcrowding and possession of a car.	Contrasts council tenants, mainly immigrant, living in crowded conditions with others.
Component 5 (7.3%; 8.3%)	
Contrasts immigrants not living with next generation, with others. They tend to be older, fewer employed, fewer with a car, than in component 4.	Contrasts immigrants with others.
Component 6 (6.7%; 5.6%)	
Contrasts the younger employed people (tendency to be male), in fairly crowded conditions, often living with next generation with others.	Contrasts younger (to a certain extent, male) employed with others.
Component 7 (5.6%; 5.1%)	
Contrasts long-stay people with a kitchen with others.	Contrasts those lacking kitchen facilities with others. (NB: 1243 out of 1268 have kitchen facilities)
Component 8 (5.0%; 4.9%)	
Contrasts women living in private accommodation with others.	Contrasts private tenants with others.
Component 9 (4.6%; 4.5%)	
Contrasts old with others.	Contrasts long-stay, mainly unemployed, individuals with others.
Component 10 (4.4%; 4.4%)	
Contrasts long-stay individuals, without a kitchen, with others.	Contrasts very old with others.
Component 11 (3.7%; 3.8,%)	
Contrasts employed (mainly female) with others.	Contrasts men with women.

* The two percentages are the percentages of variation accounted for by the relevant PC for the 65–74 and 75+ age groups, respectively.

Interpretations of the first 11 PCs for the two age groups are given in Table 4.4, together with the percentage of total variation accounted for by each PC. The variances of corresponding PCs for the two age groups differ very little, and there are similar interpretations for several pairs of PCs, for example the first, second, sixth and eighth. In other cases there are groups of PCs involving the same variables, but in different combinations for the two age groups, for example the third, fourth and fifth PCs. Similarly, the ninth and tenth PCs involve the same variables for the two age groups, but the order of the PCs is reversed.

Principal component analysis has also been found useful in other demographic studies, one of the earliest being that described by Moser and Scott (1961). In this study, there were 57 demographic variables measured for 157 British towns. A PCA of these data showed that, unlike the elderly data, dimensionality could be vastly reduced; there are 57 variables, but as few as four PCs account for 63% of the total variation. These PCs also have ready interpretations as measures of social class, population growth from 1931 to 1951, population growth after 1951, and overcrowding.

Similar studies have been done on local authority areas in the UK by Imber (1977) and Webber and Craig (1978) (see also Jolliffe et al. (1986)). In each of these studies, as well as Moser and Scott (1961) and the 'elderly at home' project, the main objective was to classify the local authorities, towns or elderly individuals, and the PCA was done as a prelude to, or as part of, cluster analysis. The use of PCA in cluster analysis is discussed further in Section 9.2, but the PCA in each study mentioned here provided useful information, separate from the results of the cluster analysis, For example, Webber and Craig (1978) used 40 variables, and they were able to interpret the first four PCs as measuring social dependence, family structure, age structure and industrial employment opportunity. These four components accounted for 29.5%, 22.7%, 12.0% and 7.4% of total variation, respectively, so that 71.6% of the total variation is accounted for in four interpretable dimensions.

4.3 Spatial and Temporal Variation in Atmospheric Science

Principal component analysis provides a widely used method of describing patterns of pressure, temperature, or other meteorological variables over a large spatial area. For example, Richman (1983) stated that, over the previous 3 years, more than 60 applications of PCA, or similar techniques, had appeared in meteorological/climatological journals. More recently, 53 out of 215 articles in the 1999 and 2000 volumes of the *International Journal of Climatology* used PCA in some form. No other statistical technique came close to this 25% rate of usage. The example considered in detail in this section is taken from Maryon (1979) and is concerned with sea level atmospheric pressure fields, averaged over half-month periods, for most of the Northern Hemisphere. There were 1440 half-months, corresponding to 60 years between 1900 and 1974, excluding the years 1916–21, 1940–48 when data were inadequate. The pressure fields are summarized by estimating average pressure at $p = 221$ grid points covering the Northern Hemisphere so that the data set consists of 1440 observations on 221 variables. Data sets of this size, or larger, are commonplace in atmospheric science, and a standard procedure is to replace the variables by a few large-variance PCs. The eigenvectors that define the PCs are often known as empirical orthogonal functions (EOFs) in the meteorological or climatological literature, and the values of the PCs (the PC scores) are sometimes referred to as amplitude time series (Rasmusson et al., 1981) or, confusingly, as coefficients (Maryon, 1979) or EOF coefficients (von Storch and Zwiers, 1999, Chapter 13). Richman (1986) distinguishes between EOF analysis and PCA, with the former having unit-length eigenvectors and the latter having eigenvectors renormalized, as in (2.3.2), to have lengths proportional to their respective eigenvalues. Other authors, such as von Storch and Zwiers (1999) treat PCA and EOF analysis as synonymous.

For each PC, there is a coefficient (in the usual sense of the word), or loading, for each variable, and because variables are gridpoints (geographical locations) it is possible to plot each loading (coefficient) on a map at its corresponding gridpoint, and then draw contours through geographical locations having the same coefficient values. The map representation can greatly aid interpretation, as is illustrated in Figure 4.1.

This figure, which comes from Maryon (1979), gives the map of coefficients, arbitrarily renormalized to give 'round numbers' on the contours, for the second PC from the pressure data set described above, and is much easier to interpret than would be the corresponding table of 221 coefficients. Half-months having large positive scores for this PC will tend to have high values of the variables, that is high pressure values, where coefficients on the map are positive, and low values of the variables (low pressure values) at gridpoints where coefficients are negative. In Figure 4.1 this corresponds to low pressure in the polar regions and high pressure in the subtropics, leading to situations where there is a strong westerly flow in high latitudes at most longitudes. This is known as strong zonal flow, a reasonably frequent meteorological phenomenon, and the second PC therefore contrasts half-months with strong zonal flow with those of opposite character. Similarly, the first PC (not shown) has one of its extremes identified as corresponding to an intense high pressure area over Asia and such situations are again a fairly frequent occurrence, although only in winter.

Several other PCs in Maryon's (1979) study can also be interpreted as corresponding to recognizable meteorological situations, especially when coefficients are plotted in map form. The use of PCs to summarize pressure fields and other meteorological or climatological fields has been found

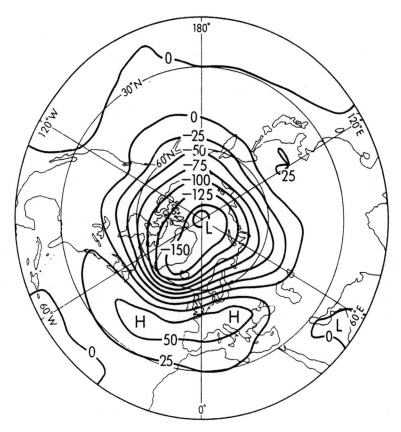

Figure 4.1. Graphical representation of the coefficients in the second PC for sea level atmospheric pressure data.

to be so valuable that it is almost routine. For example, Craddock and Flood (1969) find PCs with ready interpretations for Northern Hemispheric 500 mb geopotential surfaces, Craddock and Flintoff (1970) do the same for 1000 mb surfaces and 1000–500 mb thickness, Overland and Preisendorfer (1982) interpret the first three PCs for data on spatial distributions of cyclone frequencies in the Bering Sea, Wigley et al. (1984) discuss PCs for European precipitation data, and Folland et al. (1985) find interpretable patterns in PCs of worldwide sea surface temperature anomalies. Some patterns recur in different data sets. For example, Figure 4.1 could be interpreted as the North Atlantic Oscillation (NAO), which reflects the strength of the zonal flow in the North Atlantic and neighbouring areas, as measured by the pressure difference between the Azores and Iceland. This pattern, and a small number of others, notably ENSO (El Niño–Southern Oscillation), have been identified as major modes of climate variability in different parts of the world. They have been studied extensively (see, for example, Ambaum et al. (2001) for a discussion of the NAO).

It is not always the case that interpretation is straightforward. In atmospheric science the PCs or EOFS are often rotated in an attempt to find more clearly interpretable patterns. We return to this topic in Chapter 11.

Not only are the first few PCs readily interpreted in many meteorological and climatological examples, possibly after rotation, but they also frequently enable a considerable reduction to be made in the dimensions of the data set. In Maryon's (1979) study, for example, there are initially 221 variables, but 16 PCs account for over 90% of the total variation. Nor is this due to any disparity between variances causing a few dominant PCs; size of variance is fairly similar for all 221 variables.

Maryon's (1979) analysis was for a covariance matrix, which is reasonable since all variables are measured in the same units (see Sections 2.3 and 3.3). However, some atmospheric scientists advocate using correlation, rather than covariance, matrices so that patterns of spatial correlation can be detected without possible domination by the stations and gridpoints with the largest variances (see Wigley et al. (1984)).

It should be clear from this section that meteorologists and climatologists have played a leading role in applying PCA. In addition, they have developed many related methods to deal with the peculiarities of their data, which often have correlation structure in both time and space. A substantial part of Chapter 12 is devoted to these developments.

4.4   Properties of Chemical Compounds

The main example given in this section is based on a subset of data given by Hansch et al. (1973); the PCA was described by Morgan (1981). Seven properties (variables) were measured for each of 15 chemical substituents; the properties and substituents are listed in Table 4.5. Some of the results of a PCA based on the correlation matrix for these data are given in Table 4.6.

The aim of the work of Hansch et al. (1973), and of much subsequent research in quantitative structure–activity relationships (QSAR), is to relate aspects of the structure of chemicals to their physical properties or activities so that 'new' chemicals can be manufactured whose activities may be predicted in advance. Although PCA is less evident recently in the extensive QSAR literature, it appeared in a number of early papers on QSAR. For example, it was used in conjunction with regression (see Chapter 8 and Mager (1980a)), and as a discriminant technique (see Section 9.1 and Mager (1980b)). Here we look only at the reduction of dimensionality and interpretations obtained by Morgan (1981) in this analysis of Hansch et al.'s (1973) data. The first two PCs in Table 4.6 account for 79% of the total variation; the coefficients for each have a moderately simple structure. The first PC is essentially an average of all properties except π and MR, whereas the most important contribution to the second PC is an average of

Table 4.5. Variables and substituents considered by Hansch et al. (1973).

(a) Variables	

1. π Hansch's measure of lipophylicity

2. F
3. R   }   measures of electronic effect: F denotes 'field'; R denotes resonance

4. MR   molar refraction

5. σ_m
6. σ_p }   further measures of electronic effect

7. MW   molecular weight

(b) Substituents

1. Br	2. Cl	3. F	4. I	5. CF_3
6. CH_3	7. C_2H_5	8. C_3H_7	9. C_4H_2	10. OH
11. NH_2	12. CH_2OH	13. SO_2CH_3	14. $SOCH_3$	15. $SO_2(NH_2)$

Table 4.6. First four PCs of chemical data from Hansch et al. (1973).

Component number	1	2	3	4
π	0.15	0.49	0.70	−0.45
F	−0.42	−0.36	0.34	0.13
R	−0.37	0.30	−0.44	−0.54
MR Coefficients	−0.16	0.62	−0.23	0.49
σ_m	−0.48	−0.24	0.19	−0.03
σ_p	−0.50	0.01	−0.11	−0.30
MW	−0.40	0.30	0.31	0.40
Eigenvalue	3.79	1.73	0.74	0.59
Cumulative percentage of total variation	54.1	78.8	89.4	97.8

π and MR. Morgan (1981) also reports PCAs for a number of other similar data sets, in several of which the PCs provide useful interpretations.

4.5   Stock Market Prices

The data in this example are the only set in this chapter that previously appeared in a textbook (Press, 1972, Section 9.5.2). Both the data, and the PCs have interesting structures. The data, which were originally analysed by Feeney and Hester (1967), consist of 50 quarterly measurements between 1951 and 1963 of US stock market prices for the 30 industrial stocks making up the Dow-Jones index at the end of 1961. Table 4.7 gives, in the simplified form described for Table 4.2, the coefficients of the first two PCs, together with the percentage of variation accounted for by each PC, for both covariance and correlation matrices.

Looking at the PCs for the correlation matrix, the first is a 'size' component, similar to those discussed in Section 4.1. It reflects the fact that all stock prices rose fairly steadily during the period 1951–63, with the exception of Chrysler. It accounts for roughly two-thirds of the variation in the 30 variables. The second PC can be interpreted as a contrast between 'consumer' and 'producer' stocks. 'Consumer' companies are those that mainly supply goods or services directly to the consumer, such as AT&T, American Tobacco, General Foods, Proctor and Gamble, Sears, and Woolworth, whereas 'producer' companies sell their goods or services mainly to other companies, and include Alcoa, American Can, Anaconda, Bethlehem, Union Carbide, and United Aircraft.

The PCs for the covariance matrix can be similarly interpreted, albeit with a change of sign for the second component, but the interpretation is slightly confounded, especially for the first PC, by the different-sized variances for each variable.

Feeney and Hester (1967) also performed a number of other PCAs using these and related data. In one analysis, they removed a linear trend from the stock prices before calculating PCs, and found that they had eliminated the size (trend) PC, and that the first PC was now very similar in form to the second PC in the original analyses. They also calculated PCs based on 'rate-of-return' rather than price, for each stock, and again found interpretable PCs. Finally, PCs were calculated for subperiods of 12 years of data in order to investigate the stability of the PCs, a topic that is discussed more generally in Section 10.3.

To conclude this example, note that it is of a special type, as each variable is a time series, in which consecutive observations are not independent. Further discussion of PCA for time series data is given in Chapter 12. A possible technique for finding PCs that are free of the trend in a vector of time series, which is more general than the technique noted above for the present example, is described in Section 14.3.

Table 4.7. Simplified coefficients for the first two PCs: stock market prices.

Component number	Correlation matrix 1	Correlation matrix 2	Covariance matrix 1	Covariance matrix 2
Allied Chemical	+	(−)		
Alcoa	+	−	(+)	+
American Can	+	−		
AT&T	+	+	(+)	−
American Tobacco	+	+		
Anaconda	(+)	−		+
Bethlehem	+	−		(+)
Chrysler	(−)	(−)		
Dupont	+	(−)	+	+
Eastman Kodak	+	(+)	+	−
Esso	+	(−)		
General Electric	+		(+)	
General Foods	+	+	(+)	−
General Motors	+			
Goodyear	+			
International Harvester	+	(+)		
International Nickel	+		(+)	
International Paper	+	(−)		
Johns–Manville	+			
Owens–Illinois	+		(+)	
Proctor and Gamble	+	+	(+)	−
Sears	+	+	(+)	−
Standard Oil (Cal.)	+			
Swift	(+)	(−)		
Texaco	+	(+)	(+)	(−)
Union Carbide	+	(−)	(+)	(+)
United Aircraft	+	−		+
US Steel	+	(−)	(+)	(+)
Westinghouse	+			
Woolworth	+	+		(−)
Percentage of variation accounted for	65.7	13.7	75.8	13.9

5
Graphical Representation of Data Using Principal Components

The main objective of a PCA is to reduce the dimensionality of a set of data. This is particularly advantageous if a set of data with many variables lies, in reality, close to a two-dimensional subspace (plane). In this case the data can be plotted with respect to these two dimensions, thus giving a straightforward visual representation of what the data look like, instead of appearing as a large mass of numbers to be digested. If the data fall close to a three-dimensional subspace it is still possible to gain a good visual impression of the data using interactive computer graphics. Even with a few more dimensions it is possible, with some degree of ingenuity, to get a 'picture' of the data (see, for example, Chapters 10–12 (by Tukey and Tukey) in Barnett (1981)) although we shall concentrate almost entirely on two-dimensional representations in the present chapter.

If a good representation of the data exists in a small number of dimensions then PCA will find it, since the first q PCs give the 'best-fitting' q-dimensional subspace in the sense defined by Property G3 of Section 3.2. Thus, if we plot the values for each observation of the first two PCs, we get the best possible two-dimensional plot of the data (similarly for three or more dimensions). The first section of this chapter simply gives examples illustrating this procedure. We largely defer until the next chapter the problem of whether or not two PCs are adequate to represent most of the variation in the data, or whether we need more than two.

There are numerous other methods for representing high-dimensional data in two or three dimensions and, indeed, the book by Everitt (1978) is almost entirely on the subject, as are the conference proceedings edited by Wang (1978) and by Barnett (1981) (see also Chapter 5 of the book by Chambers et al. (1983)). A more recent thorough review of graphics for multivariate data is given by Carr (1998). A major advance has been the development of *dynamic* multivariate graphics, which Carr (1998) describes as part of 'the visual revolution in computer science.' The techniques discussed in the present chapter are almost exclusively static, although some could be adapted to be viewed dynamically. Only those graphics that have links with, or can be used in conjunction with, PCA are included.

Section 5.2 discusses principal coordinate analysis, which constructs low-dimensional plots of a set of data from information about similarities or dissimilarities between pairs of observations. It turns out that the plots given by this analysis are equivalent to plots with respect to PCs in certain special cases.

The biplot, described in Section 5.3, is also closely related to PCA. There are a number of variants of the biplot idea, but all give a simultaneous display of n observations and p variables on the same two-dimensional diagram. In one of the variants, the plot of observations is identical to a plot with respect to the first two PCs, but the biplot simultaneously gives graphical information about the relationships between variables. The relative positions of variables *and* observations, which are plotted on the same diagram, can also be interpreted.

Correspondence analysis, which is discussed in Section 5.4, again gives two-dimensional plots, but only for data of a special form. Whereas PCA and the biplot operate on a matrix of n observations on p variables, and principal coordinate analysis and other types of scaling or ordination techniques use data in the form of a similarity or dissimilarity matrix, correspondence analysis is used on contingency tables, that is, data classified according to two categorical variables. The link with PCA is less straightforward than for principal coordinate analysis or the biplot, but the ideas of PCA and correspondence analysis have some definite connections. There are many other ordination and scaling methods that give graphical displays of multivariate data, and which have increasingly tenuous links to PCA. Some of these techniques are noted in Sections 5.2 and 5.4, and in Section 5.5 some comparisons are made, briefly, between PCA and the other techniques introduced in this chapter.

Another family of techniques, *projection pursuit*, is designed to find low-dimensional representations of a multivariate data set that best display certain types of structure such as clusters or outliers. Discussion of projection pursuit will be deferred until Chapters 9 and 10, which include sections on cluster analysis and outliers, respectively.

The final section of this chapter describes some methods which have been used for representing multivariate data in two dimensions when more than two or three PCs are needed to give an adequate representation of the data. The first q PCs can still be helpful in reducing the dimensionality in such cases, even when q is much larger than 2 or 3.

Finally, we note that as well as the graphical representations described in the present chapter, we have already seen, in Section 4.3, one other type of plot that uses PCs. This type of plot is rather specialized, but is used extensively in atmospheric science. Related plots are discussed further in Chapter 12.

5.1   Plotting Data with Respect to the First Two (or Three) Principal Components

The idea here is simple: if a data set $\{\mathbf{x}_1, \mathbf{x}_2, \ldots, \mathbf{x}_n\}$ has p variables, then the observations can be plotted as points in p-dimensional space. If we wish to plot the data in a 'best-fitting' q-dimensional subspace $(q < p)$, where 'best-fitting' is defined, as in Property G3 of Section 3.2, as minimizing the sum of squared perpendicular distances of $\mathbf{x}_1, \mathbf{x}_2, \ldots, \mathbf{x}_n$ from the subspace, then the appropriate subspace is defined by the first q PCs.

Two-dimensional plots are particularly useful for detecting patterns in the data, and three-dimensional plots or models, though generally less easy to interpret quickly, can sometimes give additional insights. If the data do not lie close to a two- (or three-) dimensional subspace, then no two- (or three-) dimensional plot of the data will provide an adequate representation, although Section 5.6 discusses briefly the use of indirect ways for presenting the data in two dimensions in such cases. Conversely, if the data *are* close to a q-dimensional subspace, then most of the variation in the data will be accounted for by the first q PCs and a plot of the observations with respect to these PCs will give a realistic picture of what the data look like, unless important aspects of the data structure are concentrated in the direction of low variance PCs. Plotting data sets with respect to the first two PCs is now illustrated by two examples, with further illustrations given, in conjunction with other examples, later in this chapter and in subsequent chapters.

It should be noted that the range of structures that may be revealed by plotting PCs is limited by the fact that the PCs are uncorrelated. Hence some types of group structure or outlier patterns or non-linear relationships between PCs, may be visible, but linear relationships between PCs are impossible.

5.1.1   Examples

Two examples are given here that illustrate the sort of interpretation which may be given to plots of observations with respect to their first two PCs. These two examples do not reveal any strong, but previously unknown, structure such as clusters; examples illustrating clusters will be presented in Section 9.2. Nevertheless, useful information can still be gleaned from the plots.

Anatomical Measurements

The data presented here consist of the same seven anatomical measurements as in the data set of Section 4.1, but for a different set of students, this time comprising 11 women and 17 men. A PCA was done on the correlation matrix for all 28 observations and, as in the analyses of Section 4.1 for each sex separately, the first PC is an overall measurement of size. The second PC is a contrast between the head measurement and the other six variables, and is therefore not particularly similar to any of the first three PCs for the separate sexes found in Section 4.1, though it is closest to the third component for the women. The difference between the second PC and those from the earlier analyses may be partially due to the fact that the sexes have been combined, but it is also likely to reflect some instability in all but the first PC due to relatively small sample sizes. The first two PCs for the present data account for 69% and 11% of the total variation, respectively, so that a two-dimensional plot with respect to these PCs, representing 80% of the variation, gives a reasonably good approximation to the relative positions of the observations in seven-dimensional space.

A plot of these data with respect to the first two PCs was given in Figure 1.3, and it was noted that the first PC is successful in separating the women from the men. It can also be seen in Figure 1.3 that there is one clear outlier with respect to the second PC, seen at the bottom of the plot. A second observation, at the left of the plot, is rather extreme on the first PC. These two observations and other potential outliers will be discussed further in Section 10.1. The observation at the bottom of the diagram has such an extreme value for the second PC, roughly twice as large in absolute terms as any other observation, that it could be mainly responsible for the second PC taking the form that it does. This possibility will be discussed further in Section 10.2.

Figures 5.1(a) and (b) are the same as Figure 1.3 except that superimposed on them are convex hulls for the two groups, men and women (Figure 5.1(a)), and the minimum spanning tree (Figure 5.1(b)). Convex hulls are useful in indicating the areas of a two-dimensional plot covered by various subsets of observations. Here they confirm that, although the areas covered by men and women overlap slightly, the two sexes largely occupy different areas of the diagrams. The separation is mainly in terms of the first PC (overall size) with very little differentiation between sexes on the second PC. The plot therefore displays the unsurprising result that the two groups of observations corresponding to the two sexes differ mainly in terms of overall size.

It was noted above that the two-dimensional plot represents 80% of the total variation of the 28 observations in seven-dimensional space. Percentage of total variation is an obvious measure of how good two-dimensional representation is, but many of the other criteria that are discussed in Section 6.1 could be used instead. Alternatively, an informal way of judging the goodness-of-fit in two dimensions is to superimpose a minimum span-

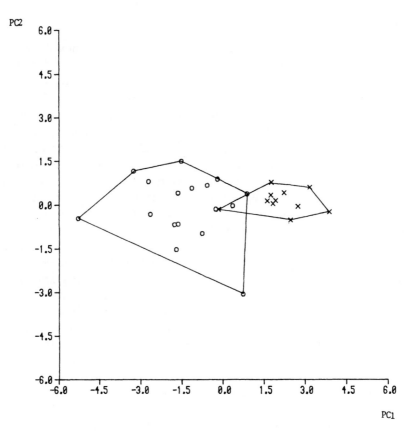

Figure 5.1. (a). Student anatomical measurements: plot of the first two PC for 28 students with convex hulls for men and women superimposed.

ning tree (MST) on the diagram, as in Figure 5.1(b). The MST is a set of lines drawn between pairs of points such that

(i) each point is connected to every other point by a sequence of lines;

(ii) there are no closed loops;

(iii) the sum of 'lengths' of lines is minimized.

If the 'lengths' of the lines are defined as distances in seven-dimensional space, then the corresponding MST will give an indication of the closeness-of-fit of the two-dimensional representation. For example, it is seen that observations 5 and 14, which are very close in two dimensions, are joined via observation 17, and so must both be closer to observation 17 in seven-dimensional space than to each other. There is therefore some distortion in the two-dimensional representation in the vicinity of observations 5 and 14. Similar remarks apply to observations 12 and 23, and to the group of observations 19, 22, 25, 27, 28. However, there appears to be little distortion for the better-separated observations.

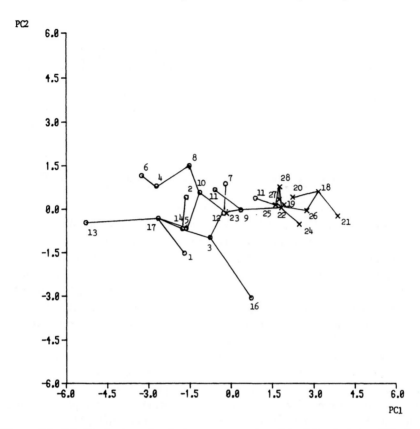

Figure 5.1. (b). Student anatomical measurements: plot of the first two PCs for 28 students with minimum spanning tree superimposed.

Artistic Qualities of Painters

The second data set described in this section was analysed by Davenport and Studdert-Kennedy (1972). It consists of a set of subjective measurements of the artistic qualities 'composition,' 'drawing,' 'colour' and 'expression' for 54 painters. The measurements, on a scale from 0 to 20, were compiled in France in 1708 by Roger de Piles for painters 'of established reputation.' Davenport and Studdert-Kennedy (1972) give data for 56 painters, but for two painters one measurement is missing, so these painters are omitted from the analysis.

Table 5.1 gives the variances and coefficients for the first two PCs based on the correlation matrix for the 54 painters with complete data. The components, and their contributions to the total variation, are very similar to those found by Davenport and Studdert-Kennedy (1972) for the covariance matrix. This strong similarity between the PCs for correlation and covariance matrices is relatively unusual (see Section 3.3) and is due to the near-equality of the variances for the four variables. The first component

Table 5.1. First two PCs: artistic qualities of painters.

		Component 1	Component 2
Composition		0.50	-0.49
Drawing	Coefficients	0.56	0.27
Colour		−0.35	−0.77
Expression		0.56	−0.31
Eigenvalue		2.27	1.04
Cumulative percentage of total variation		56.8	82.8

is interpreted by the researchers as an index of de Piles' overall assessment of the painters, although the negative coefficient for colour needs some additional explanation. The form of this first PC could be predicted from the correlation matrix. If the sign of the variable 'colour' is changed, then all correlations in the matrix are positive, so that we would expect the first PC to have positive coefficients for all variables after this redefinition of 'colour' (see Section 3.8). The second PC has its largest coefficient for colour, but the other coefficients are also non-negligible.

A plot of the 54 painters with respect to the first two components is given in Figure 5.2, and this two-dimensional display represents 82.8% of the total variation. The main feature of Figure 5.2 is that painters of the same school are mostly fairly close to each other. For example, the set of the ten 'Venetians' {Bassano, Bellini, Veronese, Giorgione, Murillo, Palma Vecchio, Palma Giovane, Pordenone, Tintoretto, Titian} are indicated on the figure, and are all in a relatively small area at the bottom left of the plot. Davenport and Studdert-Kennedy (1972) perform a cluster analysis on the data, and display the clusters on a plot of the first two PCs. The clusters dissect the data in a sensible looking manner, and none of them has a convoluted shape on the PC plot. However, there is little evidence of a strong cluster structure in Figure 5.2. Possible exceptions are a group of three isolated painters near the bottom of the plot, and four painters at the extreme left. The first group are all members of the 'Seventeenth Century School,' namely Rembrandt, Rubens, and Van Dyck, and the second group consists of three 'Venetians,' Bassano, Bellini, Palma Vecchio, together with the 'Lombardian' Caravaggio. This data set will be discussed again in Sections 5.3 and 10.2, and the numbered observations on Figure 5.2 will be referred to there. Further examples of the use of PCA in conjunction with cluster analysis are given in Section 9.2.

Throughout this section there has been the suggestion that plots of the first two PCs may reveal interesting structure in the data. This contradicts the implicit assumption that the n observations are identically distributed with a common mean and covariance matrix. Most 'structures' in the data indicate that different observations have different means, and that PCA

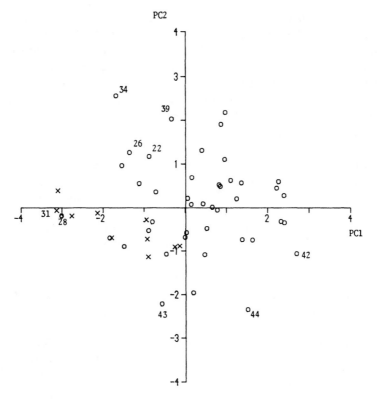

Figure 5.2. Artistic qualities of painters: plot of 54 painters with respect to their first two PCs. The symbol × denotes member of the 'Venetian' school.

is looking for major directions of variation between means rather than major directions of variation in a common distribution (de Falguerolles, personal communication). This view is in line with the fixed effect model of Section 3.9, and is discussed further in Section 5.3.

A variation of the simple plot of PC scores for the first two PCs is proposed by Tarpey (2000). Lines are added to the plot, corresponding to the directions of the first PC for two subsets of the data, derived by dividing the full data set according to the sign of the first PC for the whole data set. The idea is to indicate possible non-linearity in the data (see Section 14.1.3).

5.2 Principal Coordinate Analysis

Principal coordinate analysis is a scaling or ordination method, sometimes known as *classical scaling*. It was popularized by Gower (1966). Torgerson (1952, 1958) discussed similar ideas, but did not point out the links between principal coordinate analysis and PCA that were noted by Gower. Like the more widely known non-metric multidimensional scaling (Kruskal, 1964a,b), the technique starts with a matrix of similarities or dissimilarities between a set of observations, and aims to produce a low-dimensional graphical plot of the data in such a way that distances between points in the plot are close to the original dissimilarities. There are numerous scaling techniques; Cox and Cox (2001) provide a good overview of many of them.

The starting point (an $(n \times n)$ matrix of (dis)similarities) of principal coordinate analysis is different from that of PCA, which usually begins with the $(n \times p)$ data matrix. However, in special cases discussed below the two techniques give precisely the same low-dimensional representation. Furthermore, PCA may be used to find starting configurations for the iterative algorithms associated with non-metric multidimensional scaling (Davison, 1983, Chapters 5, 6). Before showing the equivalences between PCA and principal coordinate analysis, we need first to describe principal coordinate analysis in some detail.

Suppose that $\mathbf{T}$ is an $(n \times n)$ positive-semidefinite symmetric matrix of similarities among a set of n observations. (Note that it is fairly standard notation to use $\mathbf{A}$, rather than $\mathbf{T}$, here. However, we have avoided the use of $\mathbf{A}$ in this context, as it is consistently taken to be the matrix of PC coefficients in the current text.) From the spectral decomposition of $\mathbf{T}$ (Property A3 of Sections 2.1 and 3.1 gives the spectral decomposition of a covariance matrix, but the same idea is valid for any symmetric matrix) we have

$$\mathbf{T} = \tau_1 \mathbf{b}_1 \mathbf{b}_1' + \tau_2 \mathbf{b}_2 \mathbf{b}_2' + \cdots + \tau_n \mathbf{b}_n \mathbf{b}_n', \qquad (5.2.1)$$

where $\tau_1 \geq \tau_2 \geq \cdots \geq \tau_n$ are the eigenvalues of $\mathbf{T}$ and $\mathbf{b}_1, \mathbf{b}_2, \cdots, \mathbf{b}_n$ are the corresponding eigenvectors. Alternatively, this may be written

$$\mathbf{T} = \mathbf{c}_1 \mathbf{c}_1' + \mathbf{c}_2 \mathbf{c}_2' + \cdots + \mathbf{c}_n \mathbf{c}_n', \qquad (5.2.2)$$

where

$$\mathbf{c}_j = \tau_j^{1/2} \mathbf{b}_j, \qquad j = 1, 2, \ldots, n.$$

Now consider the n observations as points in n-dimensional space with the jth coordinate for the ith observation equal to c_{ij}, the ith element of $\mathbf{c}_j$. With this geometric interpretation of the n observations, the Euclidean distance between the hth and ith observations is

$$\Delta_{hi}^2 = \sum_{j=1}^{n} (c_{hj} - c_{ij})^2$$

$$= \sum_{j=1}^{n} c_{hj}^2 + \sum_{j=1}^{n} c_{ij}^2 - 2 \sum_{j=1}^{n} c_{hj} c_{ij}.$$

But from (5.2.2), the (h, i)th element of $\mathbf{T}$ can be written

$$t_{hi} = \sum_{j=1}^{n} c_{hj} c_{ij}, \quad h, i = 1, 2, \ldots, n,$$

so

$$\Delta_{hi}^2 = t_{hh} + t_{ii} - 2t_{hi}.$$

Principal coordinate analysis then attempts to find the 'best-fitting' q-dimensional $(q < n)$ approximation to the n-dimensional representation defined above. 'Best-fitting' is defined here in the same way as in the geometric definition of PCA (Property G3 of Section 3.2), so that 'principal components' are now found for the n 'observations' defined in n dimensions by the coordinates c_{ij}. A q-dimensional principal coordinate representation is then given by plotting the coordinates of the observations with respect to the first q 'PCs'. Principal coordinate analysis therefore consists of two stages, both of which involve finding eigenvalues and eigenvectors of $(n \times n)$ matrices:

(i) Find the eigenvectors $\mathbf{c}_1, \mathbf{c}_2, \ldots, \mathbf{c}_n$ of $\mathbf{T}$, normalized to have lengths equal to their respective eigenvalues, and represent the n observations as points in n-dimensional space with coordinate c_{ij} for the ith observation in the jth dimension.

(ii) Find the PCs for the 'data set' in n dimensions defined in (i), and calculate coordinates of the n observations with respect to the first q PCs.

If the vectors $\mathbf{c}_j$ defined in the first stage have $\sum_{i=1}^{n} c_{ij} = 0$ then the covariance matrix that is calculated in stage (ii) will be proportional to $\mathbf{C}'\mathbf{C}$ where $\mathbf{C}$ is the $(n \times n)$ matrix with jth column $\mathbf{c}_j$, $j = 1, 2, \ldots, n$. But

$$\mathbf{c}_j' \mathbf{c}_k = \begin{cases} \tau_j & j = k \\ 0 & j \neq k \end{cases},$$

as the eigenvectors in the spectral decomposition (5.2.1) have the property

$$\mathbf{b}_j' \mathbf{b}_k = \begin{cases} 1 & j = k \\ 0 & j \neq k \end{cases}$$

and

$$\mathbf{c}_j = \tau_j^{1/2} \mathbf{b}_j, \quad j = 1, 2, \ldots, n.$$

The matrix $\mathbf{C}'\mathbf{C}$ is therefore diagonal with diagonal elements τ_j, $j = 1, 2, \ldots, n$, so that the first q principal coordinates of the n observations are simply the values of c_{ij} for $i = 1, 2, \ldots, n$; $j = 1, 2, \ldots, q$. Thus when $\sum_{i=1}^{n} c_{ij} = 0$, stage (ii) is unnecessary.

In general, although a similarity matrix $\mathbf{T}$ need not lead to $\sum_{i=1}^{n} c_{ij} = 0$, this property can be readily achieved by replacing $\mathbf{T}$ by an adjusted similarity matrix. In this adjustment, t_{hi} is replaced by $t_{hi} - \bar{t}_h - \bar{t}_i + \bar{t}$ where $\bar{t}_h$ denotes the mean of the elements in the hth row (or column, since $\mathbf{T}$ is symmetric) of $\mathbf{T}$, and $\bar{t}$ is the mean of all elements in $\mathbf{T}$. This adjusted similarity matrix has $\sum_{i=1}^{n} c_{ij} = 0$, and gives the same value of Δ_{hi}^2 for each pair of observations as does $\mathbf{T}$ (Gower, 1966). Thus we can replace the second stage of principal coordinate analysis by an initial adjustment of $\mathbf{T}$, for any similarity matrix $\mathbf{T}$.

Principal coordinate analysis is equivalent to a plot with respect to the first q PCs when the measure of similarity between two points is proportional to $-d_{hi}^2$, where d_{hi}^2 is the Euclidean squared distance between the hth and ith observations, calculated from the usual $(n \times p)$ data matrix. Assume $t_{hi} = -\gamma d_{hi}^2$, where γ is a positive constant; then if stage (i) of a principal coordinate analysis is carried out, the 'distance' between a pair of points in the constructed n-dimensional space is

$$\Delta_{hi}^2 = (t_{hh} + t_{ii} - 2t_{hi})$$
$$= \gamma(-d_{hh}^2 - d_{ii}^2 + 2d_{hi}^2)$$
$$= 2\gamma d_{hi}^2,$$

as Euclidean distance from a point to itself is zero. Thus, apart from a possible rescaling if γ is taken to be a value other than $\frac{1}{2}$, the first stage of principal coordinate analysis correctly reproduces the relative positions of the n observations, which lie in a p-dimensional subspace of n-dimensional space, so that the subsequent PCA in stage (ii) gives the same result as a PCA on the original data.

Two related special cases are of interest. First, consider the situation where all variables are binary. A commonly used measure of similarity between individuals h and i is the proportion of the p variables for which h and i take the same value, and it can be easily demonstrated (Gower, 1966) that this measure is equivalent to Euclidean distance. Thus, although PCA of discrete—and in particular—binary data has its critics, it is equivalent to principal coordinate analysis with a very plausible measure of similarity. Principal component analysis for discrete data is discussed further in Section 13.1.

The second special case occurs when the elements of the similarity matrix $\mathbf{T}$ are defined as 'covariances' between observations, so that $\mathbf{T}$ is proportional to $\mathbf{XX'}$, where $\mathbf{X}$, as before, is the column-centred $(n \times p)$ matrix whose (i, j)th element is the value of the jth variable, measured about its mean $\bar{x}_j$, for the ith observation. In this case the (h, i)th similarity is, apart from a constant,

$$t_{hi} = \sum_{j=1}^{p} x_{hj} x_{ij}$$

and the distances between the points in the n-dimensional space constructed in the first stage of the principal coordinate analysis are

$$\Delta_{hi}^2 = t_{hh} + t_{ii} - 2t_{hi}$$

$$= \sum_{j=1}^{p} x_{hj}^2 + \sum_{j=1}^{p} x_{ij}^2 - 2\sum_{j=1}^{p} x_{hj} x_{ij}$$

$$= \sum_{j=1}^{p} (x_{hj} - x_{ij})^2$$

$$= d_{hi}^2,$$

the Euclidean distance between the observations using the original p variables. As before, the PCA in the second stage of principal coordinate analysis gives the same results as a PCA on the original data. Note, however, that $\mathbf{XX'}$ is not a very obvious similarity matrix. For a 'covariance matrix' between observations it is more natural to use a row-centred, rather than column-centred, version of $\mathbf{X}$.

Even in cases where PCA and principal coordinate analysis give equivalent two-dimensional plots, there is a difference, namely that in principal coordinate analysis there are no vectors of coefficients defining the axes in terms of the original variables. This means that, unlike PCA, the axes in principal coordinate analysis cannot be interpreted, unless the corresponding PCA is also done.

The equivalence between PCA and principal coordinate analysis in the circumstances described above is termed a *duality* between the two techniques by Gower (1966). The techniques are dual in the sense that PCA operates on a matrix of similarities between variables, whereas principal coordinate analysis operates on a matrix of similarities between observations (individuals), but both can lead to equivalent results.

To summarize, principal coordinate analysis gives a low-dimensional representation of data when the data are given in the form of a similarity or dissimilarity matrix. As it can be used with any form of similarity or dissimilarity matrix, it is, in one sense, 'more powerful than,' and 'extends,' PCA (Gower, 1967). However, as will be seen in subsequent chapters, PCA has many uses other than representing data graphically, which is the overriding purpose of principal coordinate analysis.

Except in the special cases discussed above, principal coordinate analysis has no direct relationship with PCA, so no examples will be given of the general application of the technique. In the case where principal coordinate analysis gives an equivalent representation to that of PCA, nothing new would be demonstrated by giving additional examples. The examples given in Section 5.1 (and elsewhere) which are presented as plots with respect to the first two PCs are, in fact, equivalent to two-dimensional principal coordinate plots if the 'dissimilarity' between observations h and i is proportional to the Euclidean squared distance between the hth and ith observations in p dimensions.

In most cases, if the data are available in the form of an $(n \times p)$ matrix of p variables measured for each of n observations, there is no advantage in doing a principal coordinate analysis instead of a PCA, unless for some reason a dissimilarity measure other than Euclidean distance is deemed to be appropriate. However, an exception occurs when $n < p$, especially if $n \ll p$ as happens for some types of chemical, meteorological and biological data. As principal coordinate analysis and PCA find eigenvectors of an $(n \times n)$ matrix and a $(p \times p)$ matrix respectively, the dual analysis based on principal coordinates will have computational advantages in such cases.

5.3   Biplots

The two previous sections describe plots of the n *observations*, usually in two dimensions. Biplots similarly provide plots of the n observations, but *simultaneously* they give plots of the relative positions of the p *variables* in two dimensions. Furthermore, superimposing the two types of plots provides additional information about relationships between variables and observations not available in either individual plot.

Since the publication of the first edition, there have been substantial developments in biplots. In particular, the monograph by Gower and Hand (1996) considerably extends the definition of biplots. As these authors note themselves, their approach to biplots is unconventional, but it is likely to become increasingly influential. The material on biplots which follows is mostly concerned with what Gower and Hand (1996) call 'classical biplots,' although correspondence analysis, which is discussed in Section 5.4, also falls under Gower and Hand's (1996) biplot umbrella. A number of other variations of biplots are discussed briefly at the end of the present section and in later chapters. As with other parts of this book, the choice of how far to stray from PCA in following interesting diversions such as these is inevitably a personal one. Some readers may prefer to go further down the biplot road; reference to Gower and Hand (1996) should satisfy their curiosity.

Classical biplots, which might also be called 'principal component biplots,' were principally developed and popularized by Gabriel (1971, and several subsequent papers), although Jolicoeur and Mosimann (1960) had earlier given an example of similar diagrams and they are periodically rediscoverd in other disciplines (see, for example, Berry et al. (1995), who refer to the same idea as 'latent semantic indexing'). The plots are based on the singular value decomposition (SVD), which was described in Section 3.5. This states that the $(n \times p)$ matrix $\mathbf{X}$ of n observations on p variables measured about their sample means can be written

$$\mathbf{X} = \mathbf{ULA}', \qquad (5.3.1)$$

where $\mathbf{U}, \mathbf{A}$ are $(n \times r)$, $(p \times r)$ matrices respectively, each with orthonormal columns, $\mathbf{L}$ is an $(r \times r)$ diagonal matrix with elements $l_1^{1/2} \geq l_2^{1/2} \geq \cdots \geq l_r^{1/2}$, and r is the rank of $\mathbf{X}$. Now define $\mathbf{L}^\alpha$, for $0 \leq \alpha \leq 1$, as the diagonal matrix whose elements are $l_1^{\alpha/2}, l_2^{\alpha/2}, \cdots, l_r^{\alpha/2}$ with a similar definition for $\mathbf{L}^{1-\alpha}$, and let $\mathbf{G} = \mathbf{U}\mathbf{L}^\alpha, \mathbf{H}' = \mathbf{L}^{1-\alpha}\mathbf{A}'$. Then

$$\mathbf{GH}' = \mathbf{U}\mathbf{L}^\alpha\mathbf{L}^{1-\alpha}\mathbf{A}' = \mathbf{ULA}' = \mathbf{X},$$

and the (i, j)th element of $\mathbf{X}$ can be written

$$x_{ij} = \mathbf{g}_i'\mathbf{h}_j, \tag{5.3.2}$$

where $\mathbf{g}_i'$, $i = 1, 2, \ldots, n$ and $\mathbf{h}_j'$, $j = 1, 2, \ldots, p$ are the rows of $\mathbf{G}$ and $\mathbf{H}$, respectively. Both the $\mathbf{g}_i$ and $\mathbf{h}_j$ have r elements, and if $\mathbf{X}$ has rank 2, all could be plotted as points in two-dimensional space. In the more general case, where $r > 2$, it was noted in Section 3.5 that (5.3.1) can be written

$$x_{ij} = \sum_{k=1}^{r} u_{ik}l_k^{1/2}a_{jk} \tag{5.3.3}$$

which is often well approximated by

$$_m\tilde{x}_{ij} = \sum_{k=1}^{m} u_{ik}l_k^{1/2}a_{jk}, \qquad \text{with} \quad m < r. \tag{5.3.4}$$

But (5.3.4) can be written

$$_m\tilde{x}_{ij} = \sum_{k=1}^{m} g_{ik}h_{jk}$$
$$= \mathbf{g}_i^{*'}\mathbf{h}_j^*,$$

where $\mathbf{g}_i^*$, $\mathbf{h}_j^*$ contain the first m elements of $\mathbf{g}_i$ and $\mathbf{h}_j$, respectively. In the case where (5.3.4) with $m = 2$ provides a good approximation to (5.3.3), $\mathbf{g}_i^*$, $i = 1, 2, \ldots, n$; $\mathbf{h}_j^*$, $j = 1, 2, \ldots, p$ together give a good two-dimensional representation of both the n observations and the p variables. This type of approximation can, of course, be used for values of $m > 2$, but the graphical representation is then less clear. Gabriel (1981) referred to the extension to $m \geq 3$ as a *bimodel*, reserving the term 'biplot' for the case where $m = 2$. However, nine years later Gabriel adopted the more common usage of 'biplot' for any value of m (see Gabriel and Odoroff (1990), which gives several examples of biplots including one with $m = 3$). Bartkowiak and Szustalewicz (1996) discuss how to display biplots in three dimensions.

In the description of biplots above there is an element of non-uniqueness, as the scalar α which occurs in the definition of $\mathbf{G}$ and $\mathbf{H}$ can take any value between zero and one and still lead to a factorization of the form (5.3.2). Two particular values of α, namely $\alpha = 0$ and $\alpha = 1$, provide especially useful interpretations for the biplot.

If $\alpha = 0$, then $\mathbf{G} = \mathbf{U}$ and $\mathbf{H}' = \mathbf{L}\mathbf{A}'$ or $\mathbf{H} = \mathbf{A}\mathbf{L}$. This means that

$$
\begin{aligned}
\mathbf{X}'\mathbf{X} &= (\mathbf{GH}')'(\mathbf{GH}') \\
&= \mathbf{HG}'\mathbf{GH}' \\
&= \mathbf{HU}'\mathbf{UH}' \\
&= \mathbf{HH}',
\end{aligned}
$$

because the columns of $\mathbf{U}$ are orthonormal. The product $\mathbf{h}'_j \mathbf{h}_k$ is therefore equal to $(n-1)$ multiplied by the covariance s_{jk} between the jth and kth variables, and $\mathbf{h}_j^{*'}\mathbf{h}_k^*$, where $\mathbf{h}_j^*$, $j = 1, 2, \cdots, p$ are as defined above, provides an approximation to $(n-1)s_{jk}$. The lengths $\mathbf{h}'_j \mathbf{h}_j$ of the vectors $\mathbf{h}_j$, $i = 1, 2, \cdots, p$ are proportional to the variances of the variables $x_1, x_2, \cdots, x_p$, and the cosines of the angles between the $\mathbf{h}_j$ represent correlations between variables. Plots of the $\mathbf{h}_j^*$ therefore provide a two-dimensional picture (usually an approximation, but often a good one) of the elements of the covariance matrix $\mathbf{S}$, and such plots are advocated by Corsten and Gabriel (1976) as a means of comparing the variance-covariance structures of several different data sets. An earlier paper by Gittins (1969), which is reproduced in Bryant and Atchley (1975), also gives plots of the $\mathbf{h}_j^*$, although it does not discuss their formal properties.

Not only do the $\mathbf{h}_j$ have a ready graphical interpretation when $\alpha = 0$, but the $\mathbf{g}_i$ also have the satisfying property that the Euclidean distance between $\mathbf{g}_h$ and $\mathbf{g}_i$ in the biplot is proportional to the Mahalanobis distance between the hth and ith observations in the complete data set. The Mahalanobis distance between two observations $\mathbf{x}_h$, $\mathbf{x}_i$, assuming that $\mathbf{X}$ has rank p so that $\mathbf{S}^{-1}$ exists, is defined as

$$
\delta_{hi}^2 = (\mathbf{x}_h - \mathbf{x}_i)'\mathbf{S}^{-1}(\mathbf{x}_h - \mathbf{x}_i), \tag{5.3.5}
$$

and is often used as an alternative to the Euclidean distance

$$
d_{hi}^2 = (\mathbf{x}_h - \mathbf{x}_i)'(\mathbf{x}_h - \mathbf{x}_i).
$$

Whereas Euclidean distance treats all variables on an equal footing, which essentially assumes that all variables have equal variances and are uncorrelated, Mahalanobis distance gives relatively less weight to variables with large variances and to groups of highly correlated variables.

To prove this Mahalanobis distance interpretation, rewrite (5.3.2) as

$$
\mathbf{x}'_i = \mathbf{g}'_i \mathbf{H}', \quad i = 1, 2, \ldots, n,
$$

and substitute in (5.3.5) to give

$$
\begin{aligned}
\delta_{hi}^2 &= (\mathbf{g}_h - \mathbf{g}_i)'\mathbf{H}'\mathbf{S}^{-1}\mathbf{H}(\mathbf{g}_h - \mathbf{g}_i) \\
&= (n-1)(\mathbf{g}_h - \mathbf{g}_i)'\mathbf{L}\mathbf{A}'(\mathbf{X}'\mathbf{X})^{-1}\mathbf{A}\mathbf{L}(\mathbf{g}_h - \mathbf{g}_i), \tag{5.3.6}
\end{aligned}
$$

as $\mathbf{H}' = \mathbf{L}\mathbf{A}'$ and $\mathbf{S}^{-1} = (n-1)(\mathbf{X}'\mathbf{X})^{-1}$.

But

$$\mathbf{X}'\mathbf{X} = (\mathbf{ULA}')'(\mathbf{ULA}')$$
$$= \mathbf{AL}(\mathbf{U}'\mathbf{U})\mathbf{LA}'$$
$$= \mathbf{AL}^2\mathbf{A}',$$

and

$$(\mathbf{X}'\mathbf{X})^{-1} = \mathbf{AL}^{-2}\mathbf{A}'.$$

Substituting in (5.3.6) gives

$$\delta_{hi}^2 = (n-1)(\mathbf{g}_h - \mathbf{g}_i)'\mathbf{L}(\mathbf{A}'\mathbf{A})\mathbf{L}^{-2}(\mathbf{A}'\mathbf{A})\mathbf{L}(\mathbf{g}_h - \mathbf{g}_i)$$
$$= (n-1)(\mathbf{g}_h - \mathbf{g}_i)'\mathbf{LL}^{-2}\mathbf{L}(\mathbf{g}_h - \mathbf{g}_i)$$
$$\text{(as the columns of } \mathbf{A} \text{ are orthonormal)},$$
$$= (n-1)(\mathbf{g}_h - \mathbf{g}_i)'(\mathbf{g}_h - \mathbf{g}_i), \quad \text{as required.}$$

An adaptation to the straightforward factorization given above for $\alpha = 0$ improves the interpretation of the plot still further. If we multiply the $\mathbf{g}_i$ by $(n-1)^{1/2}$ and correspondingly divide the $\mathbf{h}_j$ by $(n-1)^{1/2}$, then the distances between the modified $\mathbf{g}_i$ are *equal* (not just proportional) to the Mahalanobis distance and, if $m = 2 < p$, then the Euclidean distance between $\mathbf{g}_h^*$ and $\mathbf{g}_i^*$ gives an easily visualized approximation to the Mahalanobis distance between $\mathbf{x}_h$ and $\mathbf{x}_i$. Furthermore, the lengths $\mathbf{h}_j'\mathbf{h}_j$ are *equal* to variances of the variables. This adaptation was noted by Gabriel (1971), and is used in the examples below.

A further interesting property of the biplot when $\alpha = 0$ is that measures can be written down of how well the plot approximates

(a) the column-centred data matrix $\mathbf{X}$;

(b) the covariance matrix $\mathbf{S}$;

(c) the matrix of Mahalanobis distances between each pair of observations.

These measures are, respectively, (Gabriel 1971)

(a) $(l_1 + l_2) \Big/ \sum_{k=1}^{r} l_k;$

(b) $(l_1^2 + l_2^2) \Big/ \sum_{k=1}^{r} l_k^2;$

(c) $(l_1^0 + l_2^0) \Big/ \sum_{k=1}^{r} l_k^0 = 2/r.$

Because $l_1 \geq l_2 \geq \cdots \geq l_r$, these measures imply that the biplot gives a better approximation to the variances and covariances than to the (Mahalanobis) distances between observations. This is in contrast to principal coordinate plots, which concentrate on giving as good a fit as possible to interobservation dissimilarities or distances, and do not consider directly the elements of $\mathbf{X}$ or $\mathbf{S}$.

We have now seen readily interpretable properties of both the $\mathbf{g}_i^*$ and the $\mathbf{h}_j^*$ separately for the biplot when $\alpha = 0$, but there is a further property, valid for any value of α, which shows that the plots of the $\mathbf{g}_i^*$ and $\mathbf{h}_j^*$ can be usefully superimposed rather than simply considered separately.

From the relationship $x_{ij} = \mathbf{g}_i'\mathbf{h}_j$, it follows that x_{ij} is represented by the projection of $\mathbf{g}_i$ onto $\mathbf{h}_j$. Remembering that x_{ij} is the value for the ith observation of the jth variable measured about its sample mean, values of x_{ij} close to zero, which correspond to observations close to the sample mean of the jth variable, will only be achieved if $\mathbf{g}_i$ and $\mathbf{h}_j$ are nearly orthogonal. Conversely, observations for which x_{ij} is a long way from zero will have $\mathbf{g}_i$ lying in a similar direction to $\mathbf{h}_j$. The relative positions of the points defined by the $\mathbf{g}_i$ and $\mathbf{h}_j$, or their approximations in two dimensions, the $\mathbf{g}_i^*$ and $\mathbf{h}_j^*$, will therefore give information about which observations take large, average and small values on each variable.

Turning to the biplot with $\alpha = 1$, the properties relating to $\mathbf{g}_i$ and $\mathbf{h}_j$ separately are different from those for $\alpha = 0$. With $\alpha = 1$ we have

$$\mathbf{G} = \mathbf{UL}, \qquad \mathbf{H}' = \mathbf{A}',$$

and instead of $(\mathbf{g}_h - \mathbf{g}_i)'(\mathbf{g}_h - \mathbf{g}_i)$ being proportional to the Mahalanobis distance between $\mathbf{x}_h$ and $\mathbf{x}_i$, it is now equal to the Euclidean distance. This follows because

$$
\begin{aligned}
(\mathbf{x}_h - \mathbf{x}_i)'(\mathbf{x}_h - \mathbf{x}_i) &= (\mathbf{g}_h - \mathbf{g}_i)'\mathbf{H}'\mathbf{H}(\mathbf{g}_h - \mathbf{g}_i) \\
&= (\mathbf{g}_h - \mathbf{g}_i)'\mathbf{A}'\mathbf{A}(\mathbf{g}_h - \mathbf{g}_i) \\
&= (\mathbf{g}_h - \mathbf{g}_i)'(\mathbf{g}_h - \mathbf{g}_i).
\end{aligned}
$$

Therefore, if we prefer a plot on which the distance between $\mathbf{g}_h^*$ and $\mathbf{g}_i^*$ is a good approximation to Euclidean, rather than Mahalanobis, distance between $\mathbf{x}_h$ and $\mathbf{x}_i$ then the biplot with $\alpha = 1$ will be preferred to $\alpha = 0$. Note that using Mahalanobis distance emphasizes the distance between the observations in the direction of the low-variance PCs and downweights distances in the direction of high-variance PCs, when compared with Euclidean distance (see Section 10.1).

Another interesting property of the biplot with $\alpha = 1$ is that the positions of the $\mathbf{g}_i^*$ are identical to those given by a straightforward plot with respect to the first two PCs, as described in Section 5.1. It follows from equation (5.3.3) and Section 3.5 that we can write

$$x_{ij} = \sum_{k=1}^{r} z_{ik} a_{jk},$$

where $z_{ik} = u_{ik} l_k^{1/2}$ is the value of the kth PC for the ith observation. But $\alpha = 1$ implies that $\mathbf{G} = \mathbf{UL}$, so the kth element of $\mathbf{g}_i$ is $u_{ik} l_k^{1/2} = z_{ik}$.

The vector $\mathbf{g}_i^*$ consists of the first two elements of $\mathbf{g}_i$, which are simply the values of the first two PCs for the ith observation.

The properties of the $\mathbf{h}_j$ that were demonstrated above for $\alpha = 0$ will no longer be valid exactly for $\alpha = 1$, although similar interpretations can still be made, at least in qualitative terms. In fact, the coordinates of $\mathbf{h}_j^*$ are simply the coefficients of the jth variable for the first two PCs. The advantage of superimposing the plots of the $\mathbf{g}_i^*$ and $\mathbf{h}_j^*$ is preserved for $\alpha = 1$, as x_{ij} still represents the projection of $\mathbf{g}_i$ onto $\mathbf{h}_j$. In many ways, the biplot with $\alpha = 1$ is nothing new, since the $\mathbf{g}_i^*$ give PC scores and the $\mathbf{h}_j^*$ give PC coefficients, both of which are widely used on their own. The biplot, however, superimposes both the $\mathbf{g}_i^*$ and $\mathbf{h}_j^*$ to give additional information.

Other values of α could also be used; for example, Gabriel (1971) mentions $\alpha = \frac{1}{2}$, in which the sum of squares of the projections of plotted points onto either one of the axes is the same for observations as for variables (Osmond, 1985), but most applications seem to have used $\alpha = 0$, or sometimes $\alpha = 1$. For other values of α the general qualitative interpretation of the relative positions of the $\mathbf{g}_i$ and the $\mathbf{h}_j$ remains the same, but the exact properties that hold for $\alpha = 0$ and $\alpha = 1$ are no longer valid.

Another possibility is to superimpose the $\mathbf{g}_i^*$ and the $\mathbf{h}_j^*$ corresponding to different values of α. Choosing a single standard value of α for both the $\mathbf{g}_i^*$ and $\mathbf{h}_j^*$ may mean that the scales of observations and variables are so different that only one type of entity is visible on the plot. Digby and Kempton (1987, Section 3.2) choose scales for observations and variables so that both can easily be seen when plotted together. This is done rather arbitrarily, but is equivalent to using different values of α for the two types of entity. Mixing values of α in this way will, of course, lose the property that x_{ij} is the projection of $\mathbf{g}_i$ onto $\mathbf{h}_j$, but the relative positions of the $\mathbf{g}_i^*$ and $\mathbf{h}_j^*$ still give qualitative information about the size of each variable for each observation. Another way of mixing values of α is to use $\mathbf{g}_i^*$ corresponding to $\alpha = 1$ and $\mathbf{h}_j^*$ corresponding to $\alpha = 0$, so that the $\mathbf{g}_i^*$ give a PC plot, and the $\mathbf{h}_j^*$ have a direct interpretation in terms of variances and covariances. This is referred to by Gabriel (2001) as a 'correspondence analysis' (see Section 5.4) plot. Gower and Hand (1996) and Gabriel (2001), among others, have noted that different plotting positions can be chosen to give optimal approximations to two, but not all three, of the following:

(a) the elements of $\mathbf{X}$, as given by the scalar products $\mathbf{g}_i^{*'}\mathbf{h}_j^*$;

(b) Euclidean distances between the rows of $\mathbf{X}$;

(c) the covariance structure in the columns of $\mathbf{X}$.

We noted earlier that for $\alpha = 0$, (b) is fitted less well than (c). For $\alpha = 1$, (c) rather than (b) is sacrificed, while the correspondence analysis plot loses (a). Choosing $\alpha = \frac{1}{2}$ approximates (a) optimally, but is suboptimal for (b) and (c). For each of these four choices Gabriel (2001) investigates how much worse than optimal are the approximations to whichever of (a), (b), (c) are suboptimally approximated. He defines a coefficient of goodness of proportional fit equal to the squared matrix correlation between the matrix being approximated and its approximation. For example, if $\mathbf{X}$ is approximated by $\hat{\mathbf{X}}$, the matrix correlation, sometimes known as Yanai's generalized coefficient of determination (see also Section 6.3), is defined as

$$\frac{\text{tr}(\mathbf{X}'\hat{\mathbf{X}})}{\sqrt{\text{tr}(\mathbf{X}'\mathbf{X})\,\text{tr}(\hat{\mathbf{X}}'\hat{\mathbf{X}})}}.$$

By comparing this coefficient for a suboptimal choice of α with that for an optimal choice, Gabriel (2001) measures how much the approximation is degraded by the suboptimal choice. His conclusion is that the approximations are often very close to optimal, except when there is a large separation between the first two eigenvalues. Even then, the symmetric ($\alpha = \frac{1}{2}$) and correspondence analysis plots are never much inferior to the $\alpha = 0$, $\alpha = 1$ plots when one of the latter is optimal.

Another aspect of fit is explored by Heo and Gabriel (2001). They note that biplots often appear to give a better representation of patterns in the data than might be expected from simplistic interpretations of a low value for goodness-of-fit. To explain this, Heo and Gabriel (2001) invoke the special case of the unweighted version of the fixed effects model, with $\boldsymbol{\Gamma} = \mathbf{I}_p$ (see Section 3.9) and the corresponding view that we are plotting different means for different observations, rather than points from a single distribution. By simulating from the model with $q = 2$ and varying levels of σ^2 they show that the match between the biplot representation and the underlying model is often much better than that between the biplot and the data in the sample. Hence, the underlying pattern is apparent in the biplot even though the sample measure of fit is low.

5.3.1 Examples

Two examples are now presented illustrating the use of biplots. Many other examples have been given by Gabriel; in particular, see Gabriel (1981) and Gabriel and Odoroff (1990). Another interesting example, which emphasizes the usefulness of the *simultaneous* display of both rows and columns of the data matrix, is presented by Osmond (1985).

In the examples that follow, the observations are plotted as points whose coordinates are the elements of the $\mathbf{g}_i^*$, whereas variables are plotted as lines corresponding to the vectors $\mathbf{h}_j^*$, $j = 1, 2, \ldots, p$, with arrowheads at the ends of the vectors. Plots consisting of points and vectors are fairly conventional, but an alternative representation for the variables, strongly preferred by Gower and Hand (1996), is to extend the vectors $\mathbf{h}_j^*$ right across the diagram in both directions to become lines or axes. The disadvantage of this type of plot is that information about the relative sizes of the variances is lost.

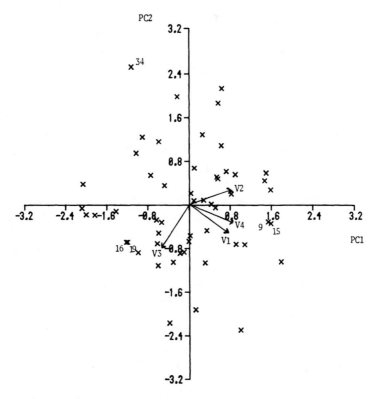

Figure 5.3. Biplot using $\alpha = 0$ for artistic qualities data.

However, what is gained is the ability to label these axes with values of the variables, so that orthogonally projecting an observation onto an axis immediately gives a prediction of the value of that variable for the chosen observation. Examples of this type of plot can be found in Gower and Hand (1996, Chapter 2).

Other variations of the plot are sometimes used. In one of their examples in which the data fall into groups, Gabriel and Odoroff (1990) replace the individual points by 'concentration ellipses' for each group. These ellipses are estimates of equal probability contours, assuming multivariate normality. Jolicoeur and Mosimann (1960) included similar ellipses on their plots.

Artistic Qualities of Painters

In Figure 5.3 a biplot is given for the data set described in Section 5.1.1 and consisting of four subjective measurements of artistic qualities for 54 painters.

The plot given uses the adapted version of $\alpha = 0$ in preference to $\alpha = 1$, because with $\alpha = 1$ the points representing the four variables are all very close to the centre of the plot, leading to difficulties in interpretation. The coordinates of the 54 painters are therefore rescaled versions of those displayed in Figure 5.2, but their relative positions are similar. For example, the group of three 'Seventeenth Century' painters at the bottom of the plot is still visible. Because of the compression of the horizontal relative to the vertical scale, the group of four painters at the left of the plot now seems to have been joined by a fifth, Murillo, who is from the same school as three of the others in this group. There is also an outlying painter, Fr. Penni, observation number 34, whose isolated position in the top left of the plot is perhaps more obvious on Figure 5.3 than Figure 5.2. The main distinguishing feature of this painter is that de Piles gave him a 0 score for composition, compared to a minimum of 4 and a maximum of 18 for all other painters.

Now consider the positions on the biplot of the vectors corresponding to the four variables. It is seen that composition and expression (V1 and V4) are close together, reflecting their relatively large positive correlation, and that drawing and colour (V2 and V3) are in opposite quadrants, confirming their fairly large negative correlation. Other correlations, and hence positions of vectors, are intermediate.

Finally, consider the simultaneous positions of painters and variables. The two painters, numbered 9 and 15, that are slightly below the positive horizontal axis are Le Brun and Domenichino. These are close to the direction defined by V4, and not far from the directions of V1 and V2, which implies that they should have higher than average scores on these three variables. This is indeed the case: Le Brun scores 16 on a scale from 0 to 20 on all three variables, and Domenichino scores 17 on V2 and V4 and 15 on V1. Their position relative to V3 suggests an average or lower score on this variable; the actual scores are 8 and 9, which confirms this suggestion. As another example consider the two painters 16 and 19 (Giorgione and Da Udine), whose positions are virtually identical, in the bottom left-hand quadrant of Figure 5.3. These two painters have high scores on V3 (18 and 16) and below average scores on V1, V2 and V4. This behaviour, but with lower scores on V2 than on V1, V4, would be predicted from the points' positions on the biplot.

100 km Running Data

The second example consists of data on times taken for each of ten 10 km sections by the 80 competitors who completed the Lincolnshire 100 km race in June 1984. There are thus 80 observations on ten variables. (I am grateful to Ron Hindley, the race organizer, for distributing the results of the race in such a detailed form.)

The variances and coefficients for the first two PCs, based on the correlation matrix for these data, are given in Table 5.2. Results for the covariance matrix are similar, though with higher coefficients in the first PC for the later sections of the race, as (means and) variances of the times taken for each section tend to increase later in the race. The first component

Table 5.2. First two PCs: 100 km running data.

		Component 1	Component 2
First 10 km		−0.30	0.45
Second 10 km		−0.30	0.45
Third 10 km		−0.33	0.34
Fourth 10 km		−0.34	0.20
Fifth 10 km	Coefficients	−0.34	−0.06
Sixth 10 km		−0.35	−0.16
Seventh 10 km		−0.31	−0.27
Eighth 10 km		−0.31	−0.30
Ninth 10 km		−0.31	−0.29
Tenth 10 km		−0.27	−0.40
Eigenvalue		72.4	1.28
Cumulative percentage of total variation		7.24	85.3

measures the overall speed of the runners, and the second contrasts those runners who slow down substantially during the course of the race with those runners who maintain a more even pace. Together, the first two PCs account for more than 85% of the total variation in the data.

The adapted $\alpha = 0$ biplot for these data is shown in Figure 5.4.

As with the previous example, the plot using $\alpha = 1$ is not very satisfactory because the vectors corresponding to the variables are all very close to the centre of the plot. Figure 5.4 shows that with $\alpha = 0$ we have the opposite extreme—the vectors corresponding to the variables and the points corresponding to the observations are completely separated. As a compromise, Figure 5.5 gives the biplot with $\alpha = \frac{1}{2}$, which at least has approximately the same degree of spread for variables and observations. As with $\alpha = 0$, the plot has been modified from the straightforward factorization corresponding to $\alpha = \frac{1}{2}$. The $\mathbf{g}_i$ have been multiplied, and the $\mathbf{h}_j$ divided, by $(n-1)^{1/4}$, so that we have a compromise between $\alpha = 1$ and the adapted version of $\alpha = 0$. The adapted plot with $\alpha = \frac{1}{2}$ is still not entirely satisfactory, but even an arbitrary rescaling of observations and/or variables, as suggested by Digby and Kempton (1987, Section 3.2), would still have all the vectors corresponding to variables within a very narrow sector of the plot. This is unavoidable for data that, as in the present case, have large correlations between all variables. The tight bunching of the vectors simply reflects large correlations, but it is interesting to note that the ordering of the vectors around their sector corresponds almost exactly to their position within the race. (The ordering is the same for both diagrams, but to avoid congestion, this fact has not been indicated on Figure 5.5.) With hindsight, this is not surprising as times in one part of the race are

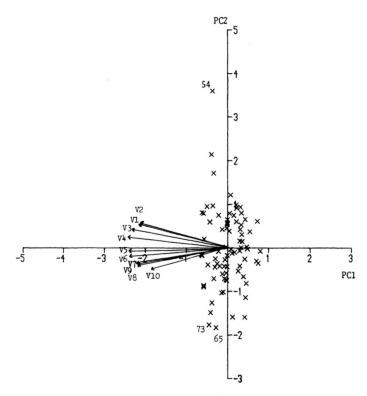

Figure 5.4. Biplot using $\alpha = 0$ for 100 km running data (V1, V2, ..., V10 indicate variables measuring times on first, second, ..., tenth sections of the race).

more likely to be similar to those in adjacent parts than to those which are more distant.

Turning to the positions of the observations, the points near the right-hand side of the diagrams correspond to the fastest athletes and those on the left to the slowest. To illustrate this, the first five and last five of the 80 finishers are indicated on Figure 5.5. Note that competitors 77 and 78 ran together throughout the race; they therefore have identical values for all ten variables and PCs, and hence identical positions on the plot. The positions of the athletes in this (horizontal) direction tally with the directions of the vectors: observations with large values on all variables, that is slow runners, will be in the direction of the vectors, namely towards the left.

Similarly, the observations near the top of the diagram are of athletes who maintained a fairly steady pace, while those at the bottom correspond to athletes who slowed down considerably during the race. Again this corresponds with the directions of the vectors: those observations at the bottom of the diagram tend to have large values of V10, V9, V8, etc. compared with V1, V2, V3, etc., meaning that these runners slow down a lot, whereas those at the top have more nearly equal values for all variables. For exam-

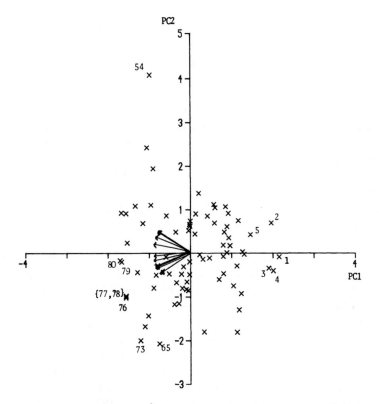

Figure 5.5. Biplot using $\alpha = \frac{1}{2}$ for 100 km running data (numbers indicate finishing position in race).

ple, consider the outlying observation at the top of Figures 5.4 and 5.5. This point corresponds to the 54th finisher, who was the only competitor to run the final 10 km faster than the first 10 km. To put this into perspective it should be noted that the average times taken for the first and last 10 km by the 80 finishers were 47.6 min, and 67.0 min respectively, showing that most competitors slowed down considerably during the race.

At the opposite extreme to the 54th finisher, consider the two athletes corresponding to the points at the bottom left of the plots. These are the 65th and 73rd finishers, whose times for the first and last 10 km were 50.0 min and 87.8 min for the 65th finisher and 48.2 min and 110.0 min for the 73rd finisher. This latter athlete therefore ran at a nearly 'average' pace for the first 10 km but was easily one of the slowest competitors over the last 10 km.

5.3.2 Variations on the Biplot

The classical biplot described above is based on the SVD of $\mathbf{X}$, the column-centred data matrix. This in turn is linked to the spectral decomposition of $\mathbf{X}'\mathbf{X}$, and hence to PCA. The variations discussed so far relate only to choices within this classical plot, for example the choice of α in defining $\mathbf{g}_i$ and $\mathbf{h}_j$ (5.3.2), the possible rescaling by a factor $(n-1)^{1/2}$ and the form of display (axes or arrowheads, concentration ellipses).

Gower and Hand (1996) describe many other variations. In particular, they look at biplots related to multivariate techniques other than PCA, including multidimensional scaling, canonical variate analysis, correspondence analysis and multiple correspondence analysis. Gabriel (1995a,b) also discusses biplots related to multivariate methods other than PCA, in particular multiple correspondence analysis and MANOVA (multivariate analysis of variance).

A key distinction drawn by Gower and Hand (1996) is between *interpolation* and *prediction* in a biplot. The former is concerned with determining where in the diagram to place an observation, given its values on the measured variables. Prediction refers to estimating the values of these variables, given the position of an observation in the plot. Both are straightforward for classical biplots—$\mathbf{g}_i^*$ is used for interpolation and $_2\tilde{x}_{ij}$ for prediction— but become more complicated for other varieties of biplot. Gower and Hand (1996, Chapter 7) describe a framework for generalized biplots that includes most other versions as special cases. One important special case is that of non-linear biplots. These will be discussed further in Section 14.1, which describes a number of non-linear modifications of PCA. Similarly, discussion of robust biplots, due to Daigle and Rivest (1992), will be deferred until Section 10.4, which covers robust versions of PCA.

The discussion and examples of the classical biplot given above use an unstandardized form of $\mathbf{X}$ and hence are related to covariance matrix PCA. As noted in Section 2.3 and elsewhere, it is more usual, and often more appropriate, to base PCA on the correlation matrix as in the examples of Section 5.3.1. Corresponding biplots can be derived from the SVD of $\tilde{\mathbf{X}}$, the column-centred data matrix whose jth column has been scaled by dividing by the standard deviation of x_j, $j = 1, 2, \ldots, p$. Many aspects of the biplot remain the same when the correlation, rather than covariance, matrix is used. The main difference is in the positions of the $\mathbf{h}_j$. Recall that if $\alpha = 0$ is chosen, together with the scaling factor $(n-1)^{1/2}$, then the length $\mathbf{h}_j^{*\prime}\mathbf{h}_j^*$ approximates the variance of x_j. In the case of a correlation-based analysis, $\mathrm{var}(x_j) = 1$ and the quality of the biplot approximation to the jth variable by the point representing $\mathbf{h}_j^*$ can be judged by the closeness of $\mathbf{h}_j^*$ to the unit circle centred at the origin. For this reason, the unit circle is sometimes drawn on correlation biplots to assist in evaluating the quality of the approximation (Besse, 1994a). Another property of correlation biplots is that the squared distance between $\mathbf{h}_j$ and $\mathbf{h}_k$ is $2(1 - r_{jk})$, where r_{jk} is the correlation between x_j and x_k. The squared distance between $\mathbf{h}_j^*$ and $\mathbf{h}_k^*$ approximates this quantity.

An alternative to the covariance and correlation biplots is the coefficient of variation biplot, due to Underhill (1990). As its name suggests, instead of dividing the jth column of $\mathbf{X}$ by the standard deviation of x_j to give a correlation biplot, here the jth column is divided by the mean of x_j. Of course, this only makes sense for certain types of non-negative variables, but Underhill (1990) shows that for such variables the resulting biplot gives a useful view of the data and variables. The cosines of the angles between the $\mathbf{h}_j^*$ still provide approximations to the correlations between variables, but the lengths of the vectors $\mathbf{h}_j^*$ now give information on the variability of the x_j *relative to their means*.

Finally, the biplot can be adapted to cope with missing values by introducing weights w_{ij} for each observation x_{ij} when approximating x_{ij} by $\mathbf{g}_i^{*'}\mathbf{h}_j^*$. A weight of zero given to missing values and a unit weight to those values which are present. The appropriate values for $\mathbf{g}_i^*, \mathbf{h}_j^*$ can be calculated using an algorithm which handles general weights, due to Gabriel and Zamir (1979). For a more general discussion of missing data in PCA see Section 13.6.

5.4 Correspondence Analysis

The technique commonly called *correspondence analysis* has been 'rediscovered' many times in several different guises with various names, such as 'reciprocal averaging' or 'dual scaling.' Greenacre (1984) provides a comprehensive treatment of the subject; in particular his Section 1.3 and Chapter 4 discuss, respectively, the history and the various different approaches to the topic. Benzécri (1992) is also comprehensive, and more recent, but its usefulness is limited by a complete lack of references to other sources. Two shorter texts, which concentrate on the more practical aspects of correspondence analysis, are Clausen (1998) and Greenacre (1993).

The name 'correspondence analysis' is derived from the French 'analyse des correspondances' (Benzécri, 1980). Although, at first sight, correspondence analysis seems unrelated to PCA it can be shown that it is, in fact, equivalent to a form of PCA for discrete (generally nominal) variables (see Section 13.1). The technique is often used to provide a graphical representation of data in two dimensions. The data are normally presented in the form of a contingency table, but because of this graphical usage the technique is introduced briefly in the present chapter. Further discussion of correspondence analysis and various generalizations of the technique, together with its connections to PCA, is given in Sections 13.1, 14.1 and 14.2.

Suppose that a set of data is presented in the form of a two-way contingency table, in which a set of n observations is classified according to its values on two discrete random variables. Thus the information available is the set of frequencies $\{n_{ij}, \ i = 1, 2, \ldots, r; \ j = 1, 2, \ldots, c\}$, where n_{ij} is the number of observations that take the ith value for the first (row) variable and the jth value for the second (column) variable. Let $\mathbf{N}$ be the $(r \times c)$ matrix whose (i, j)th element is n_{ij}.

There are a number of seemingly different approaches, all of which lead to correspondence analysis; Greenacre (1984, Chapter 4) discusses these various possibilities in some detail. Whichever approach is used, the final product is a sequence of pairs of vectors $(\mathbf{f}_1, \mathbf{g}_1), (\mathbf{f}_2, \mathbf{g}_2), \ldots, (\mathbf{f}_q, \mathbf{g}_q)$ where $\mathbf{f}_k$, $k = 1, 2, \ldots$, are r-vectors of scores or coefficients for the rows of $\mathbf{N}$, and $\mathbf{g}_k$, $k = 1, 2, \ldots$ are c-vectors of scores or coefficients for the columns of $\mathbf{N}$. These pairs of vectors are such that the first q such pairs give a 'best-fitting' representation in q dimensions, in a sense defined in Section 13.1, of the matrix $\mathbf{N}$, and of its rows and columns. It is common to take $q = 2$. The rows and columns can then be plotted on a two-dimensional diagram; the coordinates of the ith row are the ith elements of $\mathbf{f}_1, \mathbf{f}_2$, $i = 1, 2, \ldots, r$, and the coordinates of the jth column are the jth elements of $\mathbf{g}_1, \mathbf{g}_2$, $j = 1, 2, \ldots, c$.

Such two-dimensional plots cannot in general be compared in any direct way with plots made with respect to PCs or classical biplots, as $\mathbf{N}$ is a different type of data matrix from that used for PCs or their biplots. However, Greenacre (1984, Sections 9.6 and 9.10) gives examples where correspondence analysis is done with an ordinary $(n \times p)$ data matrix, $\mathbf{X}$ replacing $\mathbf{N}$. This is only possible if all variables are measured in the same units. In these circumstances, correspondence analysis produces a simultaneous two-dimensional plot of the rows and columns of $\mathbf{X}$, which is precisely what is done in a biplot, but the two analyses are not the same.

Both the classical biplot and correspondence analysis determine the plotting positions for rows and columns of $\mathbf{X}$ from the singular value decomposition (SVD) of a matrix (see Section 3.5). For the classical biplot, the SVD is calculated for the column-centred matrix $\mathbf{X}$, but in correspondence analysis, the SVD is found for a matrix of *residuals*, after subtracting 'expected values assuming independence of rows and columns' from $\mathbf{X}/n$ (see Section 13.1). The effect of looking at residual (or interaction) terms is (Greenacre, 1984, p. 288) that all the dimensions found by correspondence analysis represent aspects of the 'shape' of the data, whereas in PCA the first PC often simply represents 'size' (see Sections 4.1, 13.2). Correspondence analysis provides one way in which a data matrix may be adjusted in order to eliminate some uninteresting feature such as 'size,' before finding an SVD and hence 'PCs.' Other possible adjustments are discussed in Sections 13.2 and 14.2.3.

As with the biplot and its choice of α, there are several different ways of plotting the points corresponding to rows and columns in correspondence analysis. Greenacre and Hastie (1987) give a good description of the geometry associated with the most usual of these plots. Whereas the biplot may approximate Euclidean or Mahalanobis distances between rows, in correspondence analysis the points are often plotted to optimally approximate so-called χ^2 distances (see Greenacre (1984), Benzécri (1992)).

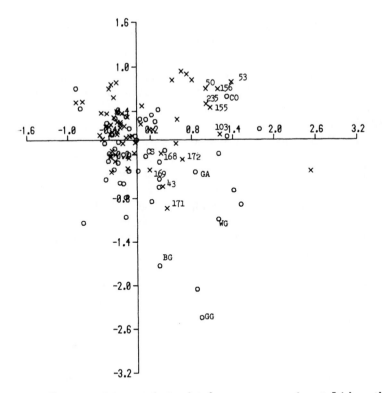

Figure 5.6. Correspondence analysis plot for summer species at Irish wetland sites. The symbol × denotes site; o denotes species.

5.4.1 Example

Figure 5.6 gives a plot obtained by correspondence analysis for a data set that recorded the presence or otherwise of 52 bird species at a number of wetland sites in Ireland. The data displayed in Figure 5.6 refer to summer sightings and are part of a much larger data set. (The larger set was kindly supplied by Dr. R.J. O'Connor of the British Trust for Ornithology, and which was analysed in various ways in two unpublished student projects/dissertations (Worton, 1984; Denham, 1985) at the University of Kent.) To avoid congestion on Figure 5.6, only a few of the points corresponding to sites and species have been labelled; these points will now be discussed. Although correspondence analysis treats the data differently from a biplot, it is still true that sites (or species) which are close to each other on the correspondence analysis plot are likely to be similar with respect to their values for the original data. Furthermore, as in a biplot, we can interpret the joint positions of sites and species.

On Figure 5.6 we first note that those sites which are close to each other on the figure also tend to be close geographically. For example, the group of sites at the top right of the plot {50, 53, 103, 155, 156, 235} are all inland

sites in the south west of Ireland, and the group {43, 168, 169, 171, 172} in
the bottom right of the diagram are all coastal sites in the south and east.

If we look at species, rather than sites, we find that similar species tend
to be located in the same part of Figure 5.6. For example, three of the
four species of goose which were recorded are in the bottom-right of the
diagram (BG, WG, GG).

Turning to the simultaneous positions of species and sites, the Grey-
lag Goose (GG) and Barnacle Goose (BG) were only recorded at site
171, among those sites which are numbered on Figure 5.6. On the plot,
site 171 is closest in position of any site to the positions of these two
species. The Whitefronted Goose (WG) is recorded at sites 171 and 172
only, the Gadwall (GA) at sites 43, 103, 168, 169, 172 among those labelled on the diagram, and the Common Sandpiper (CS) at all sites in
the coastal group {43, 168, 169, 171, 172}, but at only one of the inland
group {50, 53, 103, 155, 156, 235}. Again, these occurrences might be predicted from the relative positions of the sites and species on the plot.
However, simple predictions are not always valid, as the Coot (CO), whose
position on the plot is in the middle of the inland sites, is recorded at all
11 sites numbered on the figure.

5.5   Comparisons Between Principal Coordinates, Biplots, Correspondence Analysis and Plots Based on Principal Components

For most purposes there is little point in asking which of the graphical
techniques discussed so far in this chapter is 'best.' This is because they are
either equivalent, as is the case of PCs and principal coordinates for some
types of similarity matrix, so any comparison is trivial, or the data set is of
a type such that one or more of the techniques are not really appropriate,
and so should not be compared with the others. For example, if the data
are in the form of a contingency table, then correspondence analysis is
clearly relevant, but the use of the other techniques is more questionable.
As demonstrated by Gower and Hand (1996) and Gabriel (1995a,b), the
biplot is not restricted to 'standard' $(n \times p)$ data matrices, and could be
used on any two-way array of data. The simultaneous positions of the $\mathbf{g}_i^*$
and $\mathbf{h}_j^*$ still have a similar interpretation to that discussed in Section 5.3,
even though some of the separate properties of the $\mathbf{g}_i^*$ and $\mathbf{h}_j^*$, for instance,
those relating to variances and covariances, are clearly no longer valid. A
contingency table could also be analysed by PCA, but this is not really
appropriate, as it is not at all clear what interpretation could be given
to the results. Principal coordinate analysis needs a similarity or distance
matrix, so it is hard to see how it could be used directly on a contingency
table.

There are a number of connections between PCA and the other techniques–links with principal coordinate analysis and biplots have already been discussed, while those with correspondence analysis are deferred until Section 13.1—but for most data sets one method is more appropriate than the others. Contingency table data imply correspondence analysis, and similarity or dissimilarity matrices suggest principal coordinate analysis, whereas PCA is defined for 'standard' data matrices of n observations on p variables. Notwithstanding these distinctions, different techniques have been used on the same data sets and a number of empirical comparisons have been reported in the ecological literature. Digby and Kempton (1987, Section 4.3) compare twelve ordination methods, including principal coordinate analysis, with five different similarity measures and correspondence analysis, on both species abundances and presence/absence data. The comparison is by means of a second-level ordination based on similarities between the results of the twelve methods. Gauch (1982, Chapter 4) discusses criteria for choosing an appropriate ordination technique for ecological data, and in Gauch (1982, Chapter 3) a number of studies are described which compare PCA with other techniques, including correspondence analysis, on simulated data. The data are generated to have a similar structure to that expected in some types of ecological data, with added noise, and investigations are conducted to see which techniques are 'best' at recovering the structure. However, as with comparisons between PCA and correspondence analysis given by Greenacre (1994, Section 9.6), the relevance to the data analysed of all the techniques compared is open to debate. Different techniques implicitly assume that different types of structure or model are of interest for the data (see Section 14.2.3 for some further possibilities) and which technique is most appropriate will depend on which type of structure or model is relevant.

5.6 Methods for Graphical Display of Intrinsically High-Dimensional Data

Sometimes it will not be possible to reduce a data set's dimensionality to two or three without a substantial loss of information; in such cases, methods for displaying many variables simultaneously in two dimensions may be useful. Plots of trigonometric functions due to Andrews (1972), illustrated below, and the display in terms of faces suggested by Chernoff (1973), for which several examples are given in Wang (1978), became popular in the 1970s and 1980s. There are many other possibilities (see, for example, Tukey and Tukey (1981) and Carr(1998)) which will not be discussed here. Recent developments in the visualization of high-dimensional data using the ever-increasing power of computers have created displays which are dynamic, colourful and potentially highly informative, but there remain limitations on how many dimensions can be effectively shown simultaneously. The less sophisticated ideas of Tukey and Tukey (1981) still have a rôle to play in this respect.

Even when dimensionality cannot be reduced to two or three, a reduction to as few dimensions as possible, without throwing away too much information, is still often worthwhile before attempting to graph the data. Some techniques, such as Chernoff's faces, impose a limit on the number of variables that can be handled, although a modification due to Flury and Riedwyl (1981) increases the limit, and for most other methods a reduction in the number of variables leads to simpler and more easily interpretable diagrams. An obvious way of reducing the dimensionality is to replace the original variables by the first few PCs, and the use of PCs in this context will be particularly successful if each PC has an obvious interpretation (see Chapter 4). Andrews (1972) recommends transforming to PCs in any case, because the PCs are uncorrelated, which means that tests of significance for the plots may be more easily performed with PCs than with the original variables. Jackson (1991, Section 18.6) suggests that Andrews' curves of the residuals after 'removing' the first q PCs, that is, the sum of the last $(r - q)$ terms in the SVD of $\mathbf{X}$, may provide useful information about the behaviour of residual variability.

5.6.1   Example

In Jolliffe et al. (1986), 107 English local authorities are divided into groups or clusters, using various methods of cluster analysis (see Section 9.2), on the basis of measurements on 20 demographic variables.

The 20 variables can be reduced to seven PCs, which account for over 90% of the total variation in the 20 variables, and for each local authority an Andrews' curve is defined on the range $-\pi \le t \le \pi$ by the function

$$f(t) = \frac{z_1}{\sqrt{2}} + z_2 \sin t + z_3 \cos t + z_4 \sin 2t + z_5 \cos 2t + z_6 \sin 3t + z_7 \cos 3t,$$

where $z_1, z_2, \ldots, z_7$ are the values of the first seven PCs for the local authority. Andrews' curves may be plotted separately for each cluster. These curves are useful in assessing the homogeneity of the clusters. For example, Figure 5.7 gives the Andrews' curves for three of the clusters (Clusters 2, 11 and 12) in a 13-cluster solution, and it can be seen immediately that the shape of the curves is different for different clusters.

Compared to the variation between clusters, the curves fall into fairly narrow bands, with a few exceptions, for each cluster. Narrower bands for the curves imply greater homogeneity in the cluster.

In Cluster 12 there are two curves that are somewhat different from the remainder. These curves have three complete oscillations in the range $(-\pi, \pi)$, with maxima at 0 and $\pm 2\pi/3$. This implies that they are dominated by $\cos 3t$ and hence z_7. Examination of the seventh PC shows that

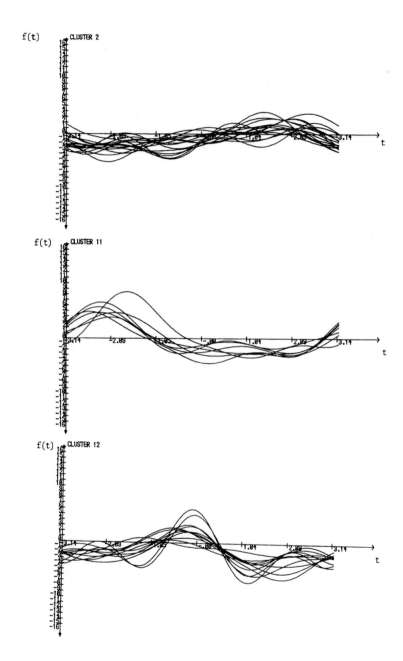

Figure 5.7. Local authorities demographic data: Andrews' curves for three clusters.

its largest coefficients are all positive and correspond to numbers of elderly persons who have recently moved to the area, numbers in privately rented accommodation, and population sparsity (Area/Population). The implication of the outlying curves for Cluster 12 is that the two local authorities corresponding to the curves (Cumbria, Northumberland) have substantially larger values for the seventh PC than do the other local authorities in the same cluster (Cornwall, Gloucestershire, Kent, Lincolnshire, Norfolk, North Yorkshire, Shropshire, Somerset and Suffolk). This is, indeed, the case and it further implies atypical values for Northumberland and Cumbria, compared to the remainder of the cluster, for the three variables having the largest coefficients for the seventh PC.

Another example of using Andrews' curves to examine the homogeneity of clusters in cluster analysis and to investigate potential outliers is given by Jolliffe et al. (1980) for the data set discussed in Section 4.2.

6

Choosing a Subset of Principal Components or Variables

In this chapter two separate, but related, topics are considered, both of which are concerned with choosing a subset of variables. In the first section, the choice to be examined is how many PCs adequately account for the total variation in $\mathbf{x}$. The major objective in many applications of PCA is to replace the p elements of $\mathbf{x}$ by a much smaller number m of PCs, which nevertheless discard very little information. It is crucial to know how small m can be taken without serious information loss. Various rules, many ad hoc, have been proposed for determining a suitable value of m, and these are discussed in Section 6.1. Examples of their use are given in Section 6.2.

Using m PCs instead of p variables considerably reduces the dimensionality of the problem when $m \ll p$, but usually the values of all p variables are still needed in order to calculate the PCs, as each PC is likely to be a function of all p variables. It might be preferable if, instead of using m PCs we could use m, or perhaps slightly more, of the original variables, to account for most of the variation in $\mathbf{x}$. The question arises of how to compare the information contained in a subset of variables with that in the full data set. Different answers to this question lead to different criteria and different algorithms for choosing the subset. In Section 6.3 we concentrate on methods that either use PCA to choose the variables or aim to reproduce the PCs in the full data set with a subset of variables, though other variable selection techniques are also mentioned briefly. Section 6.4 gives two examples of the use of variable selection methods.

All of the variable selection methods described in the present chapter are appropriate when the objective is to describe variation *within* $\mathbf{x}$ as well as possible. Variable selection when $\mathbf{x}$ is a set of regressor variables in a regression analysis, or a set of predictor variables in a discriminant analysis, is a different type of problem as criteria external to $\mathbf{x}$ must be considered. Variable selection in regression is the subject of Section 8.5. The related problem of choosing which PCs to include in a regression analysis or discriminant analysis is discussed in Sections 8.2, 9.1 respectively.

6.1 How Many Principal Components?

In this section we present a number of rules for deciding how many PCs should be retained in order to account for most of the variation in $\mathbf{x}$ (or in the standardized variables $\mathbf{x}^*$ in the case of a correlation matrix-based PCA).

In some circumstances the last few, rather than the first few, PCs are of interest, as was discussed in Section 3.4 (see also Sections 3.7, 6.3, 8.4, 8.6 and 10.1). In the present section, however, the traditional idea of trying to reduce dimensionality by replacing the p variables by the *first* m PCs $(m < p)$ is adopted, and the possible virtues of the last few PCs are ignored.

The first three types of rule for choosing m, described in Sections 6.1.1–6.1.3, are very much ad hoc rules-of-thumb, whose justification, despite some attempts to put them on a more formal basis, is still mainly that they are intuitively plausible and that they work in practice. Section 6.1.4 discusses rules based on formal tests of hypothesis. These make distributional assumptions that are often unrealistic, and they frequently seem to retain more variables than are necessary in practice. In Sections 6.1.5, 6.1.6 a number of statistically based rules, most of which do not require distributional assumptions, are described. Several use computationally intensive methods such as cross-validation and bootstrapping. Some procedures that have been suggested in the context of atmospheric science are presented briefly in Section 6.1.7, and Section 6.1.8 provides some discussion of a number of comparative studies, and a few comments on the relative merits of various rules.

6.1.1 Cumulative Percentage of Total Variation

Perhaps the most obvious criterion for choosing m, which has already been informally adopted in some of the examples of Chapters 4 and 5, is to select a (cumulative) percentage of total variation which one desires that the selected PCs contribute, say 80% or 90%. The required number of PCs is then the smallest value of m for which this chosen percentage is exceeded. It remains to define what is meant by 'percentage of variation accounted for by the first m PCs,' but this poses no real problem. Principal components are successively chosen to have the largest possible variance, and the variance of the kth PC is l_k. Furthermore, $\sum_{k=1}^{p} l_k = \sum_{j=1}^{p} s_{jj}$, that is the sum of the variances of the PCs is equal to the sum of the variances of the elements of $\mathbf{x}$. The obvious definition of 'percentage of variation accounted for by the first m PCs' is therefore

$$t_m = 100 \sum_{k=1}^{m} l_k \Big/ \sum_{j=1}^{p} s_{jj} = 100 \sum_{k=1}^{m} l_k \Big/ \sum_{k=1}^{p} l_k,$$

which reduces to

$$t_m = \frac{100}{p} \sum_{k=1}^{m} l_k$$

in the case of a correlation matrix.

Choosing a cut-off t^* somewhere between 70% and 90% and retaining m PCs, where m is the smallest integer for which $t_m > t^*$, provides a rule which in practice preserves in the first m PCs most of the information in $\mathbf{x}$. The best value for t^* will generally become smaller as p increases, or as n, the number of observations, increases. Although a sensible cutoff is very often in the range 70% to 90%, it can sometimes be higher or lower depending on the practical details of a particular data set. For example, a value greater than 90% will be appropriate when one or two PCs represent very dominant and rather obvious sources of variation. Here the less obvious structures beyond these could be of interest, and to find them a cut-off higher than 90% may be necessary. Conversely, when p is very large choosing m corresponding to 70% may give an impractically large value of m for further analyses. In such cases the threshold should be set somewhat lower.

Using the rule is, in a sense, equivalent to looking at the spectral decomposition of the covariance (or correlation) matrix $\mathbf{S}$ (see Property A3 of Sections 2.1, 3.1), or the SVD of the data matrix $\mathbf{X}$ (see Section 3.5). In either case, deciding how many terms to include in the decomposition in order to get a good fit to $\mathbf{S}$ or $\mathbf{X}$ respectively is closely related to looking at t_m, because an appropriate measure of lack-of-fit of the first m terms in either decomposition is $\sum_{k=m+1}^{p} l_k$. This follows because

$$\sum_{i=1}^{n} \sum_{j=1}^{p} (_m\tilde{x}_{ij} - x_{ij})^2 = (n-1) \sum_{k=m+1}^{p} l_k,$$

(Gabriel, 1978) and $\|_m\mathbf{S} - \mathbf{S}\| = \sum_{k=m+1}^{p} l_k$ (see the discussion of Property G4 in Section 3.2), where $_m\tilde{x}_{ij}$ is the rank m approximation to x_{ij} based on the SVD as given in equation (3.5.3), and $_m\mathbf{S}$ is the sum of the first m terms of the spectral decomposition of $\mathbf{S}$.

A number of attempts have been made to find the distribution of t_m, and hence to produce a formal procedure for choosing m, based on t_m. Mandel (1972) presents some expected values for t_m for the case where all variables are independent, normally distributed, and have the same variance. Mandel's results are based on simulation studies, and although exact results have been produced by some authors, they are only for limited special cases. For example, Krzanowski (1979a) gives exact results for $m = 1$ and $p = 3$ or 4, again under the assumptions of normality, independence and equal variances for all variables. These assumptions mean that the results can be used to determine whether or not all variables are independent, but are of little general use in determining an 'optimal' cut-off for t_m. Sugiyama and Tong (1976) describe an approximate distribution for t_m which does not assume independence or equal variances, and which can be used to test whether $l_1, l_2, \ldots, l_m$ are compatible with any given structure for $\lambda_1, \lambda_2, \ldots, \lambda_m$, the corresponding population variances. However, the test still assumes normality and it is only approximate, so it is not clear how useful it is in practice for choosing an appropriate value of m.

Huang and Tseng (1992) describe a 'decision procedure for determining the number of components' based on t_m. Given a proportion of population variance τ, which one wishes to retain, and the true minimum number of population PCs m_τ that achieves this, Huang and Tseng (1992) develop a procedure for finding a sample size n and a threshold t^* having a prescribed high probability of choosing $m = m_\tau$. It is difficult to envisage circumstances where this would be of practical value.

A number of other criteria based on $\sum_{k=m+1}^p l_k$ are discussed briefly by Jackson (1991, Section 2.8.11). In situations where some desired residual variation can be specified, as sometimes happens for example in quality control (see Section 13.7), Jackson (1991, Section 2.8.5) advocates choosing m such that the absolute, rather than percentage, value of $\sum_{k=m+1}^p l_k$ first falls below the chosen threshold.

6.1.2   Size of Variances of Principal Components

The previous rule is equally valid whether a covariance or a correlation matrix is used to compute the PCs. The rule described in this section is constructed specifically for use with correlation matrices, although it can be adapted for some types of covariance matrices. The idea behind the rule is that if all elements of $\mathbf{x}$ are independent, then the PCs are the same as the original variables and all have unit variances in the case of a correlation matrix. Thus any PC with variance less than 1 contains less information than one of the original variables and so is not worth retaining. The rule, in its simplest form, is sometimes called *Kaiser's rule* (Kaiser, 1960) and retains only those PCs whose variances l_k exceed 1. If the data set contains groups of variables having large within-group correlations, but small between group correlations, then there is one PC associated with each group whose variance is > 1, whereas any other PCs associated with the group have variances < 1 (see Section 3.8). Thus, the rule will generally retain one, and only one, PC associated with each such group of variables, which seems to be a reasonable course of action for data of this type.

As well as these intuitive justifications, Kaiser (1960) put forward a number of other reasons for a cut-off at $l_k = 1$. It must be noted, however, that most of the reasons are pertinent to factor analysis (see Chapter 7), rather than PCA, although Kaiser refers to PCs in discussing one of them.

It can be argued that a cut-off at $l_k = 1$ retains too few variables. Consider a variable which, in the population, is more-or-less independent of all other variables. In a sample, such a variable will have small coefficients in $(p - 1)$ of the PCs but will dominate one of the PCs, whose variance l_k will be close to 1 when using the correlation matrix. As the variable provides independent information from the other variables it would be unwise to delete it. However, deletion will occur if Kaiser's rule is used, and if, due to sampling variation, $l_k < 1$. It is therefore advisable to choose a cut-off l^* lower than 1, to allow for sampling variation. Jolliffe (1972) suggested, based on simulation studies, that $l^* = 0.7$ is roughly the correct level. Further discussion of this cut-off level will be given with respect to examples in Sections 6.2 and 6.4.

The rule just described is specifically designed for correlation matrices, but it can be easily adapted for covariance matrices by taking as a cut-off l^* the average value $\bar{l}$ of the eigenvalues or, better, a somewhat lower cut-off such as $l^* = 0.7\bar{l}$. For covariance matrices with widely differing variances, however, this rule and the one based on t_k from Section 6.1.1 retain very few (arguably, too few) PCs, as will be seen in the examples of Section 6.2.

An alternative way of looking at the sizes of individual variances is to use the so-called broken stick model. If we have a stick of unit length, broken at random into p segments, then it can be shown that the expected length of the kth longest segment is

$$l_k^* = \frac{1}{p} \sum_{j=k}^{p} \frac{1}{j}.$$

One way of deciding whether the proportion of variance accounted for by the kth PC is large enough for that component to be retained is to compare the proportion with l_k^*. Principal components for which the proportion exceeds l_k^* are then retained, and all other PCs deleted. Tables of l_k^* are available for various values of p and k (see, for example, Legendre and Legendre (1983, p. 406)).

6.1.3 The Scree Graph and the Log-Eigenvalue Diagram

The first two rules described above usually involve a degree of subjectivity in the choice of cut-off levels, t^* and l^* respectively. The *scree* graph, which was discussed and named by Cattell (1966) but which was already in common use, is even more subjective in its usual form, as it involves looking at a plot of l_k against k (see Figure 6.1, which is discussed in detail in Section 6.2) and deciding at which value of k the slopes of lines joining

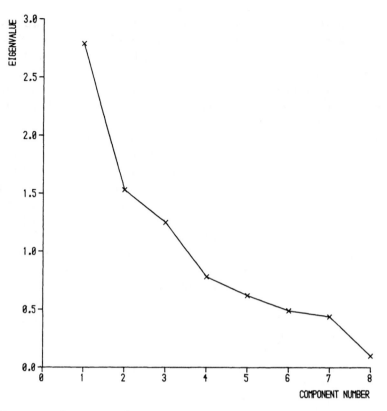

Figure 6.1. Scree graph for the correlation matrix: blood chemistry data.

the plotted points are 'steep' to the left of k, and 'not steep' to the right. This value of k, defining an 'elbow' in the graph, is then taken to be the number of components m to be retained. Its name derives from the similarity of its typical shape to that of the accumulation of loose rubble, or scree, at the foot of a mountain slope. An alternative to the scree graph, which was developed in atmospheric science, is to plot $\log(l_k)$, rather than l_k, against k; this is known as the log-eigenvalue (or LEV) diagram (see Farmer (1971), Maryon (1979)).

In introducing the scree graph, Cattell (1966) gives a somewhat different formulation from that above, and presents strong arguments that when it is used in factor analysis it is entirely objective and should produce the 'correct' number of factors (see Cattell and Vogelmann (1977) for a large number of examples). In fact, Cattell (1966) views the rule as a means of deciding upon an upper bound to the true number of factors in a factor analysis after rotation (see Chapter 7). He did not seem to envisage its use in PCA, although it has certainly been widely adopted for that purpose.

The way in which Cattell (1966) formulates the rule goes beyond a simple change of slope from 'steep' to 'shallow.' He looks for the point beyond which the scree graph defines a more-or-less straight line, not necessarily horizontal. The first point on the straight line is then taken to be the last factor/component to be retained. If there are two or more straight lines formed by the lower eigenvalues, then the cut-off is taken at the upper (left-hand) end of the left-most straight line. Cattell (1966) discusses at some length whether the left-most point on the straight line should correspond to the first excluded factor or the last factor to be retained. He concludes that it is preferable to *include* this factor, although both variants are used in practice.

The rule in Section 6.1.1 is based on $t_m = \sum_{k=1}^{m} l_k$, the rule in Section 6.1.2 looks at individual eigenvalues l_k, and the current rule, as applied to PCA, uses $l_{k-1} - l_k$ as its criterion. There is, however, no formal numerical cut-off based on $l_{k-1} - l_k$ and, in fact, judgments of when $l_{k-1} - l_k$ stops being large (steep) will depend on the *relative* values of $l_{k-1} - l_k$ and $l_k - l_{k+1}$, as well as the *absolute* value of $l_{k-1} - l_k$. Thus the rule is based subjectively on the second, as well as the first, differences among the l_k. Because of this, it is difficult to write down a formal numerical rule and the procedure has until recently remained purely graphical. Tests that attempt to formalize the procedure, due to Bentler and Yuan (1996,1998), are discussed in Section 6.1.4.

Cattell's formulation, where we look for the point at which $l_{k-1} - l_k$ becomes fairly constant for several subsequent values, is perhaps less subjective, but still requires some degree of judgment. Both formulations of the rule seem to work well in practice, provided that there is a fairly sharp 'elbow,' or change of slope, in the graph. However, if the slope gradually becomes less steep, with no clear elbow, as in Figure 6.1, then it is clearly less easy to use the procedure.

A number of methods have been suggested in which the scree plot is compared with a corresponding plot representing given percentiles, often a 95 percentile, of the distributions of each variance (eigenvalue) when PCA is done on a 'random' matrix. Here 'random' usually refers to a *correlation* matrix obtained from a random sample of n observations on p uncorrelated normal random variables, where n, p are chosen to be the same as for the data set of interest. A number of varieties of this approach, which goes under the general heading *parallel analysis*, have been proposed in the psychological literature. Parallel analysis dates back to Horn (1965), where it was described as determining the number of *factors in factor analysis*. Its ideas have since been applied, sometimes inappropriately, to PCA.

Most of its variants use simulation to construct the 95 percentiles empirically, and some examine 'significance' of loadings (eigenvectors), as well as eigenvalues, using similar reasoning. Franklin et al. (1995) cite many of the most relevant references in attempting to popularize parallel analysis amongst ecologists. The idea in versions of parallel analysis that concentrate on eigenvalues is to retain m PCs, where m is the largest integer for which the scree graph lies above the graph of upper 95 percentiles. Boot-

strap versions of these rules are used by Jackson (1993) and are discussed
further in Section 6.1.5. Stauffer et al. (1985) informally compare scree
plots from a number of ecological data sets with corresponding plots from
random data sets of the same size. They incorporate bootstrap confidence
intervals (see Section 6.1.5) but their main interest is in the stability of
the eigenvalues (see Section 10.3) rather than the choice of m. Preisendorfer and Mobley's (1988) Rule N, described in Section 6.1.7 also uses ideas
similar to parallel analysis.

Turning to the LEV diagram, an example of which is given in Section 6.2.2 below, one of the earliest published descriptions was in Craddock
and Flood (1969), although, like the scree graph, it had been used routinely
for some time before this. Craddock and Flood argue that, in meteorology,
eigenvalues corresponding to 'noise' should decay in a geometric progres-
sion, and such eigenvalues will therefore appear as a straight line on the
LEV diagram. Thus, to decide on how many PCs to retain, we should
look for a point beyond which the LEV diagram becomes, approximately,
a straight line. This is the same procedure as in Cattell's interpretation of
the scree graph, but the results are different, as we are now plotting $\log(l_k)$
rather than l_k. To justify Craddock and Flood's procedure, Farmer (1971)
generated simulated data with various known structures (or no structure).
For purely random data, with all variables uncorrelated, Farmer found that
the whole of the LEV diagram is approximately a straight line. Furthermore, he showed that if structures of various dimensions are introduced,
then the LEV diagram is useful in indicating the correct dimensionality, although real examples, of course, give much less clear-cut results than those
of simulated data.

6.1.4 The Number of Components with Unequal Eigenvalues and Other Hypothesis Testing Procedures

In Section 3.7.3 a test, sometimes known as Bartlett's test, was described
for the null hypothesis

$$H_{0,q} : \lambda_{q+1} = \lambda_{q+2} = \cdots = \lambda_p$$

against the general alternative that at least two of the last $(p-q)$ eigenvalues
are unequal. It was argued that using this test for various values of q, it
can be discovered how many of the PCs contribute substantial amounts of
variation, and how many are simply 'noise.' If m, the required number of
PCs to be retained, is defined as the number of PCs that are not noise,
then the test is used sequentially to find m.

$H_{0,p-2}$ is tested first, that is $\lambda_{p-1} = \lambda_p$, and if $H_{0,p-2}$ is not rejected then
$H_{0,p-3}$ is tested. If $H_{0,p-3}$ is not rejected, $H_{0,p-4}$ is tested next, and this
sequence continues until $H_{0,q}$ is first rejected at $q = q^*$, say. The value of
m is then taken to be $q^* + 1$ (or $q^* + 2$ if $q^* = p - 2$). There are a number of
disadvantages to this procedure, the first of which is that equation (3.7.6)

is based on the assumption of multivariate normality for $\mathbf{x}$, and is only approximately true even then. The second problem is concerned with the fact that unless $H_{0,p-2}$ is rejected, there are several tests to be done, so that the overall significance level of the sequence of tests is not the same as the individual significance levels of each test. Furthermore, it is difficult to get even an approximate idea of the overall significance level because the number of tests done is not fixed but random, and the tests are not independent of each other. It follows that, although the testing sequence suggested above can be used to estimate m, it is dangerous to treat the procedure as a formal piece of statistical inference, as significance levels are usually unknown. The reverse sequence $H_{00}, H_{01}, \ldots$ can be used instead until the first non-rejection occurs (Jackson, 1991, Section 2.6), but this suffers from similar problems.

The procedure could be added to the list of ad hoc rules, but it has one further, more practical, disadvantage, namely that in nearly all real examples it tends to retain more PCs than are really necessary. Bartlett (1950), in introducing the procedure for correlation matrices, refers to it as testing how many of the PCs are statistically significant, but 'statistical significance' in the context of these tests does not imply that a PC accounts for a substantial proportion of the total variation. For correlation matrices, Jolliffe (1970) found that the rule often corresponds roughly to choosing a cut-off l^* of about 0.1 to 0.2 in the method of Section 6.1.2. This is much smaller than is recommended in that section, and occurs because defining unimportant PCs as those with variances equal to that of the last PC is not necessarily a sensible way of finding m. If this definition is acceptable, as it may be if the model of Tipping and Bishop (1999a) (see Section 3.9) is assumed, for example, then the sequential testing procedure *may* produce satisfactory results, but it is easy to construct examples where the method gives silly answers. For instance, if there is one near-constant relationship among the elements of $\mathbf{x}$, with a much smaller variance than any other PC, then the procedure rejects $H_{0,p-2}$ and declares that all PCs need to be retained, regardless of how nearly equal are the next few eigenvalues.

The method of this section is similar in spirit to, though more formalized than, one formulation of the scree graph. Looking for the first 'shallow' slope in the graph corresponds to looking for the first of two consecutive eigenvalues that are nearly equal. The scree graph differs from the formal testing procedure in that it starts from the largest eigenvalue and compares consecutive eigenvalues two at a time, whereas the tests start with the smallest eigenvalues and compare blocks of two, three, four and so on. Another difference is that the 'elbow' point is retained in Cattell's formulation of the scree graph, but excluded in the testing procedure. The scree graph is also more subjective but, as has been stated above, the objectivity of the testing procedure is something of an illusion.

Cattell's original formulation of the scree graph differs from the above since it is differences $l_{k-1} - l_k$, rather than l_k, which must be equal beyond the cut-off point. In other words, in order to retain q PCs the last $(p - q)$ eigenvalues should have a linear trend. Bentler and Yuan (1996,1998) develop procedures for testing in the case of covariance and correlation matrices, respectively, the null hypothesis

$$H_q^* : \lambda_{q+k} = \alpha + \beta x_k, \quad k = 1, 2, \ldots, (p - q)$$

where α, β are non-negative constants and $x_k = (p - q) - k$.

For covariance matrices a maximum likelihood ratio test (MLRT) can be used straightforwardly, with the null distribution of the test statistic approximated by a χ^2 distribution. In the correlation case Bentler and Yuan (1998) use simulations to compare the MLRT, treating the correlation matrix as a covariance matrix, with a minimum χ^2 test. They show that the MLRT has a seriously inflated Type I error, even for very large sample sizes. The properties of the minimum χ^2 test are not ideal, but the test gives plausible results in the examples examined by Bentler and Yuan. They conclude that it is reliable for sample sizes of 100 or larger. The discussion section of Bentler and Yuan (1998) speculates on improvements for smaller sample sizes, on potential problems caused by possible different orderings of eigenvalues in populations and samples, and on the possibility of testing hypotheses for specific *non*-linear relationships among the last $(p - q)$ eigenvalues.

Ali et al. (1985) propose a method for choosing m based on testing hypotheses for correlations between the variables and the components. Recall from Section 2.3 that for a correlation matrix PCA and the normalization $\tilde{\alpha}_k' \tilde{\alpha}_k = \lambda_k$, the coefficients $\tilde{\alpha}_{kj}$ are precisely these correlations. Similarly, the sample coefficients $\tilde{a}_{kj}$ are correlations between the kth PC and the jth variable in the sample. The normalization constraint means that the coefficients will decrease on average as k increases. Ali et al. (1985) suggest defining m as one fewer than the index of the first PC for which none of these correlation coefficients is significantly different from zero at the 5% significance level. However, there is one immediate difficulty with this suggestion. For a fixed level of significance, the critical values for correlation coefficients decrease in absolute value as the sample size n increases. Hence for a given sample correlation matrix, the number of PCs retained depends on n. More components will be kept as n increases.

6.1.5 Choice of m Using Cross-Validatory or Computationally Intensive Methods

The rule described in Section 6.1.1 is equivalent to looking at how well the data matrix $\mathbf{X}$ is fitted by the rank m approximation based on the SVD. The idea behind the first two methods discussed in the present section is similar, except that each element x_{ij} of $\mathbf{X}$ is now predicted from an equation like the SVD, but based on a submatrix of $\mathbf{X}$ that does not include x_{ij}. In both methods, suggested by Wold (1978) and Eastment and Krzanowski

(1982), the number of terms in the estimate for $\mathbf{X}$, corresponding to the number of PCs, is successively taken as $1, 2, \ldots$, and so on, until overall prediction of the x_{ij} is no longer significantly improved by the addition of extra terms (PCs). The number of PCs to be retained, m, is then taken to be the minimum number necessary for adequate prediction.

Using the SVD, x_{ij} can be written, as in equations (3.5.2),(5.3.3),

$$x_{ij} = \sum_{k=1}^{r} u_{ik} l_k^{1/2} a_{jk},$$

where r is the rank of $\mathbf{X}$. (Recall that, in this context, $l_k, k = 1, 2, \ldots, p$ are eigenvalues of $\mathbf{X}'\mathbf{X}$, rather than of $\mathbf{S}$.)

An estimate of x_{ij}, based on the first m PCs and using all the data, is

$$_m\tilde{x}_{ij} = \sum_{k=1}^{m} u_{ik} l_k^{1/2} a_{jk}, \qquad (6.1.1)$$

but what is required is an estimate based on a subset of the data that does not include x_{ij}. This estimate is written

$$_m\hat{x}_{ij} = \sum_{k=1}^{m} \hat{u}_{ik} \hat{l}_k^{1/2} \hat{a}_{jk}, \qquad (6.1.2)$$

where $\hat{u}_{ik}, \hat{l}_k, \hat{a}_{jk}$ are calculated from suitable subsets of the data. The sum of squared differences between predicted and observed x_{ij} is then

$$\text{PRESS}(m) = \sum_{i=1}^{n} \sum_{j=1}^{p} (_m\hat{x}_{ij} - x_{ij})^2. \qquad (6.1.3)$$

The notation PRESS stands for PREdiction Sum of Squares, and is taken from the similar concept in regression, due to Allen (1974). All of the above is essentially common to both Wold (1978) and Eastment and Krzanowski (1982); they differ in how a subset is chosen for predicting x_{ij}, and in how (6.1.3) is used for deciding on m.

Eastment and Krzanowski (1982) use an estimate $\hat{a}_{jk}$ in (6.1.2) based on the data set with just the ith observation $\mathbf{x}_i$ deleted. $\hat{u}_{ik}$ is calculated with only the jth variable deleted, and $\hat{l}_k$ combines information from the two cases with the ith observation and the jth variable deleted, respectively. Wold (1978), on the other hand, divides the data into g blocks, where he recommends that g should be between four and seven and must not be a divisor of p, and that no block should contain the majority of the elements in any row or column of $\mathbf{X}$. Quantities equivalent to $\hat{u}_{ik}$, $\hat{l}_k$ and $\hat{a}_{jk}$ are calculated g times, once with each block of data deleted, and the estimates formed with the hth block deleted are then used to predict the data in the hth block, $h = 1, 2, \ldots, g$.

With respect to the choice of m, Wold (1978) and Eastment and Krzanowski (1982) each use a (different) function of PRESS(m) as a criterion for choosing m. To decide on whether to include the mth PC, Wold (1978) examines the ratio

$$R = \frac{\text{PRESS}(m)}{\sum_{i=l}^{n} \sum_{j=1}^{p} ((m-1)\tilde{x}_{ij} - x_{ij})^2}. \tag{6.1.4}$$

This compares the prediction error sum of squares after fitting m components, with the sum of squared differences between observed and estimated data points based on all the data, using $(m-1)$ components. If $R < 1$, then the implication is that a better prediction is achieved using m rather than $(m-1)$ PCs, so that the mth PC should be included.

The approach of Eastment and Krzanowski (1982) is similar to that in an analysis of variance. The reduction in prediction (residual) sum of squares in adding the mth PC to the model, divided by its degrees of freedom, is compared to the prediction sum of squares *after* fitting m PCs, divided by its degrees of freedom. Their criterion is thus

$$W = \frac{[\text{PRESS}(m-1) - \text{PRESS}(m)]/\nu_{m,1}}{\text{PRESS}(m)/\nu_{m,2}}, \tag{6.1.5}$$

where $\nu_{m,1}$, $\nu_{m,2}$ are the degrees of freedom associated with the numerator and denominator, respectively. It is suggested that if $W > 1$, then inclusion of the mth PC is worthwhile, although this cut-off at unity is to be interpreted with some flexibility. It is certainly not appropriate to stop adding PCs as soon as (6.1.5) first falls below unity, because the criterion is not necessarily a monotonic decreasing function of m. Because the ordering of the population eigenvalues may not be the same as that of the sample eigenvalues, especially if consecutive eigenvalues are close, Krzanowski (1987a) considers orders of the components different from those implied by the sample eigenvalues. For the well-known *alate adelges* data set (see Section 6.4), Krzanowski (1987a) retains components 1–4 in a straightforward implementation of W, but he keeps only components 1,2,4 when reorderings are allowed. In an example with a large number (100) of variables, Krzanowski and Kline (1995) use W in the context of factor analysis and simply take the number of components with W greater than a threshold, regardless of their position in the ordering of eigenvalues, as an indicator of the number of factors to retain. For example, the result where W exceeds 0.9 for components 1, 2, 4, 18 and no others is taken to indicate that a 4-factor solution is appropriate.

It should be noted that although the criteria described in this section are somewhat less ad hoc than those of Sections 6.1.1–6.1.3, there is still no real attempt to set up a formal significance test to decide on m. Some progress has been made by Krzanowski (1983) in investigating the sampling distribution of W using simulated data. He points out that there are two sources of variability to be considered in constructing such a distribution; namely the variability due to different sample covariance matrices $\mathbf{S}$ for a fixed population covariance matrix $\boldsymbol{\Sigma}$ and the variability due to the fact that a fixed sample covariance matrix $\mathbf{S}$ can result from different data matrices $\mathbf{X}$. In addition to this two-tiered variability, there are many parameters that can vary: n, p, and particularly the structure of $\boldsymbol{\Sigma}$. This means that simulation studies can only examine a fraction of the possible parameter values, and are therefore of restricted applicability. Krzanowski (1983) looks at several different types of structure for $\boldsymbol{\Sigma}$, and reaches the conclusion that W chooses about the right number of PCs in each case, although there is a tendency for m to be too small. Wold (1978) also found, in a small simulation study, that R retains too few PCs. This underestimation for m can clearly be overcome by moving the cut-offs for W and R, respectively, slightly below and slightly above unity. Although the cut-offs at $R = 1$ and $W = 1$ seem sensible, the reasoning behind them is not rigid, and they could be modified slightly to account for sampling variation in the same way that Kaiser's rule (Section 6.1.2) seems to work better when l^* is changed to a value somewhat below unity. In later papers (Krzanowski, 1987a; Krzanowski and Kline, 1995) a threshold for W of 0.9 is used.

Krzanowski and Kline (1995) investigate the use of W in the context of factor analysis, and compare the properties and behaviour of W with three other criteria derived from PRESS(m). Criterion P is based on the ratio

$$\frac{(\text{PRESS}(1) - \text{PRESS}(m))}{\text{PRESS}(m)},$$

P^* on

$$\frac{(\text{PRESS}(0) - \text{PRESS}(m))}{\text{PRESS}(m)},$$

and R (different from Wold's R) on

$$\frac{(\text{PRESS}(m - 1) - \text{PRESS}(m))}{(\text{PRESS}(m - 1) - \text{PRESS}(m + 1))}.$$

In each case the numerator and denominator of the ratio are divided by appropriate degrees of freedom, and in each case the value of m for which the criterion is largest gives the number of factors to be retained. On the basis of two previously analysed psychological examples, Krzanowski and Kline (1995) conclude that W and P^* select appropriate numbers of factors, whereas P and R are erratic and unreliable. As discussed later in this section, selection in factor analysis needs rather different considerations from PCA. Hence a method that chooses the 'right number' of factors may select too few PCs.

Cross-validation of PCs is computationally expensive for large data sets. Mertens et al. (1995) describe efficient algorithms for cross-validation, with applications to principal component regression (see Chapter 8) and in the investigation of influential observations (Section 10.2). Besse and Ferré (1993) raise doubts about whether the computational costs of criteria based on PRESS(m) are worthwhile. Using Taylor expansions, they show that for large n, PRESS(m) and W are almost equivalent to the much simpler quantities $\sum_{k=m+1}^{p} l_k$ and

$$\frac{l_m}{\sum_{k=m+1}^{p} l_k},$$

respectively. However, Gabriel (personal communication) notes that this conclusion holds only for large sample sizes.

In Section 3.9 we introduced the *fixed effects* model. A number of authors have used this model as a basis for constructing rules to determine m, with some of the rules relying on the resampling ideas associated with the bootstrap and jackknife. Recall that the model assumes that the rows $\mathbf{x}_i$ of the data matrix are such that $E(\mathbf{x}_i) = \mathbf{z}_i$, where $\mathbf{z}_i$ lies in a q-dimensional space F_q. If $\mathbf{e}_i$ is defined as $(\mathbf{x}_i - \mathbf{z}_i)$, then $E(\mathbf{e}_i) = \mathbf{0}$ and $\text{var}(\mathbf{e}_i) = \frac{\sigma^2}{w_i}\boldsymbol{\Gamma}$, where $\boldsymbol{\Gamma}$ is a positive definite symmetric matrix and the w_i are positive scalars whose sum is unity. For fixed q, the quantity

$$\sum_{i=1}^{n} w_i \left\| \mathbf{x}_i - \mathbf{z}_i \right\|_{\mathbf{M}}^2, \tag{6.1.6}$$

given in equation (3.9.1), is to be minimized in order to estimate σ^2, the $\mathbf{z}_i$ and F_q ($\boldsymbol{\Gamma}$ and the w_i are assumed known). The current selection problem is not only to estimate the unknown parameters, but also to find q. We wish our choice of m, the number of components retained, to coincide with the true value of q, assuming that such a value exists.

To choose m, Ferré (1990) attempts to find q so that it minimizes the loss function

$$f_q = E\left[\sum_{i=1}^{n} w_i \left\| \mathbf{z}_i - \hat{\mathbf{z}}_i \right\|_{\boldsymbol{\Gamma}^{-1}}^2\right], \tag{6.1.7}$$

where $\hat{\mathbf{z}}_i$ is the projection of $\mathbf{x}_i$ onto F_q. The criterion f_q cannot be calculated, but must be estimated, and Ferré (1990) shows that a good estimate of f_q is

$$\hat{f}_q = \sum_{k=q+1}^{p} \hat{\lambda}_k + \sigma^2 [2q(n+q-p) - np + 2(p-q) + 4\sum_{l=1}^{q}\sum_{k=q+1}^{p} \frac{\hat{\lambda}_l}{(\hat{\lambda}_l - \hat{\lambda}_k)}], \tag{6.1.8}$$

where $\hat{\lambda}_k$ is the kth largest eigenvalue of $\mathbf{V}\boldsymbol{\Gamma}^{-1}$ and

$$\mathbf{V} = \sum_{i=1}^{p} w_i (\mathbf{x}_i - \bar{\mathbf{x}})(\mathbf{x}_i - \bar{\mathbf{x}})'.$$

In the special case where $\boldsymbol{\Gamma} = \mathbf{I}_p$ and $w_i = \frac{1}{n}$, $i = 1,\ldots,n$, we have $\mathbf{V}\boldsymbol{\Gamma}^{-1} = \frac{(n-1)}{n}\mathbf{S}$, and $\hat{\lambda}_k = \frac{(n-1)}{n}l_k$, where l_k is the kth largest eigenvalue of the sample covariance matrix $\mathbf{S}$. In addition, $\hat{\mathbf{z}}_i$ is the projection of $\mathbf{x}_i$ onto the space spanned by the first q PCs. The residual variance σ^2 still needs to be estimated; an obvious estimate is the average of the $(p - q)$ smallest eigenvalues of $\mathbf{S}$.

Besse and de Falguerolles (1993) start from the same fixed effects model and concentrate on the special case just noted. They modify the loss function to become

$$L_q = \frac{1}{2} \left\| \mathbf{P}_q - \hat{\mathbf{P}}_q \right\|^2, \tag{6.1.9}$$

where $\hat{\mathbf{P}}_q = \mathbf{A}_q\mathbf{A}'_q$, $\mathbf{A_q}$ is the $(p \times q)$ matrix whose kth column is the kth eigenvalue of $\mathbf{S}$, $\mathbf{P}_q$ is the quantity corresponding to $\hat{\mathbf{P}}_q$ for the true q-dimensional subspace F_q, and $\|.\|$ denotes Euclidean norm. The loss function L_q measures the distance between the subspace F_q and its estimate $\hat{F}_q$ spanned by the columns of $\mathbf{A}_q$.

The risk function that Besse and de Falguerolles (1993) seek to minimize is $R_q = E[L_q]$. As with f_q, R_q must be estimated, and Besse and de Falguerolles (1993) compare four computationally intensive ways of doing so, three of which were suggested by Besse (1992), building on ideas from Daudin et al. (1988, 1989). Two are *bootstrap* methods; one is based on bootstrapping residuals from the q-dimensional model, while the other bootstraps the *data themselves*. A third procedure uses a *jackknife* estimate and the fourth, which requires considerably less computational effort, constructs an *approximation to the jackknife*.

Besse and de Falguerolles (1993) simulate data sets according to the fixed effects model, with $p = 10$, $q = 4$ and varying levels of the noise variance σ^2. Because q and σ^2 are known, the true value of R_q can be calculated. The four procedures outlined above are compared with the traditional scree graph and Kaiser's rule, together with boxplots of scores for each principal component. In the latter case a value m is sought such that the boxplots are much less wide for components $(m + 1), (m + 2), \ldots, p$ than they are for components $1, 2, \ldots, m$.

As the value of σ^2 increases, all of the criteria, new or old, deteriorate in their performance. Even the true value of R_q does not take its minimum value at $q = 4$, although $q = 4$ gives a local minimum in all the simulations. Bootstrapping of residuals is uninformative regarding the value of q, but the other three new procedures each have strong local minima at $q = 4$. All methods have uninteresting minima at $q = 1$ and at $q = p$, but the jackknife techniques also have minima at $q = 6, 7$ which become more pronounced as σ^2 increases. The traditional methods correctly choose $q = 4$ for small σ^2, but become less clear as σ^2 increases.

The plots of the risk estimates are very irregular, and both Besse (1992) and Besse and de Falguerolles (1993) note that they reflect the important feature of stability of the subspaces retained. Many studies of stability (see, for example, Sections 10.2, 10.3, 11.1 and Besse, 1992) show that pairs of consecutive eigenvectors are unstable if their corresponding eigenvalues are of similar size. In a similar way, Besse and de Falguerolles' (1993) risk estimates depend on the reciprocal of the difference between l_m and l_{m+1} where, as before, m is the number of PCs retained. The usual implementations of the rules of Sections 6.1.1, 6.1.2 ignore the size of gaps between eigenvalues and hence do not take stability into account. However, it is advisable when using Kaiser's rule or one of its modifications, or a rule based on cumulative variance, to treat the threshold with flexibility, and be prepared to move it, if it does not correspond to a good-sized gap between eigenvalues.

Besse and de Falguerolles (1993) also examine a real data set with $p = 16$ and $n = 60$. Kaiser's rule chooses $m = 5$, and the scree graph suggests either $m = 3$ or $m = 5$. The bootstrap and jackknife criteria behave similarly to each other. Ignoring the uninteresting minimum at $m = 1$, all four methods choose $m = 3$, although there are strong secondary minima at $m = 8$ and $m = 5$.

Another model-based rule is introduced by Bishop (1999) and, even though one of its merits is said to be that it avoids cross-validation, it seems appropriate to mention it here. Bishop (1999) proposes a Bayesian framework for Tipping and Bishop's (1999a) model, which was described in Section 3.9. Recall that under this model the covariance matrix underlying the data can be written as $\mathbf{BB'} + \sigma^2 \mathbf{I}_p$, where $\mathbf{B}$ is a $(p \times q)$ matrix. The prior distribution of $\mathbf{B}$ in Bishop's (1999) framework allows $\mathbf{B}$ to have its maximum possible value of q $(= p - 1)$ under the model. However if the posterior distribution assigns small values for all elements of a column $\mathbf{b}_k$ of $\mathbf{B}$, then that dimension is removed. The mode of the posterior distribution can be found using the EM algorithm.

Jackson (1993) discusses two bootstrap versions of 'parallel analysis,' which was described in general terms in Section 6.1.3. The first, which is a modification of Kaiser's rule defined in Section 6.1.2, uses bootstrap samples from a data set to construct confidence limits for the population eigenvalues (see Section 3.7.2). Only those components for which the corresponding 95% confidence interval lies entirely above 1 are retained. Unfortunately, although this criterion is reasonable as a means of deciding the number of factors in a factor analysis (see Chapter 7), it is inappropriate in PCA. This is because it will not retain PCs dominated by a single variable whose correlations with all the other variables are close to zero. Such variables are generally omitted from a factor model, but they provide information not available from other variables and so should be retained if most of the information in $\mathbf{X}$ is to be kept. Jolliffe's (1972) suggestion of reducing Kaiser's threshold from 1 to around 0.7 reflects the fact that we are dealing with PCA and not factor analysis. A bootstrap rule designed with PCA in mind would retain all those components for which the 95% confidence interval for the corresponding eigenvalue does not lie entirely *below* 1.

A second bootstrap approach suggested by Jackson (1993) finds 95% confidence intervals for both eigenvalues and eigenvector coefficients. To decide on m, two criteria need to be satisfied. First, the confidence intervals for λ_m and λ_{m+1} should not overlap, and second no component should be retained unless it has at least two coefficients whose confidence intervals exclude zero. This second requirement is again relevant for factor analysis, but not PCA. With regard to the first criterion, it has already been noted that avoiding small gaps between l_m and l_{m+1} is desirable because it reduces the likelihood of instability in the retained components.

6.1.6 Partial Correlation

For PCA based on a correlation matrix, Velicer (1976) suggested that the partial correlations between the p variables, given the values of the first m PCs, may be used to determine how many PCs to retain. The criterion proposed is the average of the squared partial correlations

$$V = \sum_{\substack{i=1 \\ i \neq j}}^{p} \sum_{j=1}^{p} \frac{(r_{ij}^*)^2}{p(p-1)},$$

where r_{ij}^* is the partial correlation between the ith and jth variables, given the first m PCs. The statistic r_{ij}^* is defined as the correlation between the residuals from the linear regression of the ith variable on the first m PCs, and the residuals from the corresponding regression of the jth variable on these m PCs. It therefore measures the strength of the linear relationship between the ith and jth variables after removing the common effect of the first m PCs.

The criterion V first decreases, and then increases, as m increases, and Velicer (1976) suggests that the optimal value of m corresponds to the minimum value of the criterion. As with Jackson's (1993) bootstrap rules of Section 6.1.5, and for the same reasons, this criterion is plausible as a means of deciding the number of factors in a factor analysis, but it is inappropriate in PCA. Numerous other rules have been suggested in the context of factor analysis (Reddon, 1984, Chapter 3). Many are subjective, although some, such as parallel analysis (see Sections 6.1.3, 6.1.5) attempt a more objective approach. Few are relevant to, or useful for, PCA unless they are modified in some way.

Beltrando (1990) gives a sketchy description of what appears to be another selection rule based on partial correlations. Instead of choosing m so that the average squared partial correlation is minimized, Beltrando (1990) selects m for which the number of statistically significant elements in the matrix of partial correlations is minimized.

6.1.7 Rules for an Atmospheric Science Context

As mentioned in Section 4.3, PCA has been widely used in meteorology and climatology to summarize data that vary both spatially and temporally, and a number of rules for selecting a subset of PCs have been put forward with this context very much in mind. The LEV diagram, discussed in Section 6.1.3, is one example, as is Beltrando's (1990) method in Section 6.1.6, but there are many others. In the fairly common situation where different observations correspond to different time points, Preisendorfer and Mobley (1988) suggest that important PCs will be those for which there is a clear pattern, rather than pure randomness, present in their behaviour through time. The important PCs can then be discovered by forming a time series of each PC, and testing which time series are distinguishable from white noise. Many tests are available for this purpose in the time series literature, and Preisendorfer and Mobley (1988, Sections 5g–5j) discuss the use of a number of them. This type of test is perhaps relevant in cases where the set of multivariate observations form a time series (see Chapter 12), as in many atmospheric science applications, but in the more usual (non-meteorological) situation where the observations are independent, such techniques are irrelevant, as the values of the PCs for different observations will also be independent. There is therefore no natural ordering of the observations, and if they are placed in a sequence, they should necessarily look like a white noise series.

Chapter 5 of Preisendorfer and Mobley (1988) gives a thorough review of selection rules used in atmospheric science. In Sections 5c–5e they discuss a number of rules similar in spirit to the rules of Sections 6.1.3 and 6.1.4 above. They are, however, derived from consideration of a physical model, based on spring-coupled masses (Section 5b), where it is required to distinguish signal (the important PCs) from noise (the unimportant PCs). The details of the rules are, as a consequence, somewhat different from those of Sections 6.1.3 and 6.1.4. Two main ideas are described. The first, called Rule A_4 by Preisendorfer and Mobley (1988), has a passing resemblance to Bartlett's test of equality of eigenvalues, which was defined and discussed in Sections 3.7.3 and 6.1.4. Rule A_4 assumes that the last $(p-q)$ population eigenvalues are equal, and uses the asymptotic distribution of the average of the last $(p-q)$ sample eigenvalues to test whether the common population value is equal to λ_0. Choosing an appropriate value for λ_0 introduces a second step into the procedure and is a weakness of the rule.

Rule N, described in Section 5d of Preisendorfer and Mobley (1988) is popular in atmospheric science. It is similar to the techniques of *parallel analysis*, discussed in Sections 6.1.3 and 6.1.5, and involves simulating a large number of *uncorrelated* sets of data of the same size as the real data set which is to be analysed, and computing the eigenvalues of each simulated data set. To assess the significance of the eigenvalues for the real data set, the eigenvalues are compared to percentiles derived empirically from the simulated data. The suggested rule keeps any components whose eigenvalues lie above the 95% level in the cumulative distribution of the simulated data. A disadvantage is that if the first eigenvalue for the data is very large, it makes it difficult for later eigenvalues to exceed their own

95% thresholds. It may therefore be better to look at the size of second and subsequent eigenvalues only with respect to smaller, not larger, eigenvalues. This could be achieved by removing the first term in the singular value decomposition (SVD) (3.5.2), and viewing the original second eigenvalue as the first eigenvalue in the analysis of this residual matrix. If the second eigenvalue is above its 95% threshold in this analysis, we subtract a second term from the SVD, and so on. An alternative idea, noted in Preisendorfer and Mobley (1988, Section 5f), is to simulate from a given covariance or correlation structure in which not all the variables are uncorrelated.

If the data are time series, with autocorrelation between successive observations, Preisendorfer and Mobley (1988) suggest calculating an 'equivalent sample size', n^*, allowing for the autocorrelation. The simulations used to implement Rule N are then carried out with sample size n^*, rather than the actual sample size, n. They also note that both Rules A_4 and N tend to retain too few components, and therefore recommend choosing a value for m that is the larger of the two values indicated by these rules. In Section 5k Preisendorfer and Mobley (1988) provide rules for the case of vector-valued fields.

Like Besse and de Falguerolles (1993) (see Section 6.1.5) North et al. (1982) argue strongly that a set of PCs with similar eigenvalues should either all be retained or all excluded. The size of gaps between successive eigenvalues is thus an important consideration for any decision rule, and North et al. (1982) provide a rule-of-thumb for deciding whether gaps are too small to split the PCs on either side of the gap.

The idea of using simulated data to assess significance of eigenvalues has also been explored by other authors, for example, Farmer (1971) (see also Section 6.1.3 above), Cahalan (1983) and, outside the meteorological context, Mandel (1972), Franklin et al. (1995) and the parallel analysis literature.

Other methods have also been suggested in the atmospheric science literature. For example, Jones et al. (1983), Briffa et al. (1986) use a criterion for correlation matrices, which they attribute to Guiot (1981). In this method PCs are retained if their cumulative eigenvalue product exceeds one. This technique retains more PCs than most of the other procedures discussed earlier, but Jones et al. (1983) seem to be satisfied with the results it produces. Preisendorfer and Mobley (1982, Part IV) suggest a rule that considers retaining subsets of m PCs not necessarily restricted to the first m. This is reasonable if the PCs are to be used for an external purpose, such as regression or discriminant analysis (see Chapter 8, Section 9.1), but is not really relevant if we are merely interested in accounting for as much of the variation in $\mathbf{x}$ as possible. Richman and Lamb (1987) look specifically at the case where PCs are rotated (see Section 11.1), and give a rule for choosing m based on the patterns in rotated eigenvectors.

North and Wu (2001), in an application of PCA to climate change detection, use a modification of the percentage of variation criterion of

Section 6.1.1. They use instead the percentage of 'signal' accounted for, although the PCA is done on a covariance matrix other than that associated with the signal (see Section 12.4.3). Buell (1978) advocates stability with respect to different degrees of approximation of a continuous spatial field by discrete points as a criterion for choosing m. Section 13.3.4 of von Storch and Zwiers (1999) is dismissive of selection rules.

6.1.8 Discussion

Although many rules have been examined in the last seven subsections, the list is by no means exhaustive. For example, in Section 5.1 we noted that superimposing a minimum spanning tree on a plot of the observations with respect to the first two PCs gives a subjective indication of whether or not a *two*-dimensional representation is adequate. It is not possible to give *definitive* guidance on which rules are best, but we conclude this section with a few comments on their relative merits. First, though, we discuss a small selection of the many comparative studies that have been published.

Reddon (1984, Section 3.9) describes nine such studies, mostly from the psychological literature, but all are concerned with factor analysis rather than PCA. A number of later studies in the ecological, psychological and meteorological literatures have examined various rules on both real and simulated data sets. Simulation of multivariate data sets can always be criticized as unrepresentative, because they can never explore more than a tiny fraction of the vast range of possible correlation and covariance structures. Several of the published studies, for example Grossman et al. (1991), Richman (1988), are particularly weak in this respect, looking only at simulations where all p of the variables are uncorrelated, a situation which is extremely unlikely to be of much interest in practice. Another weakness of several psychology-based studies is their confusion between PCA and factor analysis. For example, Zwick and Velicer (1986) state that 'if PCA is used to summarize a data set each retained component must contain at least two substantial loadings.' If the word 'summarize' implies a descriptive purpose the statement is nonsense, but in the simulation study that follows all their 'components' have three or more large loadings. With this structure, based on factor analysis, it is no surprise that Zwick and Velicer (1986) conclude that some of the rules they compare, which were designed with descriptive PCA in mind, retain 'too many' factors.

Jackson (1993) investigates a rather broader range of structures, including up to 12 variables in up to 3 correlated groups, as well as the completely uncorrelated case. The range of stopping rules is also fairly wide, including: Kaiser's rule; the scree graph; the broken stick rule; the proportion of total variance; tests of equality of eigenvalues; and Jackson's two bootstrap procedures described in Section 6.1.5. Jackson (1993) concludes that the broken stick and bootstrapped eigenvalue-eigenvector rules give the best results in his study. However, as with the reasoning used to develop his bootstrap rules, the results are viewed from a factor analysis rather than a PCA perspective.

Franklin et al. (1995) compare 39 published analyses from ecology. They seem to start from the unproven premise that 'parallel analysis' (see Section 6.1.3) selects the 'correct' number of components or factors to retain, and then investigate in how many of the 39 analyses 'too many' or 'too few' dimensions are chosen. Franklin et al. (1995) claim that $\frac{2}{3}$ of the 39 analyses retain too many dimensions. However, as with a number of other references cited in this chapter, they fail to distinguish between what is needed for PCA and factor analysis. They also stipulate that PCAs require normally distributed random variables, which is untrue for most purposes. It seems difficult to instil the message that PCA and factor analysis require different rules. Turner (1998) reports a large study of the properties of parallel analysis, using simulation, and notes early on that 'there are important differences between principal component analysis and factor analysis.' He then proceeds to ignore the differences, stating that the 'term *factors* will be used throughout [the] article [to refer] to either factors or components.'

Ferré (1995b) presents a comparative study which is extensive in its coverage of selection rules, but very limited in the range of data for which the techniques are compared. A total of 18 rules are included in the study, as follows:

- From Section 6.1.1 the cumulative percentage of variance with four cut-offs: 60%, 70%, 80%, 90%.

- From Section 6.1.2 Kaiser's rule with cut-off 1, together with modifications whose cut-offs are 0.7 and 2; the broken stick rule.

- From Section 6.1.3 the scree graph.

- From Section 6.1.4 Bartlett's test and an approximation due to Mardia.

- From Section 6.1.5 four versions of Eastment and Krzanowski's cross-validation methods, where two cut-offs, 1 and 0.9, are used and, for each threshold, the stopping rule can be based on either the first or last occasion that the criterion dips below the threshold; Ferré's $\hat{f}_q$; Besse and de Falguerolles's approximate jackknife criterion.

- From Section 6.1.6 Velicer's test.

The simulations are based on the fixed effects model described in Section 6.1.5. The sample size is 20, the number of variables is 10, and each simulated data matrix is the sum of a fixed (20×10) matrix $\mathbf{Z}$ of rank 8 and a matrix of independent Gaussian noise with two levels of the noise variance. This is a fixed effects model with $q = 8$, so that at first sight we might aim to choose $m = 8$. For the smaller value of noise, Ferre (1995b) considers this to be appropriate, but the higher noise level lies between the second and third largest eigenvalues of the fixed matrix $\mathbf{Z}$, so he argues that $m = 2$ should be chosen. This implies a movement away from the objective 'correct' choice given by the model, back towards what seems to be the inevitable subjectivity of the area.

The simulations are replicated 100 times for each of the two noise levels, and give results which are consistent with other studies. Kaiser's modified rule with a threshold at 2, the broken stick rule, Velicer's test, and cross-validation rules that stop after the first fall below the threshold—all retain relatively few components. Conversely, Bartlett's test, cumulative variance with a cut-off of 90%, $\hat{f}_q$ and the approximate jackknife retain greater numbers of PCs. The approximate jackknife displays the strange behaviour of retaining more PCs for larger than for smaller noise levels. If we consider $m = 8$ to be 'correct' for both noise levels, all rules behave poorly for the high noise level. For the low noise level, $\hat{f}_q$ and Bartlett's tests do best. If $m = 2$ is deemed correct for the high noise level, the best procedures are Kaiser's modified rule with threshold 2, the scree graph, and all four varieties of cross-validation. Even within this restricted study no rule is consistently good.

Bartkowiak (1991) gives an empirical comparison for some meteorological data of: subjective rules based on cumulative variance and on the scree and LEV diagrams; the rule based on eigenvalues greater than 1 or 0.7; the broken stick rule; Velicer's criterion. Most of the rules lead to similar decisions, except for the broken stick rule, which retains too few components, and the LEV diagram, which is impossible to interpret unambiguously. The conclusion for the broken stick rule is the opposite of that in Jackson's (1993) study.

Throughout our discussion of rules for choosing m we have emphasized the descriptive rôle of PCA and contrasted it with the model-based approach of factor analysis. It is usually the case that the number of components needed to achieve the objectives of PCA is greater than the number of factors in a factor analysis of the same data. However, this need not be the case when a model-based approach is adopted for PCA (see Sections 3.9, 6.1.5). As Heo and Gabriel (2001) note in the context of biplots (see Section 5.3), the fit of the first few PCs to an underlying population pattern (model) may be much better than their fit to a sample. This implies that a smaller value of m may be appropriate for model-based PCA than for descriptive purposes. In other circumstances, too, fewer PCs may be sufficient for the objectives of the analysis. For example, in atmospheric science, where p can be very large, interest may be restricted only to the first few dominant and physically interpretable patterns of variation, even though their number is fewer than that associated with most PCA-based rules. Conversely, sometimes very dominant PCs are predictable and hence of less interest than the next few. In such cases more PCs will be retained than indicated by most rules. The main message is that different objectives for a PCA lead to different requirements concerning how many PCs

Table 6.1. First six eigenvalues for the correlation matrix, blood chemistry data.

Component number	1	2	3	4	5	6
Eigenvalue, l_k	2.79	1.53	1.25	0.78	0.62	0.49
$t_m = 100 \sum_{k=1}^{m} l_k/p$	34.9	54.1	69.7	79.4	87.2	93.3
$l_{k-1} - l_k$		1.26	0.28	0.47	0.16	0.13

to retain. In reading the concluding paragraph that follows, this message should be kept firmly in mind.

Some procedures, such as those introduced in Sections 6.1.4 and 6.1.6, are usually inappropriate because they retain, respectively, too many or too few PCs in most circumstances. Some rules have been derived in particular fields of application, such as atmospheric science (Sections 6.1.3, 6.1.7) or psychology (Sections 6.1.3, 6.1.6) and may be less relevant outside these fields than within them. The simple rules of Sections 6.1.1 and 6.1.2 seem to work well in many examples, although the recommended cut-offs must be treated flexibly. Ideally the threshold should not fall between two PCs with very similar variances, and it may also change depending on the values on the values of n and p, and on the presence of variables with dominant variances (see the examples in the next section). A large amount of research has been done on rules for choosing m since the first edition of this book appeared. However it still remains true that attempts to construct rules having more sound statistical foundations seem, at present, to offer little advantage over the simpler rules in most circumstances.

6.2 Choosing m, the Number of Components: Examples

Two examples are given here to illustrate several of the techniques described in Section 6.1; in addition, the examples of Section 6.4 include some relevant discussion, and Section 6.1.8 noted a number of comparative studies.

6.2.1 Clinical Trials Blood Chemistry

These data were introduced in Section 3.3 and consist of measurements of eight blood chemistry variables on 72 patients. The eigenvalues for the correlation matrix are given in Table 6.1, together with the related information that is required to implement the ad hoc methods described in Sections 6.1.1–6.1.3.

Looking at Table 6.1 and Figure 6.1, the three methods of Sections 6.1.1–6.1.3 suggest that between three and six PCs should be retained, but the decision on a single best number is not clear-cut. Four PCs account for

Table 6.2. First six eigenvalues for the covariance matrix, blood chemistry data.

Component number	1	2	3	4	5	6
Eigenvalue, l_k	1704.68	15.07	6.98	2.64	0.13	0.07
$l_k/\bar{l}$		7.88	0.07	0.03	0.01	0.0006	0.0003
$t_m = 100 \dfrac{\sum_{k=1}^{m} l_k}{\sum_{k=1}^{p} l_k}$	98.6	99.4	99.8	99.99	99.995	99.9994
$l_{k-1} - l_k$		1689.61	8.09	4.34	2.51	0.06

nearly 80% of the total variation, but it takes six PCs to account for 90%. A cut-off at $l^* = 0.7$ for the second criterion retains four PCs, but the next eigenvalue is not very much smaller, so perhaps five should be retained. In the scree graph the slope actually increases between $k = 3$ and 4, but then falls sharply and levels off, suggesting that perhaps only four PCs should be retained. The LEV diagram (not shown) is of little help here; it has no clear indication of constant slope after any value of k, and in fact has its steepest slope between $k = 7$ and 8.

Using Cattell's (1966) formulation, there is no strong straight-line behaviour after any particular point, although perhaps a cut-off at $k = 4$ is most appropriate. Cattell suggests that the first point on the straight line (that is, the 'elbow' point) should be retained. However, if we consider the scree graph in the same light as the test of Section 6.1.4, then all eigenvalues after, and including, the elbow are deemed roughly equal and so all corresponding PCs should be deleted. This would lead to the retention of only three PCs in the present case.

Turning to Table 6.2, which gives information for the covariance matrix, corresponding to that presented for the correlation matrix in Table 6.1, the three ad hoc measures all conclusively suggest that one PC is sufficient. It is undoubtedly true that choosing $m = 1$ accounts for the vast majority of the variation in $\mathbf{x}$, but this conclusion is not particularly informative as it merely reflects that one of the original variables accounts for nearly all the variation in $\mathbf{x}$. The PCs for the covariance matrix in this example were discussed in Section 3.3, and it can be argued that it is the use of the covariance matrix, rather than the rules of Sections 6.1.1–6.1.3, that is inappropriate for these data.

6.2.2   Gas Chromatography Data

These data, which were originally presented by McReynolds (1970), and which have been analysed by Wold (1978) and by Eastment and Krzanowski (1982), are concerned with gas chromatography retention indices. After removal of a number of apparent outliers and an observation with a missing value, there remain 212 (Eastment and Krzanowski) or 213 (Wold) measurements on ten variables. Wold (1978) claims that his method indicates

Table 6.3. First six eigenvalues for the covariance matrix, gas chromatography data.

Component number	1	2	3	4	5	6
Eigenvalue, l_k	312187	2100	768	336	190	149
$l_k/\bar{l}$	9.88	0.067	0.024	0.011	0.006	0.005
$t_m = 100\dfrac{\sum_{k=1}^{m} l_k}{\sum_{k=1}^{p} l_k}$	98.8	99.5	99.7	99.8	99.9	99.94
$l_{k-1} - l_k$		310087	1332	432	146	51
R	0.02	0.43	0.60	0.70	0.83	0.99
W	494.98	4.95	1.90	0.92	0.41	0.54

the inclusion of five PCs in this example but, in fact, he slightly modifies his criterion for retaining PCs. His nominal cut-off for including the kth PC is $R < 1$; the sixth PC has $R = 0.99$ (see Table 6.3) but he nevertheless chooses to exclude it. Eastment and Krzanowski (1982) also modify their nominal cut-off but in the opposite direction, so that an *extra* PC is included. The values of W for the third, fourth and fifth PCs are 1.90, 0.92, 0.41 (see Table 6.3) so the formal rule, excluding PCs with $W < 1$, would retain three PCs. However, because the value of W is fairly close to unity, Eastment and Krzanowski (1982) suggest that it is reasonable to retain the fourth PC as well.

It is interesting to note that this example is based on a covariance matrix, and has a very similar structure to that of the previous example when the covariance matrix was used. Information for the present example, corresponding to Table 6.2, is given in Table 6.3, for 212 observations. Also given in Table 6.3 are Wold's R (for 213 observations) and Eastment and Krzanowski's W.

It can be seen from Table 6.3, as with Table 6.2, that the first two of the ad hoc methods retain only one PC. The scree graph, which cannot be sensibly drawn because $l_1 \gg l_2$, is more equivocal; it is clear from Table 6.3 that the slope drops very sharply after $k = 2$, indicating $m = 2$ (or 1), but each of the slopes for $k = 3, 4, 5, 6$ is substantially smaller than the previous slope, with no obvious levelling off. Nor is there any suggestion, for any cut-off, that the later eigenvalues lie on a straight line. There is, however, an indication of a straight line, starting at $m = 4$, in the LEV plot, which is given in Figure 6.2.

It would seem, therefore, that the cross-validatory criteria R and W differ considerably from the ad hoc rules (except perhaps the LEV plot) in the way in which they deal with covariance matrices that include a very dominant PC. Whereas most of the ad hoc rules will invariably retain only one PC in such situations, the present example shows that the cross-validatory criteria may retain several more. Krzanowski (1983) suggests that W looks for large gaps among the ordered eigenvalues, which is a similar aim to that of the scree graph, and that W can therefore be viewed as an objective ana-

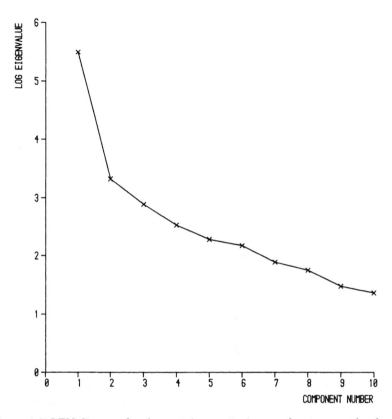

Figure 6.2. LEV diagram for the covariance matrix: gas chromatography data.

logue of the scree diagram. However, although this interpretation may be valid for the correlation matrices in his simulations, it does not seem to hold for the dominant variance structures exhibited in Tables 6.2 and 6.3.

For correlation matrices, and presumably for covariance matrices with less extreme variation among eigenvalues, the ad hoc methods and the cross-validatory criteria are likely to give more similar results. This is illustrated by a simulation study in Krzanowski (1983), where W is compared with the first two ad hoc rules with cut-offs at $t^* = 75\%$ and $l^* = 1$, respectively. Bartlett's test, described in Section 6.1.4, is also included in the comparison but, as expected from the earlier discussion, it retains too many PCs in most circumstances. The behaviour of W compared with the two ad hoc rules is the reverse of that observed in the example above. W retains fewer PCs than the $t_m > 75\%$ criterion, despite the fairly low cut-off of 75%. Similar numbers of PCs are retained for W and for the rule based on $l_k > 1$. The latter rule retains more PCs if the cut-off is lowered to 0.7 rather than 1.0, as suggested in Section 6.1.2. It can also be argued that the cut-off for W should be reduced below unity (see Section 6.1.5), in which case all three rules will give similar results.

Krzanowski (1983) examines the gas chromatography example further by generating six different artificial data sets with the same sample covariance matrix as the real data. The values of W are fairly stable across the replicates and confirm the choice of four PCs obtained above by slightly decreasing the cut-off for W. For the full data set, with outliers not removed, the replicates give some different, and useful, information from that in the original data.

6.3   Selecting a Subset of Variables

When p, the number of variables observed, is large it is often the case that a subset of m variables, with $m \ll p$, contains virtually all the information available in all p variables. It is then useful to determine an appropriate value of m, and to decide which subset or subsets of m variables are best.

Solution of these two problems, the choice of m and the selection of a good subset, depends on the purpose to which the subset of variables is to be put. If the purpose is simply to preserve most of the variation in $\mathbf{x}$, then the PCs of $\mathbf{x}$ can be used fairly straightforwardly to solve both problems, as will be explained shortly. A more familiar variable selection problem is in multiple regression, and although PCA can contribute in this context (see Section 8.5), it is used in a more complicated manner. This is because external considerations, namely the relationships of the predictor (regressor) variables with the dependent variable, as well as the internal relationships between the regressor variables, must be considered. External considerations are also relevant in other variable selection situations, for example in discriminant analysis (Section 9.1); these situations will not be considered in the present chapter. Furthermore, practical considerations, such as ease of measurement of the selected variables, may be important in some circumstances, and it must be stressed that such considerations, as well as the purpose of the subsequent analysis, can play a prominent role in variable selection, Here, however, we concentrate on the problem of finding a subset of $\mathbf{x}$ in which the sole aim is to represent the internal variation of $\mathbf{x}$ as well as possible.

Regarding the choice of m, the methods of Section 6.1 are all relevant. The techniques described there find the number of PCs that account for most of the variation in $\mathbf{x}$, but they can also be interpreted as finding the effective dimensionality of $\mathbf{x}$. If $\mathbf{x}$ can be successfully described by only m PCs, then it will often be true that $\mathbf{x}$ can be replaced by a subset of m (or perhaps slightly more) variables, with a relatively small loss of information.

Moving on to the choice of m variables, Jolliffe (1970, 1972, 1973) discussed a number of methods for selecting a subset of m variables that preserve most of the variation in $\mathbf{x}$. Some of the methods compared, and indeed some of those which performed quite well, are based on PCs. Other methods, including some based on cluster analyses of variables (see Section 9.2) were also examined but, as these do not use the PCs to select variables, they are not described here. Three main types of method using PCs were examined.

(i) Associate one variable with each of the last $m_1^*(= p - m_1)$ PCs and delete those m_1^* variables. This can either be done once only or iteratively. In the latter case a second PCA is performed on the m_1 remaining variables, and a further set of m_2^* variables is deleted, if appropriate. A third PCA can then be done on the $p - m_1^* - m_2^*$ variables, and the procedure is repeated until no further deletions are considered necessary. The choice of $m_1^*, m_2^*, \ldots$ is based on a criterion determined by the size of the eigenvalues l_k.

The reasoning behind this method is that small eigenvalues correspond to near-constant relationships among a subset of variables. If one of the variables involved in such a relationship is deleted (a fairly obvious choice for deletion is the variable with the highest coefficient in absolute value in the relevant PC) little information is lost. To decide on how many variables to delete, the criterion l_k is used as described in Section 6.1.2. The criterion t_m of Section 6.1.1 was also tried by Jolliffe (1972), but shown to be less useful.

(ii) Associate a set of m^* variables en bloc with the last m^* PCs, and then delete these variables. Jolliffe (1970, 1972) investigated this type of method, with the m^* variables either chosen to maximize sums of squares of coefficients in the last m^* PCs or to be those m^* variables that are best predicted by regression on the first $m = p - m^*$ PCs. Choice of m^* is again based on the sizes of the l_k. Such methods were found to be unsatisfactory, as they consistently failed to select an appropriate subset for some simple correlation structures.

(iii) Associate one variable with each of the first m PCs, namely the variable not already chosen with the highest coefficient in absolute value in each successive PC. These m variables are retained, and the remaining $m^* = p - m$ are deleted. The arguments leading to this approach are twofold. First, it is an obvious complementary approach to (i) and, second, in cases where there are groups of highly correlated variables it is designed to select just one variable from each group. This will happen because there will be exactly one high-variance PC associated with each group (see Section 3.8). The approach is a plausible one, as a single variable from each group should preserve most of the information given by that group when all variables in the group are highly correlated.

In Jolliffe (1972) comparisons were made, using simulated data, between non-iterative versions of method (i) and method (iii), called methods B2, B4 respectively, and with several other subset selection methods that did not use the PCs. The results showed that the PC methods B2, B4 retained the 'best' subsets more often than the other methods considered, but they also selected 'bad,' as opposed to 'good' or 'moderate', subsets more frequently than the other methods. Method B4 was most extreme in this respect; it selected 'best' and 'bad' subsets more frequently than any other method, and 'moderate' or 'good' subsets less frequently.

Similarly, for various real data sets Jolliffe (1973) found that none of the variable selection methods was uniformly best, but several of them, including B2 and B4, found reasonable subsets in most cases.

McCabe (1984) adopted a somewhat different approach to the variable selection problem. He started from the fact that, as has been seen in Chapters 2 and 3, PCs satisfy a number of different optimality criteria. A subset of the original variables that optimizes one of these criteria is termed a set of *principal variables* by McCabe (1984). Property A1 of Sections 2.1, 3.1, is uninteresting as it simply leads to a subset of variables whose variances are largest, but other properties lead to one of these four criteria:

(a) Minimize $\displaystyle\prod_{j=1}^{m^*} \theta_j$

(b) Minimize $\displaystyle\sum_{j=1}^{m^*} \theta_j$

(c) Minimize $\displaystyle\sum_{j=1}^{m^*} \theta_j^2$

(d) Minimize $\displaystyle\sum_{j=1}^{m^-} \rho_j^2$

where θ_j, $j = 1, 2, \ldots, m^*$ are the eigenvalues of the conditional covariance (or correlation) matrix of the m^* deleted variables, given the values of the m selected variables, and ρ_j, $j = 1, 2, \ldots, m^- = \min(m, m^*)$ are the canonical correlations between the set of m^* deleted variables and the set of m selected variables.

Consider, for example, Property A4 of Sections 2.1 and 3.1, where $\det(\Sigma_y)$ (or $\det(S_y)$ for samples) is to be maximized. In PCA, y consists of orthonormal linear functions of x; for principal variables y is a subset of x.

From a well-known result concerning partitioned matrices, $\det(\Sigma) = \det(\Sigma_y) \det(\Sigma_{y \cdot y})$, where $\Sigma_{y \cdot y}$ is the matrix of conditional covariances for those variables not in y, given the value of y. Because Σ, and hence $\det(\Sigma)$, is fixed for a given random vector x, maximizing $\det(\Sigma_y)$ is equivalent to minimizing $\det(\Sigma_{y \cdot y})$. Now $\det(\Sigma_{y \cdot y}) = \prod_{j=1}^{m^*} \theta_j$, so that Property A4 becomes McCabe's criterion (a) when deriving principal variables. Other properties of Chapters 2 and 3 can similarly be shown to be equivalent to one of McCabe's four criteria when dealing with principal variables.

Of the four criteria, McCabe (1984) argues that only for the first is it computationally feasible to explore all possible subsets, although the second can be used to define a stepwise variable-selection procedure; Bhargava and Ishizuka (1991) describe such a procedure. The third and fourth criteria are not explored further in McCabe's paper.

Several of the methods for selecting subsets of variables that preserve most of the information in the data associate variables with individual PCs. Cadima and Jolliffe (2001) extend the ideas of Cadima and Jolliffe (1995) for individual PCs, and look for subsets of variables that best approximate the *subspace* spanned by a subset of q PCs, in the the sense that the subspace spanned by the chosen variables is close to that spanned by the PCs of interest. A similar comparison of subspaces is the starting point for Besse and de Falguerolles's (1993) procedures for choosing the number of components to retain (see Section 6.1.5). In what follows we restrict attention to the *first* q PCs, but the reasoning extends easily to any set of q PCs.

Cadima and Jolliffe (2001) argue that there are two main ways of assessing the quality of the subspace spanned by a subset of m variables. The first compares the subspace directly with that spanned by the first q PCs; the second compares the data with its configuration when projected onto the m-variable subspaces.

Suppose that we wish to approximate the subspace spanned by the first q PCs using a subset of m variables. The matrix of orthogonal projections onto that subspace is given by

$$\mathbf{P}_q = \frac{1}{(n-1)} \mathbf{X} \mathbf{S}_q^- \mathbf{X}', \tag{6.3.1}$$

where $\mathbf{S}_q = \sum_{k=1}^{q} l_k \mathbf{a}_k \mathbf{a}_k'$ is the sum of the first q terms in the spectral decomposition of $\mathbf{S}$, and $\mathbf{S}_q^- = \sum_{k=1}^{q} l_k^{-1} \mathbf{a}_k \mathbf{a}_k'$ is a generalized inverse of $\mathbf{S}_q$. The corresponding matrix of orthogonal projections onto the space spanned by a subset of m variables is

$$\mathbf{P}_m = \frac{1}{(n-1)} \mathbf{X} \mathbf{I}_m \mathbf{S}_m^{-1} \mathbf{I}_m' \mathbf{X}', \tag{6.3.2}$$

where $\mathbf{I}_m$ is the identity matrix of order m and $\mathbf{S}_m^{-1}$ is the inverse of the $(m \times m)$ submatrix of $\mathbf{S}$ corresponding to the m selected variables.

The first measure of closeness for the two subspaces considered by Cadima and Jolliffe (2001) is the matrix correlation between $\mathbf{P}_q$ and $\mathbf{P}_m$, defined by

$$\text{corr}(\mathbf{P}_q, \mathbf{P}_m) = \frac{\text{tr}(\mathbf{P}_q' \mathbf{P}_m)}{\sqrt{\text{tr}(\mathbf{P}_q' \mathbf{P}_q) \text{tr}(\mathbf{P}_m' \mathbf{P}_m)}}. \tag{6.3.3}$$

This measure is also known as Yanai's generalized coefficient of determination (Yanai, 1980). It was used by Tanaka (1983) as one of four criteria for variable selection in factor analysis. Cadima and Jolliffe (2001) show that Yanai's coefficient can be written as

$$\text{corr}(\mathbf{P}_q, \mathbf{P}_m) = \frac{1}{\sqrt{qm}} \sum_{k=1}^{q} r_{km}^2 \qquad (6.3.4)$$

where r_{km} is the multiple correlation between the kth PC and the set of m selected variables.

The second indicator examined by Cadima and Jolliffe (2001) is again a matrix correlation, this time between the data matrix $\mathbf{X}$ and the matrix formed by orthogonally projecting $\mathbf{X}$ onto the space spanned by the m selected variables. It can be written

$$\text{corr}(\mathbf{X}, \mathbf{P}_m \mathbf{X}) = \sqrt{\frac{\sum_{k=1}^{p} \lambda_k r_{km}^2}{\sum_{k=1}^{p} \lambda_k}}. \qquad (6.3.5)$$

It turns out that this measure is equivalent to the second of McCabe's (1984) criteria defined above (see also McCabe (1986)). Cadima and Jolliffe (2001) discuss a number of other interpretations, and relationships between their measures and previous suggestions in the literature. Both indicators (6.3.4) and (6.3.5) are weighted averages of the squared multiple correlations between each PC and the set of selected variables. In the second measure, the weights are simply the eigenvalues of $\mathbf{S}$, and hence the variances of the PCs. For the first indicator the weights are positive and equal for the first q PCs, but zero otherwise. Thus the first indicator ignores PCs outside the chosen q-dimensional subspace when assessing closeness, but it also gives less weight than the second indicator to the PCs with the very largest variances relative to those with intermediate variances.

Cadima and Jolliffe (2001) discuss algorithms for finding good subsets of variables and demonstrate the use of the two measures on three examples, one of which is large ($p = 62$) compared to those typically used for illustration. The examples show that the two measures can lead to quite different optimal subsets, implying that it is necessary to know what aspect of a subspace it is most desirable to preserve before choosing a subset of variables to achieve this. They also show that

- the algorithms usually work efficiently in cases where numbers of variables are small enough to allow comparisions with an exhaustive search;

- as discussed elsewhere (Section 11.3), choosing variables on the basis of the size of coefficients or loadings in the PCs' eigenvectors can be inadvisable;

- to match the information provided by the first q PCs it is often only necessary to keep $(q + 1)$ or $(q + 2)$ variables.

For data sets in which p is too large to conduct an exhaustive search for the *optimal* subset, algorithms that can find a *good* subset are needed.

Cadima et al. (2002) compare various algorithms for finding good subsets according to the measures (6.3.4) and (6.3.5), and also with respect to the RV-coefficient, which is discussed briefly below (see also Section 3.2). Two versions of simulated annealing, a genetic algorithm, and a restricted improvement algorithm, are compared with a number of stepwise algorithms, on a total of fourteen data sets. The results show a general inferiority of the stepwise methods, but no single algorithm outperforms all the others. Cadima et al. (2002) recommend using simulated annealing or a genetic algorithm to provide a starting point for a restricted improvement algorithm, which then refines the solution. They make the interesting point that for large p the number of candidate subsets is so large that, for criteria whose range of values is bounded, it is almost inevitable that there are many solutions that are very close to optimal. For instance, in one of their examples, with $p = 62$, they find 800 solutions corresponding to a population size of 800 in their genetic algorithm. The best of these has a value 0.8079 for the criterion (6.3.5), but the worst is 0.8060, less than 0.3% smaller. Of course, it is possible that the global optimum is much greater than the best of these 800, but it seems highly unlikely.

Al-Kandari (1998) provides an extensive study of a large number of variable selection methods. The ideas of Jolliffe (1972, 1973) and McCabe (1984) are compared with a variety of new methods, based on loadings in the PCs, on correlations of the PCs with the variables, and on versions of McCabe's (1984) principal variables that are constructed from correlation, rather than covariance, matrices. The methods are compared on simulated data with a wide range of covariance or correlation structures, and on various real data sets that are chosen to have similar covariance/correlation structures to those of the simulated data. On the basis of the results of these analyses, it is concluded that few of the many techniques considered are uniformly inferior to other methods, and none is uniformly superior. The 'best' method varies, depending on the covariance or correlation structure of a data set. It also depends on the 'measure of efficiency' used to determine how good is a subset of variables, as noted also by Cadima and Jolliffe (2001). In assessing which subsets of variables are best, Al-Kandari (1998) additionally takes into account the interpretability of the PCs based on the subset, relative to the PCs based on all p variables (see Section 11.3).

Al-Kandari (1998) also discusses the distinction between criteria used to *choose* subsets of variables and criteria used to *evaluate* how good a chosen subset is. The latter are her 'measures of efficiency' and ideally these same criteria should be used to choose subsets in the first place. However, this may be computationally infeasible so that a suboptimal but computationally straightforward criterion is used to do the choosing instead. Some of Al-Kandari's (1998) results are reported in Al-Kandari and Jolliffe (2001) for covariance, but not correlation, matrices.

King and Jackson (1999) combine some of the ideas of the present Section with some from Section 6.1. Their main objective is to select a subset of m variables, but rather than treating m as fixed they also consider how to choose m. They use methods of variable selection due to Jolliffe (1972, 1973), adding a new variant that was computationally infeasible in 1972. To choose m, King and Jackson (1999) consider the rules described in Sections 6.1.1 and 6.1.2, including the broken stick method, together with a rule that selects the largest value of m for which $n/m > 3$. To assess the quality of a chosen subset of size m, King and Jackson (1999) compare plots of scores on the first two PCs for the full data set and for the data set containing only the m selected variables. They also compute a Procrustes measure of fit (Krzanowski, 1987a) between the m-dimensional configurations given by PC scores in the full and reduced data sets, and a weighted average of correlations between PCs in the full and reduced data sets.

The data set analyzed by King and Jackson (1999) has $n = 37$ and $p = 36$. The results of applying the various selection procedures to these data confirm, as Jolliffe (1972, 1973) found, that methods B2 and B4 do reasonably well. The results also confirm that the broken stick method generally chooses smaller values of m than the other methods, though its subsets do better with respect to the Procrustes measure of fit than some much larger subsets. The small number of variables retained by the broken stick implies a corresponding small proportion of total variance accounted for by the subsets it selects. King and Jackson's (1999) recommendation of method B4 with the broken stick could therefore be challenged.

We conclude this section by briefly describing a number of other possible methods for variable selection. None uses PCs directly to select variables, but all are related to topics discussed more fully in other sections or chapters. Bartkowiak (1991) uses a method described earlier in Bartkowiak (1982) to select a set of 'representative' variables in an example that also illustrates the choice of the number of PCs (see Section 6.1.8). Variables are added sequentially to a 'representative set' by considering each variable currently outside the set as a candidate for inclusion. The maximum residual sum of squares is calculated from multiple linear regressions of each of the other excluded variables on all the variables in the set plus the candidate variable. The candidate for which this maximum sum of squares is minimized is then added to the set. One of Jolliffe's (1970, 1972, 1973) rules uses a similar idea, but in a non-sequential way. A set of m variables is chosen if it maximizes the minimum multiple correlation between each of the $(p - m)$ non-selected variables and the set of m selected variables.

The RV-coefficient, due to Robert and Escoufier (1976), was defined in Section 3.2. To use the coefficient to select a subset of variables, Robert and Escoufier suggest finding $\mathbf{X}_1$ which maximizes $\mathrm{RV}(\mathbf{X}, \mathbf{M}'\mathbf{X}_1)$, where $\mathrm{RV}(\mathbf{X}, \mathbf{Y})$ is defined by equation (3.2.2) of Section 3.2. The matrix $\mathbf{X}_1$ is the $(n \times m)$ submatrix of $\mathbf{X}$ consisting of n observations on a subset of m variables, and $\mathbf{M}$ is a specific $(m \times m)$ orthogonal matrix, whose construction is described in Robert and Escoufier's paper. It is interesting to compare what is being optimized here with the approaches described earlier.

- The RV-coefficient compares *linear combinations of subsets of variables* with the full set of variables.

- Some methods, such as those of Jolliffe (1970, 1972, 1973), compare *principal components of subsets of variables* with *principal components* from the full set.

- Some approaches, such as McCabe's (1984) principal variables, simply compare subsets of the variables with the full set of variables.

- Some criteria, such as Yanai's generalized coefficient of determination, compare *subspaces spanned by a subset of variables* with *subspaces spanned by a subset of PCs*, as in Cadima and Jolliffe (2001).

No examples are presented by Robert and Escoufier (1976) of how their method works in practice. However, Gonzalez et al. (1990) give a stepwise algorithm for implementing the procedure and illustrate it with a small example ($n = 49$; $p = 6$). The example is small enough for *all* subsets of each size to be evaluated. Only for $m = 1, 2, 3$ does the stepwise algorithm give the best subset with respect to RV, as identified by the full search. Escoufier (1986) provides further discussion of the properties of the RV-coefficient when used in this context.

Tanaka and Mori (1997) also use the RV-coefficient, as one of two criteria for variable selection. They consider the same linear combinations $\mathbf{M'X_1}$ of a given set of variables as Robert and Escoufier (1976), and call these linear combinations *modified principal components*. Tanaka and Mori (1997) assess how well a subset reproduces the full set of variables by means of the RV-coefficient. They also have a second form of 'modified' principal components, constructed by minimizing the trace of the residual covariance matrix obtained by regressing $\mathbf{X}$ on $\mathbf{M'X_1}$. This latter formulation is similar to Rao's (1964) PCA of instrumental variables (see Section 14.3). The difference between Tanaka and Mori's (1997) instrumental variable approach and that of Rao (1964) is that Rao attempts to predict $\mathbf{X_2}$, the ($n \times (p - m)$) complementary matrix to $\mathbf{X_1}$ using linear functions of $\mathbf{X_1}$, whereas Tanaka and Mori try to predict the full matrix $\mathbf{X}$.

Both of Tanaka and Mori's modified PCAs solve the same eigenequation

$$(\mathbf{S}_{11}^2 + \mathbf{S}_{12}\mathbf{S}_{21})\mathbf{a} = l\mathbf{S}_{11}\mathbf{a}, \tag{6.3.6}$$

with obvious notation, but differ in the way that the quality of a subset is measured. For the instrumental variable approach, the criterion is proportional to $\sum_{k=1}^{m} l_k$, whereas for the components derived *via* the RV-coefficient, quality is based on $\sum_{k=1}^{m} l_k^2$, where l_k is the kth largest eigenvalue in the solution of (6.3.6). A backward elimination method is used to delete variables until some threshold is reached, although in the examples given by Tanaka and Mori (1997) the decision on when to stop deleting variables appears to be rather subjective.

Mori et al. (1999) propose that the subsets selected in modified PCA are also assessed by means of a PRESS criterion, similar to that defined in equation (6.1.3), except that $_m\tilde{x}_{ij}$ is replaced by the prediction of x_{ij} found from modified PCA with the ith observation omitted. Mori et al. (2000) demonstrate a procedure in which the PRESS citerion is used directly to select variables, rather than as a supplement to another criterion. Tanaka and Mori (1997) show how to evaluate the influence of variables on parameters in a PCA (see Section 10.2 for more on influence), and Mori et al. (2000) implement and illustrate a backward-elimination variable selection algorithm in which variables with the smallest influence are successively removed.

Hawkins and Eplett (1982) describe a method which can be used for selecting a subset of variables in regression; their technique and an earlier one introduced by Hawkins (1973) are discussed in Sections 8.4 and 8.5. Hawkins and Eplett (1982) note that their method is also potentially useful for selecting a subset of variables in situations other than multiple regression, but, as with the RV-coefficient, no numerical example is given in the original paper. Krzanowski (1987a,b) describes a methodology, using principal components together with Procrustes rotation for selecting subsets of variables. As his main objective is preserving 'structure' such as groups in the data, we postpone detailed discussion of his technique until Section 9.2.2.

6.4 Examples Illustrating Variable Selection

Two examples are presented here; two other relevant examples are given in Section 8.7.

6.4.1 *Alate adelges* (Winged Aphids)

These data were first presented by Jeffers (1967) and comprise 19 different variables measured on 40 winged aphids. A description of the variables, together with the correlation matrix and the coefficients of the first four PCs based on the correlation matrix, is given by Jeffers (1967) and will not be reproduced here. For 17 of the 19 variables all of the correlation coefficients are positive, reflecting the fact that 12 variables are lengths or breadths of parts of each individual, and some of the other (discrete) variables also measure aspects of the size of each aphid. Not surprisingly, the first PC based on the correlation matrix accounts for a large proportion (73.0%) of the total variation, and this PC is a measure of overall size of each aphid. The second PC, accounting for 12.5% of total variation, has its

Table 6.4. Subsets of selected variables, *Alate adelges*.

(Each row corresponds to a selected subset with × denoting a selected variable.)

				Variables					
	5	8	9	11	13	14	17	18	19
McCabe, using criterion (a)									
Three variables { best				×			×		×
Three variables { second best			×	×			×		
Four variables { best				×	×		×		×
Four variables { second best	×			×	×				×
Jolliffe, using criteria B2, B4									
Three variables { B2			×		×	×			
Three variables { B4				×	×		×		
Four variables { B2	×	×		×		×			
Four variables { B4	×			×	×		×		
Criterion (6.3.4)									
Three variables				×	×		×		
Four variables	×			×		×		×	
Criterion (6.3.5)									
Three variables	×				×		×		
Four variables	×			×	×		×		

largest coefficients on five of the seven discrete variables, and the third PC (3.9%) is almost completely dominated by one variable, number of antennal spines. This variable, which is one of the two variables negatively correlated with size, has a coefficient in the third PC that is five times as large as any other variable.

Table 6.4 gives various subsets of variables selected by Jolliffe (1973) and by McCabe (1982) in an earlier version of his 1984 paper that included additional examples. The subsets given by McCabe (1982) are the best two according to his criterion (a), whereas those from Jolliffe (1973) are selected by the criteria B2 and B4 discussed above. Only the results for $m = 3$ are given in Jolliffe (1973), but Table 6.4 also gives results for $m = 4$ using his methods. In addition, the table includes the 'best' 3- and 4-variable subsets according to the criteria (6.3.4) and (6.3.5).

There is considerable overlap between the various subsets selected. In particular, variable 11 is an almost universal choice and variables 5, 13 and 17 also appear in subsets selected by at least three of the four methods. Conversely, variables {1–4, 6, 7, 10, 12, 15, 16} appear in none of subsets of Table 6.4. It should be noted the variable 11 is 'number of antennal spines,' which, as discussed above, dominates the third PC. Variables 5 and 17, measuring number of spiracles and number of ovipositor spines, respectively, are both among the group of dominant variables for the second PC, and variable 13 (tibia length 3) has the largest coefficient of any variable for PC1.

Comparisons can be made regarding how well Jolliffe's and McCabe's selections perform with respect to the criteria (6.3.4) and (6.3.5). For (6.3.5), Jolliffe's choices are closer to optimality than McCabe's, achieving values of 0.933 and 0.945 for four variables, compared to 0.907 and 0.904 for McCabe, whereas the optimal value is 0.948. Discrepancies are generally larger but more variable for criterion (6.3.4). For example, the B2 selection of three variables achieves a value of only 0.746 compared the optimal value of 0.942, which is attained by B4. Values for McCabe's selections are intermediate (0.838, 0.880).

Regarding the choice of m, the l_k criterion of Section 6.1.2 was found by Jolliffe (1972), using simulation studies, to be appropriate for methods B2 and B4, with a cut-off close to $l^* = 0.7$. In the present example the criterion suggests $m = 3$, as $l_3 = 0.75$ and $l_4 = 0.50$. Confirmation that m should be this small is given by the criterion t_m of Section 6.1.1. Two PCs account for 85.4% of the variation, three PCs give 89.4% and four PCs contribute 92.0%, from which Jeffers (1967) concludes that two PCs are sufficient to account for most of the variation. However, Jolliffe (1973) also looked at how well other aspects of the structure of data are reproduced for various values of m. For example, the form of the PCs and the division into four distinct groups of aphids (see Section 9.2 for further discussion of this aspect) were both examined and found to be noticeably better reproduced for $m = 4$ than for $m = 2$ or 3, so it seems that the criteria of Sections 6.1.1 and 6.1.2 might be relaxed somewhat when very small values of m are indicated, especially when coupled with small values of n, the sample size. McCabe (1982) notes that four or five of the original variables are necessary in order to account for as much variation as the first two PCs, confirming that $m = 4$ or 5 is probably appropriate here.

Tanaka and Mori (1997) suggest, on the basis of their two criteria and using a backward elimination algorithm, that seven or nine variables should be kept, rather more than Jolliffe (1973) or McCabe (1982). If only four variables are retained, Tanaka and Mori's (1997) analysis keeps variables $5, 6, 14, 19$ according to the RV-coefficient, and variables $5, 14, 17, 18$ using residuals from regression. At least three of the four variables overlap with choices made in Table 6.4. On the other hand, the selection rule based on influential variables suggested by Mori et al. (2000) retains variables $2, 4, 12, 13$ in a 4-variable subset, a quite different selection from those of the other methods.

6.4.2   Crime Rates

These data were given by Ahamad (1967) and consist of measurements of the crime rate in England and Wales for 18 different categories of crime (the variables) for the 14 years, 1950–63. The sample size $n = 14$ is very

Table 6.5. Subsets of selected variables, crime rates.

(Each row corresponds to a selected subset with × denoting a selected variable.)

		1	3	4	5	7	8	10	13	14	16	17
McCabe, using criterion (a)												
Three variables	best	×									×	×
	second best	×								×		×
Four variables	best	×							×	×		×
	second best	×						×	×	×		
Jolliffe, using criteria B2, B4												
Three variables	B2	×				×			×			
	B4	×	×		×							
Four variables	B2	×				×		×	×			
	B4	×	×		×							×
Criterion (6.3.4)												
Three variables		×					×		×			
Four variables		×							×	×		×
Criterion (6.3.5)												
Three variables				×			×		×			
Four variables		×					×		×	×		

small, and is in fact smaller than the number of variables. Furthermore, the data are time series, and the 14 observations are not independent (see Chapter 12), so that the effective sample size is even smaller than 14. Leaving aside this potential problem and other criticisms of Ahamad's analysis (Walker, 1967), subsets of variables that are selected using the correlation matrix by the same methods as in Table 6.4 are shown in Table 6.5.

There is a strong similarity between the correlation structure of the present data set and that of the previous example. Most of the variables considered increased during the time period considered, and the correlations between these variables are large and positive. (Some elements of the correlation matrix given by Ahamad (1967) are incorrect; Jolliffe (1970) gives the correct values.)

The first PC based on the correlation matrix therefore has large coefficients on all these variables; it measures an 'average crime rate' calculated largely from 13 of the 18 variables, and accounts for 71.7% of the total variation. The second PC, accounting for 16.1% of the total variation, has large coefficients on the five variables whose behaviour over the 14 years is 'atypical' in one way or another. The third PC, accounting for 5.5% of the total variation, is dominated by the single variable 'homicide,' which stayed almost constant compared with the trends in other variables over the period of study. On the basis of t_m only two or three PCs are necessary, as they account for 87.8%, 93.3%, respectively, of the total variation. The third and fourth eigenvalues are 0.96, 0.68 so that a cut-off of $l^* = 0.70$ gives $m = 3$, but l_4 is so close to 0.70 that caution suggests $m = 4$. Such conservatism is particularly appropriate for small sample sizes, where sampling variation may be substantial. As in the previous example, Jolliffe (1973) found that the inclusion of a fourth variable produced a marked improvement in reproducing some of the results given by all 18 variables. McCabe (1982) also indicated that $m = 3$ or 4 is appropriate.

The subsets chosen in Table 6.5 overlap less than in the previous example, and McCabe's subsets change noticeably in going from $m = 3$ to $m = 4$. However, there is still substantial agreement; for example, variable 1 is a member of all but one of the selected subsets and variable 13 is also selected by all four methods, whereas variables $\{2, 6, 9, 11, 12, 15, 18\}$ are not selected at all.

Of the variables that are chosen by all four methods, variable 1 is 'homicide,' which dominates the third PC and is the only crime whose occurrence shows no evidence of serial correlation during the period 1950–63. Because its behaviour is different from that of all the other variables, it is important that it should be retained in any subset that seeks to account for most of the variation in $\mathbf{x}$. Variable 13 (assault) is also atypical of the general upward trend—it actually decreased between 1950 and 1963.

The values of the criteria (6.3.4) and (6.3.5) for Jolliffe's and McCabe's subsets are closer to optimality and less erratic than in the earlier example. No chosen subset does worse with respect to (6.3.5) than 0.925 for 3 variables and 0.964 for 4 variables, compared to optimal values of 0.942, 0.970 respectively. The behaviour with respect to (6.3.4) is less good, but far less erratic than in the previous example.

In addition to the examples given here, Al-Kandari (1998), Cadima and Jolliffe (2001), Gonzalez et al. (1990), Jolliffe (1973), King and Jackson (1999) and McCabe (1982, 1984) all give further illustrations of variable selection based on PCs. Krzanowski (1987b) looks at variable selection for the *alate adelges* data set of Section 6.4.1, but in the context of preserving group structure. We discuss this further in Chapter 9.

7
Principal Component Analysis and Factor Analysis

Principal component analysis has often been dealt with in textbooks as a special case of factor analysis, and this practice is continued by some widely used computer packages, which treat PCA as one option in a program for factor analysis. This view is misguided since PCA and factor analysis, as usually defined, are really quite distinct techniques. The confusion may have arisen, in part, because of Hotelling's (1933) original paper, in which principal components were introduced in the context of providing a small number of 'more fundamental' variables that determine the values of the p original variables. This is very much in the spirit of the factor model introduced in Section 7.1, although Girschick (1936) indicates that there were soon criticisms of Hotelling's PCs as being inappropriate for factor analysis. Further confusion results from the fact that practitioners of 'factor analysis' do not always have the same definition of the technique (see Jackson, 1991, Section 17.1). In particular some authors, for example Reyment and Jöreskog (1993), Benzécri (1992, Section 4.3) use the term to embrace a wide spectrum of multivariate methods. The definition adopted in this chapter is, however, fairly standard.

Both PCA and factor analysis aim to reduce the dimensionality of a set of data, but the approaches taken to do so are different for the two techniques. Principal component analysis has been extensively used as part of factor analysis, but this involves 'bending the rules' that govern factor analysis and there is much confusion in the literature over the similarities and differences between the techniques. This chapter attempts to clarify the issues involved, and starts in Section 7.1 with a definition of the basic model for factor analysis. Section 7.2 then discusses how a factor model may be estimated and how PCs are, but should perhaps not be, used in this estimation process. Section 7.3 contains further discussion of differences and similarities between PCA and factor analysis, and Section 7.4 gives a numerical example, which compares the results of PCA and factor analysis. Finally, in Section 7.5, a few concluding remarks are made regarding the 'relative merits' of PCA and factor analysis, and the possible use of rotation with PCA. The latter is discussed further in Chapter 11.

7.1 Models for Factor Analysis

The basic idea underlying factor analysis is that p observed random variables, $\mathbf{x}$, can be expressed, except for an error term, as linear functions of $m\ (<p)$ hypothetical (random) variables or *common factors*, that is if $x_1, x_2, \ldots, x_p$ are the variables and $f_1, f_2, \ldots, f_m$ are the factors, then

$$x_1 = \lambda_{11}f_1 + \lambda_{12}f_2 + \ldots + \lambda_{1m}f_m + e_1 \qquad (7.1.1)$$
$$x_2 = \lambda_{21}f_1 + \lambda_{22}f_2 + \ldots + \lambda_{2m}f_m + e_2$$
$$\vdots$$
$$x_p = \lambda_{p1}f_1 + \lambda_{p2}f_2 + \ldots + \lambda_{pm}f_m + e_p$$

where λ_{jk}, $j = 1, 2, \ldots, p$; $k = 1, 2, \ldots, m$ are constants called the *factor loadings*, and e_j, $j = 1, 2, \ldots, p$ are error terms, sometimes called *specific factors* (because e_j is 'specific' to x_j, whereas the f_k are 'common' to several x_j). Equation (7.1.1) can be rewritten in matrix form, with obvious notation, as

$$\mathbf{x} = \Lambda\mathbf{f} + \mathbf{e}. \qquad (7.1.2)$$

One contrast between PCA and factor analysis is immediately apparent. Factor analysis attempts to achieve a reduction from p to m dimensions by invoking a *model* relating $x_1, x_2, \ldots, x_p$ to m hypothetical or latent variables. We have seen in Sections 3.9, 5.3 and 6.1.5 that models have been postulated for PCA, but for most practical purposes PCA differs from factor analysis in having *no explicit model*.

The form of the basic model for factor analysis given in (7.1.2) is fairly standard, although some authors give somewhat different versions. For example, there could be *three* terms on the right-hand side corresponding to contributions from common factors, specific factors *and* measurement errors (Reyment and Jöreskog, 1993, p. 36), or the model could be made non-linear. There are a number of assumptions associated with the factor model, as follows:

(i) $\qquad\qquad\qquad E[\mathbf{e}] = \mathbf{0}, \quad E[\mathbf{f}] = \mathbf{0}, \quad E[\mathbf{x}] = \mathbf{0}.$

Of these three assumptions, the first is a standard assumption for error terms in most statistical models, and the second is convenient and loses no generality. The third may not be true, but if it is not, (7.1.2) can be simply adapted to become $\mathbf{x} = \boldsymbol{\mu} + \boldsymbol{\Lambda}\mathbf{f} + \mathbf{e}$, where $E[\mathbf{x}] = \boldsymbol{\mu}$. This modification introduces only a slight amount of algebraic complication compared with (7.1.2), but (7.1.2) loses no real generality and is usually adopted.

(ii) $E[\mathbf{ee'}] = \boldsymbol{\Psi}$ (diagonal)

$E[\mathbf{fe'}] = \mathbf{0}$ (a matrix of zeros)

$E[\mathbf{ff'}] = \mathbf{I}_m$ (an identity matrix)

The first of these three assumptions is merely stating that the error terms are uncorrelated which is a basic assumption of the factor model, namely that all of $\mathbf{x}$ which is attributable to common influences is contained in $\boldsymbol{\Lambda}\mathbf{f}$, and e_j, e_k, $j \neq k$ are therefore uncorrelated. The second assumption, that the common factors are uncorrelated with the specific factors, is also a fundamental one. However, the third assumption can be relaxed so that the common factors may be correlated (oblique) rather than uncorrelated (orthogonal). Many techniques in factor analysis have been developed for finding orthogonal factors, but some authors, such as Cattell (1978, p. 128), argue that oblique factors are almost always necessary in order to get a correct factor structure. Such details will not be explored here as the present objective is to compare factor analysis with PCA, rather than to give a full description of factor analysis, and for convenience all three assumptions will be made.

(iii) For some purposes, such as hypothesis tests to decide on an appropriate value of m, it is necessary to make distributional assumptions. Usually the assumption of multivariate normality is made in such cases but, as with PCA, many of the results of factor analysis do not depend on specific distributional assumptions.

(iv) Some restrictions are generally necessary on $\boldsymbol{\Lambda}$, because without any restrictions there will be a multiplicity of possible $\boldsymbol{\Lambda}$s that give equally good solutions. This problem will be discussed further in the next section.

7.2   Estimation of the Factor Model

At first sight, the factor model (7.1.2) looks like a standard regression model such as that given in Property A7 of Section 3.1 (see also Chapter 8). However, closer inspection reveals a substantial difference from the standard regression framework, namely that neither $\boldsymbol{\Lambda}$ nor $\mathbf{f}$ in (7.1.2) is known, whereas in regression $\boldsymbol{\Lambda}$ would be known and $\mathbf{f}$ would contain the only unknown parameters. This means that different estimation techniques must be used, and it also means that there is indeterminacy in the solutions—the 'best-fitting' solution is not unique.

Estimation of the model is usually done initially in terms of the parameters in $\mathbf{\Lambda}$ and $\mathbf{\Psi}$, while estimates of $\mathbf{f}$ are found at a later stage. Given the assumptions of the previous section, the covariance matrix can be calculated for both sides of (7.1.2) giving

$$\mathbf{\Sigma} = \mathbf{\Lambda}\mathbf{\Lambda}' + \mathbf{\Psi}. \tag{7.2.1}$$

In practice, we have the sample covariance (or correlation) matrix $\mathbf{S}$, rather than $\mathbf{\Sigma}$, and $\mathbf{\Lambda}$ and $\mathbf{\Psi}$ are found so as to satisfy

$$\mathbf{S} = \mathbf{\Lambda}\mathbf{\Lambda}' + \mathbf{\Psi},$$

(which does not involve the unknown vector of factor scores $\mathbf{f}$) as closely as possible. The indeterminacy of the solution now becomes obvious; if $\mathbf{\Lambda}$, $\mathbf{\Psi}$ is a solution of (7.2.1) and $\mathbf{T}$ is an orthogonal matrix, then $\mathbf{\Lambda}^*$, $\mathbf{\Psi}$ is also a solution, where $\mathbf{\Lambda}^* = \mathbf{\Lambda}\mathbf{T}$. This follows since

$$\mathbf{\Lambda}^*\mathbf{\Lambda}^{*'} = (\mathbf{\Lambda}\mathbf{T})(\mathbf{\Lambda}\mathbf{T})'$$
$$= \mathbf{\Lambda}\mathbf{T}\mathbf{T}'\mathbf{\Lambda}'$$
$$= \mathbf{\Lambda}\mathbf{\Lambda}',$$

as $\mathbf{T}$ is orthogonal.

Because of the indeterminacy, estimation of $\mathbf{\Lambda}$ and $\mathbf{\Psi}$ typically proceeds in two stages. In the first, some restrictions are placed on $\mathbf{\Lambda}$ in order to find a unique initial solution. Having found an initial solution, other solutions which can be found by *rotation* of $\mathbf{\Lambda}$, that is, multiplication by an orthogonal matrix $\mathbf{T}$, are explored. The 'best' of these rotated solutions is chosen according to some particular criterion. There are several possible criteria, but all are designed to make the structure of $\mathbf{\Lambda}$ as simple as possible in some sense, with most elements of $\mathbf{\Lambda}$ either 'close to zero' or 'far from zero,' and with as few as possible of the elements taking intermediate values. Most statistical computer packages provide options for several different rotation criteria, such as varimax, quartimax and promax. Cattell (1978, p. 136), Richman (1986) give non-exhaustive lists of eleven and nineteen automatic rotation methods, respectively, including some like oblimax that enable the factors to become oblique by allowing $\mathbf{T}$ to be not necessarily orthogonal. For illustration, we give the formula for what is probably the most popular rotation criterion, varimax. It is the default in several of the best known software packages. For details of other rotation criteria see Cattell (1978, p. 136), Lawley and Maxwell (1971, Chapter 6), Lewis-Beck (1994, Section II.3), Richman (1986) or Rummel (1970, Chapters 16 and 17) An example illustrating the results of using two rotation criteria is given in Section 7.4.

Suppose that $\mathbf{B} = \mathbf{\Lambda}\mathbf{T}$ and that $\mathbf{B}$ has elements b_{jk}, $j = 1, 2, \ldots, p$; $k = 1, 2, \ldots, m$. Then for varimax rotation the orthogonal rotation matrix $\mathbf{T}$ is chosen to maximize

$$Q = \sum_{k=1}^{m} \left[\sum_{j=1}^{p} b_{jk}^4 - \frac{1}{p} \left(\sum_{j=1}^{p} b_{jk}^2 \right)^2 \right].$$ (7.2.2)

The terms in the square brackets are proportional to the variances of squared loadings for each rotated factor. In the usual implementations of factor analysis the loadings are necessarily between -1 and 1, so the criterion tends to drive squared loadings towards the end of the range 0 to 1, and hence loadings towards -1, 0 or 1 and away from intermediate values, as required. The quantity Q in equation (7.2.2) is the raw varimax criterion. A normalized version is also used in which b_{jk} is replaced by

$$\frac{b_{jk}}{\sqrt{\sum_{k=1}^{m} b_{jk}^2}}$$

in (7.2.2).

As discussed in Section 11.1, rotation can be applied to principal component coefficients in order to simplify them, as is done with factor loadings. The simplification achieved by rotation can help in interpreting the factors or rotated PCs. This is illustrated nicely using diagrams (see Figures 7.1 and 7.2) in the simple case where only $m = 2$ factors or PCs are retained. Figure 7.1 plots the loadings of ten variables on two factors. In fact, these loadings are the coefficients $\mathbf{a}_1$, $\mathbf{a}_2$ for the first two PCs from the example presented in detail later in the chapter, normalized so that $\mathbf{a}_k' \mathbf{a}_k = l_k$, where l_k is the kth eigenvalue of $\mathbf{S}$, rather than $\mathbf{a}_k' \mathbf{a}_k = 1$. When an orthogonal rotation method (varimax) is performed, the loadings for the rotated factors (PCs) are given by the projections of each plotted point onto the axes represented by dashed lines in Figure 7.1.

Similarly, rotation using an oblique rotation method (direct quartimin) gives loadings after rotation by projecting onto the new axes shown in Figure 7.2. It is seen that in Figure 7.2 all points lie close to one or other of the axes, and so have near-zero loadings on the factor represented by the other axis, giving a very simple structure for the loadings. The loadings implied for the rotated factors in Figure 7.1, whilst having simpler structure than the original coefficients, are not as simple as those for Figure 7.2, thus illustrating the advantage of oblique, compared to orthogonal, rotation.

Returning to the first stage in the estimation of $\mathbf{\Lambda}$ and $\mathbf{\Psi}$, there is sometimes a problem with *identifiability*, meaning that the size of the data set is too small compared to the number of parameters to allow those parameters to be estimated (Jackson, 1991, Section 17.2.6; Everitt and Dunn, 2001, Section 12.3)). Assuming that identifiability is not a problem, there are a number of ways of constructing initial estimates (see, for example, Lewis-Beck (1994, Section II.2); Rencher (1998, Section 10.3); Everitt and Dunn (2001, Section 12.2)). Some, such as the centroid method (see Cattell, 1978, Section 2.3), were developed before the advent of computers and

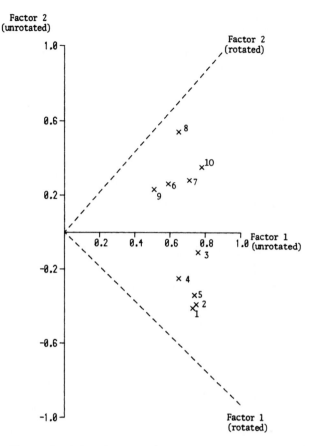

Figure 7.1. Factor loadings for two factors with respect to original and orthogonally rotated factors.

were designed to give quick computationally feasible results. Such methods do a reasonable job of getting a crude factor model, but have little or no firm mathematical basis for doing so. This, among other aspects of factor analysis, gave it a 'bad name' among mathematicians and statisticians. Chatfield and Collins (1989, Chapter 5), for example, treat the topic rather dismissively, ending with the recommendation that factor analysis 'should not be used in most practical situations.'

There are more 'statistically respectable' approaches, such as the Bayesian approach outlined by Press (1972, Section 10.6.2) and the widely implemented idea of maximum likelihood estimation of Ψ and Λ, assuming multivariate normality of $\mathbf{f}$ and $\mathbf{e}$. Finding maximum likelihood estimates of Ψ and Λ leads to an iterative procedure involving a moderate amount of algebra, which will not be repeated here (see, for example, Lawley and Maxwell (1971)).

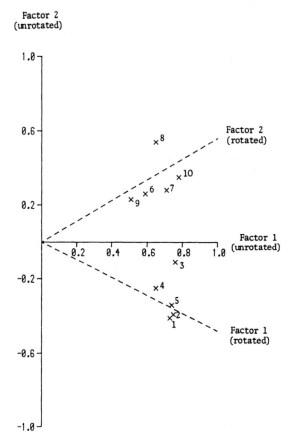

Figure 7.2. Factor loadings for two factors with respect to original and obliquely rotated factors.

An interesting point is that factor loadings found by maximum likelihood for a correlation matrix are equivalent to those for the corresponding covariance matrix, that is, they are scale invariant. This is in complete contrast to what happens for PCA (see Sections 2.3 and 3.3).

A potential problem with the maximum likelihood approach is that it relies on the assumption of multivariate normality, which may not be justified, and Everitt and Dunn (2001, Section 12.7) caution against using such estimates when the data are categorical. However, it can be shown (Morrison, 1976, Section 9.8; Rao, 1955, which is also reproduced in Bryant and Atchley (1975)) that the maximum likelihood estimators (MLEs) also optimize two criteria that make no direct distributional assumptions. If the factor model (7.1.2) holds exactly, then the partial correlations between the elements of $\mathbf{x}$, given the value of $\mathbf{f}$, are zero (see also Section 6.1.6), as $\mathbf{f}$ accounts for all the common variation in the elements of $\mathbf{x}$. To derive the criterion described by Morrison (1976), a sample estimate of the matrix of partial correlations is calculated. The determinant of this matrix will attain its maximum value of unity when all its off-diagonal elements are zero, so that maximizing this determinant is one way of attempting to minimize the absolute values of the partial correlations. This maximization problem leads to the MLEs, but here they appear regardless of whether or not multivariate normality holds.

The procedure suggested by Rao (1955) is based on canonical correlation analysis (see Section 9.3) between $\mathbf{x}$ and $\mathbf{f}$. He looks, successively, for pairs of linear functions $\{\mathbf{a}'_{k1}\mathbf{x}, \mathbf{a}'_{k2}\mathbf{f}\}$ that have maximum correlation subject to being uncorrelated with previous pairs. The factor loadings are then proportional to the elements of the $\mathbf{a}_{k2}$, $k = 1, 2, \ldots, m$, which in turn leads to the same loadings as for the MLEs based on the assumption of multivariate normality (Rao, 1955). As with the criterion based on partial correlations, no distributional assumptions are necessary for Rao's canonical analysis.

In a way, the behaviour of the partial correlation and canonical correlation criteria parallels the phenomenon in regression where the least squares criterion is valid regardless of the distribution of error terms, but if errors are normally distributed then least squares estimators have the added attraction of maximizing the likelihood function.

An alternative but popular way of getting initial estimates for $\mathbf{\Lambda}$ is to use the first m PCs. If $\mathbf{z} = \mathbf{A}'\mathbf{x}$ is the vector consisting of all p PCs, with $\mathbf{A}$ defined to have $\boldsymbol{\alpha}_k$, the kth eigenvector of $\mathbf{\Sigma}$, as its kth column as in (2.1.1), then $\mathbf{x} = \mathbf{A}\mathbf{z}$ because of the orthogonality of $\mathbf{A}$. If $\mathbf{A}$ is partitioned into its first m and last $(p - m)$ columns, with a similar partitioning of the rows of $\mathbf{z}$, then

$$\mathbf{x} = (\mathbf{A}_m \mid \mathbf{A}^*_{p-m}) \left(\frac{\mathbf{z}_m}{\mathbf{z}^*_{p-m}} \right) \tag{7.2.3}$$
$$= \mathbf{A}_m \mathbf{z}_m + \mathbf{A}^*_{p-m} \mathbf{z}^*_{p-m}$$
$$= \mathbf{\Lambda}\mathbf{f} + \mathbf{e},$$

where

$$\mathbf{\Lambda} = \mathbf{A}_m, \quad \mathbf{f} = \mathbf{z}_m \quad \text{and} \quad \mathbf{e} = \mathbf{A}^*_{p-m} \mathbf{z}^*_{p-m}.$$

Equation (7.2.3) looks very much like the factor model (7.1.2) but it violates a basic assumption of the factor model, because the elements of $\mathbf{e}$ in (7.2.3) are not usually uncorrelated. Despite the apparently greater sophistication of using the sample version of $\mathbf{A}_m$ as an initial estimator, compared with crude techniques such as centroid estimates, its theoretical justification is really no stronger.

As well as the straightforward use of PCs to estimate $\mathbf{\Lambda}$, many varieties of factor analysis use modifications of this approach; this topic will be discussed further in the next section.

7.3   Comparisons and Contrasts Between Factor Analysis and Principal Component Analysis

As mentioned in Section 7.1 a major distinction between factor analysis and PCA is that there is a definite model underlying factor analysis, but for most purposes no model is assumed in PCA. Section 7.2 concluded by describing the most common way in which PCs are used in factor analysis. Further connections and contrasts between the two techniques are discussed in the present section, but first we revisit the 'models' that have been proposed for PCA. Recall from Section 3.9 that Tipping and Bishop (1999a) describe a model in which $\mathbf{x}$ has covariance matrix $\mathbf{BB'} + \sigma^2\mathbf{I}_p$, where $\mathbf{B}$ is a $(p \times q)$ matrix. Identifying $\mathbf{B}$ with $\mathbf{\Lambda}$, and q with m, it is clear that this model is equivalent to a special case of equation (7.2.1) in which $\mathbf{\Psi} = \sigma^2\mathbf{I}_p$, so that all p specific variances are equal.

De Leeuw (1986) refers to a generalization of Tipping and Bishop's (1999a) model, in which $\sigma^2\mathbf{I}_p$ is replaced by a general covariance matrix for the error terms in the model, as the (random factor score) *factor analysis* model. This model is also discussed by Roweis (1997). A related model, in which the factors are assumed to be fixed rather than random, corresponds to Caussinus's (1986) fixed effects model, which he also calls the 'fixed factor scores model.' In such models, variability amongst individuals is mainly due to different means rather than to individuals' covariance structure, so they are distinctly different from the usual factor analysis framework.

Both factor analysis and PCA can be thought of as trying to represent some aspect of the covariance matrix $\mathbf{\Sigma}$ (or correlation matrix) as well as possible, but PCA concentrates on the diagonal elements, whereas in factor analysis the interest is in the off-diagonal elements. To justify this statement, consider first PCA. The objective is to maximize $\sum_{k=1}^{m} \text{var}(z_k)$ or, as $\sum_{k=1}^{p} \text{var}(z_k) = \sum_{j=1}^{p} \text{var}(x_j)$, to account for as much as possible of the sum of diagonal elements of $\mathbf{\Sigma}$. As discussed after Property A3 in Section 2.1, the first m PCs will in addition often do a good job of explaining the off-diagonal elements of $\mathbf{\Sigma}$, which means that PCs can frequently provide an adequate initial solution in a factor analysis. However, this is not the stated purpose of PCA and will not hold universally. Turning now to factor analysis, consider the factor model (7.1.2) and the corresponding equation (7.2.1) for $\mathbf{\Sigma}$. It is seen that, as $\mathbf{\Psi}$ is diagonal, the common factor term $\mathbf{\Lambda f}$ in (7.1.2) accounts *completely* for the *off*-diagonal elements of $\mathbf{\Sigma}$ in the perfect factor model, but there is no compulsion for the diagonal elements to be well explained by the common factors. The elements, ψ_j, $j = 1, 2, \ldots, p$, of $\mathbf{\Psi}$ will *all* be low if *all* of the variables have considerable common variation, but if a variable x_j is almost independent of all other variables, then $\psi_j = \text{var}(e_j)$ will be almost as large as $\text{var}(x_j)$. Thus, factor analysis concentrates on explaining only the off-diagonal elements of

Σ by a small number of factors, whereas, conversely, PCA concentrates on the diagonal elements of Σ.

This leads to another difference between the two techniques concerning the number of dimensions m which give an adequate representation of the p dimensional variable $\mathbf{x}$. In PCA, if any individual variables are almost independent of all other variables, then there will be a PC corresponding to each such variable, and that PC will be almost equivalent to the corresponding variable. Such 'single variable' PCs are generally included if an adequate representation of $\mathbf{x}$ is required, as was discussed in Section 6.1.5. In contrast, a common factor in factor analysis must contribute to *at least two* of the variables, so it is not possible to have a 'single variable' common factor. Instead, such factors appear as specific factors (error terms) and do not contribute to the dimensionality of the model. Thus, for a given set of data, the number of factors required for an adequate factor model is no larger—and may be strictly smaller—than the number of PCs required to account for most of the variation in the data. If PCs are used as initial factors, then the ideal choice of m is often less than that determined by the rules of Section 6.1, which are designed for descriptive PCA. As noted several times in that Section, the different objectives underlying PCA and factor analysis have led to confusing and inappropriate recommendations in some studies with respect to the best choice of rules.

The fact that a factor model concentrates on accounting for the off-diagonal elements, but not the diagonal elements, of Σ leads to various modifications of the idea of using the first m PCs to obtain initial estimates of factor loadings. As the covariance matrix of the common factors' contribution to $\mathbf{x}$ is $\Sigma - \Psi$, it seems reasonable to use 'PCs' calculated for $\Sigma - \Psi$ rather than Σ to construct initial estimates, leading to so-called *principal factor analysis*. This will, of course, require estimates of Ψ, which can be found in various ways (see, for example, Rencher, 1998, Section 10.3; Rummel, 1970, Chapter 13), either once-and-for-all or iteratively, leading to many different factor estimates. Many, though by no means all, of the different varieties of factor analysis correspond to simply using different estimates of Ψ in this type of 'modified PC' procedure. None of these estimates has a much stronger claim to absolute validity than does the use of the PCs of Σ, although arguments have been put forward to justify various different estimates of Ψ.

Another difference between PCA and (after rotation) factor analysis is that changing m, the dimensionality of the model, can have much more drastic effects on factor analysis than it does on PCA. In PCA, if m is increased from m_1 to m_2, then an additional $(m_2 - m_1)$ PCs are included, but the original m_1 PCs are still present and unaffected. However, in factor analysis an increase from m_1 to m_2 produces m_2 factors, none of which need bear any resemblance to the original m_1 factors.

A final difference between PCs and common factors is that the former can be calculated exactly from $\mathbf{x}$, whereas the latter typically cannot. The

PCs are exact linear functions of $\mathbf{x}$ and have the form

$$\mathbf{z} = \mathbf{A}'\mathbf{x}.$$

The factors, however, are not exact linear functions of $\mathbf{x}$; instead $\mathbf{x}$ is defined as a linear function of $\mathbf{f}$ apart from an error term, and when the relationship is reversed, it certainly does not lead to an exact relationship between $\mathbf{f}$ and $\mathbf{x}$. Indeed, the fact that the expected value of $\mathbf{x}$ is a linear function of $\mathbf{f}$ need not imply that the expected value of $\mathbf{f}$ is a linear function of $\mathbf{x}$ (unless multivariate normal assumptions are made). Thus, the use of PCs as initial factors may force the factors into an unnecessarily restrictive linear framework. Because of the non-exactness of the relationship between $\mathbf{f}$ and $\mathbf{x}$, the values of $\mathbf{f}$, the *factor scores*, must be *estimated*, and there are several possible ways of doing this (see, for example, Bartholomew and Knott (1999, 3.23–3.25); Jackson (1991, Section 17.7); Lawley and Maxwell (1971, Chapter 8); Lewis-Beck (1994, Section II.6)).

To summarize, there are many ways in which PCA and factor analysis differ from one another. Despite these differences, they both have the aim of reducing the dimensionality of a vector of random variables. The use of PCs to find initial factor loadings, though having no firm justification in theory (except when $\Psi = \sigma^2 \mathbf{I}_p$ as in Tipping and Bishop's (1999a) model) will often not be misleading in practice. In the special case where the elements of Ψ are proportional to the diagonal elements of Σ, Gower (1966) shows that the configuration of points produced by factor analysis will be similar to that found by PCA. In principal factor analysis, the results are equivalent to those of PCA if all (non-zero) elements of Ψ are identical (Rao, 1955). More generally, the coefficients found from PCA and the loadings found from (orthogonal) factor analysis will often be very similar, although this will not hold unless all the elements of Ψ are of approximately the same size (Rao, 1955), which again relates to Tipping and Bishop's (1999a) model.

Schneeweiss and Mathes (1995) provide detailed theoretical comparisons between factor analysis and PCA. Assuming the factor model (7.2.1), they compare Λ with estimates of Λ obtained from PCA and from factor analysis. Comparisons are also made between $\mathbf{f}$, the PC scores, and estimates of $\mathbf{f}$ using factor analysis. General results are given, as well as comparisons for the special cases where $m = 1$ and where $\Sigma = \sigma^2 \mathbf{I}$. The theorems, lemmas and corollaries given by Schneeweiss and Mathes provide conditions under which PCs and their loadings can be used as adequate surrogates for the common factors and *their* loadings. One simple set of conditions is that p is large and that the elements of Ψ are small, although, unlike the conventional factor model, Ψ need not be diagonal. Additional conditions for closeness of factors and principal components are given by Schneeweiss (1997). Further, mainly theoretical, discussion of relationships between factor analysis and PCA appears in Ogasawara (2000).

The results derived by Schneeweiss and Mathes (1995) and Schneeweiss (1997) are 'population' results, so that the 'estimates' referred to above are actually derived from correlation matrices corresponding exactly to the underlying model. In practice, the model itself is unknown and must be estimated from a data set. This allows more scope for divergence between the results from PCA and from factor analysis. There have been a number of studies in which PCA and factor analysis are compared empirically on data sets, with comparisons usually based on a subjective assessment of how well and simply the results can be interpreted. A typical study of this sort from atmospheric science is Bärring (1987). There have also been a number of comparative simulation studies, such as Snook and Gorsuch (1989), in which, unsurprisingly, PCA is inferior to factor analysis in finding underlying structure in data simulated from a factor model.

There has been much discussion in the behavioural science literature of the similarities and differences between PCA and factor analysis. For example, 114 pages of the first issue in 1990 of *Multivariate Behavioral Research* was devoted to a lead article by Velicer and Jackson (1990) on 'Component analysis versus common factor analysis ...,' together with 10 shorter discussion papers by different authors and a rejoinder by Velicer and Jackson. Widaman (1993) continued this debate, and concluded that '...principal component analysis should not be used if a researcher wishes to obtain parameters reflecting latent constructs or factors.' This conclusion reflects the fact that underlying much of the 1990 discussion is the assumption that unobservable factors are being sought from which the observed behavioural variables can be derived. Factor analysis is clearly designed with this objective in mind, whereas PCA does not directly address it. Thus, at best, PCA provides an approximation to what is truly required.

PCA and factor analysis give similar numerical results for many examples. However PCA should only be used as a surrogate for factor analysis with full awareness of the differences between the two techniques, and even then caution is necessary. Sato (1990), who, like Schneeweiss and Mathes (1995) and Schneeweiss (1997), gives a number of theoretical comparisons, showed that for $m = 1$ and small p the loadings given by factor analysis and by PCA can sometimes be quite different.

7.4 An Example of Factor Analysis

The example that follows is fairly typical of the sort of data that are often subjected to a factor analysis. The data were originally discussed by Yule et al. (1969) and consist of scores for 150 children on ten subtests of the Wechsler Pre-School and Primary Scale of Intelligence (WPPSI); there are thus 150 observations on ten variables. The WPPSI tests were designed to measure 'intelligence' of children aged $4\frac{1}{2}$–6 years, and the 150 children tested in the Yule et al. (1969) study were a sample of children who entered school in the Isle of Wight in the autumn of 1967, and who were tested during their second term in school. Their average age at the time of testing was 5 years, 5 months. Similar data sets are analysed in Lawley and Maxwell (1971).

Table 7.1 gives the variances and the coefficients of the first four PCs, when the analysis is done on the correlation matrix. It is seen that the first four components explain nearly 76% of the total variation, and that the variance of the fourth PC is 0.71. The fifth PC, with a variance of 0.51, would be discarded by most of the rules described in Section 6.1 and, indeed, in factor analysis it would be more usual to keep only two, or perhaps three, factors in the present example. Figures 7.1, 7.2 earlier in the chapter showed the effect of rotation in this example when only two PCs are considered; here, where four PCs are retained, it is not possible to easily represent the effect of rotation in the same diagrammatic way.

All of the correlations between the ten variables are positive, so the first PC has the familiar pattern of being an almost equally weighted 'average' of all ten variables. The second PC contrasts the first five variables with the final five. This is not unexpected as these two sets of variables are of different types, namely 'verbal' tests and 'performance' tests, respectively. The third PC is mainly a contrast between variables 6 and 9, which interestingly were at the time the only two 'new' tests in the WPSSI battery, and the fourth does not have a very straightforward interpretation.

Table 7.2 gives the factor loadings when the first four PCs are rotated using an orthogonal rotation method (varimax), and an oblique method (direct quartimin). It would be counterproductive to give more varieties of factor analysis for this single example, as the differences in detail tend to obscure the general conclusions that are drawn below. Often, results are far less sensitive to the choice of rotation criterion than to the choice of how many factors to rotate. Many further examples can be found in texts on factor analysis such as Cattell (1978), Lawley and Maxwell (1971), Lewis-Beck (1994) and Rummel (1970).

In order to make comparisons between Table 7.1 and Table 7.2 straightforward, the sum of squares of the PC coefficients and factor loadings are normalized to be equal to unity for each factor. Typically, the output from computer packages that implement factor analysis uses the normalization in which the sum of squares of coefficients in each PC before rotation is equal to the variance (eigenvalue) associated with that PC (see Section 2.3). The latter normalization is used in Figures 7.1 and 7.2. The choice of normalization constraints is important in rotation as it determines the properties of the rotated factors. Detailed discussion of these properties in the context of rotated PCs is given in Section 11.1.

The correlations between the oblique factors in Table 7.2 are given in Table 7.3 and it can be seen that there is a non-trivial degree of correlation between the factors given by the oblique method. Despite this, the structure of the factor loadings is very similar for the two factor rotation methods.

Table 7.1. Coefficients for the first four PCs: children's intelligence tests.

Component number		1	2	3	4
	1	0.34	−0.39	0.09	−0.08
	2	0.34	−0.37	−0.08	−0.23
	3	0.35	−0.10	0.05	0.03
	4	0.30	−0.24	−0.20	0.63
Variable	5	0.34	−0.32	0.19	−0.28
number	6	0.27	0.24	−0.57	0.30
	7	0.32	0.27	−0.27	−0.34
	8	0.30	0.51	0.19	−0.27
	9	0.23	0.22	0.69	0.43
	10	0.36	0.33	−0.03	0.02
Eigenvalue		4.77	1.13	0.96	0.71
Cumulative percentage of total variation		47.7	59.1	68.6	75.7

Table 7.2. Rotated factor loadings–four factors: children's intelligence tests.

Factor number		1	2	3	4
		Varimax			
	1	0.48	0.09	0.17	0.14
	2	0.49	0.15	0.18	−0.03
	3	0.35	0.22	0.24	0.22
	4	0.26	−0.00	0.64	0.20
Variable	5	0.49	0.16	0.02	0.15
number	6	0.05	0.34	0.60	−0.09
	7	0.20	0.51	0.18	−0.07
	8	0.10	0.54	−0.02	0.32
	9	0.10	0.13	0.07	0.83
	10	0.17	0.46	0.28	0.26
		Direct quartimin			
	1	0.51	−0.05	0.05	0.05
	2	0.53	0.04	0.05	−0.14
	3	0.32	0.13	0.16	0.15
	4	0.17	−0.19	0.65	0.20
Variable	5	0.54	0.06	−0.13	0.05
number	6	−0.07	0.28	0.67	−0.12
	7	0.16	0.53	0.13	−0.17
	8	0.03	0.62	−0.09	0.26
	9	0.00	0.09	0.02	0.87
	10	0.08	0.45	0.24	0.21

Table 7.3. Correlations between four direct quartimin factors: children's intelligence tests.

		Factor number		
		1	2	3
Factor	2	0.349		
number	3	0.418	0.306	
	4	0.305	0.197	0.112

Table 7.4. Factor loadings—three factors, varimax rotation: children's intelligence tests.

		Factor number		
		1	2	3
	1	0.47	0.09	0.14
	2	0.47	0.17	0.05
	3	0.36	0.23	0.24
	4	0.37	0.23	0.00
Variable	5	0.45	0.08	0.23
number	6	0.12	0.55	−0.05
	7	0.17	0.48	0.17
	8	0.05	0.36	0.52
	9	0.13	−0.01	0.66
	10	0.18	0.43	0.36

The first factor in both methods has its highest loadings in variables 1, 2, 3 and 5, with the next highest loadings on variables 4 and 7. In factors 2, 3, 4 there is the same degree of similarity in the position of the highest loadings: for factor 2, the loadings for variables 7, 8, 10 are highest, with an intermediate value on variable 6; factor 3 has large loadings on variables 4 and 6 and an intermediate value on variable 10; and factor 4 is dominated by variable 9 with intermediate values on variables 8 and 10. The only notable difference between the results for the two methods is that obliqueness allows the second method to achieve slightly higher values on the highest loadings and correspondingly lower values on the low loadings, as indeed it is meant to.

By contrast, the differences between the loadings before and after rotation are more substantial. After rotation, the 'general factor' with coefficients of similar size on all variables disappears, as do most negative coefficients, and the structure of the loadings is simplified. Again, this is precisely what rotation is meant to achieve.

To illustrate what happens when different numbers of factors are retained, Table 7.4 gives factor loadings for three factors using varimax rotation. The loadings for direct quartimin (not shown) are again very similar. Before rotation, changing the number of PCs simply adds or deletes PCs, leaving the remaining PCs unchanged. After rotation, however, deletion or addition of factors will usually change all of the factor loadings. In the present example, deletion of the fourth unrotated factor leaves the first rotated factor almost unchanged, except for a modest increase in the loading for variable 4. Factor 2 here is also similar to factor 2 in the four-factor analysis, although the resemblance is somewhat less strong than for factor 1. In particular, variable 6 now has the largest loading in factor 2, whereas previously it had only the fourth largest loading. The third factor in the three-factor solution is in no way similar to factor 3 in the four-factor analysis. In fact, it is quite similar to the original factor 4, and the original factor 3 has disappeared, with its highest loadings on variables 4 and 6 partially 'transferred' to factors 1 and 2, respectively.

The behaviour displayed in this example, when a factor is deleted, is not untypical of what happens in factor analysis generally, although the 'mixing-up' and 'rearrangement' of factors can be much more extreme than in the present case.

7.5 Concluding Remarks

Factor analysis is a large subject, and this chapter has concentrated on aspects that are most relevant to PCA. The interested reader is referred to one of the many books on the subject such as Cattell (1978), Lawley and Maxwell (1971), Lewis-Beck (1994) or Rummell (1970) for further details. Factor analysis is one member of the class of latent variable models (see Bartholomew and Knott (1999)) which have been the subject of much recent research. Mixture modelling, discussed in Section 9.2.3, is another of the many varieties of latent variable models.

It should be clear from the discussion of this chapter that it does not really make sense to ask whether PCA is 'better than' factor analysis or vice versa, because they are not direct competitors. If a model such as (7.1.2) seems a reasonable assumption for a data set, then factor analysis, rather than PCA, is appropriate. If no such model can be assumed, then factor analysis should not really be used.

Despite their different formulations and objectives, it can be informative to look at the results of both techniques on the same data set. Each technique gives different insights into the data structure, with PCA concentrating on explaining the diagonal elements, and factor analysis the off-diagonal elements, of the covariance matrix, and both may be useful. Furthermore, one of the main ideas of factor analysis, that of rotation, can be 'borrowed' for PCA without any implication that a factor model is being assumed. Once PCA has been used to find an m-dimensional subspace that contains most of the variation in the original p variables, it is possible to redefine, by rotation, the axes (or derived variables) that form a basis for this subspace. The rotated variables will together account for the same amount of variation as the first few PCs, but will no longer *successively* account for the maximum possible variation. This behaviour is illustrated by Tables 7.1 and 7.2; the four rotated PCs in Table 7.2 together account for 75.7% of the total variation, as did the unrotated PCs in Table 7.1. However, the percentages of total variation accounted for by individual factors (rotated PCs) are 27.4, 21.9, 14.2 and 12.1, compared with 47.7, 11.3, 9.6 and 7.1 for the unrotated PCs. The rotated PCs, when expressed in terms of the original variables, may be easier to interpret than the PCs themselves because their coefficients will typically have a simpler structure. This is discussed in more detail in Chapter 11. In addition, rotated PCs offer advantages compared to unrotated PCs in some types of analysis based on PCs (see Sections 8.5 and 10.1).

8
Principal Components in Regression Analysis

As illustrated elsewhere in this book, principal components are used in conjunction with a variety of other statistical techniques. One area in which this activity has been extensive is regression analysis.

In multiple regression, one of the major difficulties with the usual least squares estimators is the problem of multicollinearity, which occurs when there are near-constant linear functions of two or more of the predictor, or regressor, variables. A readable review of the multicollinearity problem is given by Gunst (1983). Multicollinearities are often, but not always, indicated by large correlations between subsets of the variables and, if multicollinearities exist, then the variances of some of the estimated regression coefficients can become very large, leading to unstable and potentially misleading estimates of the regression equation. To overcome this problem, various approaches have been proposed. One possibility is to use only a subset of the predictor variables, where the subset is chosen so that it does not contain multicollinearities. Numerous subset selection methods are available (see, for example, Draper and Smith, 1998, Chapter 15; Hocking, 1976; Miller, 1984, 1990), and among the methods are some based on PCs. These methods will be dealt with later in the chapter (Section 8.5), but first some more widely known uses of PCA in regression are described.

These uses of PCA follow from a second class of approaches to overcoming the problem of multicollinearity, namely the use of biased regression estimators. This class includes ridge regression, shrinkage estimators, partial least squares, the so-called LASSO, and also approaches based on PCA. The best-known such approach, generally known as PC regression, simply starts by using the PCs of the predictor variables in place of the predictor variables. As the PCs are uncorrelated, there are no multicollinearities between them, and the regression calculations are also simplified. If all the PCs are included in the regression, then the resulting model is equivalent to that obtained by least squares, so the large variances caused by multicollinearities have not gone away. However, calculation of the least squares estimates via PC regression may be numerically more stable than direct calculation (Flury and Riedwyl, 1988, p. 212).

If some of the PCs are deleted from the regression equation, estimators are obtained for the coefficients in the original regression equation. These estimators are usually biased, but can simultaneously greatly reduce any large variances for regression coefficient estimators caused by multicollinearities. Principal component regression is introduced in Section 8.1, and strategies for deciding which PCs to delete from the regression equation are discussed in Section 8.2; some connections between PC regression and other forms of biased regression are described in Section 8.3.

Variations on the basic idea of PC regression have also been proposed. One such variation, noted in Section 8.3, allows the possibility that a PC may be only 'partly deleted' from the regression equation. A rather different approach, known as *latent root regression*, finds the PCs of the predictor variables *together with* the dependent variable. These PCs can then be used to construct biased regression estimators, which differ from those derived from PC regression. Latent root regression in various forms, together with its properties, is discussed in Section 8.4. A widely used alternative to PC regression is partial least squares (PLS). This, too, is included in Section 8.4, as are a number of other regression-related techniques that have connections with PCA. One omission is the use of PCA to detect outliers. Because the detection of outliers is important in other areas as well as regression, discussion of this topic is postponed until Section 10.1.

A topic which is related to, but different from, regression analysis is that of functional and structural relationships. The idea is, like regression analysis, to explore relationships between variables but, unlike regression, the predictor variables as well as the dependent variable may be subject to error. Principal component analysis can again be used in investigating functional and structural relationships, and this topic is discussed in Section 8.6.

Finally in this chapter, in Section 8.7 two detailed examples are given of the use of PCs in regression, illustrating many of the ideas discussed in earlier sections.

8.1 Principal Component Regression

Consider the standard regression model, as defined in equation (3.1.5), that is,

$$\mathbf{y} = \mathbf{X}\boldsymbol{\beta} + \boldsymbol{\epsilon}, \tag{8.1.1}$$

where $\mathbf{y}$ is a vector of n observations on the dependent variable, measured about their mean, $\mathbf{X}$ is an $(n \times p)$ matrix whose (i, j)th element is the value of the jth predictor (or regressor) variable for the ith observation, again measured about its mean, $\boldsymbol{\beta}$ is a vector of p regression coefficients and $\boldsymbol{\epsilon}$ is a vector of error terms; the elements of $\boldsymbol{\epsilon}$ are independent, each with the same variance σ^2. It is convenient to present the model (8.1.1) in 'centred' form, with all variables measured about their means. Furthermore, it is conventional in much of the literature on PC regression to assume that the predictor variables have been standardized so that $\mathbf{X}'\mathbf{X}$ is proportional to the correlation matrix for the predictor variables, and this convention is followed in the present chapter. Similar derivations to those below are possible if the predictor variables are in uncentred or non-standardized form, or if an alternative standardization has been used, but to save space and repetition, these derivations are not given. Nor do we discuss the controversy that surrounds the choice of whether or not to centre the variables in a regression analysis. The interested reader is referred to Belsley (1984) and the discussion which follows that paper.

The values of the PCs for each observation are given by

$$\mathbf{Z} = \mathbf{XA}, \tag{8.1.2}$$

where the (i, k)th element of $\mathbf{Z}$ is the value (score) of the kth PC for the ith observation, and $\mathbf{A}$ is a $(p \times p)$ matrix whose kth column is the kth eigenvector of $\mathbf{X}'\mathbf{X}$.

Because $\mathbf{A}$ is orthogonal, $\mathbf{X}\boldsymbol{\beta}$ can be rewritten as $\mathbf{XAA}'\boldsymbol{\beta} = \mathbf{Z}\boldsymbol{\gamma}$, where $\boldsymbol{\gamma} = \mathbf{A}'\boldsymbol{\beta}$. Equation (8.1.1) can therefore be written as

$$\mathbf{y} = \mathbf{Z}\boldsymbol{\gamma} + \boldsymbol{\epsilon}, \tag{8.1.3}$$

which has simply replaced the predictor variables by their PCs in the regression model. Principal component regression can be defined as the use of the model (8.1.3) or of the reduced model

$$\mathbf{y} = \mathbf{Z}_m \boldsymbol{\gamma}_m + \boldsymbol{\epsilon}_m, \tag{8.1.4}$$

where $\boldsymbol{\gamma}_m$ is a vector of m elements that are a subset of elements of $\boldsymbol{\gamma}$, $\mathbf{Z}_m$ is an $(n \times m)$ matrix whose columns are the corresponding subset of columns of $\mathbf{Z}$, and $\boldsymbol{\epsilon}_m$ is the appropriate error term. Using least squares to estimate $\boldsymbol{\gamma}$ in (8.1.3) and then finding an estimate for $\boldsymbol{\beta}$ from the equation

$$\hat{\boldsymbol{\beta}} = \mathbf{A}\hat{\boldsymbol{\gamma}} \tag{8.1.5}$$

is equivalent to finding $\hat{\boldsymbol{\beta}}$ by applying least squares directly to (8.1.1).

The idea of using PCs rather than the original predictor variables is not new (Hotelling, 1957; Kendall, 1957), and it has a number of advantages. First, calculating $\hat{\boldsymbol{\gamma}}$ from (8.1.3) is more straightforward than finding $\hat{\boldsymbol{\beta}}$ from (8.1.1) as the columns of $\mathbf{Z}$ are orthogonal. The vector $\hat{\boldsymbol{\gamma}}$ is

$$\hat{\boldsymbol{\gamma}} = (\mathbf{Z}'\mathbf{Z})^{-1}\mathbf{Z}'\mathbf{y} = \mathbf{L}^{-2}\mathbf{Z}'\mathbf{y}, \tag{8.1.6}$$

where $\mathbf{L}$ is the diagonal matrix whose kth diagonal element is $l_k^{1/2}$, and l_k is defined here, as in Section 3.5, as the kth largest eigenvalue of $\mathbf{X'X}$, rather than $\mathbf{S}$. Furthermore, if the regression equation is calculated for PCs instead of the predictor variables, then the contributions of each transformed variable (PC) to the equation can be more easily interpreted than the contributions of the original variables. Because of uncorrelatedness, the contribution and estimated coefficient of a PC are unaffected by which other PCs are also included in the regression, whereas for the original variables both contributions and coefficients can change dramatically when another variable is added to, or deleted from, the equation. This is especially true when multicollinearity is present, but even when multicollinearity is not a problem, regression on the PCs, rather than the original predictor variables, may have advantages for computation and interpretation. However, it should be noted that although interpretation of the separate contributions of each transformed variable is improved by taking PCs, the interpretation of the regression equation itself may be hindered if the PCs have no clear meaning.

The main advantage of PC regression occurs when multicollinearities are present. In this case, by deleting a subset of the PCs, especially those with small variances, much more stable estimates of $\boldsymbol{\beta}$ can be obtained. To see this, substitute (8.1.6) into (8.1.5) to give

$$\hat{\boldsymbol{\beta}} = \mathbf{A}(\mathbf{Z'Z})^{-1}\mathbf{Z'y} \tag{8.1.7}$$
$$= \mathbf{AL}^{-2}\mathbf{Z'y}$$
$$= \mathbf{AL}^{-2}\mathbf{A'X'y}$$
$$= \sum_{k=1}^{p} l_k^{-1}\mathbf{a}_k\mathbf{a}_k'\mathbf{X'y}, \tag{8.1.8}$$

where l_k is the kth diagonal element of $\mathbf{L}^2$ and $\mathbf{a}_k$ is the kth column of $\mathbf{A}$. Equation (8.1.8) can also be derived more directly from $\hat{\boldsymbol{\beta}} = (\mathbf{X'X})^{-1}\mathbf{X'y}$, by using the spectral decomposition (see Property A3 of Sections 2.1 and 3.1) of the matrix $(\mathbf{X'X})^{-1}$, which has eigenvectors $\mathbf{a}_k$ and eigenvalues l_k^{-1}, $k = 1, 2, \ldots, p$.

Making the usual assumption that the elements of $\mathbf{y}$ are uncorrelated, each with the same variance σ^2 (that is the variance-covariance matrix of $\mathbf{y}$ is $\sigma^2\mathbf{I}_n$), it is seen from (8.1.7) that the variance-covariance matrix of $\hat{\boldsymbol{\beta}}$ is

$$\sigma^2\mathbf{A}(\mathbf{Z'Z})^{-1}\mathbf{Z'Z}(\mathbf{Z'Z})^{-1}\mathbf{A'} = \sigma^2\mathbf{A}(\mathbf{Z'Z})^{-1}\mathbf{A'}$$
$$= \sigma^2\mathbf{AL}^{-2}\mathbf{A'}$$
$$= \sigma^2\sum_{k=1}^{p} l_k^{-1}\mathbf{a}_k\mathbf{a}_k'. \tag{8.1.9}$$

This expression gives insight into how multicollinearities produce large variances for the elements of $\hat{\boldsymbol{\beta}}$. If a multicollinearity exists, then it appears as a PC with very small variance (see also Sections 3.4 and 10.1); in other words, the later PCs have very small values of l_k (the variance of the kth PC is $l_k/(n-1)$ in the present notation), and hence very large values of l_k^{-1}. Thus (8.1.9) shows that any predictor variable having moderate or large coefficients in any of the PCs associated with very small eigenvalues will have a very large variance.

One way of reducing this effect is to delete the terms from (8.1.8) that correspond to very small l_k, leading to an estimator

$$\tilde{\boldsymbol{\beta}} = \sum_{k=1}^{m} l_k^{-1} \mathbf{a}_k \mathbf{a}_k' \mathbf{X}' \mathbf{y}, \qquad (8.1.10)$$

where $l_{m+1}, l_{m+2}, \ldots, l_p$ are the very small eigenvalues. This is equivalent to setting the last $(p-m)$ elements of $\boldsymbol{\gamma}$ equal to zero.

Then the variance-covariance matrix $V(\tilde{\boldsymbol{\beta}})$ for $\tilde{\boldsymbol{\beta}}$ is

$$\sigma^2 \sum_{j=1}^{m} l_j^{-1} \mathbf{a}_j \mathbf{a}_j' \mathbf{X}' \mathbf{X} \sum_{k=1}^{m} l_k^{-1} \mathbf{a}_k \mathbf{a}_k'.$$

Substituting

$$\mathbf{X}' \mathbf{X} = \sum_{h=1}^{p} l_h \mathbf{a}_h \mathbf{a}_h'$$

from the spectral decomposition of $\mathbf{X}'\mathbf{X}$, we have

$$V(\tilde{\boldsymbol{\beta}}) = \sigma^2 \sum_{h=1}^{p} \sum_{j=1}^{m} \sum_{k=1}^{m} l_h l_j^{-1} l_k^{-1} \mathbf{a}_j \mathbf{a}_j' \mathbf{a}_h \mathbf{a}_h' \mathbf{a}_k \mathbf{a}_k'.$$

Because the vectors $\mathbf{a}_h$, $h = 1, 2, \ldots, p$ are orthonormal, the only non-zero terms in the triple summation occur when $h = j = k$, so that

$$V(\tilde{\boldsymbol{\beta}}) = \sigma^2 \sum_{k=1}^{m} l_k^{-1} \mathbf{a}_k \mathbf{a}_k' \qquad (8.1.11)$$

If none of the first m eigenvalues l_k is very small, then none of the variances given by the diagonal elements of (8.1.11) will be large.

The decrease in variance for the estimator $\tilde{\boldsymbol{\beta}}$ given by (8.1.10), compared with the variance of $\hat{\boldsymbol{\beta}}$, is achieved at the expense of introducing bias into the estimator $\tilde{\boldsymbol{\beta}}$. This follows because

$$\tilde{\boldsymbol{\beta}} = \hat{\boldsymbol{\beta}} - \sum_{k=m+1}^{p} l_k^{-1} \mathbf{a}_k \mathbf{a}_k' \mathbf{X}' \mathbf{y}, \quad E(\hat{\boldsymbol{\beta}}) = \boldsymbol{\beta},$$

and

$$E\left[\sum_{k=m+1}^{p} l_k^{-1} \mathbf{a}_k \mathbf{a}_k' \mathbf{X}'\mathbf{y}\right] = \sum_{k=m+1}^{p} l_k^{-1} \mathbf{a}_k \mathbf{a}_k' \mathbf{X}'\mathbf{X}\boldsymbol{\beta}$$

$$= \sum_{k=m+1}^{p} \mathbf{a}_k \mathbf{a}_k' \boldsymbol{\beta}.$$

This last term is, in general, non-zero so that $E(\tilde{\boldsymbol{\beta}}) \neq \boldsymbol{\beta}$. However, if multicollinearity is a serious problem, the reduction in variance can be substantial, whereas the bias introduced may be comparatively small. In fact, if the elements of $\boldsymbol{\gamma}$ corresponding to deleted components are actually zero, then no bias will be introduced.

As well as, or instead of, deleting terms from (8.1.8) corresponding to small eigenvalues, it is also possible to delete terms for which the corresponding elements of $\boldsymbol{\gamma}$ are not significantly different from zero. The question of which elements are significantly non-zero is essentially a variable selection problem, with PCs rather than the original predictor variables as variables. Any of the well-known methods of variable selection for regression (see, for example, Draper and Smith, 1998, Chapter 15) can be used. However, the problem is complicated by the desirability of also deleting high-variance terms from (8.1.8).

The definition of PC regression given above in terms of equations (8.1.3) and (8.1.4) is equivalent to using the linear model (8.1.1) and estimating $\boldsymbol{\beta}$ by

$$\tilde{\boldsymbol{\beta}} = \sum_{M} l_k^{-1} \mathbf{a}_k \mathbf{a}_k' \mathbf{X}'\mathbf{y}, \tag{8.1.12}$$

where M is some subset of the integers $1, 2, \ldots, p$. A number of authors consider only the special case (8.1.10) of (8.1.12), in which $M = \{1, 2, \ldots, m\}$, but this is often too restrictive, as will be seen in Section 8.2. In the general definition of PC regression, M can be any subset of the first p integers, so that any subset of the coefficients of $\boldsymbol{\gamma}$, corresponding to the complement of M, can be set to zero. The next section will consider various strategies for choosing M, but we first note that once again the singular value decomposition (SVD) of $\mathbf{X}$ defined in Section 3.5 can be a useful concept (see also Sections 5.3, 6.1.5, 13.4, 13.5, 13.6, 14.2 and Appendix A1). In the present context it can be used to provide an alternative formulation of equation (8.1.12) and to help in the interpretation of the results of a PC regression. Assuming that $n \geq p$ and that $\mathbf{X}$ has rank p, recall that the SVD writes $\mathbf{X}$ in the form

$$\mathbf{X} = \mathbf{ULA}',$$

where

(i) $\mathbf{A}$ and $\mathbf{L}$ are as defined earlier in this section;

(ii) the columns of $\mathbf{U}$ are those eigenvectors of $\mathbf{XX'}$ that correspond to non-zero eigenvalues, normalized so that $\mathbf{U'U} = \mathbf{I}_p$.

Then $\mathbf{X}\beta$ can be rewritten $\mathbf{ULA'}\beta = \mathbf{U}\delta$, where $\delta = \mathbf{LA'}\beta$, so that $\beta = \mathbf{AL}^{-1}\delta$. The least squares estimator for δ is

$$\hat{\delta} = (\mathbf{U'U})^{-1}\mathbf{U'y} = \mathbf{U'y},$$

leading to $\hat{\beta} = \mathbf{AL}^{-1}\hat{\delta}$.

The relationship between γ, defined earlier, and δ is straightforward, namely

$$\gamma = \mathbf{A'}\beta = \mathbf{A'}(\mathbf{AL}^{-1}\delta) = (\mathbf{A'A})\mathbf{L}^{-1}\delta = \mathbf{L}^{-1}\delta,$$

so that setting a subset of elements of δ equal to zero is equivalent to setting the same subset of elements of γ equal to zero. This result means that the SVD can provide an alternative computational approach for estimating PC regression equations, which is an advantage, as efficient algorithms exist for finding the SVD of a matrix (see Appendix A1).

Interpretation of the results of a PC regression can also be aided by using the SVD, as illustrated by Mandel (1982) for artificial data (see also Nelder (1985)).

8.2   Strategies for Selecting Components in Principal Component Regression

When choosing the subset M in equation (8.1.12) there are two partially conflicting objectives. In order to eliminate large variances due to multi-collinearities it is essential to delete all those components whose variances are very small but, at the same time, it is undesirable to delete components that have large correlations with the dependent variable y. One strategy for choosing M is simply to delete all those components whose variances are less than l^*, where l^* is some cut-off level. The choice of l^* is rather arbitrary, but when dealing with correlation matrices, where the average value of the eigenvalues is 1, a value of l^* somewhere in the range 0.01 to 0.1 seems to be useful in practice.

An apparently more sophisticated way of choosing l^* is to look at so-called variance inflation factors (VIFs) for the p predictor variables. The VIF for the jth variable when using standardized variables is defined as c_{jj}/σ^2 (which equals the jth diagonal element of $(\mathbf{X'X})^{-1}$—Marquardt, 1970), where c_{jj} is the variance of the jth element of the least squares estimator for β. If all the variables are uncorrelated, then all the VIFs are equal to 1, but if severe multicollinearities exist then the VIFs for $\hat{\beta}$ will be very large for those variables involved in the multicollinearities. By successively deleting the last few terms in (8.1.8), the VIFs for the resulting biased estimators will be reduced; deletion continues until all VIFs are

Table 8.1. Variation accounted for by PCs of predictor variables in monsoon data for (a) predictor variables, (b) dependent variable.

Component number		1	2	3	4	5	6	7	8	9	10
Percentage	(a) Predictor										
variation	variables	26	22	17	11	10	7	4	3	1	< 1
accounted	(b) Dependent										
for	variable	3	22	< 1	1	3	3	6	24	5	20

below some desired level. The original VIF for a variable is related to the squared multiple correlation R^2 between that variable and the other $(p-1)$ predictor variables by the formula VIF $= (1 - R^2)^{-1}$. Values of VIF > 10 correspond to $R^2 > 0.90$, and VIF > 4 is equivalent to $R^2 > 0.75$, so that values of R^2 can be considered when choosing how small a level of VIF is desirable. However, the choice of this desirable level is almost as arbitrary as the choice of l^* above.

Deletion based solely on variance is an attractive and simple strategy, and Property A7 of Section 3.1 gives it, at first sight, an added respectability. However, low variance for a component does not necessarily imply that the corresponding component is unimportant in the regression model. For example, Kung and Sharif (1980) give an example from meteorology where, in a regression of monsoon onset dates on all of the (ten) PCs, the most important PCs for prediction are, in decreasing order of importance, the eighth, second and tenth (see Table 8.1). The tenth component accounts for less than 1% of the total variation in the predictor variables, but is an important predictor of the dependent variable, and the most important PC in the regression accounts for 24% of the variation in y but only 3% of the variation in $\mathbf{x}$. Further examples of this type are presented in Jolliffe (1982). Thus, the two objectives of deleting PCs with small variances and of retaining PCs that are good predictors of the dependent variable may not be simultaneously achievable.

Some authors (for example, Hocking, 1976; Mosteller and Tukey, 1977, pp. 397–398; Gunst and Mason, 1980, pp. 327–328) argue that the choice of PCs in the regression should be made entirely, or mainly, on the basis of variance reduction but, as can be seen from the examples cited by Jolliffe (1982), such a procedure can be dangerous if low-variance components have predictive value. Jolliffe (1982) notes that examples where this occurs seem to be not uncommon in practice. Berk's (1984) experience with six data sets indicates the opposite conclusion, but several of his data sets are of a special type, in which strong positive correlations exist between all the regressor variables *and* between the dependent variable and the regressor variables. In such cases the first PC is a (weighted) average of the regressor variables, with all weights positive (see Section 3.8), and as y is also positively correlated with each regressor variable it is strongly correlated with the first PC. Hadi and Ling (1998) (see also Cuadras (1998)) *define* PC regression in terms of equation (8.1.10), and argue that the technique is flawed because predictive low-variance PCs may be excluded. With the more general definition of PC regression, based on (8.1.12), this criticism disappears.

In contrast to selection based solely on size of variance, the opposite extreme is to base selection only on values of t-statistics measuring the (independent) contribution of each PC to the regression equation. This, too, has its pitfalls. Mason and Gunst (1985) showed that t-tests for low-variance PCs have reduced power compared to those for high-variance components, and so are less likely to be selected. A compromise between selection on the basis of variance and on the outcome of t-tests is to delete PCs sequentially starting with the smallest variance, then the next smallest variance and so on; deletion stops when the first significant t-value is reached. Such a strategy is likely to retain more PCs than are really necessary.

Hill et al. (1977) give a comprehensive discussion of various, more sophisticated, strategies for deciding which PCs to delete from the regression equation. Their criteria are of two main types, depending on whether the primary objective is to get $\tilde{\boldsymbol{\beta}}$ close to $\boldsymbol{\beta}$, or to get $\mathbf{X}\tilde{\boldsymbol{\beta}}$, the estimate of $\mathbf{y}$, close to $\mathbf{y}$ or to $E(\mathbf{y})$. In the first case, estimation of $\boldsymbol{\beta}$ is the main interest; in the second it is prediction of $\mathbf{y}$ which is the chief concern. Whether or not $\tilde{\boldsymbol{\beta}}$ is an improvement on $\hat{\boldsymbol{\beta}}$ is determined for several of the criteria by looking at mean square error (MSE) so that variance and bias are both taken into account.

More specifically, two criteria are suggested of the first type, the 'weak' and 'strong' criteria. The weak criterion, due to Wallace (1972), prefers $\tilde{\boldsymbol{\beta}}$ to $\hat{\boldsymbol{\beta}}$ if $\text{tr}[\text{MSE}(\tilde{\boldsymbol{\beta}})] \leq \text{tr}[\text{MSE}(\hat{\boldsymbol{\beta}})]$, where $\text{MSE}(\tilde{\boldsymbol{\beta}})$ is the matrix $E[(\tilde{\boldsymbol{\beta}} - \boldsymbol{\beta})(\tilde{\boldsymbol{\beta}} - \boldsymbol{\beta})']$, with a similar definition for the matrix $\text{MSE}(\hat{\boldsymbol{\beta}})$. This simply means that $\tilde{\boldsymbol{\beta}}$ is preferred when the expected Euclidean distance between $\tilde{\boldsymbol{\beta}}$ and $\boldsymbol{\beta}$ is smaller than that between $\hat{\boldsymbol{\beta}}$ and $\boldsymbol{\beta}$.

The strong criterion insists that

$$\text{MSE}(\mathbf{c}'\tilde{\boldsymbol{\beta}}) \leq \text{MSE}(\mathbf{c}'\hat{\boldsymbol{\beta}})$$

for every non-zero p-element vector $\mathbf{c}$, where

$$\text{MSE}(\mathbf{c}'\tilde{\boldsymbol{\beta}}) = E[(\mathbf{c}'\tilde{\boldsymbol{\beta}} - \mathbf{c}'\boldsymbol{\beta})^2],$$

with, again, a similar definition for $\text{MSE}(\mathbf{c}'\hat{\boldsymbol{\beta}})$.

Among those criteria of the second type (where prediction of $\mathbf{y}$ rather than estimation of $\boldsymbol{\beta}$ is the main concern) that are considered by Hill et al. (1977), there are again two which use MSE. The first is also due to Wallace (1972) and is again termed a 'weak' criterion. It prefers $\tilde{\boldsymbol{\beta}}$ to $\hat{\boldsymbol{\beta}}$ if

$$E[(\mathbf{X}\tilde{\boldsymbol{\beta}} - \mathbf{X}\boldsymbol{\beta})'(\mathbf{X}\tilde{\boldsymbol{\beta}} - \mathbf{X}\boldsymbol{\beta})] \leq E[(\mathbf{X}\hat{\boldsymbol{\beta}} - \mathbf{X}\boldsymbol{\beta})'(\mathbf{X}\hat{\boldsymbol{\beta}} - \mathbf{X}\boldsymbol{\beta})],$$

so that $\tilde{\beta}$ is preferred to $\hat{\beta}$ if the expected Euclidean distance between $\mathbf{X}\tilde{\beta}$ (the estimate of $\mathbf{y}$) and $\mathbf{X}\beta$ (the expected value of $\mathbf{y}$) is smaller than the corresponding distance between $\mathbf{X}\hat{\beta}$ and $\mathbf{X}\beta$. An alternative MSE criterion is to look at the distance between each estimate of $\mathbf{y}$ and the actual, rather than expected, value of $\mathbf{y}$. Thus $\tilde{\beta}$ is preferred to $\hat{\beta}$ if

$$E[(\mathbf{X}\tilde{\beta} - \mathbf{y})'(\mathbf{X}\tilde{\beta} - \mathbf{y})] \leq E[(\mathbf{X}\hat{\beta} - \mathbf{y})'(\mathbf{X}\hat{\beta} - \mathbf{y})].$$

Substituting $\mathbf{y} = \mathbf{X}\beta + \boldsymbol{\epsilon}$ it follows that

$$E[(\mathbf{X}\tilde{\beta} - \mathbf{y})'(\mathbf{X}\tilde{\beta} - \mathbf{y})] = E[(\mathbf{X}\tilde{\beta} - \mathbf{X}\beta)'(\mathbf{X}\tilde{\beta} - \mathbf{X}\beta)] + n\sigma^2,$$

with a similar expression for $\hat{\beta}$. At first sight, it seems that this second criterion is equivalent to the first. However σ^2 is unknown and, although it can be estimated, we may get different estimates when the equation is fitted using $\tilde{\beta}$, $\hat{\beta}$, respectively.

Hill et al. (1977) consider several other criteria; further details may be found in their paper, which also describes connections between the various decision rules for choosing M and gives illustrative examples. They argue that the choice of PCs should not be based solely on the size of their variance, but little advice is offered on which of their criteria gives an overall 'best' trade-off between variance and bias; rather, separate circumstances are identified in which each may be the most appropriate.

Gunst and Mason (1979) also consider integrated MSE of predictions as a criterion for comparing different regression estimators. Friedman and Montgomery (1985) prefer to use the predictive ability for individual observations, rather than averaging this ability over a distribution of potential observations as is done by Gunst and Mason (1979).

Another way of comparing predicted and observed values of $\mathbf{y}$ is by means of cross-validation. Mertens et al. (1995) use a version of PRESS, defined in equation (6.1.3), as a criterion for deciding how many PCs to retain in PC regression. Their criterion is

$$\sum_{i=1}^{n}(y_i - \hat{y}_{M(i)})^2,$$

where $\hat{y}_{M(i)}$ is the estimate of y_i obtained from a PC regression based on a subset M and using the data matrix $\mathbf{X}_{(i)}$, which is $\mathbf{X}$ with its ith row deleted. They have an efficient algorithm for computing all PCAs with each observation deleted in turn, though the algebra that it uses is applicable only to covariance, not correlation, matrices. Mainly for reasons of convenience, they also restrict their procedure to implementing (8.1.10), rather than the more general (8.1.12).

Yet another approach to deletion of PCs that takes into account both variance and bias is given by Lott (1973). This approach simply calculates the adjusted multiple coefficient of determination,

$$\bar{R}^2 = 1 - \frac{(n-1)}{(n-p-1)}(1-R^2),$$

where R^2 is the usual multiple coefficient of determination (squared multiple correlation) for the regression equation obtained from each subset M of interest. The 'best' subset is then the one that maximizes $\bar{R}^2$. Lott demonstrates that this very simple procedure works well in a limited simulation study. Soofi (1988) uses a Bayesian approach to define the gain of information from the data about the ith element γ_i of $\boldsymbol{\gamma}$. The subset M is chosen to consist of integers corresponding to components with the largest values of this measure of information. Soofi shows that the measure combines the variance accounted for by a component with its correlation with the dependent variable.

It is difficult to give any general advice regarding the choice of a decision rule for determining M. It is clearly inadvisable to base the decision entirely on the size of variance; conversely, inclusion of highly predictive PCs can also be dangerous if they also have very small variances, because of the resulting instability of the estimated regression equation. Use of MSE criteria provides a number of compromise solutions, but they are essentially arbitrary.

What PC regression *can* do, which least squares cannot, is to indicate explicitly whether a problem exists with respect to the removal of multicollinearity, that is whether instability in the regression coefficients can only be removed by simultaneously losing a substantial proportion of the predictability of **y**. An extension of the cross-validation procedure of Mertens et al. (1995) to general subsets M would provide a less arbitrary way than most of deciding which PCs to keep, but the choice of M for PC regression remains an open question.

8.3 Some Connections Between Principal Component Regression and Other Biased Regression Methods

Using the expressions (8.1.8), (8.1.9) for $\hat{\boldsymbol{\beta}}$ and its variance-covariance matrix, it was seen in the previous section that deletion of the last few terms from the summation for $\hat{\boldsymbol{\beta}}$ can dramatically reduce the high variances of elements of $\hat{\boldsymbol{\beta}}$ caused by multicollinearities. However, if any of the elements of $\boldsymbol{\gamma}$ corresponding to deleted components are non-zero, then the PC estimator $\tilde{\boldsymbol{\beta}}$ for $\boldsymbol{\beta}$ is biased. Various other methods of biased estimation that aim to remove collinearity-induced high variances have also been proposed. A full description of these methods will not be given here as several do not involve

PCs directly, but there are various relationships between PC regression and other biased regression methods which will be briefly discussed.

Consider first ridge regression, which was described by Hoerl and Kennard (1970a,b) and which has since been the subject of much debate in the statistical literature. The estimator of β using the technique can be written, among other ways, as

$$\hat{\beta}_R = \sum_{k=1}^{p} (l_k + \kappa)^{-1} \mathbf{a}_k \mathbf{a}_k' \mathbf{X}' \mathbf{y},$$

where κ is some fixed positive constant and the other terms in the expression have the same meaning as in (8.1.8). The variance-covariance matrix of $\hat{\beta}_R$, is equal to

$$\sigma^2 \sum_{k=1}^{p} l_k (l_k + \kappa)^{-2} \mathbf{a}_k \mathbf{a}_k'.$$

Thus, ridge regression estimators have rather similar expressions to those for least squares and PC estimators, but variance reduction is achieved not by deleting components, but by reducing the weight given to the later components. A generalization of ridge regression has p constants κ_k, $k = 1, 2, \ldots, p$ that must be chosen, rather than a single constant κ.

A modification of PC regression, due to Marquardt (1970) uses a similar, but more restricted, idea. Here a PC regression estimator of the form (8.1.10) is adapted so that M includes the first m integers, excludes the integers $m + 2, m + 3, \ldots, p$, but includes the term corresponding to integer $(m + 1)$ with a weighting less than unity. Detailed discussion of such estimators is given by Marquardt (1970).

Ridge regression estimators 'shrink' the least squares estimators towards the origin, and so are similar in effect to the shrinkage estimators proposed by Stein (1960) and Sclove (1968). These latter estimators start with the idea of shrinking some or all of the elements of $\hat{\gamma}$ (or $\hat{\beta}$) using arguments based on loss functions, admissibility and prior information; choice of shrinkage constants is based on optimization of MSE criteria. Partial least squares regression is sometimes viewed as another class of shrinkage estimators. However, Butler and Denham (2000) show that it has peculiar properties, shrinking some of the elements of $\hat{\gamma}$ but inflating others.

All these various biased estimators have relationships between them. In particular, all the present estimators, as well as latent root regression, which is discussed in the next section along with partial least squares, can be viewed as optimizing $(\tilde{\beta} - \beta)' \mathbf{X}' \mathbf{X} (\tilde{\beta} - \beta)$, subject to different constraints for different estimators (see Hocking (1976)). If the data set is augmented by a set of dummy observations, and least squares is used to estimate β from the augmented data, Hocking (1976) demonstrates further that ridge, generalized ridge, PC regression, Marquardt's modification and shrinkage estimators all appear as special cases for particular choices of the dummy observations and their variances. In a slightly different approach to the same topic, Hocking et al. (1976) give a broad class of biased estimators, which includes all the above estimators, including those derived from PC regression, as special cases. Oman (1978) shows how several biased regression methods, including PC regression, can be fitted into a Bayesian framework by using different prior distributions for β; Leamer and Chamberlain (1976) also look at a Bayesian approach to regression, and its connections with PC regression. Other biased estimators have been suggested and compared with PC regression by Iglarsh and Cheng (1980) and Trenkler (1980), and relationships between ridge regression and PC regression are explored further by Hsuan (1981). Trenkler and Trenkler (1984) extend Hsuan's (1981) results, and examine circumstances in which ridge and other biased estimators can be made close to PC regression estimators, where the latter are defined by the restrictive equation (8.1.10).

Hoerl et al. (1986) describe a simulation study in which PC regression is compared with other biased estimators and variable selection methods, and found to be inferior. However, the comparison is not entirely fair. Several varieties of ridge regression are included in the comparison, but only one way of choosing M is considered for PC regression. This is the restrictive choice of M consisting of $1, 2, \ldots, m$, where m is the largest integer for which a t-test of the PC regression coefficient γ_m gives a significant result. Hoerl et al. (1986) refer to a number of other simulation studies comparing biased regression methods, some of which include PC regression. Theoretical comparisons between PC regression, least squares and ridge regression with respect to the predictive ability of the resulting regression equations are made by Gunst and Mason (1979) and Friedman and Montgomery (1985), but only for $p = 2$.

Essentially the same problem arises for all these biased methods as occurred in the choice of M for PC regression, namely, the question of which compromise should be chosen in the trade-off between bias and variance. In ridge regression, this compromise manifests itself in the choice of κ, and for shrinkage estimators the amount of shrinkage must be determined. Suggestions have been made regarding rules for making these choices, but the decision is usually still somewhat arbitrary.

8.4 Variations on Principal Component Regression

Marquardt's (1970) fractional rank estimator, which was described in the previous section, is one modification of PC regression as defined in Section 8.1, but it is a fairly minor modification. Another approach, suggested by Oman (1991), is to use shrinkage estimators, but instead of shrinking the least squares estimators towards zero or some other constant, the shrinkage is towards the first few PCs. This tends to downweight the contribution of the less stable low-variance PC but does not ignore them. Oman (1991) demonstrates considerable improvements over least squares with these estimators.

A rather different type of approach, which, nevertheless, still uses PCs in a regression problem, is provided by latent root regression. The main difference between this technique and straightforward PC regression is that the PCs are not calculated for the set of p predictor variables alone. Instead, they are calculated for a set of $(p + 1)$ variables consisting of the p predictor variables *and* the dependent variable. This idea was suggested independently by Hawkins (1973) and by Webster et al. (1974), and termed 'latent root regression' by the latter authors. Subsequent papers (Gunst et al., 1976; Gunst and Mason, 1977a) investigated the properties of latent root regression, and compared it with other biased regression estimators. As with the biased estimators discussed in the previous section, the latent root regression estimator can be derived by optimizing a quadratic function of β, subject to constraints (Hocking, 1976). Latent root regression, as defined in Gunst and Mason (1980, Section 10.2), will now be described; the technique introduced by Hawkins (1973) has slight differences and is discussed later in this section.

In latent root regression, a PCA is done on the set of $(p + 1)$ variables described above, and the PCs corresponding to the smallest eigenvalues are examined. Those for which the coefficient of the dependent variable y is also small are called *non-predictive multicollinearities*, and are deemed to be of no use in predicting y. However, any PC with a small eigenvalue *will be* of predictive value if its coefficient for y is large. Thus, latent root regression deletes those PCs which indicate multicollinearities, but only if the multicollinearities appear to be useless for predicting y.

Let δ_k be the vector of the p coefficients on the p predictor variables in the kth PC for the enlarged set of $(p + 1)$ variables; let δ_{0k} be the corresponding coefficient of y, and let $\tilde{l}_k$ be the corresponding eigenvalue. Then the latent root estimator for β is defined as

$$\hat{\beta}_{LR} = \sum_{M_{LR}} f_k \delta_k, \qquad (8.4.1)$$

where M_{LR} is the subset of the integers $1, 2, \ldots, p + 1$, in which integers corresponding to the non-predictive multicollinearities defined above, and no others, are deleted; the f_k are coefficients chosen to minimize residual sums of squares among estimators of the form (8.4.1).

The f_k can be determined by first using the kth PC to express $\mathbf{y}$ as a linear function of $\mathbf{X}$ to provide an estimator $\hat{\mathbf{y}}_k$. A weighted average, $\hat{\mathbf{y}}_{LR}$, of the $\hat{\mathbf{y}}_k$ for $k \in M_{LR}$ is then constructed, where the weights are chosen so as to minimize the residual sum of squares $(\hat{\mathbf{y}}_{LR} - \mathbf{y})'(\hat{\mathbf{y}}_{LR} - \mathbf{y})$. The vector $\hat{\mathbf{y}}_{LR}$ is then the latent root regression predictor $\mathbf{X}\hat{\beta}_{LR}$, and the f_k

are given by

$$f_k = -\delta_{0k}\eta_y \tilde{l}_k^{-1} \left(\sum_{M_{LR}} \delta_{0k}^2 \tilde{l}_k^{-1} \right)^{-1}, \qquad (8.4.2)$$

where $\eta_y^2 = \sum_{i=1}^n (y_i - \bar{y})^2$, and δ_{0k}, $\tilde{l}_k$ are as defined above. Note that the least squares estimator $\hat{\beta}$ can also be written in the form (8.4.1) if M_{LR} in (8.4.1) and (8.4.2) is taken to be the full set of PCs.

The full derivation of this expression for f_k is fairly lengthy, and can be found in Webster et al. (1974). It is interesting to note that f_k is proportional to the size of the coefficient of y in the kth PC, and inversely proportional to the variance of the kth PC; both of these relationships are intuitively reasonable.

In order to choose the subset $\mathbf{M}_{LR}$ it is necessary to decide not only how small the eigenvalues must be in order to indicate multicollinearities, but also how large the coefficient of y must be in order to indicate a *predictive* multicollinearity. Again, these are arbitrary choices, and ad hoc rules have been used, for example, by Gunst et al. (1976). A more formal procedure for identifying non-predictive multicollinearities is described by White and Gunst (1979), but its derivation is based on *asymptotic* properties of the statistics used in latent root regression.

Gunst et al. (1976) compared $\hat{\beta}_{LR}$ and $\hat{\beta}$ in terms of MSE, using a simulation study, for cases of only one multicollinearity, and found that $\hat{\beta}_{LR}$ showed substantial improvement over $\hat{\beta}$ when the multicollinearity is non-predictive. However, in cases where the single multicollinearity had some predictive value, the results were, unsurprisingly, less favourable to $\hat{\beta}_{LR}$. Gunst and Mason (1977a) reported a larger simulation study, which compared PC, latent root, ridge and shrinkage estimators, again on the basis of MSE. Overall, latent root estimators did well in many, but not all, situations studied, as did PC estimators, but no simulation study can ever be exhaustive, and different conclusions might be drawn for other types of simulated data.

Hawkins (1973) also proposed finding PCs for the enlarged set of $(p+1)$ variables, but he used the PCs in a rather different way from that of latent root regression as defined above. The idea here is to use the PCs themselves, or rather a rotated version of them, to decide upon a suitable regression equation. Any PC with a small variance gives a relationship between y and the predictor variables whose sum of squared residuals orthogonal to the fitted plane is small. Of course, in regression it is squared residuals in the y-direction, rather than orthogonal to the fitted plane, which are to be minimized (see Section 8.6), but the low-variance PCs can nevertheless be used to suggest low-variability relationships between y and the predictor variables. Hawkins (1973) goes further by suggesting that it may be more fruitful to look at rotated versions of the PCs, instead of the PCs themselves, in order to indicate low-variance relationships. This is done by rescaling and then using varimax rotation (see Chapter 7), which has the effect of transforming the PCs to a different set of uncorrelated variables. These variables are, like the PCs, linear functions of the original $(p + 1)$ variables, but their coefficients are mostly close to zero or a long way from zero, with relatively few intermediate values. There is no guarantee, in general, that any of the new variables will have particularly large or particularly small variances, as they are chosen by simplicity of structure of their coefficients, rather than for their variance properties. However, if only one or two of the coefficients for y are large, as should often happen with varimax rotation, then Hawkins (1973) shows that the corresponding transformed variables will have very small variances, and therefore suggest low-variance relationships between y and the predictor variables. Other possible regression equations may be found by substitution of one subset of predictor variables in terms of another, using any low-variability relationships between predictor variables that are suggested by the other rotated PCs.

The above technique is advocated by Hawkins (1973) and by Jeffers (1981) as a means of selecting which variables should appear in the regression equation (see Section 8.5), rather than as a way of directly estimating their coefficients in the regression equation, although the technique could be used for the latter purpose. Daling and Tamura (1970) also discussed rotation of PCs in the context of variable selection, but their PCs were for the predictor variables only.

In a later paper, Hawkins and Eplett (1982) propose another variant of latent root regression one which can be used to efficiently find low-variability relationships between y and the predictor variables, and which also can be used in variable selection. This method replaces the rescaling and varimax rotation of Hawkins' earlier method by a sequence of rotations leading to a set of relationships between y and the predictor variables that are simpler to interpret than in the previous method. This simplicity is achieved because the matrix of coefficients defining the relationships has non-zero entries only in its lower-triangular region. Despite the apparent complexity of the new method, it is also computationally simple to implement. The covariance (or correlation) matrix $\tilde{\Sigma}$ of y and all the predictor variables is factorized using a Cholesky factorization

$$\tilde{\Sigma} = DD',$$

where D is lower-triangular. Then the matrix of coefficients defining the relationships is proportional to D^{-1}, which is also lower-triangular. To find D it is not necessary to calculate PCs based on $\tilde{\Sigma}$, which makes the links between the method and PCA rather more tenuous than those between PCA and latent root regression. The next section discusses variable selection in regression using PCs, and because all three variants of latent root regression described above can be used in variable selection, they will all be discussed further in that section.

Another variation on the idea of PC regression has been used in several meteorological examples in which a multivariate (rather than multiple) regression analysis is appropriate, that is, where there are several dependent variables as well as regressor variables. Here PCA is performed on the dependent variables and, separately, on the predictor variables. A number of PC regressions are then carried out with, as usual, PCs of predictor variables in place of the predictor variables but, in each regression, the dependent variable is now one of the high variance PCs of the original set of dependent variables. Preisendorfer and Mobley (1988, Chapter 9) discuss this set-up in some detail, and demonstrate links between the results and those of canonical correlation analysis (see Section 9.3) on the two sets of variables. Briffa et al. (1986) give an example in which the dependent variables are mean sea level pressures at 16 grid-points centred on the UK and France, and extending from 45°–60°N, and from 20°W–10°E. The predictors are tree ring widths for 14 oak ring width chronologies from the UK and Northern France. They transform the relationships found between the two sets of PCs back into relationships between the original sets of variables and present the results in map form.

The method is appropriate if the prediction of high-variance PCs of the dependent variables is really of interest, in which case another possibility is to regress PCs of the dependent variables on the original predictor variables. However, if overall optimal prediction of linear functions of dependent variables from linear functions of predictor variables is required, then canonical correlation analysis (see Section 9.3; Mardia et al., 1979, Chapter 10; Rencher, 1995, Chapter 11) is more suitable. Alternatively, if interpretable relationships between the original sets of dependent and predictor variables are wanted, then multivariate regression analysis or a related technique (see Section 9.3; Mardia et al., 1979, Chapter 6; Rencher, 1995, Chapter10) may be the most appropriate technique.

The so-called PLS (partial least squares) method provides yet another biased regression approach with links to PC regression. The method has a long and complex history and various formulations (Geladi, 1988; Wold, 1984). It has often been expressed only in terms of algorithms for its implementation, which makes it difficult to understand exactly what it does. A number of authors, for example, Garthwaite (1994) and Helland (1988, 1990), have given interpretations that move away from algorithms towards a more model-based approach, but perhaps the most appealing interpretation from a statistical point of view is that given by Stone and Brooks (1990). They show that PLS is equivalent to successively finding linear functions of the predictor variables that have maximum *covariance* with the dependent variable, subject to each linear function being uncorrelated with previous ones. Whereas PC regression in concerned with variances derived from $\mathbf{X}$, and least squares regression maximizes correlations between $\mathbf{y}$ and $\mathbf{X}$, PLS combines correlation and variance to consider covariance. Stone and Brooks (1990) introduce a general class of regression procedures, called *continuum regression*, in which least squares and PC regression are two extremes of the class, and PLS lies halfway along the continuum in between them. As well as different algorithms and interpretations, PLS is sometimes known by a quite different name, albeit with the same acronym, in the field of statistical process control (see Section 13.7). Martin et al. (1999), for example, refer to it as 'projection to latent structure.'

Lang et al. (1998) define another general class of regression estimates, called *cyclic subspace regression*, which includes both PC regression and PLS as special cases. The nature of the special cases within this framework shows that PLS uses information from the directions of *all* the eigenvectors of $\mathbf{X'X}$, whereas PC regression, by definition, uses information from only a chosen subset of these directions.

Naes and Helland (1993) propose a compromise between PC regression and PLS, which they call *restricted principal component regression* (RPCR). The motivation behind the method lies in the idea of components (where 'component' means any linear function of the predictor variables $\mathbf{x}$) or subspaces that are 'relevant' for predicting y. An m-dimensional subspace $\mathcal{M}$ in the space of the predictor variables is *strongly relevant* if the linear functions of $\mathbf{x}$ defining the $(p - m)$-dimensional subspace $\bar{\mathcal{M}}$, orthogonal to $\mathcal{M}$, are uncorrelated with y and with the linear functions of $\mathbf{x}$ defining $\mathcal{M}$. Using this definition, if an m-dimensional relevant subspace exists it can be obtained by taking the first component found by PLS as the first component in this subspace, followed by $(m - 1)$ components, which can be considered as PCs in the space orthogonal to the first PLS component. Naes and Helland (1993) show, in terms of predictive ability, that when PC regression and PLS differ considerably in performance, RPCR tends to be close to the better of the two. Asymptotic comparisons between PLS, RPCR and PC regression (with M restricted to contain the *first m* integers) are made by Helland and Almøy (1994). Their conclusions are that PLS is preferred in many circumstances, although in some cases PC regression is a better choice.

A number of other comparisons have been made between least squares, PC regression, PLS and other biased regression techniques, and adaptations involving one or more of the biased methods have been suggested. A substantial proportion of this literature is in chemometrics, in particular concentrating on the analysis of spectroscopic data. Naes et al. (1986) find that PLS tends to be superior to PC regression, although only the rule based on (8.1.10) is considered for PC regression. For near infrared spectroscopy data, the researchers also find that results are improved by pre-processing the data using an alternative technique which they call *multiple scatter correction*, rather than simple centering. Frank and Friedman (1993) give an extensive comparative discussion of PLS and PC regression, together with other strategies for overcoming the problems caused by multicollinearity. From simulations and other considerations they conclude that the two techniques are superior to variable selection but inferior to ridge regression, although this latter conclusion is disputed by S. Wold in the published discussion that follows the article.

Naes and Isaksson (1992) use a locally weighted version of PC regression in the calibration of spectroscopic data. PCA is done on the predictor variables, and to form a predictor for a particular observation only the k observations closest to the chosen observation in the space of the first m PCs are used. These k observations are given weights in a regression of the dependent variable on the first m PCs whose values decrease as distance from the chosen observation increases. The values of m and k are chosen by cross-validation, and the technique is shown to outperform both PC regression and PLS.

Bertrand et al. (2001) revisit latent root regression, and replace the PCA of the matrix of $(p + 1)$ variables formed by $\mathbf{y}$ together with $\mathbf{X}$ by the equivalent PCA of $\mathbf{y}$ together with the PC scores $\mathbf{Z}$. This makes it easier to identify predictive and non-predictive multicollinearities, and gives a simple expression for the MSE of the latent root estimator. Bertrand et al. (2001) present their version of latent root regression as an alternative to PLS or PC regression for near infrared spectroscopic data.

Marx and Smith (1990) extend PC regression from linear models to generalized linear models. Straying further from ordinary PCA, Li et al. (2000) discuss *principal Hessian directions*, which utilize a variety of generalized PCA (see Section 14.2.2) in a regression context. These directions are used to define splits in a regression tree, where the objective is to find directions along which the regression surface 'bends' as much as possible. A weighted covariance matrix $\mathbf{S}_W$ is calculated for the predictor variables, where the weights are residuals from a multiple regression of y on all the predictor variables. Given the (unweighted) covariance matrix $\mathbf{S}$, their derivation of the first principal Hessian direction is equivalent to finding the first eigenvector in a generalized PCA of $\mathbf{S}_W$ with metric $\mathbf{Q} = \mathbf{S}^{-1}$ and $\mathbf{D} = \frac{1}{n}\mathbf{I}_n$, in the notation of Section 14.2.2.

8.5 Variable Selection in Regression Using Principal Components

Principal component regression, latent root regression, and other biased regression estimates keep all the predictor variables in the model, but change the estimates from least squares estimates in a way that reduces the effects of multicollinearity. As mentioned in the introductory section of this chapter, an alternative way of dealing with multicollinearity problems is to use only a subset of the predictor variables. Among the very many possible methods of selecting a subset of variables, a few use PCs.

As noted in the previous section, the procedures due to Hawkins (1973) and Hawkins and Eplett (1982) can be used in this way. Rotation of the PCs produces a large number of near-zero coefficients for the rotated variables, so that in low-variance relationships involving y (if such low-variance relationships exist) only a subset of the predictor variables will have coefficients substantially different from zero. This subset forms a plausible selection of variables to be included in a regression model. There may be other low-variance relationships between the predictor variables alone, again with relatively few coefficients far from zero. If such relationships exist, and involve some of the same variables as are in the relationship involving y, then substitution will lead to alternative subsets of predictor variables. Jeffers (1981) argues that in this way it is possible to identify all good subregressions using Hawkins' (1973) original procedure. Hawkins and Eplett (1982) demonstrate that their newer technique, incorporating Cholesky factorization, can do even better than the earlier method. In particular, for an example that is analysed by both methods, two subsets of variables selected by the first method are shown to be inappropriate by the second.

Principal component regression and latent root regression may also be used in an iterative manner to select variables. Consider, first, PC regression and suppose that $\tilde{\beta}$ given by (8.1.12) is the proposed estimator for β. Then it is possible to test whether or not subsets of the elements of $\tilde{\beta}$ are significantly different from zero, and those variables whose coefficients are found to be not significantly non-zero can then be deleted from the model. Mansfield et al. (1977), after a moderate amount of algebra, construct the appropriate tests for estimators of the form (8.1.10), that is, where the PCs deleted from the regression are restricted to be those with the smallest variances. Provided that the true coefficients of the deleted PCs are zero and that normality assumptions are valid, the appropriate test statistics are F-statistics, reducing to t-statistics if only one variable is considered at a time. A corresponding result will also hold for the more general form of estimator (8.1.12).

Although the variable selection procedure could stop at this stage, it may be more fruitful to use an iterative procedure, similar to that suggested by Jolliffe (1972) for variable selection in another (non-regression) context (see Section 6.3, method (i)). The next step in such a procedure is to perform a PC regression on the reduced set of variables, and then see if any further variables can be deleted from the reduced set, using the same reasoning as before. This process is repeated, until eventually no more variables are deleted. Two variations on this iterative procedure are described by Mansfield et al. (1977). The first is a stepwise procedure that first looks for the best single variable to delete, then the best pair of variables, one of which is the best single variable, then the best triple of variables, which includes the best pair, and so on. The procedure stops when the test for zero regression coefficients on the subset of excluded variables first gives a significant result. The second variation is to delete only one variable at each stage, and then recompute the PCs using the reduced set of variables, rather than allowing the deletion of several variables before the PCs are recomputed. According to Mansfield et al. (1977) this second variation gives, for several examples, an improved performance for the selected variables compared with subsets selected by the other possibilities. Only one example is described in detail in their paper, and this will be discussed further in the final section of the present chapter. In this example, they adapt their method still further by discarding a few low variance PCs before attempting any selection of variables.

A different iterative procedure is described by Boneh and Mendieta (1994). The method works on standardized variables, and hence on the correlation matrix. The first step is to do a PC regression and choose M to contain those PCs that contribute significantly to the regression. Significance is judged mainly by the use of t-tests. However, a modification is used for the PCs with the smallest variance, as Mason and Gunst (1985) have shown that t-tests have reduced power for such PCs.

Each of the p predictor variables is then regressed on the PCs in M, and the variable with the smallest residual sum of squares is selected. At subsequent stages in the iteration, suppose that a set Q of q variables has been selected and that $\bar{Q}$ is the complement of Q, consisting of $(p - q)$ variables. The variables in $\bar{Q}$ are individually regressed on all the variables in Q, and a vector of residuals is found for each variable in $\bar{Q}$. Principal components are then found for the $(p - q)$ residual variables and the dependent variable y is regressed on these $(p - q)$ PCs, together with the q variables in Q. If none of the PCs contributes significantly to this regression, the procedure stops. Otherwise, each of the residual variables is regressed on the significant PCs, and the variable is selected for which the residual sum of squares is smallest. As well as these forward selection steps, Boneh and Mendieta's (1994) procedure includes backward looks, in which previously selected variables can be deleted from (and never allowed to return to) Q. Deletion of a variable occurs if its contribution is sufficiently diminished by the later inclusion of other variables. Boneh and Mendieta (1994) claim that, using cross-validation, their method often does better than its competitors with respect to prediction error.

A similar procedure to that of Mansfield et al. (1977) for PC regression can be constructed for latent root regression, this time leading to *approximate F*-statistics (see Gunst and Mason (1980, p. 339)). Such a procedure is described and illustrated by Webster et al. (1974) and Gunst et al. (1976).

Baskerville and Toogood (1982) also suggest that the PCs appearing in latent root regression can be used to select subsets of the original predictor variables. Their procedure divides the predictor variables into four groups on the basis of their coefficients in the PCs, where each of the groups has a different degree of potential usefulness in the regression equation. The first group of predictor variables they define consists of 'isolated' variables, which are virtually uncorrelated with y and with all other predictor variables; such variables can clearly be deleted. The second and third groups contain variables that are involved in nonpredictive and predictive multicollinearities, respectively; those variables in the second group can usually be excluded from the regression analysis, whereas those in the third group certainly cannot. The fourth group simply consists of variables that do not fall into any of the other three groups. These variables may or may not be important in the regression, depending on the purpose of the analysis (for example, prediction or identification of structure) and each must be examined individually (see Baskerville and Toogood (1982) for an example).

A further possibility for variable selection is based on the idea of associating a variable with each of the first few (last few) components and then retaining (deleting) those variables associated with the first few (last few) PCs. This procedure was described in a different context in Section 6.3, and it is clearly essential to modify it in some way for use in a regression context. In particular, when there is not a single clear-cut choice of which variable to associate with a particular PC, the choice should be determined by looking at the strength of the relationships between the candidate variables and the dependent variable. Great care is also necessary to avoid deletion of variables that occur in a predictive multicollinearity.

Daling and Tamura (1970) adopt a modified version of this type of approach. They first delete the last few PCs, then rotate the remaining PCs using varimax, and finally select one variable associated with each of those rotated PCs which has a 'significant' correlation with the dependent variable. The method therefore takes into account the regression context of the problem at the final stage, and the varimax rotation increases the chances of an unambiguous choice of which variable to associate with each rotated PC. The main drawback of the approach is in its first stage, where deletion of the low-variance PCs may discard substantial information regarding the relationship between y and the predictor variables, as was discussed in Section 8.2.

8.6   Functional and Structural Relationships

In the standard regression framework, the predictor variables are implicitly assumed to be measured without error, whereas any measurement error in the dependent variable y can be included in the error term ε. If all the variables are subject to measurement error the problem is more complicated, even when there is only one predictor variable, and much has been written on how to estimate the so-called functional or structural relationships between the variables in such cases (see, for example, Kendall and Stuart (1979, Chapter 29); Anderson (1984); Cheng and van Ness (1999)). The term 'functional and structural relationships' seems to have gone out of fashion, but there are close connections to the 'errors-in-variables' models from econometrics (Darnell, 1994) and to some of the approaches of Section 9.3.

Consider the case where there are $(p+1)$ variables $x_0, x_1, x_2, \ldots, x_p$ that have a *linear functional relationship* (Kendall and Stuart, 1979, p. 416)

$$\sum_{j=0}^{p} \beta_j x_j = \text{const} \tag{8.6.1}$$

between them, but which are all subject to measurement error, so that we actually have observations on $\xi_0, \xi_1, \xi_2, \ldots, \xi_p$, where

$$\xi_j = x_j + e_j, \quad j = 0, 1, 2, \ldots, p,$$

and e_j is a measurement error term. The distinction between 'functional' and 'structural' relationships is that $x_1, x_2, \ldots, x_p$ are taken as fixed in the former but are random variables in the latter. We have included $(p + 1)$ variables in order to keep a parallel with the case of linear regression with dependent variable y and p predictor variables $x_1, x_2, \ldots, x_p$, but there is no reason here to treat any one variable differently from the remaining p. On the basis of n observations on ξ_j, $j = 0, 1, 2, \ldots, p$, we wish to estimate the coefficients $\beta_0, \beta_1, \ldots \beta_p$ in the relationship (8.6.1). If the e_j are assumed to be normally distributed, and (the ratios of) their variances are known, then maximum likelihood estimation of $\beta_0, \beta_1, \ldots, \beta_p$ leads to the coefficients of the last PC from the covariance matrix of $\xi_0/\sigma_0, \xi_1/\sigma_1, \ldots, \xi_p/\sigma_p$, where $\sigma_j^2 = \text{var}(e_j)$. This holds for both functional and structural relationships. If there is no information about the variances of the e_j, and the x_j are distinct, then no formal estimation procedure is possible, but if it is expected that the measurement errors of all $(p + 1)$ variables are of similar variability, then a reasonable procedure is to use the last PC of $\xi_0, \xi_1, \ldots, \xi_p$.

If replicate observations are available for each x_j, they can be used to estimate $\text{var}(e_j)$. In this case, Anderson (1984) shows that the maximum likelihood estimates for functional, but not structural, relationships are given by solving an eigenequation, similar to a generalized PCA in which the PCs of between-x_j variation are found with respect to a metric based on within-x_j variation (see Section 14.2.2). Even if there is no formal requirement to estimate a relationship such as (8.6.1), the last few PCs are still of interest in finding near-constant linear relationships among a set of variables, as discussed in Section 3.4.

When the last PC is used to estimate a 'best-fitting' relationship between a set of $(p + 1)$ variables, we are finding the p-dimensional hyperplane for which the sum of squares of *perpendicular* distances of the observations from the hyperplane is minimized. This was, in fact, one of the objectives of Pearson's (1901) original derivation of PCs (see Property G3 in Section 3.2). By contrast, if one of the $(p + 1)$ variables y is a dependent variable and the remaining p are predictor variables, then the 'best-fitting' hyperplane, in the least squares sense, minimizes the sum of squares of the distances *in the y direction* of the observations from the hyperplane and leads to a different relationship.

A different way of using PCs in investigating structural relationships is illustrated by Rao (1964). In his example there are 20 variables corresponding to measurements of 'absorbance' made by a spectrophotometer at 20 different wavelengths. There are 54 observations of the 20 variables, corresponding to nine different spectrophotometers, each used under three conditions on two separate days. The aim is to relate the absorbance measurements to wavelengths; both are subject to measurement error, so that a structural relationship, rather than straightforward regression analysis, is of interest. In this example, the first PCs, rather than the last, proved to be useful in investigating aspects of the structural relationship. Examination of the values of the first two PCs for the 54 observations identified systematic differences between spectrophotometers in the measurement errors for wavelength. Other authors have used similar, but rather more complicated, ideas based on PCs for the same type of data. Naes (1985) refers to the problem as one of multivariate calibration (see also Martens and Naes (1989)) and investigates an estimate (which uses PCs) of some chemical or physical quantity, given a number of spectrophotometer measurements. Sylvestre et al. (1974) take as their objective the identification and estimation of mixtures of two or more overlapping curves in spectrophotometry, and again use PCs in their procedure.

8.7   Examples of Principal Components in Regression

Early examples of PC regression include those given by Kendall (1957, p. 71), Spurrell (1963) and Massy (1965). Examples of latent root regression in one form or another, and its use in variable selection, are given by Gunst et al. (1976), Gunst and Mason (1977b), Hawkins (1973), Baskerville and Toogood (1982) and Hawkins and Eplett (1982). In Gunst and Mason (1980, Chapter 10) PC regression, latent root regression and ridge regression are all illustrated, and can therefore be compared, for the same data set. In the present section we discuss two examples illustrating some of the techniques described in this chapter.

8.7.1   Pitprop Data

No discussion of PC regression would be complete without the example given originally by Jeffers (1967) concerning strengths of pitprops, which has since been analysed by several authors. The data consist of 14 variables which were measured for each of 180 pitprops cut from Corsican pine timber. The objective is to construct a prediction equation for one of the variables (compressive strength y) using the values of the other 13 variables. These other 13 variables are physical measurements on the pitprops that could be measured fairly straightforwardly without destroying the props. The variables are listed by Jeffers (1967, 1981) and the correlation matrix for all 14 variables is reproduced in Table 8.2. In his original paper, Jeffers (1967) used PC regression to predict y from the 13 variables. The coefficients of the variables for each of the PCs are given in Table 8.3. The pattern of correlations in Table 8.2 is not easy to interpret; nor is it simple to deduce the form of the first few PCs from the correlation matrix. However, Jeffers (1967) was able to interpret the first six PCs.

Also given in Table 8.3 are variances of each component, the percentage of total variation accounted for by each component, the coefficients γ_k in a regression of y on the PCs, and the values of t-statistics measuring the importance of each PC in the regression.

Judged solely on the basis of size of variance it appears that the last three, or possibly four, PCs should be deleted from the regression. However, looking at values of γ_k and the corresponding t-statistics, it can be seen that the twelfth component is relatively important as a predictor of y, despite the fact that it accounts for only 0.3% of the total variation in the predictor variables. Jeffers (1967) only retained the first, second, third, fifth and sixth PCs in his regression equation, whereas Mardia et al. (1979, p. 246) suggest that the seventh, eighth and twelfth PCs should also be included.

This example has been used by various authors to illustrate techniques of variable selection, and some of the results are given in Table 8.4. Jeffers (1981) used Hawkins' (1973) variant of latent root regression to select subsets of five, six or seven regressor variables. After varimax rotation, only one of the rotated components has a substantial coefficient for compressive strength, y. This rotated component has five other variables that have large coefficients, and it is suggested that these should be included in the regression equation for y; two further variables with moderate coefficients might also be included. One of the five variables definitely selected by this method is quite difficult to measure, and one of the other rotated components suggests that it can be replaced by another, more readily measured, variable. However, this substitution causes a substantial drop in the squared multiple correlation for the five-variable regression equation, from 0.695 to 0.581.

Mansfield et al. (1977) used an iterative method based on PC regression and described above in Section 8.5, to select a subset of variables for these data. The procedure is fairly lengthy as only one variable is deleted at each iteration, but the F-criterion used to decide whether to delete an extra variable jumps from 1.1 to 7.4 between the fifth and sixth iterations, giving a clear-cut decision to delete five variables, that is to retain eight variables. The iterative procedure of Boneh and Mendieta (1994) also selects eight variables. As can be seen from Table 8.4, these eight-variable subsets have a large degree of overlap with the subsets found by Jeffers (1981).

Jolliffe (1973) also found subsets of the 13 variables, using various methods, but the variables in this case were chosen to reproduce the relationships between the regressor variables, rather than to predict y as well as possi-

Table 8.2. Correlation matrix for the pitprop data.

	TOPDIAM	LENGTH	MOIST	TESTSG	OVENSG	RINGTOP	RINGBUT	BOWMAX	BOWDIST	WHORLS	CLEAR	KNOTS	DIAKNOT
LENGTH	0.954												
MOIST	0.364	0.297											
TESTSG	0.342	0.284	0.882										
OVENSG	−0.129	−0.118	−0.148	0.220									
RINGTOP	0.313	0.291	0.153	0.381	0.364								
RINGBUT	0.496	0.503	−0.029	0.174	0.296	0.813							
BOWMAX	0.424	0.419	−0.054	−0.059	0.004	0.090	0.372						
BOWDIST	0.592	0.648	0.125	0.137	−0.039	0.211	0.465	0.482					
WHORLS	0.545	0.569	−0.081	−0.014	0.037	0.274	0.679	0.557	0.526				
CLEAR	0.084	0.076	0.162	0.097	0.091	−0.036	−0.113	0.061	0.085	−0.319			
KNOTS	−0.019	−0.036	0.220	0.169	−0.145	0.024	−0.232	−0.357	−0.127	−0.368	0.029		
DIAKNOT	0.134	0.144	0.126	0.015	−0.208	−0.329	−0.424	−0.202	−0.076	−0.291	0.007	0.184	
STRENGTH	−0.419	−0.338	−0.728	−0.543	0.247	0.117	0.110	−0.253	−0.235	−0.101	−0.055	−0.117	−0.153

Table 8.3. Principal component regression for the pitprop data: coefficients, variances, regression coefficients and t-statistics for each component.

	Principal component												
	1	2	3	4	5	6	7	8	9	10	11	12	13
x_1	−0.40	0.22	−0.21	−0.09	−0.08	0.12	−0.11	0.14	0.33	−0.31	0.00	0.39	−0.57
x_2	−0.41	0.19	−0.24	−0.10	−0.11	0.16	−0.08	0.02	0.32	−0.27	−0.05	−0.41	0.58
x_3	−0.12	0.54	0.14	0.08	0.35	−0.28	−0.02	0.00	−0.08	0.06	0.12	0.53	0.41
x_4	−0.17	0.46	0.35	0.05	0.36	−0.05	0.08	−0.02	−0.01	0.10	−0.02	−0.59	−0.38
x_5	−0.06	−0.17	0.48	0.05	0.18	0.63	0.42	−0.01	0.28	−0.00	0.01	0.20	0.12
x_6	−0.28	−0.01	0.48	−0.06	−0.32	0.05	−0.30	0.15	−0.41	−0.10	−0.54	0.08	0.06
x_7	−0.40	−0.19	0.25	−0.07	−0.22	0.00	−0.23	0.01	−0.13	0.19	0.76	−0.04	0.00
x_8	−0.29	−0.19	−0.24	0.29	0.19	−0.06	0.40	0.64	−0.35	−0.08	0.03	−0.05	0.02
x_9	−0.36	0.02	−0.21	0.10	−0.10	0.03	0.40	−0.70	−0.38	−0.06	−0.05	0.05	−0.06
x_{10}	−0.38	−0.25	−0.12	−0.21	0.16	−0.17	0.00	−0.01	0.27	0.71	−0.32	0.06	0.00
x_{11}	0.01	0.21	−0.07	0.80	−0.34	0.18	−0.14	0.01	0.15	0.34	−0.05	0.00	−0.01
x_{12}	0.12	0.34	0.09	−0.30	−0.60	−0.17	0.54	0.21	0.08	0.19	0.05	0.00	0.00
x_{13}	0.11	0.31	−0.33	−0.30	0.08	0.63	−0.16	0.11	−0.38	0.33	0.04	0.01	−0.01
Variance	4.22	2.38	1.88	1.11	0.91	0.82	0.58	0.44	0.35	0.19	0.05	0.04	0.04
% of total variance	32.5	18.3	14.4	8.5	7.0	6.3	4.4	3.4	2.7	1.5	0.4	0.3	0.3
Regression coefficient γ_k	0.13	−0.37	0.13	−0.05	−0.39	0.27	−0.24	−0.17	0.03	0.00	−0.12	−1.05	0.00
t-value	6.86	14.39	4.38	1.26	9.23	6.19	4.50	2.81	0.46	0.00	0.64	5.26	0.01

The x_1–x_{13} rows are grouped under the label **Coefficients**.

Table 8.4. Variable selection using various techniques on the pitprop data. (Each row corresponds to a selected subset with × denoting a selected variable.)

	\multicolumn{13}{Variables}											
	1	2	3	4	5	6	7	8	9	10	11	12	13
Five variables													
Jeffers (1981)	×		×			×	×	×					
	×			×		×	×	×					
McCabe (1982)				×					×		×	×	×
					×				×		×	×	×
	×				×						×	×	×
			×		×				×		×		×
Six variables													
Jeffers (1981)	×		×			×	×	×	×				
McCabe (1982)			×	×					×		×	×	×
Jolliffe (1973)	×		×		×			×			×		×
	×		×		×						×	×	×
{ McCabe (1982)			×		×				×		×	×	×
Jolliffe (1973)		×	×		×						×	×	×
			×		×			×			×	×	×
Eight variables													
Mansfield et al. (1977)	×		×	×	×	×	×	×			×		
Boneh and Mendieta (1994)	×	×	×	×		×		×			×	×	

ble. McCabe (1982), using a technique related to PCA (see Section 6.3), and with a similar purpose to Jolliffe's (1973) methods, chose subsets of various sizes. McCabe's subsets in Table 8.4 are the best few with respect to a single criterion, whereas Jolliffe gives the single best subset but for several different methods. The best subsets due to Jolliffe (1973) and McCabe (1982) have considerable overlap with each other, but there are substantial differences from the subsets of Jeffers (1981) and Mansfield et al. (1977). This reflects the different aims of the different selection methods. It shows again that substantial variation within the set of regressor variables does not necessarily imply any relationship with y and, conversely, that variables having little correlation with the first few PCs can still be important in predicting y. Interestingly, the subset chosen by Boneh and Mendieta (1994) overlaps less with Jeffers' (1981) selection than that of Mansfield et al. (1977), but more than do those of Jolliffe (1973) or McCabe (1982).

Table 8.5. Variables used in the household formation example.

No.	Description
1.	Population in non-private establishments
2.	Population age 0–14
3.	Population age 15–44
4.	Population age 60/65+
5.	Females currently married
6.	Married males 15–29
7.	Persons born ex UK
8.	Average population increase per annum (not births and deaths)
9.	Persons moved in previous 12 months
10.	Households in owner occupation
11.	Households renting from Local Authority
12.	Households renting private unfurnished
13.	Vacant dwellings
14.	Shared dwellings
15.	Households over one person per room
16.	Households with all exclusive amenities
17.	Ratio households to rateable units
18.	Domestic rateable value (£) per head
19.	Rateable units with rateable value < £100
20.	Students age 15+
21.	Economically active married females
22.	Unemployed males seeking work
23.	Persons employed in agriculture
24.	Persons employed in mining and manufacturing
25.	Males economically active or retired in socio-economic group 1, 2, 3, 4, 13
26.	Males economically active or retired in socio-economic group 5, 6, 8, 9, 12, 14
27.	With degrees (excluding students with degree)
28.	Economically active males socio-economic group 3, 4
29.	Average annual total income (£) per adult

8.7.2   Household Formation Data

This example uses part of a data set that arose in a study of household formation. The subset of data used here has 29 demographic variables measured in 1971 for 168 local government areas in England and Wales. The variables are listed in Table 8.5. All variables, except numbers 17, 18 and 29, are expressed as numbers per 1000 of population; precise definitions of each variable are given in Appendix B of Bassett et al. (1980).

Table 8.6. Eigenvalues of the correlation matrix and order of importance in predicting y for the household formation data.

PC number	Eigenvalue	Order of importance in predicting y	PC number	Eigenvalue	Order of importance in predicting y
1	8.62	1	15	0.24	17
2	6.09	4	16	0.21	25
3	3.40	2	17	0.18	16
4	2.30	8	18	0.14	10
5	1.19	9	19	0.14	7
6	1.06	3	20	0.10	21
7	0.78	13	21	0.10	28
8	0.69	22	22	0.07	6
9	0.58	20	23	0.07	18
10	0.57	5	24	0.05	12
11	0.46	11	25	0.04	14
12	0.36	15	26	0.03	27
13	0.27	24	27	0.02	19
14	0.25	23	28	0.003	26

Although this was not the purpose of the original project, the objective considered here is to predict the final variable (average annual total income per adult) from the other 28. This objective is a useful one, as information on income is often difficult to obtain accurately, and predictions from other, more readily available, variables would be valuable. The results presented below were given by Garnham (1979) in an unpublished M.Sc. dissertation, and further details of the regression analysis can be found in that source. A full description of the project from which the data are taken is available in Bassett et al. (1980). Most regression problems with as many as 28 regressor variables have multicollinearities, and the current example is no exception. Looking at the list of variables in Table 8.5 it is clear, even without detailed definitions, that there are groups of variables that are likely to be highly correlated. For example, several variables relate to type of household, whereas another group of variables considers rates of employment in various types of job. Table 8.6, giving the eigenvalues of the correlation matrix, confirms that there are multicollinearities; some of the eigenvalues are very small.

Consider now PC regression and some of the strategies that can be used to select a subset of PCs to be included in the regression. Deleting components with small variance, with a cut-off of about $l^* = 0.10$, implies that between seven and nine components can be left out. Sequential deletion of PCs with the smallest variances using t-statistics at each stage suggests that only six PCs can be deleted. However, from the point of view of R^2, the squared multiple correlation coefficient, deletion of eight or more might be acceptable; R^2 is 0.874 for the full model including all 28 variables, and it is reduced to 0.865, 0.851, respectively, when five and eight components are deleted.

It is interesting to examine the ordering of size of correlations between y and the PCs, or equivalently the ordering of the individual t-values, which is also given in Table 8.6. It is seen that those PCs with small variances do not necessarily have small correlations with y. The 18th, 19th and 22nd in size of variance are in the first ten in order of importance for predicting y; in particular, the 22nd PC with variance 0.07, has a highly significant t-value, and should almost certainly be retained.

An approach using stepwise deletion based solely on the size of correlation between y and each PC produces, because of the zero correlations between PCs, the subset whose value of R^2 is maximized for any given subset size. Far fewer PCs need to be retained using this approach than the 20 to 23 indicated when only small-variance components are rejected. In particular, if the 10 PCs are retained that best predict y, then R^2 is 0.848, compared with 0.874 for the full model and 0.851 using the first 20 PCs. It would appear that a strategy based solely on size of variance is unsatisfactory.

The two 'weak MSE' criteria described in Section 8.2 were also tested, in a limited way, on these data. Because of computational constraints it was not possible to find the overall 'best' subset M, so a stepwise approach was adopted, deleting PCs according to either size of variance, or correlation with y. The first criterion selected 22 PCs when selection was based on size of variance, but only 6 PCs when correlation with y was the basis for stepwise selection. The corresponding results for the second (predictive) criterion were 24 and 12 PCs, respectively. It is clear, once again, that selection based solely on order of size of variance retains more components than necessary but may still miss predictive components.

The alternative approach of Lott (1973) was also investigated for these data in a stepwise manner using correlation with y to determine order of selection, with the result that $\bar{R}^2$ was maximized for 19 PCs. This is a substantially larger number than indicated by those other methods that use correlation with y to define order of selection and, given the concensus from the other methods, suggests that Lott's (1973) method is not ideal.

When PCs are found for the augmented set of variables, including y and all the regressor variables, as required for latent root regression, there is remarkably little change in the PCs, apart from the addition of an extra one. All of the coefficients on the regressor variables are virtually unchanged, and the PCs that have largest correlation with y are in very nearly the same order as in the PC regression.

It may be of more interest to select a subset of variables, rather than a subset of PCs, to be included in the regression, and this was also attempted, using various methods, for the household formation data. Variable selection based on PC regression, deleting just one variable at a time before recomputing the PCs as suggested by Mansfield et al. (1977), indicated that only 12, and possibly fewer, variables need to be retained. R^2 for the 12-variable subset given by this method is 0.862, and it only drops to 0.847 for the 8-variable subset, compared with 0.874 for the full model and 0.851 using the first 20 PCs in the regression. Other variable selection methods, described by Jolliffe (1972) and in Section 6.3, were also tried, but these did not produce quite such good results as the Mansfield et al. (1977) method. This is not surprising since, as noted in the previous example, they are not specifically tailored for variable selection in the context of regression. However, they did confirm that only eight to ten variables are really necessary in order to provide an adequate prediction of 'income' for these data.

9
Principal Components Used with Other Multivariate Techniques

Principal component analysis is often used as a dimension-reducing technique within some other type of analysis. For example, Chapter 8 described the use of PCs as regressor variables in a multiple regression analysis. The present chapter discusses three classes of multivariate techniques, namely discriminant analysis, cluster analysis and canonical correlation analysis; for each these three there are examples in the literature that use PCA as a dimension-reducing technique.

Discriminant analysis is concerned with data in which each observation comes from one of several well-defined groups or populations. Assumptions are made about the structure of the populations, and the main objective is to construct rules for assigning future observations to one of the populations so as to minimize the probability of misclassification or some similar criterion. As with regression, there can be advantages in replacing the variables in a discriminant analysis by their principal components. The use of PCA in this way in linear discriminant analysis is discussed in Section 9.1. In addition, the section includes brief descriptions of other discriminant techniques that use PCs, and discussion of links between PCA and canonical discriminant analysis.

Cluster analysis is one of the most frequent contexts in which PCs are derived in order to reduce dimensionality prior to the use of a different multivariate technique. Like discriminant analysis, cluster analysis deals with data sets in which the observations are to be divided into groups. However, in cluster analysis little or nothing is known a priori about the groups, and the objective is to divide the given observations into groups or clusters in a 'sensible' way. There are two main ways in which PCs are employed within cluster analysis: to construct distance measures or to provide a graphical representation of the data; the latter is often called *ordination* or *scaling* (see also Section 5.1) and is useful in detecting or verifying a cluster structure. Both rôles are described and illustrated with examples in Section 9.2. The idea of clustering variables rather than observations is sometimes useful, and a connection between PCA and this idea is described. Also discussed in Section 9.2 are projection pursuit, which searches for clusters using techniques bearing some resemblance to PCA, and the construction of models for clusters which are mixtures of the PC model introduced in Section 3.9.

'Discriminant analysis' and 'cluster analysis' are standard statistical terms, but the techniques may be encountered under a variety of other names. For example, the word 'classification' is sometimes used in a broad sense, including both discrimination and clustering, but it also has more than one specialized meaning. Discriminant analysis and cluster analysis are prominent in both the pattern recognition and neural network literatures, where they fall within the areas of supervised and unsupervised learning, respectively (see, for example, Bishop (1995)). The relatively new, but large, field of data mining (Hand et al. 2001; Witten and Frank, 2000) also includes 'clustering methods... [and] supervised classification methods in general...' (Hand, 1998).

The third, and final, multivariate technique discussed in this chapter, in Section 9.3, is canonical correlation analysis. This technique is appropriate when the vector of random variables $\mathbf{x}$ is divided into two parts, $\mathbf{x}_{p_1}, \mathbf{x}_{p_2}$, and the objective is to find pairs of linear functions of $\mathbf{x}_{p_1}$ and $\mathbf{x}_{p_2}$, respectively, such that the correlation between the linear functions within each pair is maximized. In this case the replacement of $\mathbf{x}_{p_1}, \mathbf{x}_{p_2}$ by some or all of the PCs of $\mathbf{x}_{p_1}, \mathbf{x}_{p_2}$, respectively, has been suggested in the literature. A number of other techniques linked to PCA that are used to investigate relationships between two groups of variables are also discussed in Section 9.3. Situations where more than two groups of variables are to be analysed are left to Section 14.5.

9.1   Discriminant Analysis

In discriminant analysis, observations may be taken from any of $G \geq 2$ populations or groups. Assumptions are made regarding the structure of these groups, namely that the random vector $\mathbf{x}$ associated with each observation is assumed to have a particular (partly or fully specified) distribution depending on its group membership. Information may also be available about the overall relative frequencies of occurrence of each group. In addition, there is usually available a set of data $\mathbf{x}_1, \mathbf{x}_2, \ldots, \mathbf{x}_n$ (the training set) for which the group membership of each observation is known. Based on the assumptions about group structure and on the training set if one is available, rules are constructed for assigning future observations to one of the G groups in some 'optimal' way, for example, so as to minimize the probability or cost of misclassification.

The best-known form of discriminant analysis occurs when there are only two populations, and x is assumed to have a multivariate normal distribution that differs between the two populations with respect to its mean but not its covariance matrix. If the means μ_1, μ_2 and the common covariance matrix Σ are known, then the optimal rule (according to several different criteria) is based on the linear discriminant function $x'\Sigma^{-1}(\mu_1 - \mu_2)$. If μ_1, μ_2, Σ are estimated from a 'training set' by $\bar{x}_1, \bar{x}_2, S_w$, respectively, then a rule based on the sample linear discriminant function $x'S_w^{-1}(\bar{x}_1 - \bar{x}_2)$ is often used. There are many other varieties of discriminant analysis (McLachlan, 1992), depending on the assumptions made regarding the population structure, and much research has been done, for example, on discriminant analysis for discrete data and on non-parametric approaches (Goldstein and Dillon, 1978; Hand, 1982).

The most obvious way of using PCA in a discriminant analysis is to reduce the dimensionality of the analysis by replacing x by the first m (high variance) PCs in the derivation of a discriminant rule. If the first two PCs account for a high proportion of the variance, they can also be used to provide a two-dimensional graphical representation of the data showing how good, or otherwise, the separation is between the groups.

The first point to be clarified is exactly what is meant by the PCs of x in the context of discriminant analysis. A common assumption in many forms of discriminant analysis is that the covariance matrix is the same for all groups, and the PCA may therefore be done on an estimate of this common *within-group* covariance (or correlation) matrix. Unfortunately, this procedure may be unsatisfactory for two reasons. First, the within-group covariance matrix may be different for different groups. Methods for comparing PCs from different groups are discussed in Section 13.5, and later in the present section we describe techniques that use PCs to discriminate between populations when equal covariance matrices are not assumed. For the moment, however, we make the equal covariance assumption.

The second, more serious, problem encountered in using PCs based on a common within-group covariance matrix to discriminate between groups is that there is no guarantee that the separation between groups will be in the direction of the high-variance PCs. This point is illustrated diagramatically in Figures 9.1 and 9.2 for two variables. In both figures the two groups are well separated, but in the first the separation is in the direction of the first PC (that is parallel to the major axis of within-group variation), whereas in the second the separation is orthogonal to this direction. Thus, the first few PCs will only be useful for discriminating between groups in the case where within- and between-group variation have the same dominant directions. If this does not occur (and in general there is no particular reason for it to

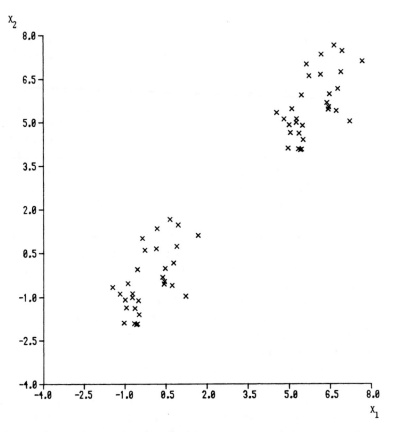

Figure 9.1. Two data sets whose direction of separation is the same as that of the first (within-group) PC.

do so) then omitting the low-variance PCs may actually throw away most of the information in **x** concerning between-group variation.

The problem is essentially the same one that arises in PC regression where, as discussed in Section 8.2, it is inadvisable to look only at high-variance PCs, as the low-variance PCs can also be highly correlated with the dependent variable. That the same problem arises in both multiple regression and discriminant analysis is hardly surprising, as linear discriminant analysis can be viewed as a special case of multiple regression in which the dependent variable is a dummy variable defining group membership (Rencher, 1995, Section 8.3).

An alternative to finding PCs from the within-group covariance matrix is mentioned by Rao (1964) and used by Chang (1983), Jolliffe et al. (1996) and Mager (1980b), among others. It ignores the group structure and calculates an overall covariance matrix based on the raw data. If the between-group variation is much larger than within-group variation, then the first few PCs for the overall covariance matrix will define directions in

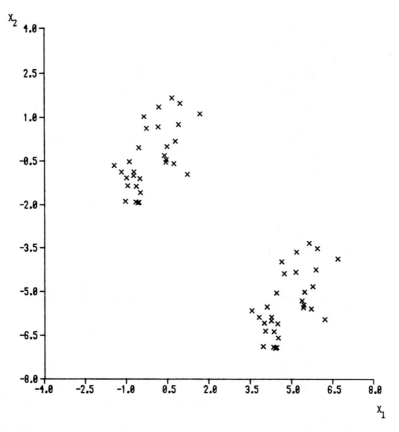

Figure 9.2. Two data sets whose direction of separation is orthogonal to that of the first (within-group) PC.

which there are large between-group differences. Such PCs therefore seem more useful than those based on within-group covariance matrices, but the technique should be used with some caution, as it will work well only if between-group variation dominates within-group variation.

It is well known that, for two completely specified normal populations, differing only in mean, the probability of misclassification using the linear discriminant function is a monotonically decreasing function of the squared Mahalanobis distance Δ^2 between the two populations (Rencher, 1998, Section 6.4). Here Δ^2 is defined as

$$\Delta^2 = (\mu_1 - \mu_2)'\Sigma^{-1}(\mu - \mu_2). \tag{9.1.1}$$

Note that we meet a number of other varieties of Mahalanobis distance elsewhere in the book. In equation (5.3.5) of Section 5.3, Mahalanobis distance *between two observations in a sample* is defined, and there is an obvious similarity between (5.3.5) and the definition given in (9.1.1) for Mahalanobis distance *between two populations*. Further modifications define Mahalanobis distance between two samples (see equation (9.1.3)), between an observation and a sample mean (see Section 10.1, below equation (10.1.2)), and between an observation and a population mean.

If we take a subset of the original p variables, then the discriminatory power of the subset can be measured by the Mahalanobis distance between the two populations in the subspace defined by the subset of variables. Chang (1983) shows that this is also true if $\boldsymbol{\Sigma}^{-1}$ is replaced in (9.1.1) by $\boldsymbol{\Psi}^{-1}$, where $\boldsymbol{\Psi} = \boldsymbol{\Sigma} + \pi(1-\pi)(\boldsymbol{\mu}_1 - \boldsymbol{\mu}_2)(\boldsymbol{\mu}_1 - \boldsymbol{\mu}_2)'$ and π is the probability of an observation coming from the ith population, $i = 1, 2$. The matrix $\boldsymbol{\Psi}$ is the overall covariance matrix for $\mathbf{x}$, ignoring the group structure. Chang (1983) shows further that the Mahalanobis distance based on the kth PC of $\boldsymbol{\Psi}$ is a monotonic, increasing function of $\theta_k = [\boldsymbol{\alpha}_k'(\boldsymbol{\mu}_1 - \boldsymbol{\mu}_2)]^2/\lambda_k$, where $\boldsymbol{\alpha}_k$, λ_k are, as usual, the vector of coefficients in the kth PC and the variance of the kth PC, respectively. Therefore, the PC with the largest discriminatory power is the one that maximizes θ_k; this will not necessarily correspond to the first PC, which maximizes λ_k. Indeed, if $\boldsymbol{\alpha}_1$ is orthogonal to $(\boldsymbol{\mu}_1 - \boldsymbol{\mu}_2)$, as in Figure 9.2, then the first PC has no discriminatory power at all. Chang (1983) gives an example in which low-variance PCs are important discriminators, and he also demonstrates that a change of scaling for the variables, for example in going from a covariance to a correlation matrix, can change the relative importance of the PCs. Townshend (1984) has an example from remote sensing in which there are seven variables corresponding to seven spectral bands. The seventh PC accounts for less than 0.1% of the total variation, but is important in discriminating between different types of land cover.

The quantity θ_k is also identified as an important parameter for discriminant analysis by Dillon et al. (1989) and Kshirsagar et al. (1990) but, like Chang (1983), they do not examine the properties of its sample analogue $\hat{\theta}_k$, where $\hat{\theta}_k$ is defined as $[\mathbf{a}_k'(\bar{\mathbf{x}}_1 - \bar{\mathbf{x}}_2)]^2/l_k$, with obvious notation. Jolliffe et al. (1996) show that a statistic which is a function of $\hat{\theta}_k$ has a t-distribution under the null hypothesis that the kth PC has equal means for the two populations.

The results of Chang (1983) and Jolliffe et al. (1996) are for two groups only, but Devijver and Kittler (1982, Section 9.6) suggest that a similar quantity to $\hat{\theta}_k$ should be used when more groups are present. Their statistic is $\tilde{\theta}_k = \mathbf{a}_k'\mathbf{S}_b\mathbf{a}_k/l_k$, where $\mathbf{S}_b$ is proportional to a quantity that generalizes $(\bar{\mathbf{x}}_1 - \bar{\mathbf{x}}_2)(\bar{\mathbf{x}}_1 - \bar{\mathbf{x}}_2)'$, namely $\sum_{g=1}^{G} n_g(\bar{\mathbf{x}}_g - \bar{\mathbf{x}})(\bar{\mathbf{x}}_g - \bar{\mathbf{x}})'$, where n_g is the number of observations in the gth group and $\bar{\mathbf{x}}$ is the overall mean of all the observations. A difference between $\hat{\theta}_k$ and $\tilde{\theta}_k$ is seen in the following: $\mathbf{a}_k$, l_k are eigenvectors and eigenvalues, respectively, for the *overall* covariance matrix in the formula for $\hat{\theta}_k$, but for the *within-group* covariance matrix $\mathbf{S}_w$ for $\tilde{\theta}_k$. Devijver and Kittler (1982) advocate ranking the PCs in terms of $\tilde{\theta}_k$ and deleting those components for which $\tilde{\theta}_k$ is smallest. Once again, this ranking will typically diverge from the ordering based on size of variance.

Corbitt and Ganesalingam (2001) also examine an extension to more than two groups based on the ideas of Dillon et al. (1989), but most of their paper consists of case studies in which only two groups are present. Corbitt and Ganesalingam (2001) show that a two-group version of their interpretation of Dillon et al.'s methodology is inferior to Jolliffe et al.'s (1996) t-tests with respect to correct classification. However, both are beaten in several of the examples studied by selecting a subset of the original variables.

Friedman (1989) demonstrates that a quantity similar to $\hat{\theta}_k$ is also relevant in the case where the within-group covariance matrices are not necessarily equal. In these circumstances a discriminant score is formed for each group, and an important part of that score is a term corresponding to $\hat{\theta}_k$, with l_k, $\mathbf{a}_k$ replaced by the eigenvalues and eigenvectors of the covariance matrix *for that group*. Friedman (1989) notes that sample estimates of large eigenvalues are biased upwards, whereas estimates of small eigenvalues are biased downwards and, because the reciprocals of the eigenvalues appear in the discriminant scores, this can lead to an exaggerated influence of the low-variance PCs in the discrimination. To overcome this, he proposes a form of 'regularized' discriminant analysis in which sample covariance matrices for each group are shrunk towards the pooled estimate $\mathbf{S}_w$. This has the effect of decreasing large eigenvalues and increasing small ones.

We return to regularized discriminant analysis later in this section, but we first note that Takemura (1985) also describes a bias in estimating eigenvalues in the context of one- and two-sample tests for multivariate normal means, based on Hotelling T^2. For two groups, the question of whether or not it is worth calculating a discriminant function reduces to testing the null hypothesis $H_0 : \boldsymbol{\mu}_1 = \boldsymbol{\mu}_2$. This is often done using Hotelling's T^2. Läuter (1996) suggests a statistic based on a subset of the PCs of the overall covariance matrix. He concentrates on the case where only the first PC is used, for which one-sided, as well as global, alternatives to H_0 may be considered.

Takemura (1985) proposes a decomposition of T^2 into contributions due to individual PCs. In the two-sample case this is equivalent to calculating t-statistics to decide which PCs discriminate between the groups corresponding to the samples, although in Takemura's case the PCs are calculated from the within-group, rather than overall, covariance matrix. Takemura (1985) suggests that later PCs might be deemed significant, and hence selected, too often. However, Jolliffe et al. (1996) dispel these worries for their tests by conducting a simulation study which shows no tendency for over-selection of the low-variance PCs in the null case, and which also gives indications of the power of the t-test when the null hypothesis of equal means in the two populations is false. Interestingly, Mason and Gunst (1985) noted bias in the opposite direction in PC regression, namely that low-variance PCs are selected less, rather than more, often than the high-variance components. Given the links between regression and discrimination, it may be that the opposite effects described by Mason and Gunst (1985) and by Takemura (1985) roughly balance in the case of using PCs in discriminant analysis.

The fact that separation between populations may be in the directions of the last few PCs does not mean that PCs should not be used at all in discriminant analysis. They can still provide a reduction of dimensionality and, as in regression, their uncorrelatedness implies that in linear discriminant analysis each PC's contribution can be assessed independently. This is an advantage compared to using the original variables $\mathbf{x}$, where the contribution of one of the variables depends on which other variables are also included in the analysis, unless all elements of $\mathbf{x}$ are uncorrelated. The main point to bear in mind when using PCs in discriminant analysis is that the best subset of PCs does not necessarily simply consist of those with the largest variances. It is easy to see, because of their uncorrelatedness, which of the PCs are best at discriminating between the populations. However, as in regression, some caution is advisable in using PCs with very low variances, because at least some of the estimated coefficients in the discriminant function will have large variances if low variance PCs are included. Many of the comments made in Section 8.2 regarding strategies for selecting PCs in regression are also relevant in linear discriminant analysis.

Some of the approaches discussed so far have used PCs from the overall covariance matrix, whilst others are based on the pooled within-group covariance matrix. This latter approach is valid for types of discriminant analysis in which the covariance structure is assumed to be the same for all populations. However, it is not always realistic to make this assumption, in which case some form of non-linear discriminant analysis may be necessary. If the multivariate normality assumption is kept, the most usual approach is *quadratic discrimination* (Rencher, 1998, Section 6.2.2). With an assumption of multivariate normality and G groups with sample means and covariance matrices $\bar{\mathbf{x}}_g, \mathbf{S}_g$, $g = 1, 2, \ldots, G$, the usual discriminant rule assigns a new vector of observations $\mathbf{x}$ to the group for which

$$(\mathbf{x} - \bar{\mathbf{x}}_g)'\mathbf{S}_g^{-1}(\mathbf{x} - \bar{\mathbf{x}}_g) + \ln(|\mathbf{S}_g|) \qquad (9.1.2)$$

is minimized. When the equal covariance assumption is made, $\mathbf{S}_g$ is replaced by the pooled covariance matrix $\mathbf{S}_w$ in (9.1.2), and only the linear part of the expression is different for different groups. In the general case, (9.1.2) is a genuine *quadratic* function of $\mathbf{x}$, leading to quadratic discriminant analysis. Flury (1995) suggests two other procedures based on his common principal component (CPC) framework, whose assumptions are intermediate compared to those of linear and quadratic discrimination. Further details will be given when the CPC framework is discussed in Section 13.5.

Alternatively, the convenience of looking only at linear functions of $\mathbf{x}$ can be kept by computing PCs separately for each population. In a number of papers (see, for example, Wold, 1976; Wold et al., 1983), Wold and others have described a method for discriminating between populations that adopts this approach. The method, called SIMCA (Soft Independent Modelling of Class Analogy), does a separate PCA for each group, and retains sufficient PCs in each to account for most of the variation within that group. The number of PCs retained will typically be different for different populations. The square of this distance for a particular population is simply the sum of squares of the values of the omitted PCs for that population, evaluated for the observation in question (Mertens, et al., 1994). The same type of quantity is also used for detecting outliers (see Section 10.1, equation (10.1.1)).

To classify a new observation, a 'distance' of the observation from the hyperplane defined by the retained PCs is calculated for each population. If future observations are to be assigned to one and only one population, then assignment is to the population from which the distance is minimized. Alternatively, a firm decision may not be required and, if all the distances are large enough, the observation can be left unassigned. As it is not close to any of the existing groups, it may be an outlier or come from a new group about which there is currently no information. Conversely, if the groups are not all well separated, some future observations may have small distances from more than one population. In such cases, it may again be undesirable to decide on a single possible class; instead two or more groups may be listed as possible 'homes' for the observation.

According to Wold et al. (1983), SIMCA works with as few as five objects from each population, although ten or more is preferable, and there is no restriction on the number of variables. This is important in many chemical problems where the number of variables can greatly exceed the number of observations. SIMCA can also cope with situations where one class is very diffuse, simply consisting of all observations that do not belong in one of a number of well-defined classes. Frank and Freidman (1989) paint a less favourable picture of SIMCA. They use a number of data sets and a simulation study to compare its performance with that of linear and quadratic discriminant analyses, with regularized discriminant analysis and with a technique called DASCO (*discriminant analysis with shrunken covariances*).

As already explained, Friedman's (1989) regularized discriminant analysis shrinks the individual within-group covariance matrices S_g towards the pooled estimate S_w in an attempt to reduce the bias in estimating eigenvalues. DASCO has a similar objective, and Frank and Freidman (1989) show that SIMCA also has a similar effect. They note that in terms of expression (9.1.2), SIMCA ignores the log-determinant term and replaces S_g^{-1} by a weighted and truncated version of its spectral decomposition, which in its full form is $S_g^{-1} = \sum_{k=1}^{p} l_{gk}^{-1} \mathbf{a}_{gk} \mathbf{a}_{gk}'$, with fairly obvious notation. If q_g PCs are retained in the gth group, then SIMCA's replacement for S_g^{-1} is $\sum_{k=q_g+1}^{p} \bar{l}_g^{-1} \mathbf{a}_{gk} \mathbf{a}_{gk}'$, where $\bar{l}_g = \dfrac{\sum_{k=q_g+1}^{p} l_{gk}}{(p-q_g)}$. DASCO treats the last $(p-q_g)$ PCs in the same way as SIMCA, but adds the terms from the spectral decomposition corresponding to the first q_g PCs. Thus it replaces $\mathbf{S}_g^{-1}$ in (9.1.2) by

$$\sum_{k=1}^{q_g} l_{gk}^{-1} \mathbf{a}_{gk} \mathbf{a}_{gk}' + \sum_{k=q_g+1}^{p} \bar{l}_g^{-1} \mathbf{a}_{gk} \mathbf{a}_{gk}'.$$

Another difference between DASCO and SIMCA is that DASCO retains the log-determinant term in (9.1.2).

In Frank and Freidman's (1989) simulation studies and data analyses, SIMCA is outperformed by both DASCO and regularized discriminant analysis in many circumstances, especially when the covariance structures are different in different groups. This is perhaps not surprising, given its omission of the log-determinant term from (9.1.2). The absence of the first q_g PCs from SIMCA's measure of discrepancy of an observation from a group also means that it is unlikely to do well when the groups differ in the directions of these PCs (Mertens et al., 1994). These latter authors treat SIMCA's measure of discrepancy between an observation and a group as an indication of the outlyingness of the observation with respect to the group, and suggest modifications of SIMCA in which other outlier detection measures are used (see Section 10.1).

A similar idea to SIMCA is suggested by Asselin de Beauville (1995). As with SIMCA, separate PCAs are done for each group, but here an observation is assigned on the basis of a measure that combines the smallest distance of the observation from an axis defining a PC for a group and its score on that PC.

It has been noted above that discriminant analysis can be treated as a multiple regression problem, with dummy variables, corresponding to the group membership, as dependent variables. Other regression techniques, as well as PC regression, can therefore be adapted for the discriminant problem. In particular, partial least squares (PLS — see Section 8.4), which is, in a sense, a compromise between least squares and PC regression, can be used in the discriminant context (Vong et al. (1990)).

SIMCA calculates PCs separately within each group, compared with the more usual practice of finding PCs from the overall covariance matrix, or from a pooled within-group covariance matrix. Yet another type of PC can be derived from the between-group covariance matrix $\mathbf{S}_b$. However, if the dominant direction of $\mathbf{S}_b$ coincides with that of the within-group covariance matrix $\mathbf{S}_w$, there may be better discriminatory power in a different direction, corresponding to a low-variance direction for $\mathbf{S}_w$. This leads Devijver and Kittler (1982, Section 9.7) to 'prewhiten' the data, using $\mathbf{S}_w$, before finding PCs with respect to $\mathbf{S}_b$. This is equivalent to the well-known procedure of canonical variate analysis or canonical discriminant analysis, in which uncorrelated linear functions are found that discriminate as well as possible between the groups. Canonical variates are defined as $\boldsymbol{\gamma}_1'\mathbf{x}, \boldsymbol{\gamma}_2'\mathbf{x}, \ldots, \boldsymbol{\gamma}_{g-1}'\mathbf{x}$ where $\boldsymbol{\gamma}_k'\mathbf{x}$ maximizes the ratio $\frac{\boldsymbol{\gamma}'\mathbf{S}_b\boldsymbol{\gamma}}{\boldsymbol{\gamma}'\mathbf{S}_w\boldsymbol{\gamma}}$, of between- to within-group variance of $\boldsymbol{\gamma}'\mathbf{x}$, subject to being uncorrelated with $\boldsymbol{\gamma}_1'\mathbf{x}$, $\boldsymbol{\gamma}_2'\mathbf{x}, \ldots, \boldsymbol{\gamma}_{k-1}'\mathbf{x}$. For more details see, for example, McLachlan (1992, Section 9.2) or Mardia et al. (1979, Section 12.5).

A variation on prewhitening using $\mathbf{S}_w$ is to 'standardize' $\mathbf{S}_b$ by dividing each variable by its within-group standard deviation and then calculate a between-group covariance matrix $\mathbf{S}_b^*$ from these rescaled variables. Finding PCs based on $\mathbf{S}_b^*$ is called *discriminant principal components analysis* by Yendle and MacFie (1989). They compare the results of this analysis with those from a PCA based on $\mathbf{S}_b$ and also with canonical discriminant analysis in which the variables $\mathbf{x}$ are replaced by their PCs. In two examples, Yendle and MacFie (1989) find, with respect to misclassification probabilities, the performance of PCA based on $\mathbf{S}_b^*$ to be superior to that based on $\mathbf{S}_b$, and comparable to that of canonical discriminant analysis using PCs. However, in the latter analysis only the restrictive special case is examined, in which the *first* q PCs are retained. It is also not made explicit whether the PCs used are obtained from the overall covariance matrix $\mathbf{S}$ or from $\mathbf{S}_w$, though it seems likely that $\mathbf{S}$ is used.

To conclude this section, we note some relationships between PCA and canonical discriminant analysis, via principal coordinate analysis (see Section 5.2), which are described by Gower (1966). Suppose that a principal coordinate analysis is done on a distance matrix whose elements are Mahalanobis distances between the samples from the G populations. These distances are defined as the square roots of

$$\delta_{hi}^2 = (\bar{\mathbf{x}}_h - \bar{\mathbf{x}}_i)'\mathbf{S}_w^{-1}(\bar{\mathbf{x}}_h - \bar{\mathbf{x}}_i); \qquad h, i = 1, 2, \ldots, G. \qquad (9.1.3)$$

Gower (1966) then shows that the configuration found in $m(< G)$ dimensions is the same as that provided by the first m canonical variates from canonical discriminant analysis. Furthermore, the same results may be found from a PCA with $\mathbf{X}'\mathbf{X}$ replaced by $(\bar{\mathbf{X}}\mathbf{W})'(\bar{\mathbf{X}}\mathbf{W})$, where $\mathbf{W}\mathbf{W}' = \mathbf{S}_w^{-1}$ and $\bar{\mathbf{X}}$ is the $(G \times p)$ matrix whose hth row gives the sample means of the p variables for the hth population, $h = 1, 2, \ldots, G$. Yet another, related, way of finding canonical variates is via a two-stage PCA, as described by Campbell and Atchley (1981). At the first stage PCs are found, based on the within-group covariance matrix $\mathbf{S}_w$, and standardized to have unit variance. The values of the means of these standardised PCs for each of the G groups are then subjected to a weighted PCA (see Section 14.2.1), with weights proportional to the sample sizes n_i in each group. The PC scores at this second stage are the values of the group means with respect to the canonical variates. Krzanowski (1990) generalizes canonical discriminant analysis, based on the common PCA model due to Flury (1988), using this two-stage derivation. Bensmail and Celeux (1996) also describe an approach to discriminant analysis based on the common PCA framework; this will be discussed further in Section 13.5. Campbell and Atchley (1981) note the possibility of an alternative analysis, different from canonical discriminant analysis, in which the PCA at the second of their two stages is unweighted.

Because of the connections between PCs and canonical variates, Mardia et al. (1979, p. 344) refer to canonical discriminant analysis as the analogue for grouped data of PCA for ungrouped data. Note also that with an appropriate choice of metric, generalized PCA as defined in Section 14.2.2 is equivalent to a form of discriminant analysis.

No examples have been given in detail in this section, although some have been mentioned briefly. An interesting example, in which the objective is to discriminate between different carrot cultivars, is presented by Horgan et al. (2001). Two types of data are available, namely the positions of 'landmarks' on the outlines of the carrots and the brightness of each pixel in a cross-section of each carrot. The two data sets are subjected to separate PCAs, and a subset of PCs taken from both analyses is used to construct a discriminant function.

9.2   Cluster Analysis

In cluster analysis, it is required to divide a set of observations into groups or clusters in such a way that most pairs of observations that are placed in the same group are more similar to each other than are pairs of observations that are placed into two different clusters. In some circumstances, it may be expected or hoped that there is a clear-cut group structure underlying the data, so that each observation comes from one of several distinct populations, as in discriminant analysis. The objective then is to determine this group structure where, in contrast to discriminant analysis, there is little or no prior information about the form that the structure takes. Cluster analysis can also be useful when there is no clear group structure in the data. In this case, it may still be desirable to segment or *dissect* (using the terminology of Kendall (1966)) the observations into relatively homogeneous groups, as observations within the same group may be sufficiently similar to be treated identically for the purpose of some further analysis, whereas this would be impossible for the whole heterogeneous data set. There are very many possible methods of cluster analysis, and several books have appeared on the subject, for example Aldenderfer and Blashfield (1984), Everitt et al. (2001), Gordon (1999). Most methods can be used either for detection of clear-cut groups or for dissection/segmentation, although there is increasing interest in mixture models, which explicitly assume the existence of clusters (see Section 9.2.3).

The majority of cluster analysis techniques require a measure of similarity or dissimilarity between each pair of observations, and PCs have been used quite extensively in the computation of one type of dissimilarity. If the p variables that are measured for each observation are quantitative and in similar units, then an obvious measure of dissimilarity between two observations is the Euclidean distance between the observations in the p-

dimensional space defined by the variables. If the variables are measured in non-compatible units, then each variable can be standardized by dividing by its standard deviation, and an arbitrary, but obvious, measure of dissimilarity is then the Euclidean distance between a pair of observations in the p-dimensional space defined by the standardized variables.

Suppose that a PCA is done based on the covariance or correlation matrix, and that m $(< p)$ PCs account for most of the variation in $\mathbf{x}$. A possible alternative dissimilarity measure is the Euclidean distance between a pair of observations in the m-dimensional subspace defined by the first m PCs; such dissimilarity measures have been used in several published studies, for example Jolliffe et al. (1980). There is often no real advantage in using this measure, rather than the Euclidean distance in the original p-dimensional space, as the Euclidean distance calculated using all p PCs from the covariance matrix is identical to that calculated from the original variables. Similarly, the distance calculated from all p PCs for the correlation matrix is the same as that calculated from the p standardized variables. Using m instead of p PCs simply provides an approximation to the original Euclidean distance, and the extra calculation involved in finding the PCs far outweighs any saving which results from using m instead of p variables in computing the distance. However, if, as in Jolliffe et al. (1980), the PCs are being calculated in any case, the reduction from p to m variables may be worthwhile.

In calculating Euclidean distances, the PCs have the usual normalization, so that the sample variance of $\mathbf{a}'_k\mathbf{x}$ is l_k, $k = 1, 2, \ldots, p$ and $l_1 \geq l_2 \geq \cdots \geq l_p$, using the notation of Section 3.1. As an alternative, a distance can be calculated based on PCs that have been renormalized so that each PC has the same variance. This renormalization is discussed further in the context of outlier detection in Section 10.1. In the present setting, where the objective is the calculation of a dissimilarity measure, its use is based on the following idea. Suppose that one of the original variables is almost independent of all the others, but that several of the remaining variables are measuring essentially the same property as each other. Euclidean distance will then give more weight to this property than to the property described by the 'independent' variable. If it is thought desirable to give equal weight to each property then this can be achieved by finding the PCs and then giving equal weight to each of the first m PCs.

To see that this works, consider a simple example in which four meteorological variables are measured. Three of the variables are temperatures, namely air temperature, sea surface temperature and dewpoint, and the fourth is the height of the cloudbase. The first three variables are highly correlated with each other, but nearly independent of the fourth. For a sample of 30 measurements on these variables, a PCA based on the correlation matrix gave a first PC with variance 2.95, which is a nearly equally weighted average of the three temperature variables. The second PC, with variance 0.99 is dominated by cloudbase height, and together the first two PCs account for 98.5% of the total variation in the four variables.

Euclidean distance based on the first two PCs gives a very close approximation to Euclidean distance based on all four variables, but it gives roughly three times as much weight to the first PC as to the second. Alternatively, if the first two PCs are renormalized to have equal weight, this implies that we are treating the *one* measurement of cloudbase height as being equal in importance to the *three* measurements of temperature.

In general, if Euclidean distance is calculated using all p renormalized PCs, then this is equivalent to calculating the Mahalanobis distance for the original variables (see Section 10.1, below equation (10.1.2), for a proof of the corresponding property for Mahalanobis distances of observations from sample means, rather than between pairs of observations). Mahalanobis distance is yet another plausible dissimilarity measure, which takes into account the variances and covariances between the elements of **x**. Naes and Isaksson (1991) give an example of (fuzzy) clustering in which the distance measure is based on Mahalanobis distance, but is truncated to exclude the last few PCs when the variances of these are small and unstable.

Regardless of the similarity or dissimilarity measure adopted, PCA has a further use in cluster analysis, namely to give a two-dimensional representation of the observations (see also Section 5.1). Such a two-dimensional representation can give a simple visual means of either detecting or verifying the existence of clusters, as noted by Rao (1964), provided that most of the variation, and in particular the between-cluster variation, falls in the two-dimensional subspace defined by the first two PCs.

Of course, the same problem can arise as in discriminant analysis, namely that the between-cluster variation may be in directions other than those of the first two PCs, even if these two PCs account for nearly all of the total variation. However, this behaviour is generally less likely in cluster analysis, as the PCs are calculated for the whole data set, not within-groups. As pointed out in Section 9.1, if between-cluster variation is much greater than within-cluster variation, such PCs will often successfully reflect the cluster structure. It is, in any case, frequently impossible to calculate within-group PCs in cluster analysis as the group structure is usually unknown a priori.

It can be argued that there are often better directions than PCs in which to view the data in order to 'see' structure such as clusters. Projection pursuit includes a number of ideas for finding such directions, and will be discussed in Section 9.2.2. However, the examples discussed below illustrate that plots with respect to the first two PCs can give suitable two-dimensional representations on which to view the cluster structure if a clear structure exists. Furthermore, in the case where there is no clear structure, but it is required to dissect the data using cluster analysis, there can be no real objection to the use of a plot with respect to the first two PCs. If we wish to view the data in two dimensions in order to see whether a set of clusters given by some procedure 'looks sensible,' then the first two PCs give the best possible representation in two dimensions in the sense defined by Property G3 of Section 3.2.

Before looking at examples of the uses just described of PCA in cluster analysis, we discuss a rather different way in which cluster analysis can be used and its connections with PCA. So far we have discussed cluster analysis on observations or individuals, but in some circumstances it is desirable to divide variables, rather than observations, into groups. In fact, by far the earliest book on cluster analysis (Tryon, 1939) was concerned with this type of application. Provided that a suitable measure of similarity between variables can be defined—the correlation coefficient is an obvious candidate—methods of cluster analysis used for observations can be readily adapted for variables.

One connection with PCA is that when the variables fall into well-defined clusters, then, as discussed in Section 3.8, there will be one high-variance PC and, except in the case of 'single-variable' clusters, one or more low-variance PCs associated with each cluster of variables. Thus, PCA will identify the presence of clusters among the variables, and can be thought of as a competitor to standard cluster analysis of variables. The use of PCA in this way in fairly common in climatology (see, for example, Cohen (1983), White et al. (1991), Romero et al. (1999)). In an analysis of a climate variable recorded at stations over a large geographical area, the loadings of the PCs at the various stations can be used to divide the area into regions with high loadings on each PC. In fact, this regionalization procedure is usually more effective if the PCs are rotated (see Section 11.1) so that most analyses are done using rotated loadings.

Identifying clusters of variables may be of general interest in investigating the structure of a data set but, more specifically, if we wish to reduce the number of variables without sacrificing too much information, then we could retain one variable from each cluster. This is essentially the idea behind some of the variable selection techniques based on PCA that were described in Section 6.3.

Hastie et al. (2000) describe a novel clustering procedure for 'variables' which uses PCA applied in a genetic context. They call their method 'gene shaving.' Their data consist of $p = 4673$ gene expression measurements for $n = 48$ patients, and the objective is to classify the 4673 genes into groups that have coherent expressions. The first PC is found for these data and a proportion of the genes (typically 10%) having the smallest absolute inner products with this PC are deleted (shaved). PCA followed by shaving is repeated for the reduced data set, and this procedure continues until ultimately only one gene remains. A nested sequence of subsets of genes results from this algorithm and an optimality criterion is used to decide which set in the sequence is best. This gives the first cluster of genes. The whole procedure is then repeated after centering the data with respect to the 'average gene expression' in the first cluster, to give a second cluster and so on.

Another way of constructing clusters of variables, which simultaneously finds the first PC within each cluster, is proposed by Vigneau and Qannari

(2001). Suppose that the p variables are divided into G groups or clusters, and that $\mathbf{x}_g$ denotes the vector of variables in the gth group, $g = 1, 2, \ldots, G$. Vigneau and Qannari (2001) seek vectors $\mathbf{a}_{11}, \mathbf{a}_{21}, \ldots, \mathbf{a}_{G1}$ that maximize $\sum_{g=1}^{G} \text{var}(\mathbf{a}_{g1}'\mathbf{x}_g)$, where $\text{var}(\mathbf{a}_{g1}'\mathbf{x}_g)$ is the sample variance of the linear function $\mathbf{a}_{g1}'\mathbf{x}_g$. This sample variance is clearly maximized by the first PC for the variables in the gth group, but simultaneously we wish to find the partition of the variables into G groups for which the sum of these variances is maximized. An iterative procedure is presented by Vigneau and Qannari (2001) for solving this problem.

The formulation of the problem assumes that variables with large squared correlations with the first PC in a cluster should be assigned to that cluster. Vigneau and Qannari consider two variations of their technique. In the first, the signs of the correlations between variables and PCs are important; only those variables with large *positive* correlations with a PC should be in its cluster. In the second, relationships with external variables are taken into account.

9.2.1 Examples

Only one example will be described in detail here, although a number of other examples that have appeared elsewhere will be discussed briefly. In many of the published examples where PCs have been used in conjunction with cluster analysis, there is no clear-cut cluster structure, and cluster analysis has been used as a dissection technique. An exception is the well-known example given by Jeffers (1967), which was discussed in the context of variable selection in Section 6.4.1. The data consist of 19 variables measured on 40 aphids and, when the 40 observations are plotted with respect to the first two PCs, there is a strong suggestion of four distinct groups; refer to Figure 9.3, on which convex hulls (see Section 5.1) have been drawn around the four suspected groups. It is likely that the four groups indicated on Figure 9.3 correspond to four different species of aphids; these four species cannot be readily distinguished using only one variable at a time, but the plot with respect to the first two PCs clearly distinguishes the four populations.

The example introduced in Section 1.1 and discussed further in Section 5.1.1, which has seven physical measurements on 28 students, also shows (in Figures 1.3, 5.1) how a plot with respect to the first two PCs can distinguish two groups, in this case men and women. There is, unlike the aphid data, a small amount of overlap between groups and if the PC plot is used to identify, rather than verify, a cluster structure, then it is likely that some misclassification between sexes will occur. A simple but specialized use of PC scores, one PC at a time, to classify seabird communities is described by Huettmann and Diamond (2001).

In the situation where cluster analysis is used for dissection, the aim of a two-dimensional plot with respect to the first two PCs will almost always be

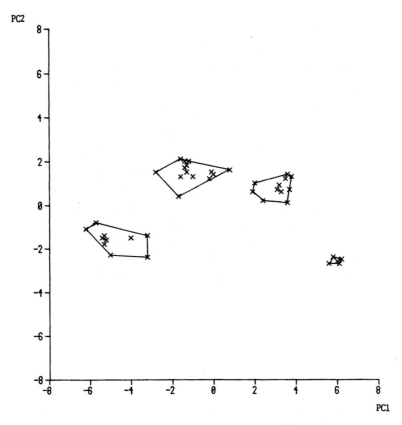

Figure 9.3. Aphids: plot with respect to the first two PCs showing four groups corresponding to species.

to verify that a given dissection 'looks' reasonable, rather than to attempt to identify clusters. An early example of this type of use was given by Moser and Scott (1961), in their Figure 9.2. The PCA in their study, which has already been mentioned in Section 4.2, was a stepping stone on the way to a cluster analysis of 157 British towns based on 57 variables. The PCs were used both in the construction of a distance measure, and as a means of displaying the clusters in two dimensions.

Principal components are used in cluster analysis in a similar manner in other examples discussed in Section 4.2, details of which can be found in Jolliffe et al. (1980, 1982a, 1986), Imber (1977) and Webber and Craig (1978). Each of these studies is concerned with demographic data, as is the example described next in detail.

Demographic Characteristics of English Counties

In an unpublished undergraduate dissertation, Stone (1984) considered a cluster analysis of 46 English counties. For each county there were 12

Table 9.1. Demographic variables used in the analysis of 46 English counties.

1.	Population density—numbers per hectare
2.	Percentage of population aged under 16
3.	Percentage of population above retirement age
4.	Percentage of men aged 16–65 who are employed
5.	Percentage of men aged 16–65 who are unemployed
6.	Percentage of population owning their own home
7.	Percentage of households which are 'overcrowded'
8.	Percentage of employed men working in industry
9.	Percentage of employed men working in agriculture
10.	(Length of public roads)/(area of county)
11.	(Industrial floor space)/(area of county)
12.	(Shops and restaurant floor space)/(area of county)

Table 9.2. Coefficients and variances for the first four PCs: English counties data.

Component number		1	2	3	4
	1	0.35	−0.19	0.29	0.06
	2	0.02	0.60	−0.03	0.22
	3	−0.11	−0.52	−0.27	−0.36
	4	−0.30	0.07	0.59	−0.03
	5	0.31	0.05	−0.57	0.07
Variable	6	−0.29	0.09	−0.07	−0.59
	7	0.38	0.04	0.09	0.08
	8	0.13	0.50	−0.14	−0.34
	9	−0.25	−0.17	−0.28	0.51
	10	0.37	−0.09	0.09	−0.18
	11	0.34	0.02	−0.00	−0.24
	12	0.35	−0.20	0.24	0.07
Eigenvalue		6.27	2.53	1.16	0.96
Cumulative percentage of total variation		52.3	73.3	83.0	90.9

demographic variables, which are listed in Table 9.1.

The objective of Stone's analysis, namely dissection of local authority areas into clusters, was basically the same as that in other analyses by Imber (1977), Webber and Craig (1978) and Jolliffe et al. (1986), but these various analyses differ in the variables used and in the local authorities considered. For example, Stone's list of variables is shorter than those of the other analyses, although it includes some variables not considered by any of the others. Also, Stone's list of local authorities includes large metropolitan counties such as Greater London, Greater Manchester and Merseyside as single entities, whereas these large authorities are subdivided into smaller areas in the other analyses. A comparison of the clusters obtained from several different analyses is given by Jolliffe et al. (1986).

As in other analyses of local authorities, PCA is used in Stone's analysis in two ways: first, to summarize and explain the major sources of variation in the data, and second, to provide a visual display on which to judge the adequacy of the clustering.

Table 9.2 gives the coefficients and variances for the first four PCs using the correlation matrix for Stone's data. It is seen that the first two components account for 73% of the total variation, but that most of the relevant rules of Section 6.1 would retain four components (the fifth eigenvalue is 0.41).

There are fairly clear interpretations for each of the first three PCs. The first PC provides a contrast between urban and rural areas, with positive coefficients for variables that are high in urban areas, such as densities of population, roads, and industrial and retail floor space; negative coefficients occur for owner occupation, percentage of employed men in agriculture, and overall employment level, which at the time of the study tended to be higher in rural areas. The main contrast for component 2 is between the percentages of the population below school-leaving age and above retirement age. This component is therefore a measure of the age of the population in each county, and it identifies, at one extreme, the south coast retirement areas.

The third PC contrasts employment and unemployment rates. This contrast is also present in the first urban versus rural PC, so that the third PC is measuring variation in employment/unemployment rates within rural areas and within urban areas, rather than between the two types of area.

Turning now to the cluster analysis of the data, Stone (1984) examines several different clustering methods, and also considers the analysis with and without Greater London, which is very different from any other area, but whose omission produces surprisingly little change. Figure 9.4 shows the position of the 46 counties with respect to the first two PCs, with the four-cluster solution obtained using complete-linkage cluster analysis (see Gordon, 1999, p. 85) indicated by different symbols for different clusters. The results for complete-linkage are fairly similar to those found by several of the other clustering methods investigated.

In the four-cluster solution, the single observation at the bottom left of

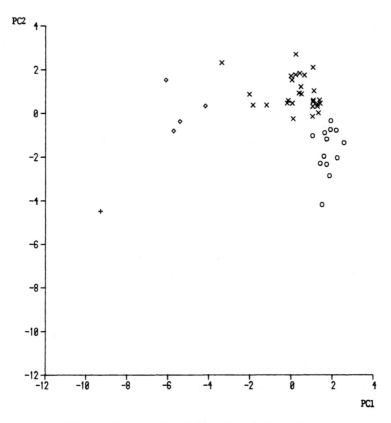

Figure 9.4. English counties: complete-linkage four-cluster solution superimposed on a plot of the first two PCs.

the diagram is Greater London and the four-county cluster at the top left consists of other metropolitan counties. The counties at the right of the diagram are more rural, confirming the interpretation of the first PC given earlier. The split between the larger groups at the right of the plot is rather more arbitrary but, as might be expected from the interpretation of the second PC, most of the retirement areas have similar values in the vertical direction; they are all in the bottom part of the diagram. Conversely, many of the counties towards the top have substantial urban areas within them, and so have somewhat lower values on the first PC as well.

The clusters are rather nicely located in different areas of the figure, although the separation between them is not particularly clear-cut, except for Greater London. This behaviour is fairly similar to what occurs for other clustering methods in this example, and for different numbers of clusters. For example, in the eight-cluster solution for complete-linkage clustering, one observation splits off from each of the clusters in the top left and bottom right parts of the diagram to form single-county clusters. The large 27-

county cluster in the top right of the plot splits into three groups containing 13, 10 and 4 counties, with some overlap between them.

This example is typical of many in which cluster analysis is used for dissection. Examples like that of Jeffers' (1967) aphids, where a very clear-cut and previously unknown cluster structure is uncovered, are relatively unusual, although another illustration is given by Blackith and Reyment (1971, p. 155). In their example, a plot of the observations with respect to the second and third (out of seven) PCs shows a very clear separation into two groups. It is probable that in many circumstances 'projection-pursuit' methods, which are discussed next, will provide a better two-dimensional space in which to view the results of a cluster analysis than that defined by the first two PCs. However, if dissection rather than discovery of a clear-cut cluster structure is the objective of a cluster analysis, then there is likely to be little improvement over a plot with respect to the first two PCs.

9.2.2 Projection Pursuit

As mentioned earlier in this chapter, it may be possible to find low-dimensional representations of a data set that are better than the first few PCs at displaying 'structure' in the data. One approach to doing this is to define structure as 'interesting' and then construct an index of 'interesting-ness,' which is successively maximized. This is the idea behind projection pursuit, with different indices leading to different displays. If 'interesting' is defined as 'large variance,' it is seen that PCA is a special case of projection pursuit. However, the types of structure of interest are often clusters or outliers, and there is no guarantee that the high-variance PCs will find such features. The term 'projection pursuit' dates back to Friedman and Tukey (1974), and a great deal of work was done in the early 1980s. This is described at length in three key papers: Friedman (1987), Huber (1985), and Jones and Sibson (1987). The last two both include extensive discussion, in addition to the paper itself. Some techniques are good at finding clusters, whereas others are better at detecting outliers.

Most projection pursuit techniques start from the premise that the least interesting structure is multivariate normality, so that deviations from normality form the basis of many indices. There are measures based on skewness and kurtosis, on entropy, on looking for deviations from uniformity in transformed data, and on finding 'holes' in the data. More recently, Foster (1998) suggested looking for directions of high density, after 'sphering' the data to remove linear structure. Sphering operates by transforming the variables $\mathbf{x}$ to $\mathbf{z} = \mathbf{S}^{-\frac{1}{2}}(\mathbf{x} - \bar{\mathbf{x}})$, which is equivalent to converting to PCs, which are then standardized to have zero mean and unit variance. Friedman (1987) also advocates sphering as a first step in his version of projection pursuit. After identifying the high-density directions for the sphered data, Foster (1998) uses the inverse transformation to discover the nature of the interesting structures in terms of the original variables.

Projection pursuit indices usually seek out deviations from multivariate normality. Bolton and Krzanowski (1999) show that if normality holds then PCA finds directions for which the maximized likelihood is minimized. They interpret this result as PCA choosing interesting directions to be those for which normality is least likely, thus providing a link with the ideas of projection pursuit. A different projection pursuit technique with an implicit assumption of normality is based on the fixed effects model of Section 3.9. Recall that the model postulates that, apart from an error term $\mathbf{e}_i$ with $\text{var}(\mathbf{e}_i) = \frac{\sigma^2}{w_i}\boldsymbol{\Gamma}$, the variables $\mathbf{x}$ lie in a q-dimensional subspace. To find the best-fitting subspace, $\sum_{i=1}^{n} w_i \|\mathbf{x}_i - \mathbf{z}_i\|_{\mathbf{M}}^2$ is minimized for an appropriately chosen metric $\mathbf{M}$. For multivariate normal $\mathbf{e}_i$ the optimal choice for $\mathbf{M}$ is $\boldsymbol{\Gamma}^{-1}$. Given a structure of clusters in the data, all w_i equal, and $\mathbf{e}_i$ describing variation *within* clusters, Caussinus and Ruiz (1990) suggest a robust estimate of $\boldsymbol{\Gamma}$, defined by

$$\hat{\boldsymbol{\Gamma}} = \frac{\sum_{i=1}^{n-1} \sum_{j=i+1}^{n} K[\|\mathbf{x}_i - \mathbf{x}_j\|_{\mathbf{S}^{-1}}^2](\mathbf{x}_i - \mathbf{x}_j)(\mathbf{x}_i - \mathbf{x}_j)'}{\sum_{i=1}^{n-1} \sum_{j=i+1}^{n} K[\|\mathbf{x}_i - \mathbf{x}_j\|_{\mathbf{S}^{-1}}^2]}, \quad (9.2.1)$$

where $K[.]$ is a decreasing positive real function (Caussinus and Ruiz, 1990, use $K[d] = e^{-\frac{\beta}{2}t}$ for $\beta > 0$) and $\mathbf{S}$ is the sample covariance matrix. The best fit is then given by finding eigenvalues and eigenvectors of $\mathbf{S}\hat{\boldsymbol{\Gamma}}^{-1}$, which is a type of generalized PCA (see Section 14.2.2). There is a similarity here with canonical discriminant analysis (Section 9.1), which finds eigenvalues and eigenvectors of $\mathbf{S}_b\mathbf{S}_w^{-1}$, where $\mathbf{S}_b$, $\mathbf{S}_w$ are between and within-group covariance matrices. In Caussinus and Ruiz's (1990) form of projection pursuit, $\mathbf{S}$ is the overall covariance matrix, and $\hat{\boldsymbol{\Gamma}}$ is an estimate of the within-group covariance matrix. Equivalent results would be obtained if $\mathbf{S}$ were replaced by an estimate of between-group covariance, so that the only real difference from canonical discriminant analysis is that the groups are known in the latter case but are unknown in projection pursuit. Further theoretical details and examples of Caussinus and Ruiz's technique can be found in Caussinus and Ruiz-Gazen (1993, 1995). The choice of values for β is discussed, and values in the range 0.5 to 3.0 are recommended. There is a link between Caussinus and Ruiz-Gazen's technique and the mixture models of Section 9.2.3. In discussing theoretical properties of their technique, they consider a framework in which clusters arise from a mixture of multivariate normal distributions. The q dimensions of the underlying model correspond to q clusters and $\boldsymbol{\Gamma}$ represents 'residual' or within-group covariance.

Although not projection pursuit as such, Krzanowski (1987b) also looks for low-dimensional representations of the data that preserve structure, but in the context of variable selection. Plots are made with respect to the first two PCs calculated from only a subset of the variables. A criterion for choosing a subset of k variables is based on how closely a Procrustes rotation of the configuration of scores on the first q PCs for a selected subset of variables matches the corresponding configuration based on the first q PCs for the full set of variables. It is shown that the visibility of group structure may be enhanced by plotting with respect to PCs that are calculated from only a subset of variables. The selected variables differ from those chosen by the methods of Section 6.3 (Krzanowski, 1987b; Jmel, 1992), which illustrates again that different selection rules are needed, depending on the purpose for which the variables are chosen (Jolliffe, 1987a).

Some types of projection pursuit are far more computationally demanding than PCA, and for large data sets an initial reduction of dimension may be necessary before they can be implemented. In such cases, Friedman (1987) suggests reducing dimensionality by retaining only the high-variance PCs from a data set and conducting the projection pursuit on those. Caussinus (1987) argues that an initial reduction of dimensionality using PCA may be useful even when there are no computational problems.

9.2.3 Mixture Models

Cluster analysis traditionally consisted of a wide variety of rather ad hoc descriptive techniques, with little in the way of statistical underpinning. The consequence was that it was fine for dissection, but less satisfactory for deciding whether clusters actually existed and, if so, how many there were. An attractive alternative approach is to model the cluster structure by a mixture model, in which the probability density function (p.d.f.) for the vector of variables $\mathbf{x}$ is expressed as

$$f(\mathbf{x}; \boldsymbol{\theta}) = \sum_{g=1}^{G} \pi_g f_g(\mathbf{x}; \boldsymbol{\theta}_g), \qquad (9.2.2)$$

where G is the number of clusters, π_g is the probability of an observation coming from the gth cluster, $f_g(\mathbf{x}; \boldsymbol{\theta}_g)$ is the p.d.f. in the gth cluster, and $\boldsymbol{\theta}' = (\boldsymbol{\theta}_1', \boldsymbol{\theta}_2', \ldots, \boldsymbol{\theta}_G')$ is a vector of parameters that must be estimated. A particular form needs to be assumed for each p.d.f. $f_g(\mathbf{x}; \boldsymbol{\theta}_g)$, the most usual choice being multivariate normality in which $\boldsymbol{\theta}_g$ consists of the mean vector $\boldsymbol{\mu}_g$, and the covariance matrix $\boldsymbol{\Sigma}_g$, for the gth cluster.

The problem of fitting a model such as (9.2.2) is difficult, even for small values of p and G, so the approach was largely ignored at the time when many clustering algorithms were developed. Later advances in theory, in computational sophistication, and in computing speed made it possible for versions of (9.2.2) to be developed and fitted, especially in the univariate case (see, for example, Titterington et al. (1985); McLachlan and Bashford (1988); Böhning (1999)). However, many multivariate problems are still intractable because of the large number of parameters that need to be estimated. For example, in a multivariate normal mixture the total number of parameters in the π_g, $\boldsymbol{\mu}_g$ and $\boldsymbol{\Sigma}_g$ is $\frac{1}{2}G[p^2 + 3p + 2] - 1$. To overcome this intractability, the elements of the $\boldsymbol{\Sigma}_g$ can be constrained in some way, and a promising approach was suggested by Tipping and Bishop (1999b), based on their probabilistic model for PCA which is described in Section 3.9. In this approach, the p.d.f.s in (9.2.2) are replaced by p.d.f.s derived from Tipping and Bishop's (1999a,b) model. These p.d.f.s are multivariate normal, but instead of having general covariance matrices, the matrices take the form $\mathbf{B}_g\mathbf{B}_g' + \sigma_g^2\mathbf{I}_p$, where $\mathbf{B}_g$ is a $(p \times q)$ matrix, and q $(< p)$ is the same for all groups. This places constraints on the covariance matrices, but the constraints are not as restrictive as the more usual ones, such as equality or diagonality of matrices. Tipping and Bishop (1999b) describe a two-stage EM algorithm for finding maximum likelihood estimates of all the parameters in the model. As with Tipping and Bishop's (1999a) single population model, it turns out the columns of $\mathbf{B}_g$ define the space spanned by the first q PCs, this time within each cluster. There remains the question of the choice of q, and there is still a restriction to multivariate normal distributions for each cluster, but Tipping and Bishop (1999b) provide examples where the procedure gives clear improvements compared to the imposition of more standard constraints on the $\boldsymbol{\Sigma}_g$. Bishop (1999) outlines how a Bayesian version of Tipping and Bishop's (1999a) model can be extended to mixtures of distributions.

9.3 Canonical Correlation Analysis and Related Techniques

Canonical correlation analysis (CCA) is the central topic in this section. Here the variables are in two groups, and relationships are sought *between* these groups of variables. CCA is probably the most commonly used technique for tackling this objective. The emphasis in this section, as elsewhere in the book, is on how PCA can be used with, or related to, the technique. A number of other methods have been suggested for investigating relationships between groups of variables. After describing CCA, and illustrating it with an example, some of these alternative approaches are briefly described, with their connections to PCA again highlighted. Discussion of techniques which analyse more than two sets of variables simultaneously is largely deferred to Section 14.5.

9.3.1 Canonical Correlation Analysis

Suppose that $\mathbf{x}_{p_1}$, $\mathbf{x}_{p_2}$ are vectors of random variables with p_1, p_2 elements, respectively. The objective of canonical correlation analysis (CCA) is to find successively for $k = 1, 2, \ldots, \min[p_1, p_2]$, pairs $\{\mathbf{a}_{k1}'\mathbf{x}_{p_1}, \mathbf{a}_{k2}'\mathbf{x}_{p_2}\}$ of linear functions of $\mathbf{x}_{p_1}$, $\mathbf{x}_{p_2}$, respectively, called *canonical variates*, such that the correlation between $\mathbf{a}'_{k1}\mathbf{x}_{p_1}$ and $\mathbf{a}'_{k2}\mathbf{x}_{p_2}$ is maximized, subject to $\mathbf{a}'_{k1}\mathbf{x}_{p_1}$, $\mathbf{a}'_{k2}\mathbf{x}_{p_2}$ both being uncorrelated with $\mathbf{a}'_{jh}\mathbf{x}_{p_h}$, $j = 1, 2, \ldots, (k-1)$; $h = 1, 2$. The name of the technique is confusingly similar to 'canonical variate analysis,' which is used in discrimination (see Section 9.1). In fact, there is a link between the two techniques (see, for example, Gittins, 1985, Chapter 4; Mardia et al. 1979, Exercise 11.5.4), but this will not be discussed in detail here. Because of this link, the view of canonical discriminant analysis as a two-stage PCA, noted by Campbell and Atchley (1981) and discussed in Section 9.1, is also a valid perspective for CCA. Although CCA treats the .two sets of variables $\mathbf{x}_{p_1}$, $\mathbf{x}_{p_2}$ on an equal footing, it can still be used, as in the example of Section 9.3.2, if one set is clearly a set of responses while the other is a set of predictors. However, alternatives such as multivariate regression and other techniques discussed in Sections 8.4, 9.3.3 and 9.3.4 may be more appropriate in this case.

A number of authors have suggested that there are advantages in calculating PCs $\mathbf{z}_{p_1}, \mathbf{z}_{p_2}$ separately for $\mathbf{x}_{p_1}, \mathbf{x}_{p_2}$ and then performing the CCA on $\mathbf{z}_{p_1}, \mathbf{z}_{p_2}$ rather than $\mathbf{x}_{p_1}, \mathbf{x}_{p_2}$. Indeed, the main derivation of CCA given by Preisendorfer and Mobley (1988, Chapter 8) is in terms of the PCs for the two groups of variables. If $\mathbf{z}_{p_1}, \mathbf{z}_{p_2}$ consist of all p_1, p_2 PCs, respectively, then the results using $\mathbf{z}_{p_1}, \mathbf{z}_{p_2}$ are equivalent to those for $\mathbf{x}_{p_1}, \mathbf{x}_{p_2}$. This follows as $\mathbf{z}_{p_1}, \mathbf{z}_{p_2}$ are exact linear functions of $\mathbf{x}_{p_1}, \mathbf{x}_{p_2}$, respectively, and, conversely, $\mathbf{x}_{p_1}, \mathbf{x}_{p_2}$ are exact linear functions of $\mathbf{z}_{p_1}, \mathbf{z}_{p_2}$, respectively. We are looking for 'optimal' linear functions of $\mathbf{z}_{p_1}, \mathbf{z}_{p_2}$, but this is equivalent to searching for 'optimal' linear functions of $\mathbf{x}_{p_1}, \mathbf{x}_{p_2}$ so we have the same analysis as that based on $\mathbf{x}_{p_1}, \mathbf{x}_{p_2}$.

Muller (1982) argues that using $\mathbf{z}_{p_1}, \mathbf{z}_{p_2}$ instead of $\mathbf{x}_{p_1}, \mathbf{x}_{p_2}$ can make some of the theory behind CCA easier to understand, and that it can help in interpreting the results of such an analysis. He also illustrates the use of PCA as a preliminary dimension-reducing technique by performing CCA based on just the first few elements of $\mathbf{z}_{p_1}$ and $\mathbf{z}_{p_2}$. Von Storch and Zwiers (1999, Section 14.1.6) note computational advantages in working with the PCs and also suggest using only the first few PCs to construct the canonical variates. This works well in the example given by Muller (1982), but cannot be expected to do so in general, for reasons similar to those already discussed in the contexts of regression (Chapter 8) and discriminant analysis (Section 9.1). There is simply no reason why those linear functions of $\mathbf{x}_{p_1}$ that are highly correlated with linear functions of $\mathbf{x}_{p_2}$ should necessarily be in the subspace spanned by the first few PCs of $\mathbf{x}_{p_1}$; they could equally well be related to the last few PCs of $\mathbf{x}_{p_1}$. The fact that a linear function of $\mathbf{x}_{p_1}$ has a small variance, as do the last few PCs, in no way prevents it from having a high correlation with some linear function of $\mathbf{x}_{p_2}$. As well as suggesting the use of PCs in CCA, Muller (1982) describes the closely related topic of using canonical correlation analysis to compare sets of PCs. This will be discussed further in Section 13.5.

An interesting connection between PCA and CCA is given by considering the problem of minimizing $\text{var}[\mathbf{a}_1'\mathbf{x}_{p_1} - \mathbf{a}_2'\mathbf{x}_{p_2}]$. If constraints $\mathbf{a}_1'\boldsymbol{\Sigma}_{11}\mathbf{a}_1 = \mathbf{a}_2'\boldsymbol{\Sigma}_{22}\mathbf{a}_2 = 1$ are added to this problem, where $\boldsymbol{\Sigma}_{11}, \boldsymbol{\Sigma}_{22}$ are the covariance matrices for $\mathbf{x}_{p_1}, \mathbf{x}_{p_2}$, respectively, we obtain the first pair of canonical variates. If, instead, the constraint $\mathbf{a}_1'\mathbf{a}_1 + \mathbf{a}_2'\mathbf{a}_2 = 1$ is added, the coefficients $(\mathbf{a}_1', \mathbf{a}_2')'$ define the *last* PC for the vector of random variables $\mathbf{x} = (\mathbf{x}_{p_1}', \mathbf{x}_{p_2}')'$. There has been much discussion in the literature of a variety of connections between multivariate techniques, including PCA and CCA. Gittins (1985, Sections 4.8, 5.6, 5.7) gives numerous references. In the special case where $p_1 = p_2$ and the same variables are measured in both $\mathbf{x}_{p_1}$ and $\mathbf{x}_{p_2}$, perhaps at different time periods or for matched pairs of individuals, Flury and Neuenschwander (1995) demonstrate a theoretical equivalence between the canonical variates and a common principal component model (see Section 13.5) when the latter model holds.

9.3.2 Example of CCA

Jeffers (1978, p. 136) considers an example with 15 variables measured on 272 sand and mud samples taken from various locations in Morecambe Bay, on the north west coast of England. The variables are of two types: eight variables are chemical or physical properties of the sand or mud samples, and seven variables measure the abundance of seven groups of invertebrate species in the samples. The relationships between the two groups of variables, describing environment and species, are of interest, so that canonical correlation analysis is an obvious technique to use.

Table 9.3 gives the coefficients for the first two pairs of canonical variates, together with the correlations between each pair—the *canonical correlations*. The definitions of each variable are not given here (see Jeffers (1978, pp. 103, 107)). The first canonical variate for species is dominated by a single species. The corresponding canonical variate for the environmental variables involves non-trivial coefficients for four of the variables, but is not difficult to interpret (Jeffers, 1978, p. 138). The second pair of canonical variates has fairly large coefficients for three species and three environmental variables.

Jeffers (1978, pp. 105–109) also looks at PCs for the environmental and species variables separately, and concludes that four and five PCs, respectively, are necessary to account for most of the variation in each group. He goes on to look, informally, at the between-group correlations for each set of retained PCs.

Instead of simply looking at the individual correlations between PCs for different groups, an alternative is to do a canonical correlation analysis based only on the retained PCs, as suggested by Muller (1982). In the present example this analysis gives values of 0.420 and 0.258 for the first two canonical correlations, compared with 0.559 and 0.334 when all the variables are used. The first two canonical variates for the environmental

Table 9.3. Coefficients for the first two canonical variates in a canonical correlation analysis of species and environmental variables.

		First canonical variates	Second canonical variates
Environment variables	x_1	0.03	0.17
	x_2	0.51	0.52
	x_3	0.56	0.49
	x_4	0.37	0.67
	x_5	0.01	−0.08
	x_6	0.03	0.07
	x_7	−0.00	0.04
	x_8	0.53	−0.02
Species variables	x_9	0.97	−0.19
	x_{10}	−0.06	−0.25
	x_{11}	0.01	−0.28
	x_{12}	0.14	0.58
	x_{13}	0.19	0.00
	x_{14}	0.06	0.46
	x_{15}	0.01	0.53
Canonical correlation		0.559	0.334

variables and the first canonical variate for the species variables are each dominated by a single PC, and the second canonical variate for the species variables has two non-trivial coefficients. Thus, the canonical variates for PCs look, at first sight, easier to interpret than those based on the original variables. However, it must be remembered that, even if only one PC occurs in a canonical variate, the PC itself is not necessarily an easily interpreted entity. For example, the environmental PC that dominates the first canonical variate for the environmental variables has six large coefficients. Furthermore, the between-group relationships found by CCA of the retained PCs are different in this example from those found from CCA on the original variables.

9.3.3   Maximum Covariance Analysis (SVD Analysis), Redundancy Analysis and Principal Predictors

The first technique described in this section has been used in psychology for many years, dating back to Tucker (1958), where it is known as *inter-battery factor analysis*. This method postulates a model in which

$$\mathbf{x}_{p_1} = \boldsymbol{\mu}_1 + \boldsymbol{\Lambda}_1 \mathbf{z} + \boldsymbol{\Gamma}_1 \mathbf{y}_1 + \mathbf{e}_1 \tag{9.3.1}$$
$$\mathbf{x}_{p_2} = \boldsymbol{\mu}_2 + \boldsymbol{\Lambda}_2 \mathbf{z} + \boldsymbol{\Gamma}_2 \mathbf{y}_2 + \mathbf{e}_2,$$

where $\boldsymbol{\mu}_1, \boldsymbol{\mu}_2$ are vectors of means, $\boldsymbol{\Lambda}_1, \boldsymbol{\Lambda}_2, \boldsymbol{\Gamma}_1, \boldsymbol{\Gamma}_2$ are matrices of coefficients, $\mathbf{z}$ is a vector of latent variables common to both $\mathbf{x}_{p_1}$ and $\mathbf{x}_{p_2}$, $\mathbf{y}_1, \mathbf{y}_2$ are vectors of latent variables specific to $\mathbf{x}_{p_1}, \mathbf{x}_{p_2}$, and $\mathbf{e}_1, \mathbf{e}_2$ are vectors of errors. Tucker (1958) fits the model using the singular value decomposition of the $(p_1 \times p_2)$ matrix of correlations between two batteries of tests $\mathbf{x}_{p_1}, \mathbf{x}_{p_2}$, and notes that his procedure is equivalent to finding linear combinations of the two batteries that have maximum covariance. Browne (1979) demonstrates some algebraic connections between the results of this technique and those of CCA.

The method was popularised in atmospheric science by Bretherton et al. (1992) and Wallace et al. (1992) under the name *singular value decomposition* (SVD) analysis. This name arose because, as Tucker (1958) showed, the analysis can be conducted via an SVD of the $(p_1 \times p_2)$ matrix of covariances between $\mathbf{x}_{p_1}$ and $\mathbf{x}_{p_2}$, but the use of this general term for a specific technique is potentially very confusing. The alternative *canonical covariance analysis*, which Cherry (1997) notes was suggested in unpublished work by Muller, is a better descriptor of what the technique does, namely that it successively finds pairs of linear functions of $\mathbf{x}_{p_1}$ and $\mathbf{x}_{p_2}$ that have maximum covariance and whose vectors of loadings are orthogonal. Even better is *maximum covariance analysis*, which is used by von Storch and Zwiers (1999, Section 14.1.7) and others (Frankignoul, personal communication), and we will adopt this terminology. Maximum covariance analysis differs from CCA in two ways: covariance rather than correlation is maximized, and vectors of loadings are orthogonal instead of derived variates being uncorrelated. The rationale behind maximum covariance analysis is that it may be important to explain a large proportion of the variance in one set of variables using the other set, and a pair of variates from CCA with large correlation need not necessarily have large variance.

Bretherton et al. (1992) and Wallace et al. (1992) discuss maximum covariance analysis (SVD analysis) in some detail, make comparisons with competing techniques and give examples. Cherry (1997) and Hu (1997) point out some disadvantages of the technique, and Cherry (1997) also demonstrates a relationship with PCA. Suppose that separate PCAs are done on the two sets of variables and that the values (scores) of the n observations on the first q PCs are given by the $(n \times q)$ matrices $\mathbf{Z}_1, \mathbf{Z}_2$ for the two sets of variables. If $\mathbf{B}_1, \mathbf{B}_2$ are orthogonal matrices chosen to minimize $\|\mathbf{Z}_1 \mathbf{B}_1 - \mathbf{Z}_2 \mathbf{B}_2\|$, the resulting matrices $\mathbf{Z}_1 \mathbf{B}_1, \mathbf{Z}_2 \mathbf{B}_2$ contain the values for the n observations of the first q pairs of variates from a maximum covariance analysis. Thus, maximum covariance analysis can be viewed as two PCAs, followed by rotation to match up the results of the two analyses as closely as possible.

Like maximum covariance analysis, *redundancy analysis* attempts to incorporate variance as well as correlation in its search for relationships between two sets of variables. The *redundancy coefficient* was introduced by Stewart and Love (1968), and is an index of the average proportion of the variance of the variables in one set that is reproducible from the variables in the other set. One immediate difference from both CCA and maximum covariance analysis is that it does not view the two sets of variables symmetrically. One set is treated as response variables and the other as predictor variables, and the results of the analysis are different depending on the choice of which set contains responses. For convenience, in what follows $\mathbf{x}_{p_1}$ and $\mathbf{x}_{p_2}$ consist of responses and predictors, respectively.

Stewart and Love's (1968) redundancy index, given a pair of canonical variates, can be expressed as the product of two terms. These terms are the squared canonical correlation and the variance of the canonical variate for the response set. It is clear that a different value results if the rôles of predictor and response variables are reversed. The redundancy coefficient can be obtained by regressing each response variable on all the predictor variables and then averaging the p_1 squared multiple correlations from these regressions. This has a link to the interpretation of PCA given in the discussion of Property A6 in Chapter 2, and was used by van den Wollenberg (1977) and Thacker (1999) to introduce two slightly different techniques.

In van den Wollenberg's (1977) redundancy analysis, linear functions $\mathbf{a}'_{k2}\mathbf{x}_{p_2}$ of $\mathbf{x}_{p_2}$ are found that successively maximize their average squared correlation with the elements of the response set $\mathbf{x}_{p_1}$, subject to the vectors of loadings $\mathbf{a}_{12}$, $\mathbf{a}_{22}$, ... being orthogonal. It turns out (van den Wollenberg, 1977) that finding the required linear functions is achieved by solving the equation

$$\mathbf{R}_{xy}\mathbf{R}_{yx}\mathbf{a}_{k2} = l_k\mathbf{R}_{xx}\mathbf{a}_{k2}, \qquad (9.3.2)$$

where $\mathbf{R}_{xx}$ is the correlation matrix for the predictor variables, $\mathbf{R}_{xy}$ is the matrix of correlations between the predictor and response variables, and $\mathbf{R}_{yx}$ is the transpose of $\mathbf{R}_{xy}$. A linear function of $\mathbf{x}_{p_1}$ can be found by reversing the rôles of predictor and response variables, and hence replacing x by y and vice versa, in equation (9.3.2).

Thacker (1999) also considers a linear function $\mathbf{z}_1 = \mathbf{a}'_{12}\mathbf{x}_{p_2}$ of the predictors $\mathbf{x}_{p_2}$. Again $\mathbf{a}_{12}$ is chosen to maximize $\sum_{j=1}^{p_1} r_{1j}^2$, where r_{1j} is the correlation between $\mathbf{z}_1$ and the jth response variable. The variable $\mathbf{z}_1$ is called the first *principal predictor* by Thacker (1999). Second, third, ... principal predictors are defined by maximizing the same quantity, subject to the constraint that each principal predictor must be uncorrelated with all previous principal predictors. Thacker (1999) shows that the vectors of loadings $\mathbf{a}_{12}$, $\mathbf{a}_{22}$, ... are solutions of the equation

$$\mathbf{S}_{xy}[\mathrm{diag}(\mathbf{S}_{yy})]^{-1}\mathbf{S}_{yx}\mathbf{a}_{k2} = l_k\mathbf{S}_{xx}\mathbf{a}_{k2}, \qquad (9.3.3)$$

where $\mathbf{S}_{xx}$, $\mathbf{S}_{yy}$, $\mathbf{S}_{xy}$ and $\mathbf{S}_{yx}$ are covariance matrices defined analogously to the correlation matrices $\mathbf{R}_{xx}$, $\mathbf{R}_{yy}$, $\mathbf{R}_{xy}$ and $\mathbf{R}_{yx}$ above. The eigenvalue l_k corresponding to $\mathbf{a}_{k2}$ is equal to the sum of squared correlations $\sum_{j=1}^{p_1} r_{kj}^2$

between $\mathbf{a}'_{k2}\mathbf{x}_{p2}$ and each of the variables x_j. The difference between principal predictors and redundancy analysis is that the principal predictors are uncorrelated, whereas the derived variables in redundancy analysis are correlated but have vectors of loadings that are orthogonal. The presence of correlation in redundancy analysis may be regarded as a drawback, and van den Wollenberg (1977) suggests using the first few derived variables from redundancy analysis as input to CCA. This will then produce uncorrelated canonical variates whose variances are unlikely to be small. The possibility of using the first few PCs from each set as input to CCA was mentioned above, as was the disadvantage that excluded low-variance PCs might contain strong inter-set correlations. As low-variance directions are unlikely to be of interest in redundancy analysis, using the first few PCs as input seems to be far safer in this case and is another option.

It is of interest to note the similarity between equations (9.3.2), (9.3.3) and the eigenequation whose solution gives the loadings $\mathbf{a}_{k2}$ on $\mathbf{x}_{p2}$ for canonical correlation analysis, namely

$$\mathbf{S}_{xy}\mathbf{S}_{yy}^{-1}\mathbf{S}_{yx}\mathbf{a}_{k2} = l_k\mathbf{S}_{xx}\mathbf{a}_{k2}, \qquad (9.3.4)$$

using the present notation. Wang and Zwiers (2001) solve a version of (9.3.2) with covariance matrices replacing correlation matrices, by first solving the eigenequation

$$\mathbf{S}_{yx}\mathbf{S}_{xx}^{-1}\mathbf{S}_{xy}\mathbf{b}_{k2} = l_k\mathbf{b}_{k2}, \qquad (9.3.5)$$

and then setting $\mathbf{a}_{k2} = l_k^{-\frac{1}{2}}\mathbf{S}_{xx}^{-1}\mathbf{S}_{xy}\mathbf{b}_{k2}$. This is equivalent to a PCA of the covariance matrix $\mathbf{S}_{yx}\mathbf{S}_{xx}^{-1}\mathbf{S}_{xy}$ of the predicted values of the response variables obtained from a multivariate regression on the predictor variables. Multivariate regression is discussed further in Section 9.3.4.

Van den Wollenberg (1977) notes that PCA is a special case of redundancy analysis (and principal predictors, but not CCA) when $\mathbf{x}_{p_1}$ and $\mathbf{x}_{p_2}$ are the same (see also Property A6 in Chapter 2). Muller (1981) shows that redundancy analysis is equivalent to orthogonally rotating the results of a multivariate regression analysis. DeSarbo and Jedidi (1986) give a number of other properties, together with modifications and extensions, of redundancy analysis.

9.3.4   Other Techniques for Relating Two Sets of Variables

A number of other techniques for relating two sets of variables were noted in Section 8.4. They include separate PCAs on the two groups of variables, followed by the calculation of a regression equation to predict, again separately, each PC from one set from the PCs in the other set. Another way of using PCA is to concatenate the two sets of variables and find PCs for the combined set of $(p_1 + p_2)$ variables. This is sometimes known as *combined PCA*, and is one of the methods that Bretherton et al. (1992) compare with maximum covariance analysis. The reasoning behind the analysis, apart from being very easy to implement, is that two variables are more likely to simultaneously have large loadings in the same high-variance combined PC if they are strongly correlated. Thus, by looking at which variables from different groups appear together in the same high-variance components, some ideas can be gained about the relationships between the two groups. This is true to some extent, but the combined components do not directly quantify the precise form of the relationships, nor their strength, in the way that CCA or maximum covariance analysis does. One other PCA-based technique considered by Bretherton et al. (1992) is to look at correlations between PCs of one set of variables and the variables themselves from the other set. This takes us back to a collection of simple PC regressions.

Another technique from Section 8.4, partial least squares (PLS), can be generalized to the case of more than one response variable (Wold, 1984). Like single-response PLS, multiresponse PLS is often defined in terms of an algorithm, but Frank and Friedman (1993) give an interpretation showing that multiresponse PLS successively maximizes the covariance between linear functions of the two sets of variables. It is therefore similar to maximum covariance analysis, which was discussed in Section 9.3.3, but differs from it in not treating response and predictor variables symmetrically. Whereas in maximum covariance analysis the vectors of coefficients of the linear functions are orthogonal within each set of variables, no such restriction is placed on the response variables in multiresponse PLS. For the predictor variables there is a restriction, but it is that the linear functions are uncorrelated, rather than having orthogonal vectors of coefficients.

The standard technique when one set of variables consists of responses and the other is made up of predictors is *multivariate linear regression*. Equation (8.1.1) generalizes to

$$\mathbf{Y} = \mathbf{XB} + \mathbf{E}, \qquad (9.3.6)$$

where $\mathbf{Y}$, $\mathbf{X}$ are $(n \times p_1)$, $(n \times p_2)$ matrices of n observations on p_1 response variables and p_2 predictor variables, respectively, $\mathbf{B}$ is a $(p_2 \times p_1)$ matrix of unknown parameters, and $\mathbf{E}$ is an $(n \times p_1)$ matrix of errors. The number of parameters to be estimated is at least $p_1 p_2$ (as well as those in $\mathbf{B}$, there are usually some associated with the covariance matrix of the error term). Various attempts have been made to reduce this number by simplifying the model. The reduced rank models of Davies and Tso (1982) form a general class of this type. In these models $\mathbf{B}$ is assumed to be of rank $m < p_2$. There is more than one way to estimate the unknown parameters in a reduced rank regression model. That recommended by Davies and Tso (1982) first finds the least squares estimate $\hat{\mathbf{B}}$ of $\mathbf{B}$ in (9.3.6) and uses this to obtain predicted values $\hat{\mathbf{Y}} = \mathbf{X}\hat{\mathbf{B}}$ for $\mathbf{Y}$. Next a singular value decomposition (with its usual meaning) is done on $\hat{\mathbf{Y}}$ and $\mathbf{B}$ is then projected onto the subspace spanned by the first m terms in the SVD. This is equivalent to projecting the rows of $\hat{\mathbf{Y}}$ onto the subspace spanned by the first m PCs of $\hat{\mathbf{Y}}$.

Two further equivalences are noted by ter Braak and Looman (1994), namely that the reduced rank regression model estimated in this way is equivalent to redundancy analysis, and also to PCA of instrumental variables, as introduced by Rao (1964) (see Section 14.3). Van den Brink and ter Braak (1999) also refer to redundancy analysis as 'PCA in which sample scores are constrained to be linear combinations of the explanatory [predictor] variables.' They extend redundancy analysis to the case where the variables in $\mathbf{X}$ and $\mathbf{Y}$ are observed over several time periods and the model changes with time. This extension is discussed further in Section 12.4.2. Because of the link with PCA, it is possible to construct biplots (see Section 5.3) of the regression coefficients in the reduced rank regression model (ter Braak and Looman, 1994).

Aldrin (2000) proposes a modification of reduced rank regression, called *softly shrunk reduced-rank regression (SSRRR)*, in which the terms in the SVD of $\hat{\mathbf{Y}}$ are given varying non-zero weights, rather than the all-or-nothing inclusion/exclusion of terms in reduced rank regression. Aldrin (2000) also suggests that a subset of PCs of the predictor variables may be used as input for a reduced rank regression or SSRRR instead of the predictor variables themselves. In a simulation study comparing least squares with a number of biased multivariate regression procedures, SSRRR with PCs as input seems to be the best method overall.

Reduced rank regression models essentially assume a latent structure underlying the predictor variables, so that their dimensionality can be reduced below p_2. Burnham et al. (1999) describe so-called *latent variable multivariate regression models*, which take the idea of reduced rank regression further by postulating overlapping latent structures underlying both the response and predictor variables. The model can be written

$$\mathbf{X} = \mathbf{Z}_X \boldsymbol{\Gamma}_X + \mathbf{E}_X$$
$$\mathbf{Y} = \mathbf{Z}_Y \boldsymbol{\Gamma}_Y + \mathbf{E}_Y,$$

where $\mathbf{Z}_X$, $\mathbf{Z}_Y$ are of dimension $(n \times m)$ and contain values of m latent variables for the n observations; $\boldsymbol{\Gamma}_X$, $\boldsymbol{\Gamma}_Y$ are $(m \times p_1)$, $(m \times p_2)$ matrices of unknown parameters, and $\mathbf{E}_X$, $\mathbf{E}_Y$ are matrices of errors.

To fit this model, Burnham et al. (1999) suggest carrying out PCAs on the data in $\mathbf{X}$, on that in $\mathbf{Y}$, and on the combined $(n \times (p_1 + p_2))$ matrix containing both response and predictor variables. In each PCA, a judgment is made of how many PCs seem to represent common underlying structure and how many represent error or noise. Suppose that the numbers of non-noisy PCs in the three analyses are m_X, m_Y and m_C, with obvious notation. The implication is then that the overlapping part of the latent structures has dimension $m_X + m_Y - m_C$. If $m_X = m_Y = m_C$ there is complete overlap, whereas if $m_C = m_X + m_Y$ there is none. This model is very similar to that of (9.3.1). The difference is that the separation into latent variables common to both sets of measured variables and those specific to one set of measured variables is explicit in (9.3.1). Burnham et al. (1999) successfully fit their model to a number of examples from chemometrics.

10
Outlier Detection, Influential Observations, Stability, Sensitivity, and Robust Estimation of Principal Components

This chapter deals with four related topics, which are all concerned with situations where some of the observations may, in some way, be atypical of the bulk of the data.

First, we discuss the problem of detecting *outliers* in a set of data. Outliers are generally viewed as observations that are a long way from, or inconsistent with, the remainder of the data. Such observations can, but need not, have a drastic and disproportionate effect on the results of various analyses of a data set. Numerous methods have been suggested for detecting outliers (see, for example, Barnett and Lewis, 1994; Hawkins, 1980); some of the methods use PCs, and these methods are described in Section 10.1.

The techniques described in Section 10.1 are useful regardless of the type of statistical analysis to be performed, but in Sections 10.2–10.4 we look specifically at the case where a PCA is being done. Depending on their position, outlying observations may or may not have a large effect on the results of the analysis. It is of interest to determine which observations do indeed have a large effect. Such observations are called *influential observations* and are discussed in Section 10.2. Leaving out an observation is one type of perturbation to a data set. Sensitivity and stability of PCA with respect to other types of perturbation is the subject of Section 10.3.

Given that certain observations are outliers or influential, it may be desirable to adapt the analysis to remove or diminish the effects of such observations; that is, the analysis is made *robust*. Robust analyses have been developed in many branches of statistics (see, for example, Huber (1981); Hampel et al. (1986) for some of the theoretical background, and

Hoaglin et al. (1983) for a more readable approach), and robustness with respect to distributional assumptions, as well as with respect to outlying or influential observations, may be of interest. A number of techniques have been suggested for robustly estimating PCs, and these are discussed in the fourth section of this chapter; the final section presents a few concluding remarks.

10.1 Detection of Outliers Using Principal Components

There is no formal, widely accepted, definition of what is meant by an 'outlier.' The books on the subject by Barnett and Lewis (1994) and Hawkins (1980) both rely on informal, intuitive definitions, namely that outliers are observations that are in some way different from, or inconsistent with, the remainder of a data set. For p-variate data, this definition implies that outliers are a long way from the rest of the observations in the p-dimensional space defined by the variables. Numerous procedures have been suggested for detecting outliers with respect to a single variable, and many of these are reviewed by Barnett and Lewis (1994) and Hawkins (1980). The literature on multivariate outliers is less extensive, with each of these two books containing only one chapter (comprising less than 15% of their total content) on the subject. Several approaches to the detection of multivariate outliers use PCs, and these will now be discussed in some detail. As well as the methods described in this section, which use PCs in fairly direct ways to identify potential outliers, techniques for robustly estimating PCs (see Section 10.4) may also be used to detect outlying observations.

A major problem in detecting multivariate outliers is that an observation that is not extreme on any of the original variables can still be an outlier, because it does not conform with the correlation structure of the remainder of the data. It is impossible to detect such outliers by looking solely at the original variables one at a time. As a simple example, suppose that heights and weights are measured for a sample of healthy children of various ages between 5 and 15 years old. Then an 'observation' with height and weight of 175 cm (70 in) and 25 kg (55 lb), respectively, is not particularly extreme on either the height or weight variables individually, as 175 cm is a plausible height for the older children and 25 kg is a plausible weight for the youngest children. However, the combination (175 cm, 25 kg) is virtually impossible, and will be a clear outlier because it combines a large height with a small weight, thus violating the general pattern of a positive correlation between the two variables. Such an outlier is apparent on a plot of the two variables (see Figure 10.1) but, if the number of variables p is large, it is quite possible that some outliers will not be apparent on any of the $\frac{1}{2}p(p-1)$ plots of two variables at a time. Thus, for large p we need to consider the possibility

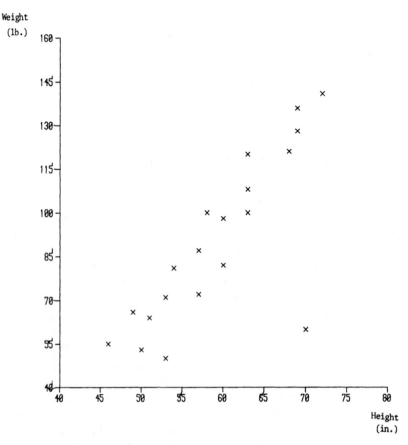

Figure 10.1. Example of an outlier that is not detectable by looking at one variable at a time.

that outliers will manifest themselves in directions other than those which are detectable from simple plots of pairs of the original variables.

Outliers can be of many types, which complicates any search for directions in which outliers occur. However, there are good reasons for looking at the directions defined by either the first few or the last few PCs in order to detect outliers. The first few and last few PCs will detect different types of outlier, and in general the last few are more likely to provide additional information that is not available in plots of the original variables.

As discussed in Gnanadesikan and Kettenring (1972), the outliers that are detectable from a plot of the first few PCs are those which *inflate* variances and covariances. If an outlier is the cause of a large increase in one or more of the variances of the original variables, then it must be extreme on those variables and thus detectable by looking at plots of single variables. Similarly, an observation that inflates a covariance or correlation between two variables will usually be clearly visible on a plot of these two variables, and will often be extreme with respect to one or both of these variables looked at individually.

By contrast, the last few PCs may detect outliers that are not apparent with respect to the original variables. A strong correlation structure between variables implies that there are linear functions of the variables with small variances compared to the variances of the original variables. In the simple height-and-weight example described above, height and weight have a strong positive correlation, so it is possible to write

$$x_2 = \beta x_1 + \varepsilon,$$

where x_1, x_2 are height and weight measured about their sample means, β is a positive constant, and ε is a random variable with a much smaller variance than x_1 or x_2. Therefore the linear function

$$x_2 - \beta x_1$$

has a small variance, and the last (in this case the second) PC in an analysis of x_1, x_2 has a similar form, namely $a_{22}x_2 - a_{12}x_1$, where $a_{12}, a_{22} > 0$. Calculation of the value of this second PC for each observation will detect observations such as (175 cm, 25 kg) that are outliers with respect to the correlation structure of the data, though not necessarily with respect to individual variables. Figure 10.2 shows a plot of the data from Figure 10.1, with respect to the PCs derived from the correlation matrix. The outlying observation is 'average' for the first PC, but very extreme for the second.

This argument generalizes readily when the number of variables p is greater than two; by examining the values of the last few PCs, we may be able to detect observations that violate the correlation structure imposed by the bulk of the data, but that are not necessarily aberrant with respect to individual variables. Of course, it is possible that, if the sample size is relatively small or if a few observations are sufficiently different from the rest, then the outlier(s) may so strongly influence the last few PCs that these PCs now reflect mainly the position of the outlier(s) rather than the structure of the majority of the data. One way of avoiding this masking or camouflage of outliers is to compute PCs leaving out one (or more) observations and then calculate for the deleted observations the values of the last PCs based on the reduced data set. To do this for each observation is a heavy computational burden, but it might be worthwhile in small samples where such camouflaging is, in any case, more likely to occur. Alternatively, if PCs are estimated robustly (see Section 10.4), then the influence of outliers on the last few PCs should be reduced and it may be unnecessary to repeat the analysis with each observation deleted.

A series of scatterplots of pairs of the first few and last few PCs may be useful in identifying possible outliers. One way of presenting each PC separately is as a set of parallel boxplots. These have been suggested as a means of deciding how many PCs to retain (see Section 6.1.5), but they may also be useful for flagging potential outliers (Besse, 1994).

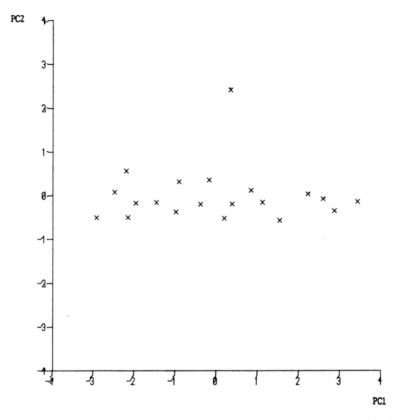

Figure 10.2. The data set of Figure 10.1, plotted with respect to its PCs.

As well as simple plots of observations with respect to PCs, it is possible to set up more formal tests for outliers based on PCs, assuming that the PCs are normally distributed. Strictly, this assumes that $\mathbf{x}$ has a multivariate normal distribution but, because the PCs are linear functions of p random variables, an appeal to the Central Limit Theorem may justify approximate normality for the PCs even when the original variables are not normal. A battery of tests is then available for each individual PC, namely those for testing for the presence of outliers in a sample of univariate normal data (see Hawkins (1980, Chapter 3) and Barnett and Lewis (1994, Chapter 6)). The latter reference describes 47 tests for univariate normal data, plus 23 for univariate gamma distributions and 17 for other distributions. Other tests, which combine information from several PCs rather than examining one at a time, are described by Gnanadesikan and Kettenring (1972) and Hawkins (1974), and some of these will now be discussed. In particular, we define four statistics, which are denoted d_{1i}^2, d_{2i}^2, d_{3i}^2 and d_{4i}.

The last few PCs are likely to be more useful than the first few in detecting outliers that are not apparent from the original variables, so one possible test statistic, d_{1i}^2, suggested by Rao (1964) and discussed further by Gnanadesikan and Kettenring (1972), is the sum of squares of the values of the last q $(< p)$ PCs, that is

$$d_{1i}^2 = \sum_{k=p-q+1}^{p} z_{ik}^2, \tag{10.1.1}$$

where z_{ik} is the value of the kth PC for the ith observation. The statistics d_{1i}^2, $i = 1, 2, \ldots, n$ should, approximately, be independent observations from a gamma distribution if there are no outliers, so that a gamma probability plot with suitably estimated shape parameter may expose outliers (Gnanadesikan and Kettenring, 1972).

A possible criticism of the statistic d_{1i}^2 is that it still gives insufficient weight to the last few PCs, especially if q, the number of PCs contributing to d_{1i}^2, is close to p. Because the PCs have decreasing variance with increasing index, the values of z_{ik}^2 will typically become smaller as k increases, and d_{1i}^2 therefore implicitly gives the PCs decreasing weight as k increases. This effect can be severe if some of the PCs have very small variances, and this is unsatisfactory as it is precisely the low-variance PCs which may be most effective in determining the presence of certain types of outlier.

An alternative is to give the components equal weight, and this can be achieved by replacing z_{ik} by $z_{ik}^* = z_{ik}/l_k^{1/2}$, where l_k is the variance of the kth sample PC. In this case the sample variances of the z_{ik}^* will all be equal to unity. Hawkins (1980, Section 8.2) justifies this particular renormalization of the PCs by noting that the renormalized PCs, in reverse order, are the uncorrelated linear functions $\tilde{a}_p'\mathbf{x}, \tilde{a}_{p-1}'\mathbf{x}, \ldots, \tilde{a}_1'\mathbf{x}$ of $\mathbf{x}$ which, when constrained to have unit variances, have coefficients $\tilde{a}_{jk}$ that successively maximize the criterion $\sum_{j=1}^{p} \tilde{a}_{jk}^2$, for $k = p, (p-1), \ldots, 1$. Maximization of this criterion is desirable because, given the fixed-variance property, linear functions that have large absolute values for their coefficients are likely to be more sensitive to outliers than those with small coefficients (Hawkins, 1980, Section 8.2). It should be noted that when $q = p$, the statistic

$$d_{2i}^2 = \sum_{k=p-q+1}^{p} \frac{z_{ik}^2}{l_k} \tag{10.1.2}$$

becomes $\sum_{k=1}^{p} z_{ik}^2/l_k$, which is simply the (squared) Mahalanobis distance D_i^2 between the ith observation and the sample mean, defined as $D_i^2 = (\mathbf{x}_i - \bar{\mathbf{x}})'\mathbf{S}^{-1}(\mathbf{x}_i - \bar{\mathbf{x}})$. This follows because $\mathbf{S} = \mathbf{AL}^2\mathbf{A}'$ where, as usual, $\mathbf{L}^2$ is the diagonal matrix whose kth diagonal element is l_k, and $\mathbf{A}$ is the matrix whose (j, k)th element is a_{jk}. Furthermore,

$$\mathbf{S}^{-1} = \mathbf{AL}^{-2}\mathbf{A}'$$
$$\mathbf{x}_i' = \mathbf{z}_i'\mathbf{A}'$$
$$\bar{\mathbf{x}}' = \bar{\mathbf{z}}'\mathbf{A}',$$

so D_i^2 is

$$(\mathbf{x}_i - \bar{\mathbf{x}})'\mathbf{S}^{-1}(\mathbf{x}_i - \bar{\mathbf{x}}) = (\mathbf{z}_i - \bar{\mathbf{z}})'\mathbf{A}'\mathbf{A}\mathbf{L}^{-2}\mathbf{A}'\mathbf{A}(\mathbf{z}_i - \bar{\mathbf{z}})$$
$$= (\mathbf{z}_i - \bar{\mathbf{z}})'\mathbf{L}^{-2}(\mathbf{z}_i - \bar{\mathbf{z}})$$
$$= \sum_{k=1}^{p} \frac{z_{ik}^2}{l_k},$$

where z_{ik} is the kth PC score for the ith observation, measured about the mean of the scores for all observations. Flury (1997, p. 609-610) suggests that a plot of $(D_i^2 - d_{2i}^2)$ versus d_{2i}^2 will reveal observations that are not well represented by the first $(p - q)$ PCs. Such observations are potential outliers.

Gnanadesikan and Kettenring (1972) consider also the statistic

$$d_{3i}^2 = \sum_{k=1}^{p} l_k z_{ik}^2, \qquad (10.1.3)$$

which emphasizes observations that have a large effect on the *first few* PCs, and is equivalent to $(\mathbf{x}_i - \bar{\mathbf{x}})'\mathbf{S}(\mathbf{x}_i - \bar{\mathbf{x}})$. As stated earlier, the first few PCs are useful in detecting some types of outlier, and d_{3i}^2 emphasizes such outliers. However, we repeat that such outliers are often detectable from plots of the original variables, unlike the outliers exposed by the last few PCs. Various types of outlier, including some that are extreme with respect to both the first few *and* and the last few PCs, are illustrated in the examples given later in this section.

Hawkins (1974) prefers to use d_{2i}^2 with $q < p$ rather than $q = p$ (again, in order to emphasize the low-variance PCs), and he considers how to choose an appropriate value for q. This is a rather different problem from that considered in Section 6.1, as we now wish to decide how many of the PCs, *starting with the last* rather than starting with the first, need to be retained. Hawkins (1974) suggests three possibilities for choosing q, including the 'opposite' of Kaiser's rule (Section 6.1.2)—that is, the retention of PCs with eigenvalues less than unity. In an example, he selects q as a compromise between values suggested by his three rules.

Hawkins (1974) also shows that outliers can be successfully detected using the statistic

$$d_{4i} = \max_{p-q+1 \le k \le p} |z_{ik}^*|, \qquad (10.1.4)$$

and similar methods for choosing q are again suggested. Fellegi (1975), too, is enthusiastic about the performance of the statistic d_{4i}. Hawkins and Fatti (1984) claim that outlier detection is improved still further by a series of transformations, including varimax rotation (see Sections 7.2 and 11.1), before computing d_{4i}. The test statistic for the ith observation then becomes the maximum absolute value of the last q renormalized and rotated PCs evaluated for that observation.

Note that d_{1i}^2, computed separately for several populations, is also used in a form of discriminant analysis (SIMCA) by Wold (1976) (see Section 9.1). Mertens et al. (1994) use this relationship to suggest modifications to SIMCA. They investigate variants in which d_{1i}^2 is replaced by d_{2i}^2, d_{3i}^2 or d_{4i} as a measure of the discrepancy between a new observation and a group. In an example they find that d_{2i}^2, but not d_{3i}^2 or d_{4i}, improves the cross-validated misclassification rate compared to that for d_{1i}^2.

The exact distributions for d_{1i}^2, d_{2i}^2, d_{3i}^2 and d_{4i} can be deduced if we assume that the observations are from a multivariate normal distribution with mean $\boldsymbol{\mu}$ and covariance matrix $\boldsymbol{\Sigma}$, where $\boldsymbol{\mu}$, $\boldsymbol{\Sigma}$ are both known (see Hawkins (1980, p. 113) for results for d_{2i}^2, d_{4i}). Both d_{3i}^2 and d_{2i}^2 when $q = p$, as well as d_{1i}^2, have (approximate) gamma distributions if no outliers are present and if normality can be (approximately) assumed (Gnanadesikan and Kettenring, 1972), so that gamma probability plots of d_{2i}^2 (with $q = p$) and d_{3i}^2 can again be used to look for outliers. However, in practice $\boldsymbol{\mu}$, $\boldsymbol{\Sigma}$ are unknown, and the data will often not have a multivariate normal distribution, so that any distributional results derived under the restrictive assumptions can only be approximations. Jackson (1991, Section 2.7.2) gives a fairly complicated function of d_{1i}^2 that has, approximately, a standard normal distribution when no outliers are present.

In order to be satisfactory, such approximations to the distributions of d_{1i}^2, d_{2i}^2, d_{3i}^2, d_{4i} often need not be particularly accurate. Although there are exceptions, such as detecting possible unusual patient behaviour in safety data from clinical trials (see Penny and Jolliffe, 2001), outlier detection is frequently concerned with finding observations that are blatantly different from the rest, corresponding to very small significance levels for the test statistics. An observation that is 'barely significant at 5%' is typically not of interest, so that there is no great incentive to compute significance levels very accurately. The outliers that we wish to detect should 'stick out like a sore thumb' provided we find the right direction in which to view the data; the problem in multivariate outlier detection is to find appropriate directions. If, on the other hand, identification of less clear-cut outliers is important and multivariate normality cannot be assumed, Dunn and Duncan (2000) propose a procedure, in the context of evaluating habitat suitability, for assessing 'significance' based on the empirical distribution of their test statistics. The statistics they use are individual terms from d_{2i}^2.

PCs can be used to detect outliers in any multivariate data set, regardless of the subsequent analysis which is envisaged for that data set. For particular types of data or analysis, other considerations come into play. For multiple regression, Hocking (1984) suggests that plots of PCs derived from $(p + 1)$ variables consisting of the p predictor variables and the dependent variable, as used in latent root regression (see Section 8.4), tend to reveal outliers together with observations that are highly influential (Section 10.2) for the regression equation. Plots of PCs derived from the predictor variables only also tend to reveal influential observations. Hocking's (1984) suggestions are illustrated with an example, but no indication is given of whether the first few or last few PCs are more likely to be useful—his example has only three predictor variables, so it is easy to look at all possible plots. Mason and Gunst (1985) refer to outliers among the predictor variables as *leverage points*. They recommend constructing scatter plots of the first few PCs normalized to have unit variance, and claim that such plots are often effective in detecting leverage points that cluster and leverage points that are extreme in two or more dimensions. In the case of *multivariate* regression, another possibility for detecting outliers (Gnanadesikan and Kettenring, 1972) is to look at the PCs of the (multivariate) residuals from the regression analysis.

Peña and Yohai (1999) propose a PCA on a matrix of regression diagnostics that is also useful in detecting outliers in multiple regression. Suppose that a sample of n observations is available for the analysis. Then an $(n \times n)$ matrix can be calculated whose (h, i)th element is the difference $\hat{y}_h - \hat{y}_{h(i)}$ between the predicted value of the dependent variable y for the hth observation when all n observations are used in the regression, and when $(n-1)$ observations are used with the ith observation omitted. Peña and Yohai (1999) refer to this as a *sensitivity matrix* and seek a unit-length vector such that the sum of squared lengths of the projections of the rows of the matrix onto that vector is maximized. This leads to the first principal component of the sensitivity matrix, and subsequent components can be found in the usual way. Peña and Yohai (1999) call these components *principal sensitivity components* and show that they also represent directions that maximize standardized changes to the vector of the regression coefficient. The definition and properties of principal sensitivity components mean that high-leverage outliers are likely to appear as extremes on at least one of the first few components.

Lu et al. (1997) also advocate the use of the PCs of a matrix of regression diagnostics. In their case the matrix is what they call the *standardized influence matrix* (**SIM**). If a regression equation has p unknown parameters and n observations with which to estimate them, a $(p \times n)$ influence matrix can be formed whose (j, i)th element is a standardized version of the theoretical influence function (see Section 10.2) for the jth parameter evaluated for the ith observation. Leaving aside the technical details, the so-called complement of the standardized influence matrix (**SIM**^{c}) can be viewed as a covariance matrix for the 'data' in the influence matrix. Lu et al. (1997) show that finding the PCs of these standardized data, and hence the eigenvalues and eigenvectors of **SIM**^{c}, can identify outliers and influential points and give insights into the structure of that influence. Sample versions of **SIM** and **SIM**^{c} are given, as are illustrations of their use.

Another specialized field in which the use of PCs has been proposed in order to detect outlying observations is that of statistical process control, which is the subject of Section 13.7. A different way of using PCs to detect outliers is proposed by Gabriel and Zamir (1979). This proposal uses the idea of weighted PCs, and will be discussed further in Section 14.2.1.

Projection pursuit was introduced in Section 9.2.2 as a family of techniques for finding clusters, but it can equally well be used to look for outliers. PCA is not specifically designed to find dimensions which best display either clusters or outliers. As with clusters, optimizing a criterion other than variance can give better low-dimensional displays in which to identify outliers. As noted in Section 9.2.2, projection pursuit techniques find directions in p-dimensional space that optimize some index of 'interestingness,' where 'uninteresting' corresponds to multivariate normality and 'interesting' implies some sort of 'structure,' such as clusters or outliers.

Some indices are good at finding clusters, whereas others are better at detecting outliers (see Friedman (1987); Huber (1985); Jones and Sibson (1987)). Sometimes the superiority in finding outliers has been observed empirically; in other cases the criterion to be optimized has been chosen with outlier detection specifically in mind. For example, if outliers rather than clusters are of interest, Caussinus and Ruiz (1990) suggest replacing the quantity in equation (9.2.1) by

$$\hat{\boldsymbol{\Gamma}} = \frac{\sum_{i=1}^{n} K[\|\mathbf{x}_i - \mathbf{x}^*\|_{\mathbf{S}^{-1}}^2](\mathbf{x}_i - \mathbf{x}^*)(\mathbf{x}_i - \mathbf{x}^*)'}{\sum_{i=1}^{n} K[\|\mathbf{x}_i - \mathbf{x}^*\|_{\mathbf{S}^{-1}}^2]}, \tag{10.1.5}$$

where $\mathbf{x}^*$ is a robust estimate of the centre of the $\mathbf{x}_i$ such as a multivariate median, and $K[.]$, $\mathbf{S}$ are defined as in (9.2.1). Directions given by the first few eigenvectors of $\mathbf{S}\hat{\boldsymbol{\Gamma}}^{-1}$ are used to identify outliers. Further theoretical details and examples of the technique are given by Caussinus and Ruiz-Gazen (1993, 1995). A mixture model is assumed (see Section 9.2.3) in which one element in the mixture corresponds to the bulk of the data, and the other elements have small probabilities of occurrence and correspond to different types of outliers. In Caussinus et al. (2001) it is assumed that if there are q types of outlier, then q directions are likely needed to detect them. The bulk of the data is assumed to have a spherical distribution, so there is no single $(q+1)$th direction corresponding to these data. The question of an appropriate choice for q needs to be considered. Using asymptotic results for the null (one-component mixture) distribution of a matrix which is closely related to $\mathbf{S}\hat{\boldsymbol{\Gamma}}^{-1}$, Caussinus et al. (2001) use simulation to derive tables of critical values for its eigenvalues. These tables can then be used to assess how many eigenvalues are 'significant,' and hence decide on an appropriate value for q. The use of the tables is illustrated by examples.

The choice of the value of β is discussed by Caussinus and Ruiz-Gazen (1995) and values in the range 0.1 to 0.5 are recommended. Caussinus et al. (2001) use somewhat smaller values in constructing their tables, which are valid for values of β in the range 0.01 to 0.1. Penny and Jolliffe (2001) include Caussinus and Ruiz-Gazen's technique in a comparative study of methods for detecting multivariate outliers. It did well compared to other methods in some circumstances, particularly when there are multiple outliers and p is not too large.

Before turning to examples, recall that an example in which outliers are detected using PCs in a rather different way was given in Section 5.6. In that example, Andrews' curves (Andrews, 1972) were computed using PCs and some of the observations stood out as different from the others when plotted as curves. Further examination of these different observations showed that they were indeed 'outlying' in some respects, compared to the remaining observations.

10.1.1 Examples

In this section one example will be discussed in some detail, while three others will be described more briefly.

Anatomical Measurements

A set of seven anatomical measurements on 28 students was discussed in Section 5.1.1 and it was found that on a plot of the first two PCs (Figures 1.3, 5.1) there was an extreme observation on the second PC. When the measurements of this individual were examined in detail, it was found that he had an anomalously small head circumference. Whereas the other 27 students all had head girths in the narrow range 21–24 cm, this student (no. 16) had a measurement of 19 cm. It is impossible to check whether this was an incorrect measurement or whether student 16 indeed had an unusually small head (his other measurements were close to average), but it is clear that this observation would be regarded as an 'outlier' according to most definitions of the term.

This particular outlier is detected on the second PC, and it was suggested above that any outliers detected by high-variance PCs are usually detectable on examination of individual variables; this is indeed the case here. Another point concerning this observation is that it is so extreme on the second PC that it may be suspected that it alone is largely responsible for the direction of this PC. This question will be investigated at the end of Section 10.2, which deals with influential observations.

Figure 1.3 indicates one other possible outlier at the extreme left of the diagram. This turns out to be the largest student in the class—190 cm (6 ft 3 in) tall, with all measurements except head girth at least as large as all other 27 students. There is no suspicion here of any incorrect measurements.

Turning now to the last few PCs, we hope to detect any observations which are 'outliers' with respect to the correlation structure of the data. Figure 10.3 gives a plot of the scores of the observations for the last two PCs, and Table 10.1 gives the values of d_{1i}^2, d_{2i}^2 and d_{4i}, defined in equations (10.1.1), (10.1.2) and (10.1.4), respectively, for the six 'most extreme'

Table 10.1. Anatomical measurements: values of d_{1i}^2, d_{2i}^2, d_{4i} for the most extreme observations.

		Number of PCs used, q					
$q = 1$					$q = 2$		
d_{1i}^2	Obs. No.	d_{1i}^2	Obs. No.	d_{2i}^2	Obs. No.	d_{4i}	Obs. No.
0.81	15	1.00	7	7.71	15	2.64	15
0.47	1	0.96	11	7.69	7	2.59	11
0.44	7	0.91	15	6.70	11	2.01	1
0.16	16	0.48	1	4.11	1	1.97	7
0.15	4	0.48	23	3.52	23	1.58	23
0.14	2	0.36	12	2.62	12	1.49	27
				$q = 3$			
		d_{1i}^2	Obs. No.	d_{2i}^2	Obs. No.	d_{4i}	Obs. No.
		1.55	20	9.03	20	2.64	15
		1.37	5	7.82	15	2.59	5
		1.06	11	7.70	5	2.59	11
		1.00	7	7.69	7	2.53	20
		0.96	1	7.23	11	2.01	1
		0.93	15	6.71	1	1.97	7

observations on each statistic, where the number of PCs included, q, is 1, 2 or 3. The observations that correspond to the most extreme values of d_{1i}^2, d_{2i}^2 and d_{4i} are identified in Table 10.1, and also on Figure 10.3.

Note that when $q = 1$ the observations have the same ordering for all three statistics, so only the values of d_{1i}^2 are given in Table 10.1. When q is increased to 2 or 3, the six most extreme observations are the same (in a slightly different order) for both d_{1i}^2 and d_{2i}^2. With the exception of the sixth most extreme observation for $q = 2$, the same observations are also identified by d_{4i}. Although the sets of the six most extreme observations are virtually the same for d_{1i}^2, d_{2i}^2 and d_{4i}, there are some differences in ordering. The most notable example is observation 15 which, for $q = 3$, is most extreme for d_{4i} but only sixth most extreme for d_{1i}^2.

Observations 1, 7 and 15 are extreme on all seven statistics given in Table 10.1, due to large contributions from the final PC alone for observation 15, the last two PCs for observation 7, and the fifth and seventh PCs for observation 1. Observations 11 and 20, which are not extreme for the final PC, appear in the columns for $q = 2$ and 3 because of extreme behaviour on the sixth PC for observation 11, and on both the fifth and sixth PCs for observation 20. Observation 16, which was discussed earlier as a clear outlier on the second PC, appears in the list for $q = 1$, but is not notably extreme for any of the last three PCs.

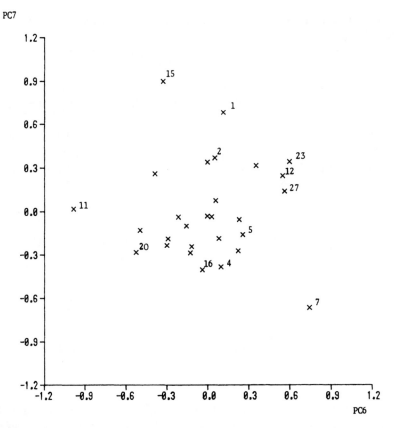

Figure 10.3. Anatomical measurements: plot of observations with respect to the last two PCs.

Most of the observations identified in Table 10.1 are near the edge of the plot given in Figure 10.3. Observations 2, 4, 5, 12, 16, 20, 23 and 27 are close to the main body of the data, but observations 7, 11, 15, and to a lesser extent 1, are sufficiently far from the remaining data to be worthy of further consideration. To roughly judge their 'significance,' recall that, if no outliers are present and the data are approximately multivariate normal, then the values of d_{4i}, are (approximately) absolute values of a normal random variable with zero mean and unit variance. The quantities given in the relevant columns of Table 10.1 are therefore the six largest among $28q$ such variables, and none of them look particularly extreme. Nevertheless, it is of interest to investigate the reasons for the outlying positions of some of the observations, and to do so it is necessary to examine the coefficients of the last few PCs. The final PC, accounting for only 1.7% of the total variation, is largely a contrast between chest and hand measurements with positive coefficients 0.55, 0.51, and waist and height measurements, which have negative coefficients −0.55, −0.32. Looking at observation 15, we find that this (male) student has the equal largest chest measurement, but that only 3 of the other 16 male students are shorter than him, and only two have a smaller waist measurement—perhaps he was a body builder? Similar analyses can be done for other observations in Table 10.1. For example, observation 20 is extreme on the fifth PC. This PC, which accounts for 2.7% of the total variation, is mainly a contrast between height and forearm length with coefficients 0.67, -0.52, respectively. Observation 20 is (jointly with one other) the shortest student of the 28, but only one of the other ten women has a larger forearm measurement. Thus, observations 15 and 20, and other observations indicated as extreme by the last few PCs, are students for whom some aspects of their physical measurements contradict the general positive correlation among all seven measurements.

Household Formation Data

These data were described in Section 8.7.2 and are discussed in detail by Garnham (1979) and Bassett *et al.* (1980). Section 8.7.2 gives the results of a PC regression of average annual total income per adult on 28 other demographic variables for 168 local government areas in England and Wales. Garnham (1979) also examined plots of the last few and first few PCs of the 28 predictor variables in an attempt to detect outliers. Two such plots, for the first two and last two PCs, are reproduced in Figures 10.4 and 10.5. An interesting aspect of these figures is that the most extreme observations with respect to the last two PCs, namely observations 54, 67, 41 (and 47, 53) are also among the most extreme with respect to the first two PCs. Some of these observations are, in addition, in outlying positions on plots of other low-variance PCs. The most blatant case is observation 54, which is among the few most extreme observations on PCs 24 to 28 inclusive, and also on PC1. This observation is 'Kensington and Chelsea,' which must be an outlier with respect to several variables individually, as well as being different in correlation structure from most of the remaining observations.

In addition to plotting the data with respect to the last few and first few PCs, Garnham (1979) examined the statistics d_{1i}^2 for $q = 1, 2, \ldots, 8$ using gamma plots, and also looked at normal probability plots of the values of various PCs. As a combined result of these analyses, he identified six likely outliers, the five mentioned above together with observation 126, which is moderately extreme according to several analyses.

The PC regression was then repeated without these six observations. The results of the regression were noticeably changed, and were better in two respects than those derived from all the observations. The number of PCs which it was necessary to retain in the regression was decreased, and the prediction accuracy was improved, with the standard error of prediction reduced to 77.3% of that for the full data set.

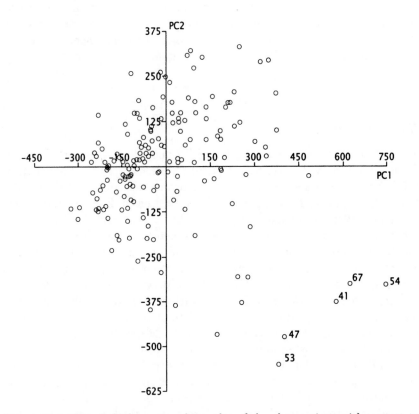

Figure 10.4. Household formation data: plot of the observations with respect to the first two PCs.

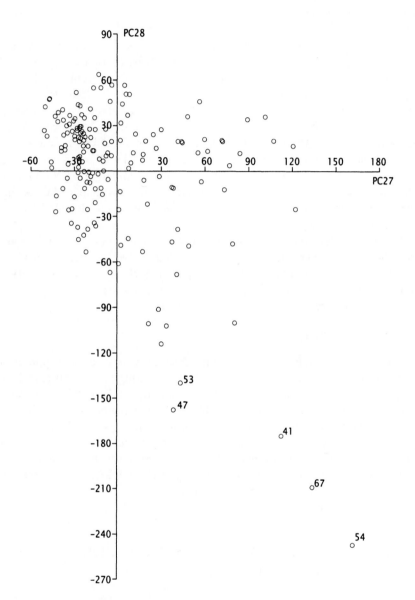

Figure 10.5. Household formation data: plot of the observations with respect to the last two PCs.

Trace Element Concentrations

These data, which are discussed by Hawkins and Fatti (1984), consist of measurements of the log concentrations of 12 trace elements in 75 rock-chip samples. In order to detect outliers, Hawkins and Fatti simply look at the values for each observation on each variable, on each PC, and on transformed and rotated PCs. To decide whether an observation is an outlier, a cut-off is defined assuming normality and using a Bonferroni bound with significance level 0.01. On the original variables, only two observations satisfy this criterion for outliers, but the number of outliers increases to seven if (unrotated) PCs are used. Six of these seven outlying observations are extreme on one of the last four PCs, and each of these low-variance PCs accounts for less than 1% of the total variation. The PCs are thus again detecting observations whose correlation structure differs from the bulk of the data, rather than those that are extreme on individual variables. Indeed, one of the 'outliers' on the original variables is not detected by the PCs.

When transformed and rotated PCs are considered, nine observations are declared to be outliers, including all those detected by the original variables and by the unrotated PCs. There is a suggestion, then, that transfomation and rotation of the PCs as advocated by Hawkins and Fatti (1984) provides an even more powerful tool for detecting outliers.

Epidemiological Data

Bartkowiak et al. (1988) use PCs in a number of ways to search for potential outliers in a large epidemiological data set consisting of 2433 observations on 7 variables. They examine the first two and last two PCs from both correlation and covariance matrices. In addition, some of the variables are transformed to have distributions closer to normality, and the PCAs are repeated after transformation. The researchers report that (unlike Garnham's (1979) analysis of the household formation data) the potential outliers found by the various analyses overlap only slightly. Different analyses are capable of identifying different potential outliers.

10.2   Influential Observations in a Principal Component Analysis

Outliers are generally thought of as observations that in some way are atypical of a data set but, depending on the analysis done, removal of an outlier may or may not have a substantial effect on the results of that analysis. Observations whose removal does have a large effect are called 'influential,' and, whereas most influential observations are outliers in some respect, outliers need not be at all influential. Also, whether or not an observation is influential depends on the analysis being done on the data set; observations that are influential for one type of analysis or parameter of interest may not be so for a different analysis or parameter. This behaviour is evident in PCA where observations that are influential for the coefficients of a PC are not necessarily influential for the variance of that PC, and vice versa. We have seen in the Section 10.1 that PCA can be used to search for influential observations in a regression analysis. The present section concentrates on looking for observations that are influential for some aspect of a PCA, either the variances, the coefficients (loadings) or the PCs themselves (the PC scores).

The intuitive definition of the influence of an observation on a statistic, such as the kth eigenvalue l_k or eigenvector $\mathbf{a}_k$ of a sample covariance matrix, is simply the change in l_k or $\mathbf{a}_k$, perhaps renormalized in some way, when the observation is deleted from the sample. For example, the *sample influence function* for the ith observation on a quantity $\hat{\theta}$, which might be l_k or $\mathbf{a}_k$, is defined by Critchley (1985) as $(n-1)(\hat{\theta} - \hat{\theta}_{(i)})$, where n is the sample size and $\hat{\theta}_{(i)}$ is the quantity corresponding to $\hat{\theta}$ when the ith observation is omitted from the sample. Gnanadesikan and Kettenring (1972) suggested similar leave-one-out statistics for the correlation coefficient and for $\sum_{k=1}^{p} l_k$. The problems with influence defined in this manner are: there may be no closed form for the influence function; the influence needs to be computed afresh for each different sample. Various other definitions of sample influence have been proposed (see, for example, Cook and Weisberg, 1982, Section 3.4; Critchley, 1985); some of these have closed-form expressions for regression coefficients (Cook and Weisberg, 1982, Section 3.4), but not for the statistics of interest in PCA. Alternatively, a theoretical influence function may be defined that can be expressed as a once-and-for-all formula and, provided that sample sizes are not too small, can be used to estimate the influence of actual and potential observations within a sample.

To define the theoretical influence function, suppose that $\mathbf{y}$ is a p-variate random vector, and let $\mathbf{y}$ have cumulative distribution function (c.d.f.) $F(\mathbf{y})$. If $\boldsymbol{\theta}$ is a vector of parameters of the distribution of $\mathbf{y}$ (such as λ_k, $\boldsymbol{\alpha}_k$, respectively, the kth eigenvalue and eigenvector of the covariance matrix of $\mathbf{y}$) then $\boldsymbol{\theta}$ can be written as a functional of F. Now let $F(\mathbf{y})$ be perturbed to become

$$\tilde{F}(\mathbf{y}) = (1 - \varepsilon)F(\mathbf{y}) + \varepsilon \delta_{\mathbf{x}},$$

where $0 < \varepsilon < 1$ and $\delta_{\mathbf{x}}$ is the c.d.f. of the random variable that takes the value $\mathbf{x}$ with certainty; let $\tilde{\boldsymbol{\theta}}$ be the value of $\boldsymbol{\theta}$ when F becomes $\tilde{F}$. Then the influence function $I(\mathbf{x}; \boldsymbol{\theta})$ for $\boldsymbol{\theta}$ evaluated at a value $\mathbf{x}$ of the random variable is defined to be (Hampel, 1974)

$$I(\mathbf{x}; \boldsymbol{\theta}) = \lim_{\varepsilon \to 0} \frac{\tilde{\boldsymbol{\theta}} - \boldsymbol{\theta}}{\varepsilon}.$$

Alternatively, if $\tilde{\theta}$ is expanded about θ as a power series in ε, that is

$$\tilde{\theta} = \theta + \mathbf{c}_1\varepsilon + \mathbf{c}_2\varepsilon^2 + \cdots, \tag{10.2.1}$$

then the influence function is the coefficient $\mathbf{c}_1$ of ε in this expansion.

Some of the above may appear somewhat abstract, but in many situations an expression can be derived for $I(\mathbf{x}; \boldsymbol{\theta})$ without too much difficulty and, as we shall see in examples below, $I(\mathbf{x}; \boldsymbol{\theta})$ can provide valuable guidance about the influence of individual observations in *samples*.

The influence functions for λ_k and $\boldsymbol{\alpha}_k$ are given by Radhakrishnan and Kshirsagar (1981) and by Critchley (1985) for the case of covariance matrices. Critchley (1985) also discusses various sample versions of the influence function and considers the coefficients of ε^2, as well as ε, in the expansion (10.2.1). Pack et al. (1988) give the main results for correlation matrices, which are somewhat different in character from those for covariance matrices. Calder (1986) can be consulted for further details.

For covariance matrices, the theoretical influence function for λ_k can be written very simply as

$$I(\mathbf{x}; \lambda_k) = z_k^2 - \lambda_k, \tag{10.2.2}$$

where z_k is the value of the kth PC for the given value of $\mathbf{x}$, that is, z_k is the kth element of $\mathbf{z}$, where $\mathbf{z} = \mathbf{A}'\mathbf{x}$, using the same notation as in earlier chapters. Thus, the influence of an observation on λ_k depends only on its score on the kth component; an observation can be extreme on any or all of the other components without affecting λ_k. This illustrates the point made earlier that outlying observations need not necessarily be influential for every part of an analysis.

For correlation matrices, $I(\mathbf{x}; \lambda_k)$ takes a different form, which can be written most conveniently as

$$I(\mathbf{x}; \lambda_k) = \sum_{\substack{i=1 \\ i\neq j}}^{p} \sum_{j=1}^{p} \alpha_{ki}\alpha_{kj} I(\mathbf{x}; \rho_{ij}), \tag{10.2.3}$$

where α_{kj} is the jth element of $\boldsymbol{\alpha}_k$, $I(\mathbf{x}; \rho_{ij}) = -\frac{1}{2}\rho_{ij}(x_i^2 + x_j^2) + x_i x_j$, and x_i, x_j are elements of $\mathbf{x}$ standardized to zero mean and unit variance. $I(\mathbf{x}; \rho_{ij})$ is the influence function for the correlation coefficient ρ_{ij}, which is given by Devlin et al. (1975). The expression (10.2.3) is relatively simple, and it shows that investigation of the influence of an observation on the correlation coefficients is useful in determining the influence of the observation on λ_k.

There is a corresponding expression to (10.2.3) for covariance matrices that expresses $I(\mathbf{x}; \lambda_k)$ in terms of influence functions for the elements of the covariance matrix. However, when $I(x; \lambda_k)$ in (10.2.3) is written in terms of $\mathbf{x}$, or the PCs, by substituting for $I(\mathbf{x}; \rho_{ij})$, it cannot be expressed in as simple a form as in (10.2.2). In particular, $I(\mathbf{x}; \lambda_k)$ now depends on z_j, $j = 1, 2, \ldots, p$, and not just on z_k. This result reflects the fact that a change to a covariance matrix may change one of the eigenvalues without affecting the others, but that this cannot happen for a correlation matrix. For a correlation matrix the sum of the eigenvalues is a constant, so that if one of them is changed there must be compensatory changes in at least one of the others.

Expressions for $I(\mathbf{x}; \boldsymbol{\alpha}_k)$ are more complicated than those for $I(\mathbf{x}; \lambda_k)$; for example, for covariance matrices we have

$$I(\mathbf{x}; \boldsymbol{\alpha}_k) = -z_k \sum_{h \neq k}^{p} z_h \boldsymbol{\alpha}_h (\lambda_h - \lambda_k)^{-1} \tag{10.2.4}$$

compared with (10.2.2) for $I(\mathbf{x}; \lambda_k)$. A number of comments can be made concerning (10.2.4) and the corresponding expression for correlation matrices, which is

$$I(\mathbf{x}; \boldsymbol{\alpha}_k) = \sum_{h \neq k}^{p} \boldsymbol{\alpha}_h (\lambda_h - \lambda_k)^{-1} \sum_{i=1}^{p} \sum_{\substack{j=1 \\ i \neq j}}^{p} \alpha_{hi} \alpha_{kj} I(\mathbf{x}; \rho_{ij}). \tag{10.2.5}$$

First, and perhaps most important, the form of the expression is completely different from that for $I(\mathbf{x}; \lambda_k)$. It is possible for an observation to be influential for λ_k but not for $\boldsymbol{\alpha}_k$, and vice versa. This behaviour is illustrated by the examples in Section 10.2.1 below.

A second related point is that for covariance matrices $I(\mathbf{x}; \boldsymbol{\alpha}_k)$ depends on all of the PCs, $z_1, z_2, \ldots, z_p$, unlike $I(\mathbf{x}; \lambda_k)$, which depends just on z_k. The dependence is quadratic, but involves only cross-product terms $z_j z_k$, $j \neq k$, and not linear or squared terms. The general shape of the influence curves $I(\mathbf{x}; \boldsymbol{\alpha}_k)$ is hyperbolic for both covariance and correlation matrices, but the details of the functions are different. The dependence of both (10.2.4) and (10.2.5) on eigenvalues is through $(\lambda_h - \lambda_k)^{-1}$. This means that influence, and hence changes to $\boldsymbol{\alpha}_k$ resulting from small perturbations to the data, tend to be large when λ_k is close to $\lambda_{(k-1)}$ or to $\lambda_{(k+1)}$.

A final point is, that unlike regression, the influence of different observations in PCA is approximately additive, that is the presence of one observation does not affect the influence of another (Calder (1986), Tanaka and Tarumi (1987)).

To show that theoretical influence functions are relevant to sample data, predictions from the theoretical influence function can be compared with the sample influence function, which measures actual changes caused by the deletion from a data set of one observation at a time. The theoretical influence function typically contains unknown parameters and these must be replaced by equivalent sample quantities in such comparisons. This gives what Critchley (1985) calls the *empirical influence function*. He also considers a third sample-based influence function, the *deleted empirical influence function* in which the unknown quantities in the theoretical influence function are estimated using a sample from which the observation whose influence is to be assessed is omitted. The first example given in Section 10.2.1 below illustrates that the empirical influence function can give a good approximation to the sample influence function for moderate sample sizes. Critchley (1985) compares the various influence functions from a more theoretical viewpoint.

A considerable amount of work was done in the late 1980s and early 1990s on influence functions in multivariate analysis, some of which extends the basic results for PCA given earlier in this section. Benasseni in France and Tanaka and co-workers in Japan were particularly active in various aspects of influence and sensitivity for a wide range of multivariate techniques. Some of their work on sensitivity will be discussed further in Section 10.3.

Tanaka (1988) extends earlier work on influence in PCA in two related ways. The first is to explicitly consider the situation where there are equal eigenvalues—equations (10.2.4) and (10.2.5) break down in this case. Secondly he considers influence functions for subspaces spanned by subsets of PCs, not simply individual PCs. Specifically, if $\mathbf{A}_q$ is a matrix whose columns are a subset of q eigenvectors, and Λ_q is the diagonal matrix of corresponding eigenvalues, Tanaka (1988) finds expressions for $I(\mathbf{x}; \mathbf{A}_q\Lambda_q\mathbf{A}_q')$ and $I(\mathbf{x}; \mathbf{A}_q\mathbf{A}_q')$. In discussing a general strategy for analysing influence in multivariate methods, Tanaka (1995) suggests that groups of observations with similar patterns of influence across a set of parameters may be detected by means of a PCA of the empirical influence functions for each parameter.

Benasseni (1990) examines a number of measures for comparing principal component subspaces computed with and without one of the observations. After eliminating some possible measures such as the RV-coefficient (Robert and Escoufier, 1976) and Yanai's generalized coefficient of determination (Yanai, 1980) for being too insensitive to perturbations, he settles on

$$\rho_{1(i)} = 1 - \sum_{k=1}^{q} \frac{\|\mathbf{a}_k - \mathbf{P}_{(i)}\mathbf{a}_k\|}{q}$$

and

$$\rho_{2(i)} = 1 - \sum_{k=1}^{q} \frac{\|\mathbf{a}_{k(i)} - \mathbf{P}\mathbf{a}_{k(i)}\|}{q}$$

where $\mathbf{a}_k, \mathbf{a}_{k(i)}$ are eigenvectors with and without the ith observation, $\mathbf{P}, \mathbf{P}_{(i)}$ are projection matrices onto the subspaces derived with and without the ith observation, and the summation is over the q eigenvectors within the subspace of interest. Benasseni (1990) goes on to find expressions for the theoretical influence functions for these two quantities, which can then be used to compute empirical influences.

Reducing the comparison of two subspaces to a single measure inevitably leads to a loss of information about the structure of the differences between the two. Ramsier (1991) introduces a graphical method, using ideas similar to those of Andrews' curves (see Section 5.6), in which curves representing subspaces with and without individual observations omitted are displayed. It is not easy to deduce exactly how a particular change in subspace structure is reflected in differences between curves. However, curves plotted with an observation missing that are close to the curve for the full data set imply negligible influence on the subspace for that observation, and similarly shaped curves for different omitted observations suggest that these observations have similar detailed effects on the subspace's structure.

Krzanowski (1987a) notes that the algorithm used by Eastment and Krzanowski (1982) to decide how many components to retain (see Section 6.1.5) calculates elements of the singular value decomposition of the data matrix $\mathbf{X}$ with individual observations or variables omitted. It can therefore be used to evaluate the influence of each observation on the subspace spanned by however many components it has been decided to keep. Mertens et al. (1995) similarly use a cross-validation algorithm to give easily computed expressions for the sample influence of observations on the eigenvalues of a covariance matrix. They also provide a closed form expression for the angle between an eigenvector of that matrix using all the data and the corresponding eigenvector when an observation is omitted. An example illustrating the use of these expressions for spectroscopic data is given by Mertens (1998), together with some discussion of the relationships between measures of influence and outlyingness.

Wang and Nyquist (1991) provide a number of algebraic results, proofs and approximations relating eigenvalues and eigenvectors of covariance matrices with and without the removal of one of the n observations. Hadi and Nyquist (1993) give improved approximations for eigenvalues, and Wang and Liski (1993) extend the results for both eigenvalues and eigenvectors to the situation where more than one observation is removed. Comparisons are made with Critchley's (1985) results for the special case of a single deleted observation.

Brooks (1994) uses simulation to address the question of when an apparently influential observation can be declared 'significant.' He points out that the sample influence function (which he confusingly refers to as the 'empirical influence function') for an observation $\mathbf{x}_i$ depends on all the other observations in the sample, so that simulation of repeated values of this function under some null hypothesis needs whole samples to be generated. The theoretical influence function can be evaluated more easily because equations (10.2.2)—(10.2.5) depend only on the eigenvalues and eigenvectors of the correlation or covariance matrix together with the *single* observation whose influence is to be assessed. Thus, if the sample correlation or covariance matrix is used as a surrogate for the corresponding population matrix, it is only necessary to simulate individual observations, rather than a whole new sample, in order to generate a value for this influence function. Of course, any simulation study requires some assumption about the distribution from which the simulated data are generated. Brooks (1994) gives an example in which 1000 simulated observations are generated from a 7-variable multivariate normal distribution whose parameters are those estimated from the available sample of data. Empirical distributions of the estimated theoretical influence functions for eigenvalues and eigenvectors of the correlation matrix are constructed from the simulated observations by computing the values of these functions for each of the 1000 observations. The actual values of the functions for the same 7 anatomical variables as were discussed in Sections 5.1.1 and 10.1, but for a different sample of 44 students, are then compared to these distributions. Observations whose values are in the upper 5% (1%) tails of the distributions can be considered to be 'significantly influential' at the 5% (1%) level. Brooks (1994) finds that, on this basis, 10 of the 44 observations are significant at the 5% level for more than one of the 7 eigenvalues and/or eigenvectors. Brooks (1994) uses the same reasoning to investigate 'significantly influential' observations in a two-dimensional principal component subspace based on Tanaka's (1988) influence functions $I(\mathbf{x}; \mathbf{A}_q \mathbf{\Lambda}_q \mathbf{A}_q')$ and $I(\mathbf{x}; \mathbf{A}_q \mathbf{A}_q')$.

10.2.1   Examples

Two examples are now given, both using data sets that have been discussed earlier. In the first example we examine the usefulness of expressions for theoretical influence in predicting the actual effect of omitting observations for the data on artistic qualities of painters described in Section 5.1.1. As a second illustration, we follow up the suggestion, made in Section 10.1, that an outlier is largely responsible for the form of the second PC in the student anatomical data.

Artistic Qualities of Painters

We consider again the set of four subjectively assessed variables for 54 painters, which was described by Davenport and Studdert-Kennedy (1972) and discussed in Section 5.1.1. Tables 10.2 and 10.3 give some comparisons between the values of the influence functions obtained from expressions such as (10.2.2), (10.2.3), (10.2.4) and (10.2.5) by substituting sample quantities l_k, a_{kj}, r_{ij} in place of the unknown λ_k, α_{kj}, ρ_{ij}, and the actual changes observed in eigenvalues and eigenvectors when individual observations are omitted. The information given in Table 10.2 relates to PCs derived from the covariance matrix; Table 10.3 gives corresponding results for the correlation matrix. Some further explanation is necessary of exactly how the numbers in these two tables are derived.

First, the 'actual' changes in eigenvalues are precisely that—the differences between eigenvalues with and without a particular observation included in the analysis. The tables give the four largest and four smallest such changes for each PC, and identify those observations for which

Table 10.2. Artistic qualities of painters: comparisons between estimated (empirical) and actual (sample) influence of individual observations for the first two PCs, based on the covariance matrix.

Component number	1						2					
	Influence						Influence					
	Eigenvalue			Eigenvector			Eigenvalue			Eigenvector		
	Estimated	Actual	Obs. no.	Estimated	Actual	Obs. no.	Estimated	Actual	Obs. no.	Estimated	Actual	Obs. no.
	2.490	2.486	31	6.791	6.558	34	1.759	1.854	44	11.877	13.767	34
	2.353	2.336	7	5.025	4.785	44	1.757	1.770	48	5.406	5.402	44
	2.249	2.244	28	3.483	4.469	42	1.478	1.491	34	3.559	4.626	42
	2.028	2.021	5	1.405	1.260	48	1.199	1.144	43	2.993	3.650	43
	...	...	...	...	...	...	...	...	...	...	...	...
	0.248	0.143	44	0.001	0.001	53	−0.073	−0.085	49	0.016	0.017	24
	−0.108	−0.117	52	0.000	0.000	17	−0.090	−0.082	20	0.009	0.009	1
	−0.092	−0.113	37	0.000	0.000	24	0.039	0.045	37	0.000	0.001	30
	0.002	−0.014	11	0.000	0.000	21	−0.003	−0.010	10	0.000	0.000	53

Table 10.3. Artistic qualities of painters: comparisons between estimated (empirical) and actual (sample) influence of individual observations for the first two PCs, based on the correlation matrix.

Component number											
	1						2				
	Influence						Influence				
	Eigenvalue			Eigenvector			Eigenvalue			Eigenvector		
	Estimated	Actual	Obs. no.	Estimated	Actual	Obs. no.	Estimated	Actual	Obs. no.	Estimated	Actual	Obs. no.
	−0.075	−0.084	34	1.406	1.539	26	0.075	0.080	34	5.037	7.546	34
	0.068	0.074	31	1.255	1.268	44	0.064	0.070	44	4.742	6.477	43
	−0.067	−0.073	43	0.589	0.698	42	0.061	0.062	43	4.026	4.333	26
	0.062	0.067	28	0.404	0.427	22	0.047	0.049	39	1.975	2.395	42
	...			...			...			...		
	−0.003	−0.003	6	0.001	0.001	33	−0.001	−0.002	45	0.011	0.012	21
	−0.001	−0.001	32	0.000	0.000	53	−0.001	−0.001	14	0.010	0.012	38
	0.001	0.001	11	0.000	0.000	30	−0.000	−0.001	42	0.001	0.001	30
	0.001	0.001	13	0.000	0.000	21	−0.000	−0.000	11	0.001	0.001	53

changes occur. The 'estimated' changes in eigenvalues given in Tables 10.2 and 10.3 are derived from multiples of $I(\cdot)$ in (10.2.2), (10.2.3), respectively, with the value of the kth PC for each individual observation substituted for z_k, and with l_k, a_{kj}, r_{ij} replacing λ_k, α_{kj}, ρ_{ij}. The multiples are required because Change = Influence $\times$ ε, and we need a replacement for ε; here we have used $1/(n-1)$, where $n = 54$ is the sample size. Thus, apart from a multiplying factor $(n-1)^{-1}$, 'actual' and 'estimated' changes are sample and empirical influence functions, respectively.

In considering changes to an eigenvector, there are changes to each of the p (= 4) coefficients in the vector. Comparing vectors is more difficult than comparing scalars, but Tables 10.2 and 10.3 give the sum of squares of changes in the individual coefficients of each vector, which is a plausible measure of the difference between two vectors. This quantity is a monotonically increasing function of the angle in p-dimensional space between the original and perturbed versions of $\mathbf{a}_k$, which further increases its plausibility. The idea of using angles between eigenvectors to compare PCs is discussed in a different context in Section 13.5.

The 'actual' changes for eigenvectors again come from leaving out one observation at a time, recomputing and then comparing the eigenvectors, while the estimated changes are computed from multiples of sample versions of the expressions (10.2.4) and (10.2.5) for $I(\mathbf{x}; \boldsymbol{\alpha}_k)$. The changes in eigenvectors derived in this way are much smaller in absolute terms than the changes in eigenvalues, so the eigenvector changes have been multiplied by 10^3 in Tables 10.2 and 10.3 in order that all the numbers are of comparable size. As with eigenvalues, apart from a common multiplier we are comparing empirical and sample influences.

The first comment to make regarding the results given in Tables 10.2 and 10.3 is that the estimated values are extremely good in terms of obtaining the correct ordering of the observations with respect to their influence. There are some moderately large discrepancies in absolute terms for the observations with the largest influences, but experience with this and several other data sets suggests that the most influential observations are correctly identified unless sample sizes become very small. The discrepancies in absolute values can also be reduced by taking multiples other than $(n-1)^{-1}$ and by including second order (ε^2) terms.

A second point is that the observations which are most influential for a particular eigenvalue need not be so for the corresponding eigenvector, and vice versa. For example, there is no overlap between the four most influential observations for the first eigenvalue and its eigenvector in either the correlation or covariance matrix. Conversely, observations can sometimes have a large influence on both an eigenvalue and its eigenvector (see Table 10.3, component 2, observation 34).

Next, note that observations may be influential for one PC only, or affect two or more. An observation is least likely to affect more than one PC in the case of eigenvalues for a covariance matrix—indeed there is no overlap between the four most influential observations for the first and second eigenvalues in Table 10.2. However, for eigenvalues in a correlation matrix, more than one value is likely to be affected by a very influential observation, because the sum of eigenvalues remains fixed. Also, large changes in an eigenvector for either correlation or covariance matrices result in at least one other eigenvector being similarly changed, because of the orthogonality constraints. These results are again reflected in Tables 10.2 and 10.3, with observations appearing as influential for both of the first two eigenvectors, and for both eigenvalues in the case of the correlation matrix.

Comparing the results for covariance and correlation matrices in Tables 10.2 and 10.3, we see that several observations are influential for both matrices. This agreement occurs because, in the present example, the original variables all have similar variances, so that the PCs for correlation and covariance matrices are similar. In examples where the PCs based on correlation and covariance matrices are very different, the sets of influential observations for the two analyses often show little overlap.

Turning now to the observations that have been identified as influential in Table 10.3, we can examine their positions with respect to the first two PCs on Figures 5.2 and 5.3. Observation 34, which is the most influential observation on eigenvalues 1 and 2 and on eigenvector 2, is the painter indicated in the top left of Figure 5.2, Fr. Penni. His position is not particularly extreme with respect to the first PC, and he does not have an unduly large influence on its direction. However, he does have a strong influence on both the direction and variance (eigenvalue) of the second PC, and to balance the increase which he causes in the second eigenvalue there is a compensatory decrease in the first eigenvalue. Hence, he is influential on that eigenvalue too. Observation 43, Rembrandt, is at the bottom of Figure 5.2 and, like Fr. Penni, has a direct influence on PC2 with an indirect but substantial influence on the first eigenvalue. The other two observations, 28 and 31, Caravaggio and Palma Vecchio, which are listed in Table 10.3 as being influential for the first eigenvalue, have a more direct effect. They are the two observations with the most extreme values on the first PC and appear at the extreme left of Figure 5.2.

Finally, the observations in Table 10.3 that are most influential on the first eigenvector, two of which also have large values of influence for the second eigenvector, appear on Figure 5.2 in the second and fourth quadrants in moderately extreme positions.

Student Anatomical Measurements

In the discussion of the data on student anatomical measurements in Section 10.1 it was suggested that observation 16 is so extreme on the second PC that it could be largely responsible for the direction of that component. Looking at influence functions for these data enables us to investigate this conjecture. Not surprisingly, observation 16 is the most influential observation for the second eigenvector and, in fact, has an influence nearly six times as large as that of the second most influential observation. It is also the most influential on the first, third and fourth eigenvectors, showing that the perturbation caused by observation 16 to the second PC has a 'knock-on' effect to other PCs in order to preserve orthogonality. Although observation 16 is very influential on the eigenvectors, its effect is less marked on the eigenvalues. It has only the fifth highest influence on the second eigenvalue, though it is highest for the fourth eigenvalue, second highest for the first, and fourth highest for the third. It is clear that values of influence on eigenvalues need not mirror major changes in the structure of the eigenvectors, at least when dealing with correlation matrices.

Having said that observation 16 is clearly the most influential, for eigenvectors, of the 28 observations in the data set, it should be noted that its influence in absolute terms is not outstandingly large. In particular, the coefficients rounded to one decimal place for the second PC when observation 16 is omitted are

$$0.2 \quad 0.1 \quad -0.4 \quad 0.8 \quad -0.1 \quad -0.3 \quad -0.2.$$

The corresponding coefficients when all 28 observations are included are

$$-0.0 \quad -0.2 \quad -0.2 \quad 0.9 \quad -0.1 \quad -0.0 \quad -0.0.$$

Thus, when observation 16 is removed, the basic character of PC2 as mainly a measure of head size is unchanged, although the dominance of head size in this component is reduced. The angle between the two vectors defining the second PCs, with and without observation 16, is about 24°, which is perhaps larger than would be deduced from a quick inspection of the simplified coefficients above. Pack et al. (1988) give a more thorough discussion of influence in the context of this data set, together with similar data sets measured on different groups of students (see also Brooks (1994)). A problem that does not arise in the examples discussed here, but which does in Pack et al.'s (1988) larger example, is the possibility that omission of an observation causes switching or rotation of eigenvectors when consecutive eigenvalues have similar magnitudes. What appear to be large changes in eigenvectors may be much smaller when a possible reordering of PCs in the modified PCA is taken into account. Alternatively, a subspace of two or more PCs may be virtually unchanged when an observation is deleted, but individual eigenvectors spanning that subspace can look quite different. Further discussion of these subtleties in the context of an example is given by Pack et al. (1988).

10.3   Sensitivity and Stability

Removing a single observation and estimating its influence explores one type of perturbation to a data set, but other perturbations are possible. In one the weights of one or more observations are reduced, without removing them entirely. This type of 'sensitivity' is discussed in general for multivariate techniques involving eigenanalyses by Tanaka and Tarumi (1985, 1987), with PCA as a special case. Benasseni (1986a) also examines the effect of differing weights for observations on the eigenvalues in a PCA. He gives bounds on the perturbed eigenvalues for any pattern of perturbations of the weights for both covariance and correlation-based analyses. The work is extended in Benasseni (1987a) to include bounds for eigenvectors as well as eigenvalues. A less structured perturbation is investigated empirically by Tanaka and Tarumi (1986). Here each element of a (4×4) 'data' matrix has an independent random perturbation added to it.

In Tanaka and Mori (1997), where the objective is to select a subset of variables reproducing all the p variables as well as possible and hence has connections with Section 6.3, the *influence of variables* is discussed. Fujikoshi et al. (1985) examine changes in the eigenvalues of a covariance matrix when additional variables are introduced. Krzanowski (1987a) indicates how to compare data configurations given by sets of retained PCs, including all the variables and with each variable omitted in turn. The calculations are done using an algorithm for computing the singular value decomposition (SVD) with a variable missing, due to Eastment and Krzanowski (1982), and the configurations are compared by means of Procrustes rotation (see Krzanowski and Marriott 1994, Chapter 5). Holmes-Junca (1985) gives an extensive discussion of the effect of omitting observations or variables from a PCA. As in Krzanowski (1987a), the SVD plays a prominent rôle, but the framework in Holmes-Junca (1985) is a more general one in which unequal weights may be associated with the observations, and a general metric may be associated with the variables (see Section 14.2.2).

A different type of stability is investigated by Benasseni (1986b). He considers replacing each of the n p-dimensional observations in a data set by a p-dimensional random variable whose probability distribution is centred on the observed value. He relates the covariance matrix in the perturbed case to the original covariance matrix and to the covariance matrices of the n random variables. From this relationship, he deduces bounds on the eigenvalues of the perturbed matrix. In a later paper, Benasseni (1987b) looks at fixed, rather than random, perturbations to one or more of the observations. Expressions are given for consequent changes to eigenvalues and eigenvectors of the covariance matrix, together with approximations to those changes. A number of special forms for the perturbation, for example where it affects only one of the p variables, are examined in detail. Corresponding results for the correlation matrix are discussed briefly.

Dudziński et al. (1975) discuss what they call 'repeatability' of principal components in samples, which is another way of looking at the stability of the components' coefficients. For each component of interest the angle is calculated between the vector of coefficients in the population and the corresponding vector in a sample. Dudziński et al. (1975) define a *repeata-*

bility index as the proportion of times in repeated samples that this angle has a cosine whose value is greater than 0.95. They conduct a simulation study to examine the dependence of this index on sample size, on the ratio of consecutive population eigenvalues, and on whether or not the data are normally distributed. To generate non-normal samples, Dudziński and researchers use the bootstrap idea of sampling *with replacement* from a data set that is clearly non-normal. This usage predates the first appearance of the term 'bootstrap.' Their simulation study is relatively small, but it demonstrates that repeatability is often greater for normal than for non-normal data with the same covariance structure, although the differences are usually small for the cases studied and become very small as the repeatability increases with sample size. Repeatability, as with other forms of stability, decreases as consecutive eigenvalues become closer.

Dudziński et al. (1975) implemented a bootstrap-like method for assessing stability. Daudin et al. (1988) use a fully-fledged bootstrap and, in fact, note that more than one bootstrapping procedure may be relevant in resampling to examine the properties of PCA, depending in part on whether a correlation-based or covariance- based analysis is done. They consider a number of measures of stability for both eigenvalues and eigenvectors, but the stability of subspaces spanned by subsets of PCs is deemed to be of particular importance. This idea is used by Besse (1992) and Besse and de Falguerolles (1993) to choose how many PCs to retain (see Section 6.1.5). Stability indices based on the jackknife are also used by the latter authors, and Daudin et al. (1989) discuss one such index in detail for both correlation- and covariance-based PCA. Besse and de Falguerolles (1993) describe the equally weighted version of the criterion (6.1.6) as a natural criterion for stability, but prefer

$$L_q = \frac{1}{2}\|\mathbf{P}_q - \hat{\mathbf{P}}_q\|^2$$

in equation (6.1.9). Either (6.1.6) or the expectation of L_q can be used as a measure of the stability of the first q PCs or the subspace spanned by them, and q is then chosen to optimize stability. Besse and de Falguerolles (1993) discuss a variety of ways of estimating their stability criterion, including some based on the bootstrap and jackknife. The bootstrap has also been used to estimate standard errors of elements of the eigenvectors $\mathbf{a}_k$ (see Section 3.7.2), and these standard errors can be viewed as further measures of stability of PCs, specifically the stability of their coefficients.

Stauffer et al. (1985) conduct a study that has some similarities to that of Dudziński et al. (1975). They take bootstrap samples from three ecological data sets and use them to construct standard errors for the eigenvalues of the correlation matrix. The stability of the eigenvalues for each data set is investigated when the full data set is replaced by subsets of variables or subsets of observations. Each eigenvalue is examined as a percentage of variation remaining after removing earlier PCs, as well as in absolute terms.

Comparisons are made with corresponding results derived from 'random,' that is uncorrelated, data sets of the same size as the real data sets. This is done with an objective similar to that of parallel analysis (see Section 6.1.3) in mind.

A different approach to influence from that described in Section 10.2 was proposed by Cook (1986) and called *local influence*. Shi (1997) develops these ideas in the context of PCA. It is included in this section rather than in 10.2 because, as with Tanaka and Tarumi (1985, 1987) and Benasseni (1986a, 1987a) weights are attached to each observation, and the weights are varied. In Shi's (1997) formulation the ith observation is expressed as

$$x_i(\mathbf{w}) = w_i(x_i - \bar{x}),$$

and $\mathbf{w} = (w_1, w_2, \ldots, w_n)'$ is a vector of weights. The vector $\mathbf{w}_0 = (1, 1, \ldots, 1)'$ gives the unperturbed data, and Shi considers perturbations of the form $\mathbf{w} = \mathbf{w}_0 + \epsilon \mathbf{h}$, where $\mathbf{h}$ is a fixed unit-length vector. The *generalized influence function* for a functional $\boldsymbol{\theta}$ is defined as

$$\mathrm{GIF}(\boldsymbol{\theta}, \mathbf{h}) = \lim_{\varepsilon \to 0} \frac{\boldsymbol{\theta}(\mathbf{w}_0 + \epsilon \mathbf{h}) - \boldsymbol{\theta}(\mathbf{w}_0)}{\epsilon}.$$

For a scalar θ, such as an eigenvalue, local influence investigates the direction $\mathbf{h}$ in which $\mathrm{GIF}(\theta, \mathbf{h})$ is maximized. For a vector $\boldsymbol{\theta}$, such as an eigenvector, $\mathrm{GIF}(.)$ is converted into a scalar (a norm) via a quadratic form, and then $\mathbf{h}$ is found for which this norm is maximized.

Yet another type of stability is the effect of changes in the PC coefficients $\mathbf{a}_k$ on their variances l_k and vice versa. Bibby (1980) and Green (1977) both consider changes in variances, and other aspects such as the non-orthogonality of the changed $\mathbf{a}_k$ when the elements of $\mathbf{a}_k$ are rounded. This is discussed further in Section 11.3.

Krzanowski (1984b) considers what could be thought of as the opposite problem to that discussed by Bibby (1980). Instead of looking at the effect of small changes in $\mathbf{a}_k$ on the value of l_k, Krzanowski (1984b) examines the effect of small changes in the value of l_k on $\mathbf{a}_k$, although he addresses the problem in the population context and hence works with $\boldsymbol{\alpha}_k$ and λ_k. He argues that this is an important problem because it gives information on another aspect of the stability of PCs: the PCs can only be confidently interpreted if their coefficients are stable with respect to small changes in the values of the λ_k.

If λ_k is *decreased* by an amount ε, then Krzanowski (1984b) looks for a vector $\boldsymbol{\alpha}_k^\varepsilon$ that is maximally different from $\boldsymbol{\alpha}_k$, subject to $\mathrm{var}(\boldsymbol{\alpha}_k^{\prime\varepsilon} \mathbf{x}) = \lambda_k - \varepsilon$. He finds that the angle θ between $\boldsymbol{\alpha}_k$ and $\boldsymbol{\alpha}_k^\varepsilon$ is given by

$$\cos \theta = [1 + \frac{\varepsilon}{(\lambda_k - \lambda_{k+1})}]^{-1/2}, \tag{10.3.1}$$

and so depends mainly on the difference between λ_k and λ_{k+1}. If λ_k, λ_{k+1} are close, then the kth PC, $\boldsymbol{\alpha}_k' \mathbf{x} = z_k$, is more unstable than if λ_k, λ_{k+1} are well separated. A similar analysis can be carried out if λ_k is *increased* by an amount ε, in which case the stability of z_k depends on the separation between λ_k and λ_{k-1}. Thus, the stability of a PC depends on the separation of its variance from the variances of adjacent PCs, an unsurprising result, especially considering the discussion of 'influence' in Section 10.2. The ideas in Section 10.2 are, as here, concerned with how perturbations affect α_k, but they differ in that the perturbations are deletions of individual observations, rather than hypothetical changes in λ_k. Nevertheless, we find in both cases that the changes in α_k are largest if λ_k is close to λ_{k+1} or λ_{k-1} (see equations (10.2.4) and (10.2.5)).

As an example of the use of the sample analogue of the expression (10.3.1), consider the PCs presented in Table 3.2. Rounding the coefficients in the PCs to the nearest 0.2 gives a change of 9% in l_1 and changes the direction of $\mathbf{a}_1$ through an angle of about 8° (see Section 11.3). We can use (10.3.1) to find the maximum angular change in $\mathbf{a}_1$ that can occur if l_1 is decreased by 9%. The maximum angle is nearly 25°, so that rounding the coefficients has certainly not moved $\mathbf{a}_1$ in the direction of maximum sensitivity.

The eigenvalues l_1, l_2 and l_3 in this example are 2.792, 1.532 and 1.250, respectively, so that the separation between l_1 and l_2 is much greater than that between l_2 and l_3. The potential change in $\mathbf{a}_2$ for a given decrease in l_2 is therefore larger than that for $\mathbf{a}_1$, given a corresponding decrease in l_1. In fact, the same percentage decrease in l_2 as that investigated above for l_1 leads to a maximum angular change of 35°; if the change is made the same in absolute (rather than percentage) terms, then the maximum angle becomes 44°.

10.4   Robust Estimation of Principal Components

It has been noted several times in this book that for PCA's main (descriptive) rôle, the form of the distribution of $\mathbf{x}$ is usually not very important. The main exception to this statement is seen in the case where outliers may occur. If the outliers are, in fact, influential observations, they will have a disproportionate effect on the PCs, and if PCA is used blindly in this case, without considering whether any observations are influential, then the results may be largely determined by such observations. For example, suppose that all but one of the n observations lie very close to a plane within p-dimensional space, so that there are two dominant PCs for these $(n-1)$ observations. If the distance of the remaining observation from the plane is greater than the variation of the $(n-1)$ observations within the plane, then the first component for the n observations will be determined solely by the direction of the single outlying observation from the plane. This, incidentally, is a case where the first PC, rather than last few, will detect the outlier (see Section 10.1), but if the distance from the plane is larger than distances within the plane, then the observation is likely to 'stick out' on at least one of the original variables as well.

To avoid such problems, it is possible to use robust estimation of the covariance or correlation matrix, and hence of the PCs. Such estimation downweights or discards the effect of any outlying observations. Five robust estimators using three different approaches are investigated by Devlin et al. (1981). The first approach robustly estimates each element of the covariance or correlation matrix separately and then 'shrinks' the elements to achieve positive-definiteness if this is not achieved with the initial estimates. The second type of approach involves robust regression of x_j on $(x_1, x_2, \ldots, x_{j-1})$ for $j = 2, 3, \ldots, p$. An illustration of the robust regression approach is presented for a two-variable example by Cleveland and Guarino (1976). Finally, the third approach has three variants; in one (multivariate trimming) the observations with the largest Mahalanobis distances D_i from a robust estimate of the mean of $\mathbf{x}$ are discarded, and in the other two they are downweighted. One of these variants is used by Coleman (1985) in the context of quality control. Both of the downweighting schemes are examples of so-called *M-estimators* proposed by Maronna (1976). One is the maximum likelihood estimator for a p-variate elliptical t distribution, which has longer tails than the multivariate normal distribution for which the usual non-robust estimate is optimal. The second downweighting scheme uses Huber weights (1964, 1981) which are constant for values of D_i up to a threshold D_i^*, and equal to D_i^*/D_i thereafter.

All but the first of these five estimates involve iteration; full details are given by Devlin et al. (1981), who also show that the usual estimator of the covariance or correlation matrix can lead to misleading PCs if outlying observations are included in the analysis. Of the five possible robust alternatives which they investigate, only one, that based on robust regression, is clearly dominated by other methods, and each of the remaining four may be chosen in some circumstances.

Robust estimation of covariance matrices, using an iterative procedure based on downweighting observations with large Mahalanobis distance from the mean, also based on M-estimation, was independently described by Campbell (1980). He uses Huber weights and, as an alternative, a so-called redescending estimate in which the weights decrease more rapidly than Huber weights for values of D_i larger than the threshold D_i^*. Campbell (1980) notes that, as a by-product of robust estimation, the weights given to each data point give an indication of potential outliers. As the weights are non-increasing functions of Mahalanobis distance, this procedure is essentially using the statistic d_{2i}^2, defined in Section 10.1, to identify outliers, except that the mean and covariance matrix of $\mathbf{x}$ are estimated robustly in the present case.

Other methods for robustly estimating covariance or correlation matrices have been suggested since Devlin et al.'s (1981) work. For example, Mehrotra (1995) proposes a new elementwise estimate, and compares it to Devlin et al.'s (1981) estimates via a simulation study whose structure is the same as that of Devlin and co-workers. Maronna and Yohai (1998) review many of the robust covariance estimators. As Croux and Haesbroek (2000) point out, every new robust covariance matrix estimator has a new robust PCA method associated with it.

Campbell (1980) also proposed a modification of his robust estimation technique in which the weights assigned to an observation, in estimating the covariance matrix, are no longer functions of the overall Mahalanobis distance of each observation from the mean. Instead, when estimating the kth PC, the weights are a decreasing function of the absolute value of the score of each observation for the kth PC. As with most of the techniques tested by Devlin et al. (1981), the procedure is iterative and the algorithm to implement it is quite complicated, with nine separate steps. However, Campbell (1980) and Matthews (1984) each give an example for which the technique works well, and as well as estimating the PCs robustly, both authors use the weights found by the technique to identify potential outliers.

The discussion in Section 10.2 noted that observations need not be particularly 'outlying' in order to be influential. Thus, robust estimation methods that give weights to each observation based on Mahalanobis distance will 'miss' any influential observations that are not extreme with respect to Mahalanobis distance. It would seem preferable to downweight observations according to their influence, rather than their Mahalanobis distance. As yet, no systematic work seems to have been done on this idea, but it should be noted that the influence function for the kth eigenvalue of a covariance matrix is an increasing function of the absolute score on the kth PC (see equation (10.2.2)). The weights used in Campbell's (1980) procedure therefore downweight observations according to their influence on eigenvalues (though not eigenvectors) of the covariance (but not correlation) matrix.

It was noted above that Devlin et al. (1981) and Campbell (1980) use M-estimators in some of their robust PCA estimates. A number of other authors have also considered M-estimation in the context of PCA. For example, Daigle and Rivest (1992) use it to construct a robust version of the biplot (see Section 5.3). Ibazizen (1986) gives a thorough discussion of an M-estimation approach to robust PCA in which the definition of the first robust PC is based on a robust version of Property A5 in Section 2.1. This property implies that the first PC minimizes the sum of residual variances arising from predicting each variable from the same single linear function of the variables. Each variance is an expected squared difference between the variable and its mean. To robustify this quantity the mean is replaced by a robust estimator of location, and the *square* of the difference is replaced by another function, namely one of those typically chosen in M-estimation. Ibazizen (1986) includes a substantial amount of underlying theory for this robust PC, together with details of an algorithm to implement the procedure for both the first and suitably defined second, third, ... robust PCs. Xu and Yuille (1995) present a robust PCA using a 'statistical physics' approach within a neural network framework (see Sections 14.1.3, 14.6.1). The function optimized in their algorithm can be regarded as a generalization of a robust redescending M-estimator.

Locantore et al. (1999) discuss robust PCA for functional data (see Section 12.3). In fact, their data are images rather than functions, but a pre-processing step turns them into functions. Both means and covariance matrices are robustly estimated, the latter by shrinking extreme observations to the nearest point on the surface of a hypersphere or hyperellipsoid centred at the robustly estimated mean, a sort of multivariate Winsorization. Locantore et al.'s (1999) paper is mainly a description of an interesting case study in opthalmology, but it is followed by contributions from 9 sets of discussants, and a rejoinder, ranging over many aspects of robust PCA in a functional context and more generally.

A different type of approach to the robust estimation of PCs is discussed by Gabriel and Odoroff (1983). The approach relies on the fact that PCs may be computed from the singular value decomposition (SVD) of the $(n \times p)$ data matrix $\mathbf{X}$ (see Section 3.5 and Appendix A1). To find the SVD, and hence the PCs, a set of equations involving weighted means of functions of the elements of $\mathbf{X}$ must be solved iteratively. Replacing the weighted means by medians, weighted trimmed means, or some other measure of location which is more robust than the mean leads to estimates of PCs that are less sensitive than the usual estimates to the presence of 'extreme' observations.

Yet another approach, based on 'projection pursuit,' is proposed by Li and Chen (1985). As with Gabriel and Odoroff (1983), and unlike Campbell (1980) and Devlin et al. (1981), the PCs are estimated directly without first finding a robustly estimated covariance matrix. Indeed, Li and Chen suggest that it may be better to estimate the covariance matrix $\boldsymbol{\Sigma}$ from the robust PCs via the spectral decomposition (see Property A3 of Sections 2.1 and 3.1), rather than estimating $\boldsymbol{\Sigma}$ directly. Their idea is to find linear functions of $\mathbf{x}$ that maximize a robust estimate of scale, rather than functions that maximize variance. Properties of their estimates are investigated, and the estimates' empirical behaviour is compared with that of Devlin et al. (1981)'s estimates using simulation studies. Similar levels of performance are found for both types of estimate, although, as noted by Croux and Haesbroeck (2000), methods based on robust estimation of covariance matrices have poor properties for large values of p and projection pursuit-based techniques may be preferred in such cases. One disadvantage of Li and Chen's (1985) approach is that it is complicated to implement. Both Xie et al. (1993) and Croux and Ruiz-Gazen (1995) give improved algorithms for doing so.

Another way of 'robustifying' PCA directly, rather than doing so via a robust covariance or correlation matrix, is described by Baccini et al.

(1996). When the unweighted version of the fixed effects model of Section 3.9 assumes a multivariate normal distribution for its error term e_i, maximum likelihood estimation of the model leads to the usual PCs. However, if the elements of e_i are instead assumed to be independent Laplace random variables with probability density functions $f(e_{ij}) = \frac{1}{2\sigma} exp(-\frac{1}{\sigma}|e_{ij}|)$, maximum likelihood estimation requires the minimization of $\sum_{i=1}^{n} \sum_{j=1}^{p} |x_{ij} - z_{ij}|$, leading to an L_1-norm variant of PCA. Baccini et al. (1996) show that the L_1-norm PCs can be estimated using a canonical correlation analysis (see Section 9.3) of the original variables and ranked version of the variables. Although by no means a robust method, it seems natural to note here the minimax approach to component analysis proposed by Bargmann and Baker (1977). Whereas PCA minimizes the sum of squared discrepancies between x_{ij} and a rank-m approximation, and Baccini et al. (1996) minimize the sum of absolute discrepencies, Bargmann and Baker (1977) suggest minimizing the *maximum* discrepancy. They provide arguments to justify the procedure, but it is clearly sensitive to extreme observations and could even be thought of as 'anti-robust.'

A different, but related, topic is robust estimation of the *distribution* of the PCs, their coefficients and their variances, rather than robust estimation of the PCs themselves. It was noted in Section 3.6 that this can be done using bootstrap estimation (Diaconis and Efron, 1983). The 'shape' of the estimated distributions should also give some indication of whether any highly influential observations are present in a data set (the distributions may be multimodal, corresponding to the presence or absence of the influential observations in each sample from the data set), although the method will not directly identify such observations.

The ideas of robust estimation and influence are brought together in Jaupi and Saporta (1993), Croux and Haesbroeck (2000) and Croux and Ruiz-Gazen (2001). Given a robust PCA, it is of interest to examine influence functions for the results of such an analysis. Jaupi and Saporta (1993) investigate influence for M-estimators, and Croux and Haesbroek (2000) extend these results to a much wider range of robust PCAs for both covariance and correlation matrices. Croux and Ruiz-Gazen (2001) derive influence functions for Li and Chen's (1985) projection pursuit-based robust PCA. Croux and Haesbroeck (2000) also conduct a simulation study using the same structure as Devlin et al. (1981), but with a greater range of robust procedures included. They recommend the S-estimator, described in Rousseeuw and Leroy (1987, p. 263) for practical use.

Naga and Antille (1990) explore the stability of PCs derived from robust estimators of covariance matrices, using the measure of stability defined by Daudin et al. (1988) (see Section 10.3). PCs derived from covariance estimators based on minimum variance ellipsoids perform poorly with respect to this type of stability on the data sets considered by Naga and Antille (1990), but those associated with M-estimated covariance matrices are much better.

Finally, we mention that Ruymgaart (1981) discusses a class of robust PC estimators. However, his discussion is restricted to bivariate distributions (that is $p = 2$) and is entirely theoretical in nature. Ibazizen (1986) suggests that it would be difficult to generalize Ruymgaart's (1981) ideas to more than two dimensions.

10.5   Concluding Remarks

The topics discussed in this chapter pose difficult problems in data analysis. Much research has been done and is continuing on all of them. It is useful to identify potentially outlying observations, and PCA provides a number of ways of doing so. Similarly, it is important to know which observations have the greatest influence on the results of a PCA.

Identifying potential outliers and influential observations is, however, only part of the problem; the next, perhaps more difficult, task is to decide whether the most extreme or influential observations are sufficiently extreme or influential to warrant further action and, if so, what that action should be. Tests of significance for outliers were discussed only briefly in Section 10.1 because they are usually only approximate, and tests of significance for influential observations in PCA have not yet been widely used. Perhaps the best advice is that observations that are much more extreme or influential than most of the remaining observations in a data set should be thoroughly investigated, and explanations sought for their behaviour. The analysis could also be repeated with such observations omitted, although it may be dangerous to act as if the deleted observations never existed. Robust estimation provides an automatic way of dealing with extreme (or influential) observations but, if at all possible, it should be accompanied by a careful examination of any observations that have been omitted or substantially downweighted by the analysis.

11
Rotation and Interpretation of Principal Components

It was noted earlier, especially in Chapter 4, that PCs are particularly useful if they can be simply interpreted. The word *reification* is sometimes used for this process of interpretation, and a number of examples have been seen in previous chapters for which this has been successful. However, the construction of PCs as linear combinations of *all* the measured variables means that interpretation is not always easy. Various suggestions have been made to simplify the process of interpretation. These are the subject of this chapter.

One way to aid interpretation is to rotate the components, as is done with the factor loadings in factor analysis (see Chapter 7). Rotation of PCs is discussed in Section 11.1. The approach can provide useful simplification in some cases, but it has a number of drawbacks, and some alternative approaches are described. In one the two steps of PCA followed by rotation are replaced by a single optimization problem which takes into account both variance maximization and simplicity. In others simplification criteria are emphasized at the expense of variance maximization.

Section 11.2 describes some alternatives to PCA which aim to provide simpler 'components.' Some techniques restrict the coefficients of the variables in each component to integer values, whilst another drives some of the coefficients to zero. The purpose of these techniques is to provide *replacements* for PCs that are simpler to interpret, but which do not sacrifice much variance. In other circumstances, the objective may be to *approximate* the PCs in a way which makes them simpler to interpret, again without much loss of variance. The most common way of doing this is to ignore (effectively set to zero) coefficients whose absolute values fall below some threshold. There are links here to variable selection (see Section 6.3). This strategy and some of the problems associated with it are discussed in Section 11.3, and the chapter concludes with a short section on the desire in some disciplines to attach *physical* interpretations to the principal components.

11.1   Rotation of Principal Components

In Chapter 7 it was seen that rotation is an integral part of factor analysis, with the objective of making the rotated factors as simple as possible to interpret. The same ideas can be used to simplify principal components. A principal component is a linear function of *all* the p original variables. If the coefficients or loadings for a PC are all of a similar size, or if a few are large and the remainder small, the component looks easy to interpret, although, as will be seen in Section 11.3, looks can sometimes be deceiving. Several examples in Chapter 4 are like this, for instance, components 1 and 2 in Section 4.1. If there are intermediate loadings, as well as large and small ones, the component can be more difficult to interpret, for example, component 4 in Table 7.1.

Suppose that it has been decided that the first m components account for most of the variation in a p-dimensional data set. It can then be argued that it is more important to interpret simply the m-dimensional space defined by these m components than it is to interpret each individual component. One way to tackle this objective is to rotate the axes within this m-dimensional space in a way that simplifies the interpretation of the axes as much as possible. More formally, suppose that $\mathbf{A}_m$ is the $(p \times m)$ matrix whose kth column is the vector of loadings for the kth PC. Following similar steps to those in factor analysis (see Section 7.2), orthogonally rotated PCs have loadings given by the columns of $\mathbf{B}_m$, where $\mathbf{B}_m = \mathbf{A}_m \mathbf{T}$, and $\mathbf{T}$ is a $(m \times m)$ orthogonal matrix. The matrix $\mathbf{T}$ is chosen so as to optimize one of many simplicity criteria available for factor analysis. Rotation of PCs is commonplace in some disciplines, such as atmospheric science, where there has been extensive discussion of its advantages and disadvantages (see, for example Richman (1986, 1987); Jolliffe (1987b); Rencher (1995, Section 12.8.2)). Oblique, instead of orthogonal, rotation is possible, and this gives extra flexibility (Cohen, 1983; Richman, 1986). As noted in Chapter 7, the choice of simplicity criterion is usually less important than the choice of m, and in the examples of the present chapter only the well-known varimax criterion is used (see equation (7.2.2)). Note, however, that Jackson (1991, Section 8.5.1) gives an example in which the results from two orthogonal rotation criteria (varimax and quartimax) have non-trivial differences. He states that neither is helpful in solving the problem that he wishes to address for his example.

It is certainly possible to simplify interpretion of PCs by using rotation. For example, Table 7.2 gives two versions of rotated loadings for the first four PCs in a study of scores on 10 intelligence tests for 150 children from the Isle of Wight. The two versions correspond to varimax rotation and to an oblique rotation criterion, direct quartimin. The unrotated loadings are given in Table 7.1. The type of simplicity favoured by almost all rotation criteria attempts to drive loadings towards zero or towards their maximum possible absolute value, which with most scalings of the loadings is 1. The idea is that it should then be clear which variables are 'important' in a (rotated) component, namely, those with large absolute values for their loadings, and those which are not important (loadings near zero). Intermediate-value loadings, which are difficult to interpret, are avoided as much as possible by the criteria. Comparing Tables 7.1 and 7.2, it is apparent that this type of simplicity has been achieved by rotation.

There are other types of simplicity. For example, the first unrotated component in Table 7.1 has all its loadings of similar magnitude and the same sign. The component is thus simple to interpret, as an average of scores on all ten tests. This type of simplicity is shunned by most rotation criteria, and it is difficult to devise a criterion which takes into account more than one type of simplicity, though Richman (1986) attempts to broaden the definition of simple structure by graphical means.

In the context of PCA, rotation has a number of drawbacks:

- A choice has to made from a large number of possible rotation criteria. Cattell (1978) and Richman (1986), respectively, give non-exhaustive lists of 11 and 19 such criteria. Frequently, the choice is made arbitrarily, for example by using the default criterion in a computer package (often varimax). Fortunately, as noted already, different choices of criteria, at least within orthogonal rotation, often make little difference to the results.

- PCA *successively maximizes* variance accounted for. When rotation is done, the total variance within the rotated m-dimensional subspace remains unchanged; it is still the maximum that can be achieved, but it is redistributed amongst the rotated components more evenly than before rotation. This means that information about the nature of any really dominant components may be lost, unless they are already 'simple' in the sense defined by the chosen rotation criterion.

- The choice of m can have a large effect on the results after rotation. This is illustrated in Tables 7.4 and 7.2 when moving from $m = 3$ to $m = 4$ for the children's intelligence example. Interpreting the 'most important dimensions' for a data set is clearly harder if those 'dimensions' appear, disappear, and possibly reappear, as m changes.

- The choice of normalization constraint used for the columns in the matrix $\mathbf{A}_m$ changes the properties of the rotated loadings. From the theory of PCA, it is natural to expect the columns of $\mathbf{A}_m$ to be normalized so as to have unit length, but it is far more common in computer packages to rotate a matrix $\tilde{\mathbf{A}}_m$ whose columns have lengths equal to their corresponding eigenvalues, as in the sample version of equation (2.3.2). The main reason for this convention is almost certainly due to the fact that it has been borrowed from factor analysis. This is discussed further later in this section, but first the properties associated with the rotated components under these two normalizations are explored.

In PCA the components possess two 'orthogonality' properties. First

$$\mathbf{A}'_m\mathbf{A}_m = \mathbf{I}_m, \tag{11.1.1}$$

where $\mathbf{I}_m$ is the identity matrix of order m. Hence, the vectors of loadings for different components are orthogonal. Second, if $\mathbf{A}$ and $\mathbf{Z}$ are, as in previous chapters, the $(p \times p)$ matrix of loadings or coefficients and the $(n \times p)$ matrix of scores, respectively, for all p PCs, and $\mathbf{L}^2$, as in Chapter 5, is the diagonal matrix whose elements are eigenvalues of $\mathbf{X}'\mathbf{X}$, then

$$\mathbf{Z}'\mathbf{Z} = \mathbf{A}'\mathbf{X}'\mathbf{X}\mathbf{A} = \mathbf{A}'\mathbf{A}\mathbf{L}^2\mathbf{A}'\mathbf{A} = \mathbf{L}^2. \tag{11.1.2}$$

The second equality in (11.1.2) follows from the algebra below equation (5.3.6), and the last equality is a result of the orthogonality of $\mathbf{A}$. The implication of this result is that all the p unrotated components, including the first m, are uncorrelated with other.

The fact that orthogonality of vectors of loadings and uncorrelatedness of component scores both hold for PCs is because the loadings are given by eigenvectors of the covariance matrix $\mathbf{S}$ corresponding to the result for $\mathbf{\Sigma}$ in Section 1.1. After rotation, one or both of these properties disappears. Let $\mathbf{Z}_m = \mathbf{X}\mathbf{A}_m$ be the $(n \times m)$ matrix of PC scores for n observations on the first m PCs, so that

$$\mathbf{Z}_m^R = \mathbf{X}\mathbf{B}_m = \mathbf{X}\mathbf{A}_m\mathbf{T} = \mathbf{Z}_m\mathbf{T} \tag{11.1.3}$$

is the corresponding matrix of rotated PC scores. Consider

$$\mathbf{B}'_m\mathbf{B}_m = \mathbf{T}'\mathbf{A}'_m\mathbf{A}_m\mathbf{T}, \tag{11.1.4}$$

and

$$\mathbf{Z}_m^{\prime R}\mathbf{Z}_m^R = \mathbf{T}'\mathbf{Z}'_m\mathbf{Z}_m\mathbf{T}. \tag{11.1.5}$$

With the usual PCA normalization $\mathbf{a}'_k\mathbf{a}_k = 1$, or $\mathbf{A}'_m\mathbf{A}_m = \mathbf{I}_m$, equation (11.1.4) becomes $\mathbf{T}'\mathbf{T}$, which equals $\mathbf{I}_m$ for orthogonal rotation. However, if $\mathbf{L}_m^2$ is the $(m \times m)$ submatrix of $\mathbf{L}^2$ consisting of its first m rows and columns then, from (11.1.2), equation (11.1.5) becomes $\mathbf{T}'\mathbf{L}_m^2\mathbf{T}$, which is not diagonal. Hence the rotated components using this normalization have orthogonal loadings, but are not uncorrelated.

Next consider the common alternative normalization, corresponding to equation (2.3.2), in which each column $\mathbf{a}_k$ of $\mathbf{A}_m$ is multiplied by the square root of the corresponding eigenvalue of $\mathbf{S}$, giving

$$\tilde{\mathbf{a}}_k = \left(\frac{l_k}{n-1}\right)^{\frac{1}{2}} \mathbf{a}_k$$

or

$$\tilde{\mathbf{A}}_m = (n-1)^{-\frac{1}{2}} \mathbf{A}_m \mathbf{L}_m$$

[Note that in this section we are using the notation of Section 5.3. Hence the matrix $\mathbf{L}$ has elements $l_k^{1/2}$ that are square roots of the eigenvalues of $\mathbf{X}'\mathbf{X}$. As $\mathbf{S} = \mathbf{X}'\mathbf{X}/(n-1)$, the square roots of its eigenvalues are $\left(\frac{l_k}{n-1}\right)^{1/2}$].

For this normalization, equation (11.1.4) becomes $\mathbf{T}'\mathbf{L}_m^2\mathbf{T}/(n-1)$ and (11.1.5) is $\mathbf{T}'\mathbf{L}_m^4\mathbf{T}/(n-1)$. Neither of these matrices is diagonal, so the rotated PCs are correlated and have loadings that are not orthogonal. To obtain uncorrelated components a different normalization is needed in which each column $\mathbf{a}_k$ of $\mathbf{A}_m$ is *divided* by the square root of the corresponding eigenvalue of $\mathbf{S}$. This normalization is used in the discussion of outlier detection in Section 10.1. Here it gives

$$\tilde{\tilde{\mathbf{A}}}_m = (n-1)^{\frac{1}{2}} \mathbf{A}_m \mathbf{L}_m^{-1}$$

and equation (11.1.5) becomes $(n-1)\mathbf{T}'\mathbf{T} = (n-1)\mathbf{I}_m$, which is diagonal. Hence, the components are uncorrelated when this normalization is used. However, equation (11.1.4) is now $(n-1)\mathbf{T}'\mathbf{L}_m^{-2}\mathbf{T}$, which is not diagonal, so the loadings of the rotated components are not orthogonal.

It would seem that the common normalization $\tilde{\mathbf{A}}_m = (n-1)^{-\frac{1}{2}} \mathbf{A}_m \mathbf{L}_m$ should be avoided, as it has neither orthogonal loadings nor uncorrelated components once rotation is done. However, it is often apparently used in atmospheric science with a claim that the components are uncorrelated. The reason for this is that rotation is interpreted in a factor analysis framework rather than in terms of rotating PCA loadings. If the loadings $\mathbf{B}_m$ are used to calculate rotated PC scores, then the properties derived above hold. However, in a factor analysis context an attempt is made to reconstruct $\mathbf{X}$ from the underlying factors. With notation based on that of Chapter 7, we have $\mathbf{X} \approx \mathbf{F}\mathbf{\Lambda}'$, and if PCs are used as factors this becomes $\mathbf{X} \approx \mathbf{Z}_m^R \mathbf{A}_m'$. If a column of $\mathbf{A}_m$ is multiplied by $\left(\frac{l_k}{n-1}\right)^{1/2}$, then a corresponding column of $\mathbf{Z}_m^R$ must be *divided* by the same quantity to preserve the approximation to $\mathbf{X}$. This is similar to what happens when using the singular value decomposition (SVD) of $\mathbf{X}$ to construct different varieties of biplot (Section 5.3). By contrast, in the PCA approach to rotation multiplication of a column of $\mathbf{A}_m$ by a constant implies that the corresponding column of $\mathbf{Z}_m^R$ is *multiplied* by the same constant. The consequence is that when the factor analysis (or SVD) framework is used, the scores are found using the normalization $\tilde{\tilde{\mathbf{A}}}_m = (n-1)^{1/2}\mathbf{A}_m\mathbf{L}_m^{-1}$ when the normalization $\tilde{\mathbf{A}}_m = (n-1)^{-\frac{1}{2}}\mathbf{A}_m\mathbf{L}_m$ is adopted for the loadings, and vice versa. Jolliffe

(1995) and Mestas-Nuñez (2000) discuss the effects of different normalizations. Mestas-Nuñez distinguishes between the PCA and factor analysis approaches as corresponding to the 'analysis' and 'synthesis' formulae, respectively. This terminology comes from Preisendorfer and Mobley (1988, Section 2b). Other terminology used by Mestas-Nuñez is less helpful. He refers to unit-length eigenvectors as defining the 'EOF model,' and eigenvectors with squared lengths equal to their eigenvalues as the 'PCA model.' This distinction, which was used by Richman (1986), is confusing—it is more usual to refer to the eigenvectors as EOFs, and the derived variables as PCs, whatever their normalization.

Von Storch and Zwiers (1999, Section 13.1.11) note a further implication of the two approaches. In PCA using a covariance matrix with all variables measured in the same units, it is clear that that the loadings in the PCs are dimensionless, and the PCs themselves have the same units as the original variables. For example, if each of the measured variables is a temperature, so are the PCs. In the factor analysis approach, dividing each PC by its standard deviation makes the resulting components dimensionless, and the *loadings* are now measured in the same units as the original variables. Things are different when PCA is done on a correlation matrix, as the standardized variables forming the input to the analysis are themselves dimensionless.

Arbuckle and Friendly (1977) describe a different way of rotating PCs. Their data consist of p measurements over time of the same quantity, for n individuals. There is an assumption that the measurements represent discrete points on an underlying smooth curve, so that the coefficients in any PC should also vary smoothly over time. Such data can be analysed using functional PCA (see Section 12.3), but Arbuckle and Friendly (1997) treat smoothness as a form of simplicity and attempt to rotate a chosen subset of PCs towards smoothness rather than towards the more usual form of simple structure. The criterion which they minimize over possible rotations is the sum of squared first differences of the coefficients when they are arranged in time order.

11.1.1   Examples

Mediterranean Sea Surface Temperatures

The data presented here were originally analysed by Bartzokas et al. (1994). They consist of values of sea surface temperatures (SST) for sixteen $5° \times 5°$ grid boxes covering most of the Mediterranean, averaged over 3-month seasons, for each of the 43 years 1946–1988. Here we consider data for autumn. Figure 11.1 shows the varimax rotated loadings of the 'first' rotated PC when three PCs are rotated. The three plots within Figure 11.1 provide results from analyses in which three different normalization constraints are used, corresponding to $\mathbf{A}_m$, $\tilde{\mathbf{A}}_m$ and $\tilde{\tilde{\mathbf{A}}}_m$. Of course, the term 'first' ro-

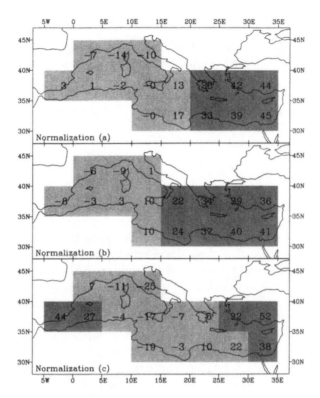

Figure 11.1. Loadings of first rotated autumn components for three normalization constraints based on (a) $\mathbf{A}_m$; (b) $\tilde{\mathbf{A}}_m$; (c) $\tilde{\tilde{\mathbf{A}}}_m$

tated PC has no clear meaning, as there is no longer a successive variance maximization property after rotation, but these three rotated components are matched in the sense that they all have large loadings in the Eastern Mediterranean. The rotated loadings are rescaled after rotation for convenience in making comparisons, so that their sums of squares are equal to unity. The numbers given in Figure 11.1 are these rescaled loadings multiplied by 100. The darker shading in the figure highlights those grid boxes for which the size of their loadings is at least 50% of the largest loading (in absolute value) for that component.

In can be seen that the loadings in the first two plots of Figure 11.1 are similar, although those corresponding to $\tilde{\mathbf{A}}_m$ emphasize a larger area of the Eastern Mediterranean than those derived from $\mathbf{A}_m$. The rotated loadings corresponding to $\tilde{\tilde{\mathbf{A}}}_m$ show more substantial differences, with a clear negative area in the centre of the Mediterranean and the extreme western grid-box having a large positive loading.

For the normalization based on $\mathbf{A}_m$, the vectors of rotated loadings are orthogonal, but the correlations of the displayed component with the other two rotated components are 0.21 and 0.35. The rotated components corresponding to $\tilde{\tilde{\mathbf{A}}}_m$ are uncorrelated, but the angles between the vector of loadings for the displayed component and those of the other two rotated components are 61° and 48°, respectively, rather than 90° as required for orthogonality. For the normalization based on $\tilde{\mathbf{A}}_m$ the respective correlations between the plotted component and the other rotated components are 0.30 and 0.25, with respective corresponding angles between vectors of loadings equal to 61° and 76°.

Artistic Qualities of Painters

In the SST example, there are quite substantial differences in the rotated loadings, depending on which normalization constraint is used. Jolliffe (1989) suggests a strategy for rotation that avoids this problem, and also alleviates two of the other three drawbacks of rotation noted above. This strategy is to move away from necessarily rotating the *first* few PCs, but instead to rotate subsets of components with similar eigenvalues. The effect of different normalizations on rotated loadings depends on the relative lengths of the vectors of loadings within the set of loadings that are rotated. If the eigenvalues are similar for all PCs to be rotated, any normalization in which lengths of loading vectors are functions of eigenvalues is similar to a normalization in which lengths are constant. The three constraints above specify lengths of 1, $\frac{l_k}{n-1}$, and $\frac{n-1}{l_k}$, and it therefore matters little which is used when eigenvalues in the rotated set are similar.

With similar eigenvalues, there is also no dominant PC or PCs within the set being rotated, so the second drawback of rotation is removed. The arbitary choice of m also disappears, although it is replaced by another arbitrary choice of which sets of eigenvalues are sufficiently similar to justify rotation. However, there is a clearer view here of what is required (close eigenvalues) than in choosing m. The latter is multifaceted, depending on what the m retained components are meant to achieve, as evidenced by the variety of possible rules described in Section 6.1. If approximate multivariate normality can be assumed, the choice of which subsets to rotate becomes less arbitrary, as tests are available for hypotheses averring that blocks of consecutive population eigenvalues are equal (Flury and Riedwyl, 1988, Section 10.7). These authors argue that when such hypotheses cannot be rejected, it is dangerous to interpret individual eigenvectors—only the subspace defined by the block of eigenvectors is well-defined, not individual eigenvectors. A possible corollary of this argument is that PCs corresponding to such subspaces should always be rotated in order to interpret the subspace as simply as possible.

Jolliffe (1989) gives three examples of rotation for PCs with similar eigenvalues. One, which we summarize here, is the four-variable artistic qualities data set described in Section 5.1. In this example the four eigenvalues are 2.27, 1.04, 0.40 and 0.29. The first two of these are well-separated, and

Table 11.1. Unrotated and rotated loadings for components 3 and 4: artistic qualities data.

	PC3	PC4	RPC3	RPC4
Composition	−0.59	−0.41	−0.27	0.66
Drawing	0.60	−0.50	0.78	−0.09
Colour	0.49	−0.22	0.53	0.09
Expression	0.23	0.73	−0.21	0.74
Percentage of total variation	10.0	7.3	9.3	8.0

together they account for 83% of the total variation. Rotating them may lose information on individual dominant sources of variation. On the other hand, the last two PCs have similar eigenvalues and are candidates for rotation. Table 11.1 gives unrotated (PC3, PC4) and rotated (RPC3, RPC4) loadings for these two components, using varimax rotation and the normalization constraints $a'_k a_k = 1$. Jolliffe (1989) shows that using alternative rotation criteria to varimax makes almost no difference to the rotated loadings. Different normalization constraints also affect the results very little. Using $\tilde{a}'_k \tilde{a}_k = l_k$ gives vectors of rotated loadings whose angles with the vectors for the constraint $a'_k a_k = 1$ are only 6° and 2°.

It can be seen from Table 11.1 that rotation of components 3 and 4 considerably simplifies their structure. The rotated version of component 4 is almost a pure contrast between expression and composition, and component 3 is also simpler than either of the unrotated components. As well as illustrating rotation of components with similar eigenvalues, this example also serves as a reminder that the last few, as well as the first few, components are sometimes of interest (see Sections 3.1, 3.7, 6.3, 8.4–8.6, 9.1 and 10.1).

In such cases, interpretation of the last few components may have as much relevance as interpretation of the first few, and for the last few components close eigenvalues are more likely.

11.1.2 One-step Procedures Using Simplicity Criteria

Kiers (1993) notes that even when a PCA solution is rotated to simple structure, the result may still not be as simple as required. It may therefore be desirable to put more emphasis on simplicity than on variance maximization, and Kiers (1993) discusses and compares four techniques that do this. Two of these explicitly attempt to maximize one of the standard simplicity criteria, namely varimax and quartimax respectively, over all possible sets of orthogonal components, for a chosen number of components m. A third method divides the variables into non-overlapping clusters and associates exactly one component with each cluster. The criterion to be optimized is the sum over variables of squared loadings for each variable for the single component associated with that variable's cluster. The final method is similar to the one that maximizes the quartimax criterion, but it relaxes the requirement of orthogonality of components.

Gains in simplicity achieved by any of the methods are paid for by a loss of variance explained compared to rotated PCA, but none of the four methods explicitly takes into account the desirability of minimizing the variance lost. Kiers (1993) investigates the trade-off between simplicity gain and variance loss for the four methods in a simulation study. He finds that neither of the first two methods offers any advantage over rotated PCA. The third and fourth methods show some improvement over rotated PCA in recovering the simple structures that are built into the simulated data, and the fourth is better than the third in terms of retaining variance.

Two of the disadvantages of standard rotation noted earlier are the loss of the successive optimization property and the possible sensitivity of the results to the choice of m. All four techniques compared by Kiers (1993) share these disadvantages. We now describe a method that avoids these drawbacks and explicitly takes into account both variance and simplicity. Like Kiers' (1993) methods it replaces the two stages of rotated PCA by a single step. Linear combinations of the p variables are successively found that maximize a criterion in which variance is combined with a penalty function that pushes the linear combination towards simplicity. The method is known as the Simplified Component Technique (SCoT) (Jolliffe and Uddin, 2000).

Let $c'_k x_i$ be the value of the kth simplified component (SC) for the ith observation. Suppose that $Sim(c_k)$ is a measure of simplicity for the vector c_k, for example the varimax criterion, and $Var(c'_k x)$ denotes the sample variance $c'_k S c_k$ of $c'_k x$. Then SCoT successively maximizes

$$(1 - \psi)Var(c'_k x) + \psi Sim(c_k) \qquad (11.1.6)$$

subject to $c'_k c_k = 1$ and (for $k \geq 2$, $h < k$) $c'_h c_k = 0$. Here ψ is a simplicity/complexity parameter, which needs to be chosen. The value $\psi = 0$ corresponds to PCA, and as ψ increases the SCs move away from the PCs towards greater simplicity. When $\psi = 1$, each SC is identical to one of the original variables, with zero loadings for all other variables.

Mediterranean SST

We return to the Mediterranean SST data, which were introduced earlier in this section, but here we describe results for both autumn and winter. Figures 11.2–11.5, reproduced with permission from Jolliffe et al. (2002b), show the loadings for the first two PCs, and the first two simplified components, for each of the two seasons. Also included in the figures are two varimax rotated PCs, using the normalization corresponding to A_m, and the first two components from the SCoTLASS and simple component techniques which are described in Section 11.2. The normalization of loadings and the shading in Figures 11.2–11.5 follow the same conventions as in Figure 11.1, except that any loadings which are *exactly* zero are unshaded in the figures.

For the autumn data, the first two SCoT components may be viewed as slightly simpler versions of PC1 and PC2. The largest loadings become larger and the smallest become smaller, as is often observed during rotation. The strong similarity between SCoT and PCA components is reflected by the fact that the first two SCoT components account for 78.0% of the total variation compared to 78.2% for the first two PCs. Turning to the winter data, the first two PCs and RPCs in Figures 11.4, 11.5 are not too different from those in autumn (Figures 11.2, 11.3), but there are bigger differences for SCoT. In particular, the second SCoT component is very much dominated by a single grid box in the Eastern Mediterranean. This extreme simplicity leads to a reduction in the total variation accounted for by the first two SCoT components to 55.8%, compared to 71.0% for the first two PCs.

Results are presented for only one value of the tuning parameter ψ, and the choice of this value is not an easy one. As ψ increases, there is often a rapid jump from components that are very close to the corresponding PC, like those in autumn, to components that are dominated by a single variable, as for SC2 in winter. This jump is usually accompanied by a large drop in the proportion of variation accounted for, compared to PCA. This behaviour as ψ varies can also be seen in a later example (Tables 11.3, 11.4), and is an unsatisfactory feature of SCoT. Choosing different values $\psi_1, \psi_2, \ldots$ of ψ for SC1, SC2, $\ldots$, only partially alleviates this problem. The cause seems to be the presence of a number of local maxima for the criterion defined by (11.1.6). As ψ changes, a different local maximum may take over as the global maximum, leading to a sudden switch between two quite different solutions. Further discussion of the properties of SCoT, together with additional examples, is given by Jolliffe and Uddin (2000) and Uddin (1999).

Filzmoser (2000) argues that sometimes there is simple structure in a plane but not in any single direction within that plane. He derives a way of finding such simply structured planes—which he calls *principal planes*. Clearly, if a data set has such structure, the methods discussed in this section and the next are unlikely to find simple single components without a large sacrifice in variance.

11.2   Alternatives to Rotation

This section describes two ideas for constructing linear functions of the p original variables having large variances; these techniques differ from PCA in imposing additional constraints on the loadings or coefficients of the

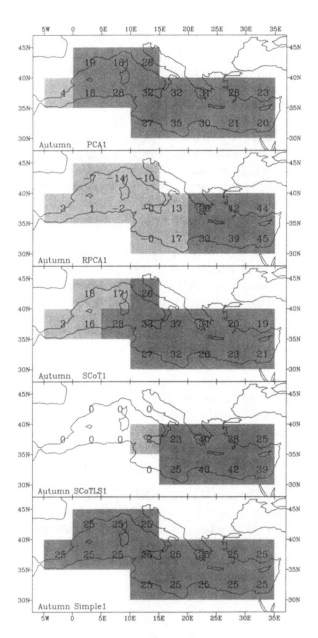

Figure 11.2. Loadings of first autumn components for PCA, RPCA, SCoT, SCoTLASS and simple component analysis.

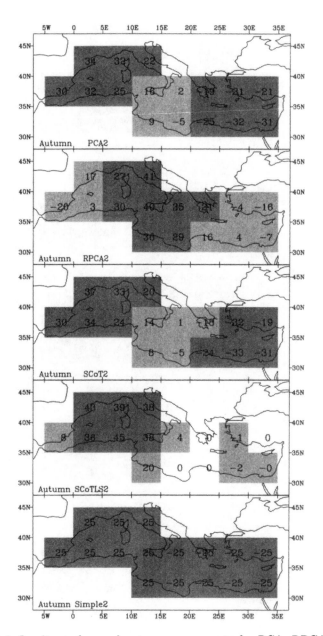

Figure 11.3. Loadings of second autumn components for PCA, RPCA, SCoT, SCoTLASS and simple component analysis.

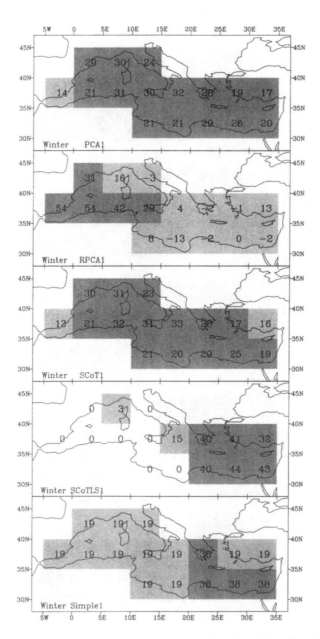

Figure 11.4. Loadings of first winter components for PCA, RPCA, SCoT, SCoTLASS and simple component analysis.

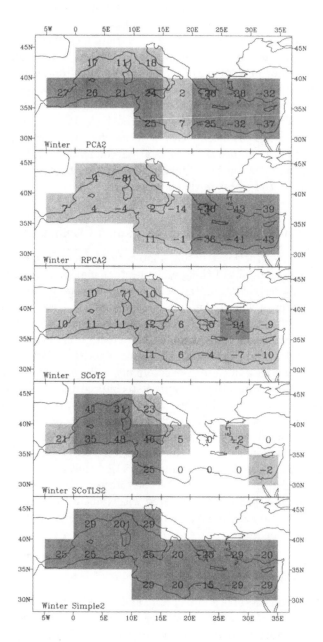

Figure 11.5. Loadings of second winter components for PCA, RPCA, SCoT, SCoTLASS and simple component analysis.

variables in the functions. The constraints are designed to make the resulting components simpler to interpret than PCs, but without sacrificing too much of the variance accounted for by the PCs. The first idea, discussed in Section 11.2.1, is a simple one, namely, restricting coefficients to a set of integers, though it is less simple to put into practice. The second type of technique, described in Section 11.2.2, borrows an idea from regression, that of the LASSO (Least Absolute Shrinkage and Selection Operator). By imposing an additional constraint in the PCA optimization problem, namely, that the sum of the absolute values of the coefficients in a component is bounded, some of the coefficients can be forced to zero. A technique from atmospheric science, empirical orthogonal teleconnections, is described in Section 11.2.3, and Section 11.2.4 makes comparisons between some of the techniques introduced so far in the chapter.

11.2.1 Components with Discrete-Valued Coefficients

A fairly obvious way of constructing simpler versions of PCs is to successively find linear functions of the p variables that maximize variance, as in PCA, but with a restriction on the values of coefficients in those functions to a small number of values. An extreme version of this was suggested by Hausmann (1982), in which the loadings are restricted to the values $+1$, -1 and 0. To implement the technique, Hausman (1982) suggests the use of a branch-and-bound algorithm. The basic algorithm does not include an orthogonality constraint on the vectors of loadings of successive 'components,' but Hausmann (1982) adapts it to impose this constraint. This improves interpretability and speeds up the algorithm, but has the implication that it may not be possible to find as many as p components. In the 6-variable example given by Hausmann (1982), after 4 orthogonal components have been found with coefficients restricted to $\{-1, 0, +1\}$ the null vector is the only vector with the same restriction that is orthogonal to all four already found. In a unpublished M.Sc. project report, Brooks (1992) discusses some other problems associated with Hausmann's algorithm.

Further information is given on Hausmann's example in Table 11.2. Here the following can be seen:

- The first component is a straightforward average or 'size' component in both analyses.

- Despite a considerable simplication, and a moderately different interpretation, for the second constrained component, there is very little loss in variance accounted by the first two constrained components compared to first two PCs.

A less restrictive method is proposed by Vines (2000), in which the coefficients are also restricted to integers. The algorithm for finding so-called *simple components* starts with a set of p particularly simple vectors of

Table 11.2. Hausmann's 6-variable example: the first two PCs and constrained components.

	First component		Second component	
Variable	PC	Constrained	PC	Constrained
Sentence structure	0.43	1	−0.28	0
Logical relationships	0.44	1	0.12	0
Essay	0.32	1	−0.66	1
Composition	0.46	1	−0.18	1
Computation	0.39	1	0.53	−1
Algebra	0.40	1	0.40	−1
Percentage of total variation	74.16	74.11	16.87	16.66

loadings chosen without worrying about the variances of the corresponding components. Typically this is the set of vectors $\mathbf{a}_k$ where $a_{kk} = 1$ and $a_{kj} = 0$ $(k = 1, 2, \ldots, p; \ j = 1, 2, \ldots, p; \ j \neq k)$, a_{kj} being the jth element of $\mathbf{a}_k$. A sequence of 'simplicity-preserving' transformations is then applied to these vectors. Each transformation chooses a pair of the vectors and rotates them orthogonally in such a way that the variance associated with the currently higher variance component of the pair is increased at the expense of the lower variance component. The algorithm stops when no non-trivial simplicity-preserving transformation leads to an improvement in variance.

Simplicity is achieved by considering a restricted set of angles for each rotation. Only angles that result in the elements of the transformed vectors being proportional to integers are allowed. Thus, 'simple' vectors are defined in this technique as those whose elements are proportional to integers. It is usually the case that the transformed vector associated with the higher variance tends to be simpler (proportional to smaller magnitude integers) than the other transformed vector. Cumulatively, this means that when the algorithm terminates, all the vectors of loadings are simple, with those for the first few components tending to be simpler than those for later components. The choice of which pair of vectors to rotate at each stage of the algorithm is that pair for which the increase in variance of the higher-variance component resulting from a simplicity-preserving rotation is maximized, although this strict requirement is relaxed to enable more than one rotation of mutually exclusive pairs to be implemented simultaneously.

The algorithm for simple components includes a tuning parameter c, which determines the number of angles considered for each simplicity-preserving transformation. This number is 2^{c+2} for $c = 0, 1, 2, \ldots$. As c increases, the simple components tend to become closer to the principal components, but simplicity is sacrificed as the elements of the vectors of loadings progressively become proportional to larger magnitude integers.

In practice, it has been found that $c = 0$ usually gives the best balance between simplicity and retention of variance. Examples of this technique's application are now given.

Mediterranean SST

Returning to the Mediterrean SST example of Section 11.1.2, Figures 11.2–11.5 show that for autumn the simple components using $c = 0$ are very simple indeed. In the figures the normalization $\mathbf{a}'_k\mathbf{a}_k = 1$ is used to aid comparisons between methods, but when converted to integers all coefficients in the first simple component are equal to 1. The second component is a straightforward contrast between east and west with all coefficients equal to $+1$ or -1. The results for winter are slightly less simple. The first simple component has four grid boxes with coefficients equal to 2 with the remaining 12 coefficients equal to 1, and the second component has coefficients proportional to $3, 4, 5$ and 6 in absolute value. The first two simple components account for 70.1%, 67.4% of total variation in autumn and winter, respectively, compared to 78.2%, 71.0% for the first two PCs.

Pitprops

Here we revisit the pitprop data, originally analysed by Jeffers (1967) and discussed in Section 8.7.1. Tables 11.3 and 11.4 give the coefficients and cumulative variance for the first and fourth simple components for these data. Also given in the tables is corresponding information for SCoT and SCoTLASS. The first simple component is, as in the SST example, very simple with all its coefficients proportional to $+1$, 0 or -1. Its loss of variance compared to the first PC is non-trivial, though not large. The second component (not shown) is also simple, the third (also not shown) is less so, and the fourth (Table 11.4) is by no means simple, reflecting the pattern that higher variance simple components are simpler than later ones. The cumulative loss of variance over 4 components compared to PCA is similar to that over 2 components in the SST example.

11.2.2   Components Based on the LASSO

Tibshirani (1996) considers the difficulties associated with interpreting multiple regression equations with many predictor variables. As discussed in Chapter 8, these problems may occur due to the instability of the regression coefficients in the presence of collinearity, or may simply be as a consequence of the large number of variables included in the regression equation. Alternatives to least squares regression that tackle the instability are of two main types. Biased regression methods such as PC regression keep all variables in the regression equation but typically shrink some of the regression coefficients towards zero (see Section 8.3). On the other hand, variable selection procedures (Section 8.5) choose a *subset* of variables and keep only

Table 11.3. Jeffers' pitprop data - coefficients and variance for the first component.

Variable	PCA	SCoT	SCoTLASS	Simple
1	0.40	0.44	0.50	1
2	0.41	0.44	0.51	1
3	0.12	0.10	0	0
4	0.17	0.14	0	0
5	0.06	0.04	0	1
6	0.28	0.25	0.05	1
7	0.40	0.40	0.35	1
8	0.29	0.27	0.23	1
9	0.36	0.35	0.39	1
10	0.38	0.38	0.41	1
11	−0.01	−0.01	0	0
12	−0.11	−0.09	0	−1
13	−0.11	−0.08	0	−1
Variance(%)	32.45	32.28	29.07	28.23

Table 11.4. Jeffers' pitprop data - coefficients and cumulative variance for the fourth component.

Variable	PCA	SCoT	SCoTLASS	Simple
1	−0.03	0.02	−0.07	−133
2	−0.02	0.02	−0.11	−133
3	0.02	−0.05	0.16	603
4	0.01	−0.00	0.08	601
5	0.25	−0.02	0	79
6	−0.15	−0.01	−0.02	−333
7	−0.13	−0.01	−0.05	−273
8	0.29	0.02	0.47	250
9	0.13	0.03	0	−68
10	−0.20	−0.04	0	224
11	0.81	1.00	0.51	20
12	−0.30	−0.00	−0.68	−308
13	−0.10	0.00	−0.04	−79
Cumulative Variance (%)	74.0	59.9	70.0	68.7

the selected variables in the equation. Tibshirani (1996) proposes a new method, the 'least absolute shrinkage and selection operator' or LASSO, which is a hybrid of variable selection and shrinkage estimators. The procedure shrinks the coefficients of some of the variables not simply *towards* zero, but *exactly* to zero, giving an implicit form of variable selection. The LASSO idea can be transferred to PCA, as will now be shown.

In standard multiple regression we have the equation

$$y_i = \beta_0 + \sum_{j=1}^{p} \beta_j x_{ij} + \epsilon_i, \quad i = 1, 2, \ldots, n,$$

where $y_1, y_2, \ldots, y_n$ are measurements on a response variable y; x_{ij}, $i = 1, 2, \ldots, n$, $j = 1, 2, \ldots, p$, are corresponding values of p predictor variables; $\beta_0, \beta_1, \beta_2, \ldots, \beta_p$ are parameters in the regression equation; and ϵ_i is an error term. In least squares regression, the parameters are estimated by minimizing the residual sum of squares,

$$\sum_{i=1}^{n} \left(y_i - \beta_0 - \sum_{j=1}^{p} \beta_j x_{ij} \right)^2.$$

The LASSO imposes an additional restriction on the coefficients, namely

$$\sum_{j=1}^{p} |\beta_j| \le t$$

for some 'tuning parameter' t. For suitable choices of t this constraint has the interesting property that it forces some of the coefficients in the regression equation to zero.

Now consider PCA, in which linear combinations $\mathbf{a}_k' \mathbf{x}$, $k = 1, 2, \ldots, p$, of the p measured variables $\mathbf{x}$ are found that successively have maximum variance $\mathbf{a}_k' \mathbf{S} \mathbf{a}_k$, subject to $\mathbf{a}_k' \mathbf{a}_k = 1$ (and, for $k \ge 2$, $\mathbf{a}_h' \mathbf{a}_k = 0$, $h < k$).

Jolliffe et al. (2002a) suggest an adaptation of PCA, which they call the Simplified Component Technique—LASSO (SCoTLASS), in which the additional constraint

$$\sum_{j=1}^{p} |a_{kj}| \le t \tag{11.2.1}$$

is added to the PCA optimization problem, where a_{kj} is the jth element of the kth vector $\mathbf{a}_k$, $k = 1, 2, \ldots, p$, and t is a 'tuning' parameter.

SCoTLASS differs from PCA in the inclusion of the constraints defined in (11.2.1), and a decision must be made on the value of the tuning parameter, t. It is easy to see that

- for $t \ge \sqrt{p}$, we get PCA;

- for $t < 1$, there is no solution;

- for $t = 1$, we must have exactly one non-zero a_{kj} for each k.

As t decreases from $\sqrt{p}$, the SCoTLASS components move progressively away from PCA and some of the loadings become zero. Eventually, for $t = 1$, a solution is reached in which only one variable has a non-zero loading on each component, as with $\psi = 1$ in SCoT (Section 11.1.2). Examples follow.

Mediterranean SST

We return to the Mediterranean SST example and Figures 11.2–11.5 once more. As with SCoT in Section 11.1.2, results are given for one value of the tuning parameter, in this case t, chosen to give a compromise between variance retention and simplicity. In the autumn, SCoTLASS behaves rather like rotated PCA, except that its patterns are more clearcut, with several grid boxes having zero rather than small loadings. Its first component concentrates on the Eastern Mediterranean, as does rotated PCA, but its second component is centred a little further west than the second rotated PC. The patterns found in winter are fairly similar to those in autumn, and again similar to those of rotated PCA except for a reversal in order. In autumn and winter, respectively, the first two SCoTLASS components account for 70.5%, 65.9% of the total variation. This compares with 78.2%, 71.0% for PCA and 65.8%, 57.9% for rotated PCA. The comparison with rotated PCA is somewhat unfair to the latter as we have chosen to display only two out of three rotated PCs. If just two PCs had been rotated, the rotated PCs would account for the same total variation as the first two PCs. However, the fact that the first two SCoTLASS components are much simplified versions of the two displayed rotated PCs, and at the same time have substantially larger variances, suggests that SCoTLASS is superior to rotated PCA in this example.

Pitprops

Tables 11.3 and 11.4 give coefficients and cumulative variances for the first and fourth SCoTLASS components for Jeffers' (1967) pitprop data. The first component sacrifices about the same amount of variance compared to PCA as the first simple component. Both achieve a greatly simplified pattern of coefficients compared to PCA, but, as in the SST example, of quite different types. For the fourth component, the cumulative variance is again similar to that of the simple components, but the SCoTLASS component is clearly simpler in this case. Further examples and discussion of the properties of SCoTLASS can be found in Jolliffe et al. (2002a) and Uddin (1999).

11.2.3   Empirical Orthogonal Teleconnections

Van den Dool et al. (2000) propose a technique, the results of which they refer to as empirical orthogonal teleconnections (EOTs). The data they consider have the standard atmospheric science form in which the p variables correspond to measurements of the same quantity at p different spatial locations and the n observations are taken at n different points in time (see Section 12.2). Suppose that r_{jk} is the correlation between the jth and kth variables (spatial locations), and s_k^2 is the sample variance at location k. Then the location j^* is found for which the criterion $\sum_{k=1}^{p} r_{jk}^2 s_k^2$ is maximized. The first EOT is then the vector whose elements are the coefficients in separate regressions of each variable on the chosen variable j^*. The first derived variable is not the linear combination of variables defined by this EOT, but simply the time series at the chosen location j^*. To find a second EOT, the residuals are calculated from the separate regressions that define the first EOT, and the original analysis is repeated on the resulting matrix of residuals. Third, fourth, ... EOTs can be found in a similar way. At this stage it should be noted that 'correlation' appears to be defined by van den Dool et al. (2000) in a non-standard way. Both 'variances' and 'covariances' are obtained from *uncentred* products, as in the uncentred version of PCA (see Section 14.2.3). This terminology is reasonable if means can be assumed to be zero. However, it does not appear that this assumption can be made in van den Dool and co-workers' examples, so that the 'correlations' and 'regressions' employed in the EOT technique cannot be interpreted in the usual way.

Leaving aside for the moment the non-standard definition of correlations, it is of interest to investigate whether there are links between the procedure for finding the optimal variable j^* that determines the first EOT and 'principal variables' as defined by McCabe (1984) (see Section 6.3). Recall that McCabe's (1984) idea is to find subsets of variables that optimize the same criteria as are optimized by the linear combinations of variables that are the PCs. Now from the discussion following Property A6 in Section 2.3 the first correlation matrix PC is the linear combination of variables that maximizes the sum of squared correlations between the linear combination and each of the variables, while the first covariance matrix PC similarly maximizes the corresponding sum of squared covariances. If 'linear combination of variables' is replaced by 'variable,' these criteria become $\sum_{k=1}^{p} r_{jk}^2$ and $\sum_{k=1}^{p} r_{jk}^2 s_k^2 s_j^2$, respectively, compared to $\sum_{k=1}^{p} r_{jk}^2 s_k^2$ for the first EOT. In a sense, then, the first EOT is a compromise between the first principal variables for covariance and correlation matrices. However, the non-standard definitions of variances and correlations make it difficult to understand exactly what the results of the analysis represent.

11.2.4   Some Comparisons

In this subsection some comparisons are made between the two main techniques described in the section, and to other methods discussed earlier in the chapter. Further discussion of the properties of SCoT and SCoT-LASS can be found in Uddin (1999). Another approach to simplification is described in Section 14.6.3.

- SCoT, SCoTLASS and Vines' (2000) simple components all have tuning parameters which must be chosen. At present there is no procedure for choosing the tuning parameters automatically, and it is advisable to try more than one value and judge subjectively when a suitable compromise between simplicity and variance retention has been achieved. For simple components $c = 0$ is often a good choice. The sudden switching between solutions as ψ varies, noted for SCoT, seems not to be a problem with respect to t or c for SCoTLASS or simple components.

- Principal components have the special property that the vectors of loadings are orthogonal *and* the component scores are uncorrelated. It was noted in Section 11.1 that rotated PCs lose at least one of these properties, depending on which normalization constraint is used. None of the new techniques is able to retain both properties either. SCoT and SCoTLASS, as defined above and implemented in the examples, retain orthogonality of the vectors of loadings but sacrifice uncorrelatedness of the components. It is straightforward, though a little more complicated computationally, to implement versions of SCoT and SCoTLASS that keep uncorrelatedness rather than orthogonality. All that is required is to substitute the conditions $c'_h Sc_k = 0$, $h < k$ (or $a'_h Sa_k = 0$, $h < k$) for $c'_h c_k = 0$, $h < k$ (or $a'_h a_k = 0$, $h < k$) in the definitions of the techniques in Sections 11.1.2 and 11.2.2. Because of the presence of orthogonal rotations at the heart of the algorithm for simple components, it is not obvious how a modification replacing orthogonality by uncorrelatedness could be constructed for this technique.

- As noted in Section 11.1, ordinary two-stage rotation of PCs has a number of drawbacks. The nature of the new techniques means that all four of the listed difficulties can be avoided, but the fourth is replaced by a choice between orthogonal vectors of loadings or uncorrelated components. The removal of the drawbacks is paid for by a loss in variation accounted for, although the reduction can be small for solutions that provide considerable simplification, as demonstrated in the examples above. One way to quantify simplicity is to calculate the value of the varimax criterion, or whatever other criterion is used in rotated PCA, for individual components derived from any of the methods. The new techniques often do better than two-stage rotation with respect to the latter's own simplicity criterion, with only a small reduction in variance, giving another reason to prefer the new techniques.

11.3   Simplified Approximations to Principal Components

The techniques of the previous section are *alternatives* to PCA that sacrifice some variance in order to enhance simplicity. A different approach to improving interpretability is to find the PCs, as usual, but then to approximate them. In Chapter 4, especially its first section, it was mentioned that there is usually no need to express the coefficients of a set of PCs to more than one or two decimal places. Rounding the coefficients in this manner is one way of approximating the PCs. The vectors of rounded coefficients will no longer be exactly orthogonal, the rounded PCs will not be uncorrelated and their variances will be changed, but typically these effects will not be very great, as demonstrated by Green (1977) and Bibby (1980). The latter paper presents bounds on the changes in the variances of the PCs (both in absolute and relative terms) that are induced by rounding coefficients, and shows that in practice the changes are quite small, even with fairly severe rounding.

To illustrate the effect of severe rounding, consider again Table 3.2, in which PCs for eight blood chemistry variables have their coefficients rounded to the nearest 0.2. Thus, the coefficients for the first PC, for example, are given as

$$0.2 \quad 0.4 \quad 0.4 \quad 0.4 \quad -0.4 \quad -0.4 \quad -0.2 \quad -0.2.$$

Their values to three decimal place are, by comparison,

$$0.195 \quad 0.400 \quad 0.459 \quad 0.430 \quad -0.494 \quad -0.320 \quad -0.177 \quad -0.171.$$

The variance of the rounded PC is 2.536, compared to an exact variance of 2.792, a change of 9%. The angle between the vectors of coefficients defining the exact and rounded PCs is about 8°. For the second, third and fourth PCs given in Table 3.2, the changes in variances are 7%, 11% and 11%, respectively, and the angles between vectors of coefficients for exact and rounded PCs are 8° in each case. The angle between the vectors of coefficients for the first two rounded PCs is 99° and their correlation is −0.15. None of these changes or angles is unduly extreme considering the severity of the rounding that is employed. However, in an example from quality control given by Jackson (1991, Section 7.3) some of correlations between PCs whose coefficients are approximated by integers are worryingly large.

Bibby (1980) and Jackson (1991, Section 7.3) also mention the possibility of using conveniently chosen integer values for the coefficients. For example, the simplified first PC from Table 3.2 is proportional to $2(x_2+x_3+x_4)+x_1-(x_7+x_8)-2(x_5+x_6)$, which should be much simpler to interpret than the exact form of the PC. This is in the same spirit as the techniques of Section 11.2.1, which restrict coefficients to be proportional to integer values. The difference is that the latter methods provide simpler *alternatives* to PCA, whereas Bibby's (1980) suggestion *approximates* the PCs. Both types of simplification increase the interpretability of individual components, but comparisons *between* components are more difficult when integers are used, because different components have different values of $\mathbf{a}_k'\mathbf{a}_k$.

It is possible to test whether a single simplified (rounded or otherwise) PC is a plausible 'population' PC using the result in equation (3.7.5) (see Jackson (1991, Section 7.4)) but the lack of orthogonality between a set of simplified PCs means that it is not possible for them to simultaneously represent population PCs.

Green (1977) investigates a different effect of rounding in PCA. Instead of looking at the direct impact on the PCs, he looks at the proportions of variance accounted for *in each individual variable* by the first m PCs, and examines by how much these proportions are reduced by rounding. He concludes that changes due to rounding are small, even for quite severe rounding, and recommends rounding to the nearest 0.1 or even 0.2, as this will increase interpretability with little effect on other aspects of the analysis.

It is fairly common practice in interpreting a PC to ignore (set to zero), either consciously or subconsciously, the variables whose coefficients have the smallest absolute values for that principal component. A second stage then focuses on these 'truncated' components to see whether the pattern of non-truncated coefficients can be interpreted as a simple weighted average or contrast for the non-ignored variables. Cadima and Jolliffe (1995) show that the first 'truncation' step does not always do what might be expected and should be undertaken with caution. In particular, this step can be considered as choosing a subset of variables (those not truncated) with which to approximate a PC. Cadima and Jolliffe (1995) show that

- for the chosen subset of variables, the linear combination given by the coefficients in the untruncated PC may be far from the optimal linear approximation to that PC using those variables;

- a different subset of the same size may provide a better approximation.

As an illustration of the first point, consider an example given by Cadima and Jolliffe (1995) using data presented by Lebart et al. (1982). In this example there are seven variables measuring yearly expenditure of groups of French families on 7 types of foodstuff. The loadings on the variables in the second PC to two decimal places are $0.58, 0.41, -0.10, -0.11, -0.24, 0.63$ and 0.14, so a truncated version of this component is

$$\hat{z}_2 = 0.58x_1 + 0.41x_2 + 0.63x_6.$$

Thus, PC2 can be interpreted as a weighted average of expenditure on bread (x_1), vegetables (x_2) and milk (x_6). However, a question that arises is whether this truncated version of PC2 is the best linear approximation to PC2 using only these three variables. The answer is an emphatic 'No.' The best linear approximation in a least squares sense is obtained by regressing PC2 on x_1, x_2 and x_6. This gives $\hat{\hat{z}}_2 = 0.55x_1 - 0.27x_2 + 0.73x_7$, which differs notably in interpretation from $\hat{z}_2$, as the expenditure on vegetables is now contrasted, rather than averaged, with expenditure on bread and milk. Furthermore $\hat{\hat{z}}_2$ has a correlation of 0.964 with PC2, whereas the correlation between PC2 and $\hat{z}_2$ is 0.766. Hence the truncated component $\hat{z}_2$ not only gives a misleading interpretation of PC2 in terms of x_1, x_2 and x_6, but also gives an inferior approximation compared to $\hat{\hat{z}}_2$.

To illustrate Cadima and Jolliffe's (1995) second point, the third PC in the same example has coefficients $0.40, -0.29, -0.34, 0.07, 0.38, -0.23$ and 0.66 on the 7 variables. If the 4 variables x_1, x_3, x_5, x_7 are kept, the best linear approximation to PC3 has coefficients $-0.06, -0.48, 0.41, 0.68$, respectively, so that x_1 looks much less important. The correlation of the approximation with PC3 is increased to 0.979, compared to 0.773 for the truncated component. Furthermore, although x_1 has the second highest coefficient in PC3, it is not a member of the 3-variable subset $\{x_3, x_5, x_7\}$ that best approximates PC3. This subset does almost as well as the best 4-variable subset, achieving a correlation of 0.975 with PC3.

Ali et al. (1985) also note that using loadings to interpret PCs can be misleading, and suggest examining correlations between variables and PCs instead. In a correlation matrix-based PCA, these PC-variable correlations are equal to the loadings when the normalization $\tilde{\mathbf{a}}_k$ is used (see Section 2.3 for the population version of this result). The use of PCA in regionalization studies in climatology, as discussed in Section 9.2, often uses these correlations to define and interpret clusters. However, Cadima and Jolliffe (1995) show that neither the absolute size of loadings in a PC nor the absolute size of correlations between variables and components gives reliable guidance on which subset of variables best approximates a PC (see also Rencher (1995, Section 12.8.3)). Cadima and Jolliffe (2001) give further examples of this phenomenon in the context of variable selection (see Section 6.3).

Richman and Gong (1999) present an extensive study, which starts from the explicit premise that loadings greater than some threshold are retained, while those below the threshold are set to zero. Their objective is to find an optimal value for this threshold in a spatial atmospheric science context. The loadings they consider are, in fact, based on the normalization $\tilde{\mathbf{a}}_k$, so for correlation matrix PCA they are correlations between PCs and variables. Rotated PCs and covariance matrix PCs are also included in the study. Richman and Gong (1999) use biserial correlation to measure how well the loadings in a truncated PC (or truncated rotated PC) match the pattern of correlations between the spatial location whose loading on the PC (rotated PC) is largest, and all other spatial locations in a data set. The optimum threshold is the one for which this match is best. Results are given that show how the optimum threshold changes as sample size increases (it decreases) for unrotated, orthogonally rotated and obliquely rotated PCs. The precision of the results is indicated by boxplots. The paper provides useful guidance on choosing a threshold in the spatial atmospheric science setting if truncation is to be done. However, it largely ignores the fact stressed above that neither the size of loadings nor that of correlations are necessarily reliable indicators of how to interpret a PC, and hence that truncation according to a fixed threshold should be used with extreme caution.

Interpretation can play a rôle in variable selection. Section 6.3 defines a number of criteria for choosing a subset of variables; these criteria are based on how well the subset represents the full data set in one sense or another. Often there will be several, perhaps many for large p, subsets that do almost as well as the best in terms of the chosen criterion. To decide between them it may be desirable to select those variables that are most easily measured or, alternatively, those that can be most easily interpreted. Taking this train of thought a little further, if PCs are calculated from a subset of variables, it is preferable that the chosen subset gives PCs that are easy to interpret. Al-Kandari and Jolliffe (2001) take this consideration into account in a study that compares the merits of a number of variable selection methods and also compares criteria for assessing the value of subsets. A much fuller discussion is given by Al-Kandari (1998).

11.3.1   Principal Components with Homogeneous, Contrast and Sparsity Constraints

Chipman and Gu (2001) present a number of related ideas that lead to results which appear similar in form to those produced by the techniques in Section 11.2. However, their reasoning is closely related to truncation so they are included here. Components with *homogeneous constraints* have their coefficients restricted to -1, 0 and $+1$ as with Hausmann (1982), but instead of solving a different optimization problem we start with the PCs and approximate them. For a threshold τ_H a vector of PC coefficients $\mathbf{a}_k$ is replaced by a vector $\mathbf{a}_k^H$ with coefficients

$$a_{kj}^H = \begin{cases} \text{sign}(a_{kj}) & \text{if } a_{kj} \geq \tau_H \\ 0 & \text{if } a_{kj} < \tau_H. \end{cases}$$

The threshold τ_H is chosen to minimize the angle between the vectors $\mathbf{a}_k^H$ and $\mathbf{a}_k$.

Contrast constraints allow coefficients a_{kj}^C to take the values $-c_1$, 0 and c_2, where $c_1, c_2 > 0$ and $\sum_{j=1}^p a_{kj}^C = 0$. Again a threshold τ_C determines the coefficients. We have

$$
a_{kj}^C = \begin{cases} -c_1 & \text{if } a_{kj} \leq -\tau_C \\ 0 & \text{if } |a_{kj}| < \tau_C \\ c_2 & \text{if } a_{kj} \geq \tau_C, \end{cases}
$$

and τ_C is chosen to minimize the angle between $\mathbf{a}_k^C$ and $\mathbf{a}_k$.

The idea behind *sparsity constraints* as defined by Chipman and Gu (2001) is different. An approximating vector $\mathbf{a}_k^S$ is simply the vector $\mathbf{a}_k$ with all except its q largest coefficients in absolute value truncated to zero. The value of q is chosen by minimizing the quantity $(\frac{2\theta}{\pi} + \frac{\eta q}{p})$ with respect to q, where θ is the angle between $\mathbf{a}_k^S$ and $\mathbf{a}_k$ and η is a tuning parameter. As η increases, so the optimal value of q decreases and the component becomes more 'sparse.'

The three ideas of homogeneous, contrast and sparsity constraints formalize and extend the informal procedure of truncation, and Chipman and Gu (2001) give interesting examples of their use. It must, however, be remembered that these techniques implicitly assume that the sizes of coefficients or loadings for a PC are a reliable guide to interpreting that PC. As we have seen, this is not necessarily true.

11.4 Physical Interpretation of Principal Components

There is often a desire to physically interpret principal components. In psychology, for example, it is frequently assumed that there are certain underlying factors that cannot be measured directly, but which can be deduced from responses to questions or scores on tests of a sample of individuals. PCA can be used to look for such factors. In psychology, factor analysis, which builds a model that includes the factors, is usually rightly preferred to PCA, although the software used may be such that PCA is actually done.

Atmospheric science is another area where it is believed that there are fundamental modes of variation underlying a data set, and PCA or some modification of it is often used to look for them. Preisendorfer and Mobley (1988, Chapters 3 and 10) discuss at length the interplay between PCA and the dominant modes of variation in physical systems, and use a version of PCA for continuous variables (functional PCA—see Section 12.3) to explore the relationships between EOFs (the vectors of coefficients) and the physical modes of the atmosphere. Recent discussions of the North Atlantic Oscillation and Arctic Oscillation (Ambaum et al., 2001) and the Indian Ocean dipole (Allan et al., 2001) illustrate the controversy that sometimes surrounds the physical interpretation of PCs.

In practice, there may be a circularity of argument when PCA is used to search for physically meaningful modes in atmospheric data. The form of these modes is often assumed known and PCA is used in an attempt to confirm them. When it fails to do so, it is 'accused' of being inadequate. However, it has very clear objectives, namely finding uncorrelated derived variables that in succession maximize variance. If the physical modes are not expected to maximize variance and/or to be uncorrelated, PCA should not be used to look for them in the first place.

One of the reasons why PCA 'fails' to find the expected physical modes in some cases in atmospheric science is because of its dependence on the size and shape of the spatial domain over which observations are taken. Buell (1975) considers various spatial correlation functions, both circular (isotropic) and directional (anisotropic), together with square, triangular and rectangular spatial domains. The resulting EOFs depend on the size of the domain in the sense that in a small domain with positive correlations between all points within it, the first EOF is certain to have all its elements of the same sign, with the largest absolute values near the centre of the domain, a sort of 'overall size' component (see Section 13.2). For larger domains, there may be negative as well as positive correlations, so that the first EOF represents a more complex pattern. This gives the impression of instability, because patterns that are present in the analysis of a large domain are not necessarily reproduced when the PCA is restricted to a subregion of this domain.

The shape of the domain also influences the form of the PCs. For example, if the first EOF has all its elements of the same sign, subsequent ones must represent 'contrasts,' with a mixture of positive and negative values, in order to satisfy orthogonality constraints. If the spatial correlation is isotropic, the contrast represented by the second PC will be between regions with the greatest geographical separation, and hence will be determined by the shape of the domain. Third and subsequent EOFs can also be predicted for isotropic correlations, given the shape of the domain (see Buell (1975) for diagrams illustrating this). However, if the correlation is anisotropic and/or non-stationary within the domain, things are less simple. In any case, the form of the correlation function is important in determining the PCs, and it is only when it takes particularly simple forms that the nature of the PCs can be easily predicted from the size and shape of the domain. PCA will often give useful information about the sources of maximum variance in a spatial data set, over and above that available from knowledge of the size and shape of the spatial domain of the data. However, in interpreting PCs derived from spatial data, it should not be forgotten that the size and shape of the domain can have a strong influence on the results. The degree of dependence of EOFs on domain shape has been, like the use of rotation and the possibility of physical interpretation, a source of controversy in atmospheric science. For an entertaining and enlightening exchange of strong views on the importance of domain shape, which also brings in rotation and interpretation, see Legates (1991, 1993) and Richman (1993).

Similar behaviour occurs for PCA of time series when the autocorrelation function (see Section 12.1) takes a simple form. Buell (1979) discusses this case, and it is well-illustrated by the road-running data which are analysed in Sections 5.3 and 12.3. The first PC has all its loadings of the same sign, with the greatest values in the middle of the race, while the second PC is a contrast between the race segments with the greatest separation in time. If there are predictable patterns in time or in spatial data, it may be of interest to examine the major sources of variation orthogonal to these predictable directions. This is related to what is done in looking for 'shape' components orthogonal to the isometric size vector for size and shape data (see Section 13.2). Rao's 'principal components uncorrelated with instrumental variables' (Section 14.3) also have similar objectives.

A final comment is that even when there is no underlying structure in a data set, sampling variation ensures that some linear combinations of the variables have larger variances than others. Any 'first PC' which has actually arisen by chance can, with some ingenuity, be 'interpreted.' Of course, completely unstructured data is a rarity but we should always try to avoid 'overinterpreting' PCs, in the same way that in other branches of statistics we should be wary of spurious regression relationships or clusters, for example.

12

Principal Component Analysis for Time Series and Other Non-Independent Data

12.1 Introduction

In much of statistics it is assumed that the n observations $\mathbf{x}_1, \mathbf{x}_2, \ldots, \mathbf{x}_n$ are independent. This chapter discusses the implications for PCA of non-independence among $\mathbf{x}_1, \mathbf{x}_2, \ldots, \mathbf{x}_n$. Much of the chapter is concerned with PCA for time series data, the most common type of non-independent data, although data where $\mathbf{x}_1, \mathbf{x}_2, \ldots, \mathbf{x}_n$ are measured at n points in space are also discussed. Such data often have dependence which is more complicated than for time series. Time series data are sufficiently different from ordinary independent data for there to be aspects of PCA that arise only for such data, for example, PCs in the frequency domain.

The results of Section 3.7, which allow formal inference procedures to be performed for PCs, rely on independence of $\mathbf{x}_1, \mathbf{x}_2, \ldots, \mathbf{x}_n$, as well as (usually) on multivariate normality. They cannot therefore be used if more than very weak dependence is present between $\mathbf{x}_1, \mathbf{x}_2, \ldots, \mathbf{x}_n$. However, when the main objective of PCA is descriptive, not inferential, complications such as non-independence do not seriously affect this objective. The effective sample size is reduced below n, but this reduction need not be too important. In fact, in some circumstances we are actively looking for dependence among $\mathbf{x}_1, \mathbf{x}_2, \ldots, \mathbf{x}_n$. For example, grouping of observations in a few small areas of the two-dimensional space defined by the first two PCs implies dependence between those observations that are grouped together. Such behaviour is actively sought in cluster analysis (see Section 9.2) and is often welcomed as a useful insight into the structure of the data, rather than decried as an undesirable feature.

We have already seen a number of examples where the data are time series, but where no special account is taken of the dependence between observations. Section 4.3 gave an example of a type that is common in atmospheric science, where the variables are measurements of the same meteorological variable made at p different geographical locations, and the n observations on each variable correspond to different times. Section 12.2 largely deals with techniques that have been developed for data of this type. The examples given in Section 4.5 and Section 6.4.2 are also illustrations of PCA applied to data for which the variables (stock prices and crime rates, respectively) are measured at various points of time. Furthermore, one of the earliest published applications of PCA (Stone, 1947) was on (economic) time series data.

In time series data, dependence between the $\mathbf{x}$ vectors is induced by their relative closeness in time, so that $\mathbf{x}_h$ and $\mathbf{x}_i$ will often be highly dependent if $|h-i|$ is small, with decreasing dependence as $|h-i|$ increases. This basic pattern may in addition be perturbed by, for example, seasonal dependence in monthly data, where decreasing dependence for increasing $|h-i|$ is interrupted by a higher degree of association for observations separated by exactly one year, two years, and so on.

Because of the emphasis on time series in this chapter, we need to introduce some of its basic ideas and definitions, although limited space permits only a rudimentary introduction to this vast subject (for more information see, for example, Brillinger (1981); Brockwell and Davis (1996); or Hamilton (1994)). Suppose, for the moment, that only a single variable is measured at equally spaced points in time. Our time series is then $\ldots x_{-1}, x_0, x_1, x_2, \ldots$. Much of time series analysis is concerned with series that are stationary, and which can be described entirely by their first- and second-order moments; these moments are

$$
\begin{aligned}
\mu &= E(x_i), & i &= \ldots, -1, 0, 1, 2, \ldots \\
\gamma_k &= E[(x_i - \mu)(x_{i+k} - \mu)], & i &= \ldots, -1, 0, 1, 2, \ldots \\
& & k &= \ldots, -1, 0, 1, 2, \ldots,
\end{aligned}
\tag{12.1.1}
$$

where μ is the mean of the series and is the same for all x_i in stationary series, and γ_k, the kth autocovariance, is the covariance between x_i and x_{i+k}, which depends on k but not i for stationary series. The information contained in the autocovariances can be expressed equivalently in terms of the power spectrum of the series

$$
f(\lambda) = \frac{1}{2\pi} \sum_{k=-\infty}^{\infty} \gamma_k e^{-ik\lambda},
\tag{12.1.2}
$$

where $i = \sqrt{-1}$ and λ denotes angular frequency. Roughly speaking, the function $f(\lambda)$ decomposes the series into oscillatory portions with different frequencies of oscillation, and $f(\lambda)$ measures the relative importance of these portions as a function of their angular frequency λ. For example, if a series is almost a pure oscillation with angular frequency λ_0, then $f(\lambda)$ is large for λ close to λ_0 and near zero elsewhere. This behaviour is signalled in the autocovariances by a large value of γ_k at $k = k_0$, where k_0 is the period of oscillation corresponding to angular frequency λ_0 (that is $k_0 = 2\pi/\lambda_0$), and small values elsewhere.

Because there are two different but equivalent functions (12.1.1) and (12.1.2) expressing the second-order behaviour of a time series, there are two different types of analysis of time series, namely in the time domain using (12.1.1) and in the frequency domain using (12.1.2).

Consider now a time series that consists not of a single variable, but p variables. The definitions (12.1.1), (12.1.2) generalize readily to

$$\boldsymbol{\Gamma}_k = E[(\mathbf{x}_i - \boldsymbol{\mu})(\mathbf{x}_{i+k} - \boldsymbol{\mu})'], \qquad (12.1.3)$$

where

$$\boldsymbol{\mu} = E[\mathbf{x}_i]$$

and

$$\mathbf{F}(\lambda) = \frac{1}{2\pi} \sum_{k=-\infty}^{\infty} \boldsymbol{\Gamma}_k e^{-ik\lambda} \qquad (12.1.4)$$

The mean $\boldsymbol{\mu}$ is now a p-element vector, and $\boldsymbol{\Gamma}_k$, $\mathbf{F}(\lambda)$ are $(p \times p)$ matrices.

Principal component analysis operates on a covariance or correlation matrix, but in time series we can calculate not only covariances between variables measured at the same time (the usual definition of covariance, which is given by the matrix $\boldsymbol{\Gamma}_0$ defined in (12.1.3)), but also covariances between variables at different times, as measured by $\boldsymbol{\Gamma}_k, k \neq 0$. This is in contrast to the more usual situation where our observations $\mathbf{x}_1, \mathbf{x}_2, \ldots$ are independent, so that any covariances between elements of $\mathbf{x}_i, \mathbf{x}_j$ are zero when $i \neq j$. In addition to the choice of which $\boldsymbol{\Gamma}_k$ to examine, the fact that the covariances have an alternative representation in the frequency domain means that there are several different ways in which PCA can be applied to time series data.

Before looking at specific techniques, we define the terms 'white noise' and 'red noise.' A white noise series is one whose terms are all identically distributed and independent of each other. Its spectrum is flat, like that of white light; hence its name. Red noise is equivalent to a series that follows a positively autocorrelated first-order autoregressive model

$$x_t = \phi x_{t-1} + \epsilon_t, \quad t = \ldots 0, 1, 2 \ldots,$$

where ϕ is a constant such that $0 < \phi < 1$ and $\{\epsilon_t\}$ is a white noise series. The spectrum of a red noise series decreases as frequency increases, like that of red light in the range of visual radiation.

The next section of this chapter describes a range of approaches based on PCA that have been used on time series data in atmospheric science. Although many are inspired by the special nature of the data commonly encountered in this area, with observations corresponding to times and variables to spatial position, they are not necessarily restricted to such data.

Time series are usually measured at discrete points in time, but sometimes the series are curves. The analysis of such data is known as *functional data analysis* (functional PCA is the subject of Section 12.3). The final section of the chapter collects together a number of largely unconnected ideas and references concerning PCA in the context of time series and other non-independent data.

12.2   PCA-Related Techniques for (Spatio-) Temporal Atmospheric Science Data

It was noted in Section 4.3 that, for a common type of data in atmospheric science, the use of PCA, more often referred to as *empirical orthogonal function* (EOF) analysis, is widespread. The data concerned consist of measurements of some variable, for example, sea level pressure, temperature, ..., at p spatial locations (usually points on a grid) at n different times. The measurements at different spatial locations are treated as variables and the time points play the rôle of observations. An example of this type was given in Section 4.3. It is clear that, unless the observations are well-separated in time, there is likely to be correlation between measurements at adjacent time points, so that we have non-independence between observations. Several techniques have been developed for use in atmospheric science that take account of correlation in both time and space, and these will be described in this section. First, however, we start with the simpler situation where there is a single time series. Here we can use a principal component-like technique, called *singular spectrum analysis* (SSA), to analyse the autocorrelation in the series. SSA is described in Section 12.2.1, as is its extension to several time series, multichannel singular spectrum analysis (MSSA).

Suppose that a set of p series follows a multivariate first-order autoregressive model in which the values of the series at time t are linearly related to the values at time $(t - 1)$, except for a multivariate white noise term. An estimate of the matrix defining the linear relationship can be subjected to an eigenanalysis, giving insight into the structure of the series. Such an analysis is known as *principal oscillation pattern* (POP) analysis, and is discussed in Section 12.2.2.

One idea underlying POP analysis is that there may be patterns in the maps comprising our data set, which travel in space as time progresses, and that POP analysis can help to find such patterns. Complex (Hilbert) empirical orthogonal functions (EOFs), which are described in Section 12.2.3, are designed to achieve the same objective. Detection of detailed oscillatory behaviour is also the aim of multitaper frequency-domain singular value decomposition, which is the subject of Section 12.2.4.

Some time series have cyclic behaviour with fixed periods, such as an annual or diurnal cycle. Modifications of POP analysis and PCA that take such cycles into account are discussed in Section 12.2.5. A brief discussion of examples and comparative studies is given in Section 12.2.6.

12.2.1 Singular Spectrum Analysis (SSA)

Like a number of other statistical techniques, SSA appears to have been 'invented' independently in a number of different fields. Elsner and Tsonis (1996) give references from the 1970s and 1980s from chaos theory, biological oceanography and signal processing, and the same idea is described in the statistical literature by Basilevsky and Hum (1979), where it is referred to as the 'Karhunen-Loève method,' a term more often reserved for continuous time series (see Section 12.3). Other names for the technique are 'Pisarenko's method' (Smyth, 2000) and 'singular systems analysis' (von Storch and Zwiers, 1999, Section 13.6); fortunately the latter has the same acronym as the most popular name. A comprehensive coverage of SSA is given by Golyandina et al. (2001).

The basic idea in SSA is simple: a principal component analysis is done with the variables analysed being lagged versions of a single time series variable. More specifically, our p variables are $x_t, x_{(t+1)}, \ldots, x_{(t+p-1)}$ and, assuming the time series is stationary, their covariance matrix is such that the (i, j)th element depends only on $|i - j|$. Such matrices are known as Töplitz matrices. In the present case the (i, j)th element is the autocovariance $\gamma_{|i-j|}$. Because of the simple structure of Töplitz matrices, the behaviour of the first few PCs, and their corresponding eigenvalues and eigenvectors (the EOFs) which are trigonometric functions (Brillinger, 1981, Section 3.7), can be deduced for various types of time series structure. The PCs are moving averages of the time series, with the EOFs providing the weights in the moving averages.

Töplitz matrices also occur when the p-element random vector $\mathbf{x}$ consists of *non-overlapping* blocks of p consecutive values of a single time series. If the time series is stationary, the covariance matrix $\mathbf{\Sigma}$ for $\mathbf{x}$ has Töplitz structure, with the well-known pattern of trigonometric functions for its eigenvalues and eigenvectors. Craddock (1965) performed an analysis of this type on monthly mean temperatures for central England for the period November 1680 to October 1963. The p ($= 12$) elements of $\mathbf{x}$ are mean temperatures for the 12 months of a particular year, where a 'year' starts in November. There is some dependence between different values of $\mathbf{x}$, but it is weaker than that between elements within a particular $\mathbf{x}$; between-year correlation was minimized by starting each year at November, when there was apparently evidence of very little continuity in atmospheric behaviour for these data. The sample covariance matrix does not, of course, have exact Töplitz structure, but several of the eigenvectors have approximately the form expected for such matrices.

Durbin (1984) uses the structure of the eigenvectors of Töplitz matrices in a different context. In regression analysis (see Chapter 8), if the dependent variable and the predictor variables are all time series, then the Töplitz structure for the covariance matrix of error terms in the regression model can be used to deduce properties of the least squares estimators of regression coefficients.

Returning to SSA, for a time series with an oscillatory component SSA has an associated pair of EOFs with identical eigenvalues. Coefficients of both EOFs have the same oscillatory pattern but are $\frac{\pi}{2}$ radians out of phase with each other. Although Elsner and Tsonis (1996) describe a number of other uses of SSA, the major application in atmospheric science has been to 'discover' dominant periodicities in a series (Allen and Smith, 1996). One advantage that SSA has in this respect over traditional spectral analysis is that the frequencies of oscillations detected by SSA can take any value in a given range rather than be restricted to a fixed set of frequencies. A disadvantage of SSA is, however, a tendency to find apparent periodicities when none exists. Allen and Smith (1996) address this problem with a carefully constructed hypothesis testing procedure for deciding whether or not apparent periodicities are statistically significant in the presence of red noise, a more realistic assumption than white noise for background variation in many climatological series. Allen and Smith (1997) discuss what they call a generalization of SSA for detecting 'signal' in the presence of red noise.

The way that SSA has been introduced above is in a 'population' context, although when talking about statistical significance in the last paragraph it is clear that we are discussing samples. A sample consists of a series $x_1, x_2, \ldots, x_n$, which is rearranged to give an $(n' \times p)$ matrix whose ith row is

$$\mathbf{x}'_i = (x_i, x_{(i+1)}, \ldots, x_{(i+p-1)}) \quad i = 1, 2, \ldots, n',$$

where $n' = n - p + 1$.

A practical consideration is the choice of p - large values of p allow longer-period oscillations to be resolved, but choosing p too large leaves too few observations, n', from which to estimate the covariance matrix of the p variables. Elsner and Tsonis (1996, Section 5.2) give some discussion of, and references to, the choice of p, and remark that choosing $p = \frac{n}{4}$ is a common practice.

Estimation of the covariance matrix raises further questions. If the n' observations on p variables are treated as an 'ordinary' data matrix, the corresponding covariance matrix will generally have all its elements unequal. If the series is stationary, the covariance between the ith and jth variables should only depend on $|i - j|$. Estimates can be constructed to be constrained to give this Töplitz structure for the covariance matrix. There is discussion of which estimates to use in Elsner and Tsonis (1996, Section 5.3). We now offer examples of SSA.

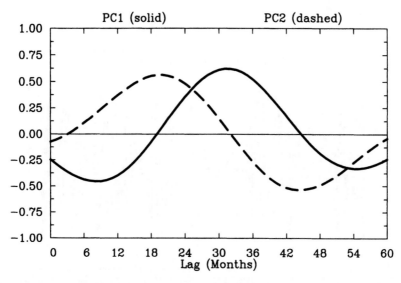

Figure 12.1. Plots of loadings for the first two components in an SSA with $p = 61$ of the Southern Oscillation Index data.

Southern Oscillation Index

The data considered here are monthly values of the Southern Oscillation Index (SOI) for the years 1876–2000, produced by the Australian Bureau of Meteorology's National Climate Centre. The number of observations in the series is therefore $n = 12 \times 125 = 1500$. The index is a measure of the East-West pressure gradient between Tahiti in the mid-Pacific Ocean and Darwin, Australia. It is a major source of climate variation. SSA was carried on the data with $p = 61$, and Figure 12.1 gives a plot of the loadings for the first two EOFs. Their eigenvalues correspond to 13.7% and 13.4% of the total variation. The closeness of the eigenvalues suggests a quasi-oscillatory pattern, and this is clearly present in the loadings of Figure 12.1. Note, however, that the relationship between the two EOFs is not a simple displacement by $\frac{\pi}{2}$. Figure 12.2 shows time series plots of the scores for the first two components (PCs) in the SSA. These again reflect the oscillatory nature of the components. A reconstruction of the series using only the first two PCs shown in Figure 12.3 captures some of the major features of the original series, but a large amount of other variability remains, reflecting the fact that the two components only account for 27.1% of the total variation.

Multichannel SSA

In multichannel SSA (MSSA) we have the more usual atmospheric science set-up of p spatial locations and n time points, but rather than finding a covariance matrix directly from the $(n \times p)$ data matrix, the data are rearranged into a larger $(n' \times p')$ matrix, where $n' = n - m + 1$, $p' = mp$

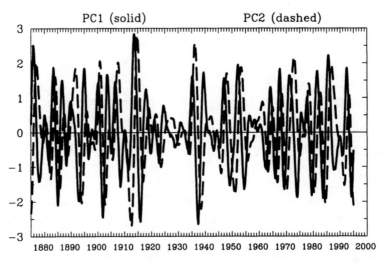

Figure 12.2. Plots of scores for the first two components in an SSA with $p = 61$ for Southern Oscillation Index data.

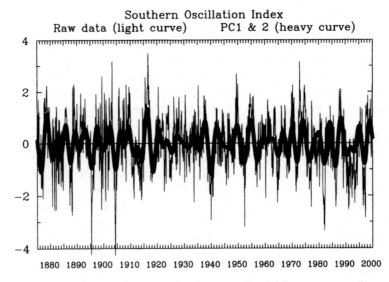

Figure 12.3. Southern Oscillation Index data together with a reconstruction using the first two components from an SSA with $p = 61$.

and a typical row of the matrix is

$$\mathbf{x}'_i = (x_{i1}, x_{(i+1)1}, \ldots, x_{(i+m-1)1}, x_{i2}, \ldots, x_{(i+m-1)2}, \ldots, x_{(i+m-1)p}),$$

$i = 1, 2, \ldots, n'$, where x_{ij} is the value of the measured variable at the ith time point and the jth spatial location, and m plays the same rôle in MSSA as p does in SSA. The covariance matrix for this data matrix has the form

$$\begin{bmatrix} \mathbf{S}_{11} & \mathbf{S}_{12} & \cdots & \mathbf{S}_{1p} \\ \mathbf{S}_{21} & \mathbf{S}_{22} & \cdots & \mathbf{S}_{2p} \\ \vdots & \vdots & & \\ \mathbf{S}_{p1} & \mathbf{S}_{p2} & \cdots & \mathbf{S}_{pp} \end{bmatrix},$$

where $\mathbf{S}_{kk}$ is an $(m \times m)$ covariance matrix at various lags for the kth variable (location), with the same structure as the covariance matrix in an SSA of that variable. The off-diagonal matrices $\mathbf{S}_{kl}$, $k \neq l$, have (i, j)th element equal to the covariance between locations k and l at time lag $|i-j|$. Plaut and Vautard (1994) claim that the 'fundamental property' of MSSA is its ability to detect oscillatory behaviour in the same manner as SSA, but rather than an oscillation of a single series the technique finds oscillatory spatial patterns. Furthermore, it is capable of finding oscillations with the same period but different spatially orthogonal patterns, and oscillations with the same spatial pattern but different periods.

The same problem of ascertaining 'significance' arises for MSSA as in SSA. Allen and Robertson (1996) tackle this problem in a similar manner to that adopted by Allen and Smith (1996) for SSA. The null hypothesis here extends one-dimensional 'red noise' to a set of p independent AR(1) processes. A general multivariate AR(1) process is not appropriate as it can itself exhibit oscillatory behaviour, as exemplified in POP analysis (Section 12.2.2).

MSSA extends SSA from one time series to several, but if the number of time series p is large, it can become unmanageable. A solution, which is used by Benzi et al. (1997), is to carry out PCA on the $(n \times p)$ data matrix, and then implement SSA separately on the first few PCs. Alternatively for large p, MSSA is often performed on the first few PCs instead of the variables themselves, as in Plaut and Vautard (1994).

Although MSSA is a natural extension of SSA, it is also equivalent to extended empirical orthogonal function (EEOF) analysis which was introduced independently of SSA by Weare and Nasstrom (1982). Barnett and Hasselmann (1979) give an even more general analysis, in which different meteorological variables, as well as or instead of different time lags, may be included at the various locations. When different variables *replace* different time lags, the temporal correlation in the data is no longer taken into account, so further discussion is deferred to Section 14.5.

The general technique, including both time lags and several variables, is referred to as multivariate EEOF (MEEOF) analysis by Mote et al.

(2000), who give an example of the technique for five variables, and compare the results to those of separate EEOFs (MSSAs) for each variable. Mote and coworkers note that it is possible that some of the dominant MEEOF patterns may not be dominant in any of the individual EEOF analyses, and this may viewed as a disadvantage of the method. On the other hand, MEEEOF analysis has the advantage of showing directly the connections between patterns for the different variables. Discussion of the properties of MSSA and MEEOF analysis is ongoing (in addition to Mote et al. (2000), see Monahan et al. (1999), for example). Compagnucci et al. (2001) propose yet another variation on the same theme. In their analysis, the PCA is done on the transpose of the matrix used in MSSA, a so-called T-mode instead of S-mode analysis (see Section 14.5). Compagnucci et al. call their technique *principal sequence pattern analysis*.

12.2.2   Principal Oscillation Pattern (POP) Analysis

SSA, MSSA, and other techniques described in this chapter can be viewed as special cases of PCA, once the variables have been defined in a suitable way. With the chosen definition of the variables, the procedures perform an eigenanalysis of a covariance matrix. POP analysis is different, but it is described briefly here because its results are used for similar purposes to those of some of the PCA-based techniques for time series included elsewhere in the chapter. Furthermore its core is an eigenanalysis, albeit not on a covariance matrix.

POP analysis was introduced by Hasselman (1988). Suppose that we have the usual $(n \times p)$ matrix of measurements on a meteorological variable, taken at n time points and p spatial locations. POP analysis has an underlying assumption that the p time series can be modelled as a multivariate first-order autoregressive process. If $\mathbf{x}'_t$ is the tth row of the data matrix, we have

$$(\mathbf{x}_{(t+1)} - \boldsymbol{\mu}) = \boldsymbol{\Upsilon}(\mathbf{x}_t - \boldsymbol{\mu}) + \boldsymbol{\epsilon}_t, \quad t = 1, 2, \ldots, (n-1), \qquad (12.2.1)$$

where $\boldsymbol{\Upsilon}$ is a $(p \times p)$ matrix of constants, $\boldsymbol{\mu}$ is a vector of means for the p variables and $\boldsymbol{\epsilon}_t$ is a multivariate white noise term. Standard results from multivariate regression analysis (Mardia et al., 1979, Chapter 6) lead to estimation of $\boldsymbol{\Upsilon}$ by $\hat{\boldsymbol{\Upsilon}} = \mathbf{S}_1 \mathbf{S}_0^{-1}$, where $\mathbf{S}_0$ is the usual sample covariance matrix for the p variables, and $\mathbf{S}_1$ has (i, j)th element equal to the sample covariance between the ith and jth variables at lag 1. POP analysis then finds the eigenvalues and eigenvectors of $\hat{\boldsymbol{\Upsilon}}$. The eigenvectors are known as principal oscillation patterns (POPs) and denoted $\mathbf{p}_1, \mathbf{p}_2, \ldots, \mathbf{p}_p$. The quantities $z_{t1}, z_{t2}, \ldots, z_{tp}$ which can be used to reconstitute $\mathbf{x}_t$ as $\sum_{k=1}^{p} z_{tk} \mathbf{p}_k$ are called the POP coefficients. They play a similar rôle in POP analysis to that of PC scores in PCA.

One obvious question is why this technique is called principal *oscillation* pattern analysis. Because $\hat{\boldsymbol{\Upsilon}}$ is not symmetric it typically has a mixture of real and complex eigenvectors. The latter occur in pairs, with each pair sharing the same eigenvalue and having eigenvectors that are complex conjugate pairs. The real eigenvectors describe non-oscillatory, non-propagating damped patterns, but the complex eigenvectors represent damped oscillations and can include standing waves and/or spatially propagating waves, depending on the relative magnitudes of the real and imaginary parts of each complex POP (von Storch et al., 1988).

As with many other techniques, the data may be pre-processed using PCA, with x in equation (12.2.1) replaced by its PCs. The description of POP analysis in Wu et al. (1994) includes this initial step, which provides additional insights.

Kooperberg and O'Sullivan (1996) introduce and illustrate a technique which they describe as a hybrid of PCA and POP analysis. The analogous quantities to POPs resulting from the technique are called Predictive Oscillation Patterns (PROPs). In their model, x_t is written as a linear transformation of a set of underlying 'forcing functions,' which in turn are linear functions of x_t. Kooperberg and O'Sullivan (1996) find an expression for an upper bound for forecast errors in their model, and PROP analysis minimizes this quantity. The criterion is such that it simultaneously attempts to account for as much as possible of x_t, as with PCA, and to reproduce as well as possible the temporal dependence in x_t, as in POP analysis.

In an earlier technical report, Kooperberg and O'Sullivan (1994) mention the possible use of canonical correlation analysis (CCA; see Section 9.3) in a time series context. Their suggestion is that a second group of variables is created by shifting the usual measurements at p locations by one time period. CCA is then used to find relationships between the original and time-lagged sets of variables.

12.2.3   Hilbert (Complex) EOFs

There is some confusion in the literature over the terminology 'complex EOFs,' or 'complex PCA.' It is perfectly possible to perform PCA on complex numbers, as well as real numbers, whether or not the measurements are made over time. We return to this general version of complex PCA in Section 13.8. Within the time series context, and especially for meteorological time series, the term 'complex EOFs' has come to refer to a special type of complex series. To reduce confusion, von Storch and Zwiers (1999) suggest (for reasons that will soon become apparent) referring to this procedure as Hilbert EOF analysis. We will follow this recommendation. Baines has suggested removing ambiguity entirely by denoting the analysis as 'complex Hilbert EOF analysis.' The technique seems to have originated with Rasmusson et al. (1981), by whom it was referred to as *Hilbert singular decomposition*.

Suppose that $\mathbf{x}_t$, $t = 1, 2, \ldots, n$ is a p-variate time series, and let

$$\mathbf{y}_t = \mathbf{x}_t + i\mathbf{x}_t^H, \qquad (12.2.2)$$

where $i = \sqrt{-1}$ and $\mathbf{x}_t^H$ is the Hilbert transform of $\mathbf{x}_t$, defined as

$$\mathbf{x}_t^H = \sum_{s=0}^{\infty} \frac{2}{(2s+1)\pi} \left(\mathbf{x}_{(t+2s+1)} - \mathbf{x}_{(t-2s-1)} \right).$$

The definition assumes that $\mathbf{x}_t$ is observed at an infinite number of times $t = \ldots, -1, 0, 1, 2, \ldots$. Estimation of $\mathbf{x}_t^H$ for finite samples, to use in equation (12.2.2), is discussed by von Storch and Zwiers (1999, Section 16.2.4) and Bloomfield and Davis (1994).

If a series is made up of oscillatory terms, its Hilbert transform advances each oscillatory term by $\frac{\pi}{2}$ radians. When $\mathbf{x}_t$ is comprised of a single periodic oscillation, $\mathbf{x}_t^H$ is identical to $\mathbf{x}_t$, except that it is shifted by $\frac{\pi}{2}$ radians. In the more usual case, where $\mathbf{x}_t$ consists of a mixture of two or more oscillations or pseudo-oscillations at different frequencies, the effect of transforming to $\mathbf{x}_t^H$ is more complex because the phase shift of $\frac{\pi}{2}$ is implemented separately for each frequency.

A Hilbert EOF (HEOF) analysis is simply a PCA based on the covariance matrix of $\mathbf{y}_t$ defined in (12.2.2). As with (M)SSA and POP analysis, HEOF analysis will find dominant oscillatory patterns, which may or may not be propagating in space, that are present in a standard meteorological data set of p spatial locations and n time points.

Similarly to POP analysis, the eigenvalues and eigenvectors (HEOFs) are complex, but for a different reason. Here a covariance matrix is analysed, unlike POP analysis, but the variables from which the covariance matrix is formed are complex-valued. Other differences exist between POP analysis and HEOF analysis, despite the similarities in the oscillatory structures they can detect, and these are noted by von Storch and Zwiers (1999, Section 15.1.7). HEOF analysis maximizes variances, has orthogonal component scores and is empirically based, all attributes shared with PCA. In direct contrast, POP analysis does not maximize variance, has non-orthogonal POP coefficients (scores) and is model-based. As in ordinary PCA, HEOFs may be simplified by rotation (Section 11.1), and Bloomfield and Davis (1994) discuss how this can be done.

There is a connection between HEOF analysis and PCA in the frequency domain which is discussed in Section 12.4.1. An example of HEOF analysis is now given.

Southern Hemisphere Sea Surface Temperature

This example and its figures are taken, with permission, from Cai and Baines (2001). The data are sea surface temperatures (SSTs) in the Southern Hemisphere. Figure 12.4 gives a shaded contour map of the coefficients in the first four (ordinary) PCs for these data (the first four EOFs), together with the variance accounted for by each PC. Figure 12.5 displays similar plots for the real and imaginary parts of the first Hilbert EOF (labelled CEOFs on the plots). It can be seen that the real part of the first Hilbert EOF in Figure 12.5 looks similar to EOF1 in Figure 12.4. There are also similarities between EOF3 and the imaginary part of the first Hilbert EOF.

Figures 12.6, 12.7 show plots of time series (scores), labelled temporal coefficients in 12.7, for the first and third ordinary PCs, and the real and imaginary parts of the first HEOF, respectively. The similarity between the first PC and the real part of the first HEOF is obvious. Both represent the same oscillatory behaviour in the series. The imaginary part of the first HEOF is also very similar, but lagged by $\frac{\pi}{2}$, whereas the scores on the third EOF show a somewhat smoother and more regular oscillation. Cai and Baines (2001) note that the main oscillations visible in Figures 12.6, 12.7 can be identified with well-known El Niño-Southern Oscillation (ENSO) events. They also discuss other physical interpretations of the results of the HEOF analysis relating them to other meteorological variables, and they provide tests for statistical significance of HEOFs.

The main advantage of HEOF analysis over ordinary PCA is its ability to identify and reconstruct propagating waves, whereas PCA only finds standing oscillations. For the first and second Hilbert EOFs, and for their sum, this propagating behaviour is illustrated in Figure 12.8 by the movement of similar-valued coefficients from west to east as time progresses in the vertical direction.

12.2.4   Multitaper Frequency Domain-Singular Value Decomposition (MTM SVD)

In a lengthy paper, Mann and Park (1999) describe MTM-SVD, developed earlier by the same authors. It combines multitaper spectrum estimation methods (MTM) with PCA using the singular value decomposition (SVD). The paper also gives a critique of several of the other techniques discussed in the present section, together with frequency domain PCA which is covered in Section 12.4.1. Mann and Park (1999) provide detailed examples, both real and artificial, in which MTM-SVD is implemented.

Like MSSA, POP analysis, HEOF analysis and frequency domain PCA, MTM-SVD looks for oscillatory behaviour in space and time. It is closest in form to PCA of the spectral matrix $\mathbf{F}(\lambda)$ (Section 12.4.1), as it operates in the frequency domain. However, in transferring from the time domain to the frequency domain MTM-SVD constructs a set of different tapered Fourier transforms (hence the 'multitaper' in its name). The frequency domain matrix is then subjected to a singular value decomposition, giving 'spatial' or 'spectral' EOFs, and 'principal modulations' which are analogous to principal component scores. Mann and Park (1999) state that the

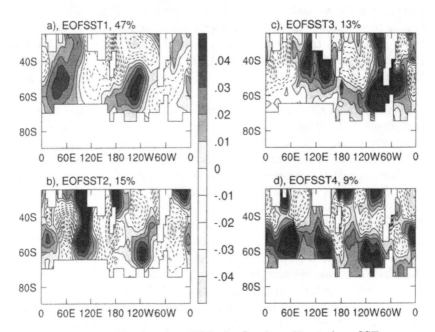

Figure 12.4. The first four EOFs for Southern Hemisphere SST.

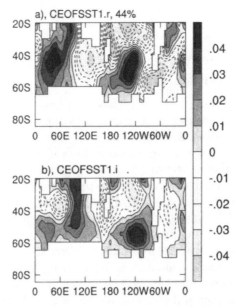

Figure 12.5. Real and imaginary parts of the first Hilbert EOF for Southern Hemisphere SST.

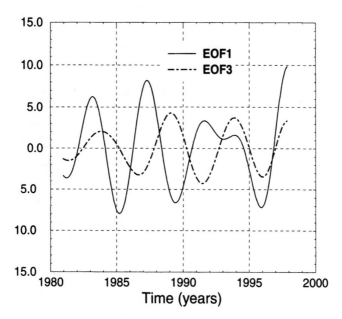

Figure 12.6. Plots of temporal scores for EOF1 and EOF3 for Southern Hemisphere SST.

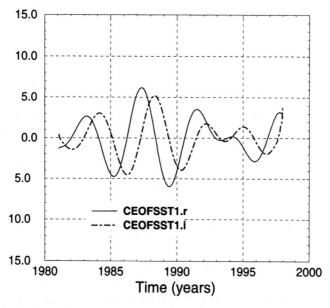

Figure 12.7. Plots of temporal scores for real and imaginary parts of the first Hilbert EOF for Southern Hemisphere SST.

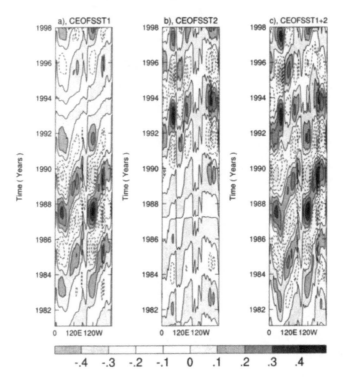

Figure 12.8. Propagation of waves in space and time in Hilbert EOF1, Hilbert EOF2, and the sum of these two Hilbert EOFs.

distinction between MTM-SVD and standard frequency domain PCA is that the former provides a *local* frequency domain decomposition of the different spectral estimates given by the multitapers, whereas the latter produces a *global* frequency domain decomposition over the spectral estimates. Mann and Park (1999) describe tests for the statistical significance of the oscillations found by MTM-SVD using a bootstrap approach, which, it is claimed, is effective for a general, smoothly varying, coloured noise background, and is not restricted to red noise as in Allen and Smith (1996) and Allen and Robertson (1996).

12.2.5 Cyclo-Stationary and Periodically Extended EOFs (and POPs)

The assumption of temporal stationarity is implicit for most of the methods described in this chapter. In meteorological data there is often a cycle with a fixed period, most commonly the annual cycle but sometimes a diurnal cycle. There may then be stationarity for times at the same point in the cycle but not across different points in the cycle. For example, for monthly data the probability distribution may be same every April but different in

April than in June. Similarly, the joint distribution for April and August, which have a four-month separation, may be the same in different years, but different from the joint distribution for June and October, even though the latter are also separated by four months. Such behaviour is known as *cyclo-stationarity*, and both PCA and POP analysis have been modified for such data.

The modification is easier to explain for POP analysis than for PCA. Suppose that τ is the length of the cycle; for example $\tau = 12$ for monthly data and an annual cycle. Then for a time series of length $n = n'\tau$ equation (12.2.1) is replaced by

$$(\mathbf{x}_{(s\tau+t+1)} - \boldsymbol{\mu}_{(t+1)}) = \boldsymbol{\Upsilon}_t(\mathbf{x}_{(s\tau+t)} - \boldsymbol{\mu}_t) + \boldsymbol{\epsilon}_{(s\tau+t)}, \qquad (12.2.3)$$

$$t = 0, 1, \ldots, (\tau - 2); \quad s = 0, 1, 2, \ldots, (n' - 1),$$

with $(t + 1)$ replaced by 0, and s replaced by $(s + 1)$ on the left-hand side of (12.2.3) when $t = (\tau - 1)$ and $s = 0, 1, 2, \ldots, (n' - 2)$. Here the mean $\boldsymbol{\mu}_t$ and the matrix $\boldsymbol{\Upsilon}_t$ can vary within cycles, but not between cycles. Cyclo-stationary POP analysis estimates $\boldsymbol{\Upsilon}_0, \boldsymbol{\Upsilon}_1, \ldots, \boldsymbol{\Upsilon}_{(\tau-1)}$ and is based on an eigenanalysis of the product of those estimates $\hat{\boldsymbol{\Upsilon}}_0\hat{\boldsymbol{\Upsilon}}_1 \ldots \hat{\boldsymbol{\Upsilon}}_{(\tau-1)}$.

The cyclo-stationary variety of PCA is summarized by Kim and Wu (1999) but is less transparent in its justification than cyclo-stationary POP analysis. First, vectors $\mathbf{a}_{0t}, \mathbf{a}_{1t}, \ldots, \mathbf{a}_{(\tau-1)t}$ are found such that $\mathbf{x}_t = \sum_{j=0}^{\tau-1} \mathbf{a}_{jt}e^{\frac{2\pi ijt}{\tau}}$, and a new vector of variables is then constructed by concatenating $\mathbf{a}_{0t}, \mathbf{a}_{1t}, \ldots, \mathbf{a}_{(\tau-1)t}$. Cyclo-stationary EOFs (CSEOFs) are obtained as eigenvectors of the covariance matrix formed from this vector of variables. Kim and Wu (1999) give examples of the technique, and also give references explaining how to calculate CSEOFs.

Kim and Wu (1999) describe an additional modification of PCA that deals with periodicity in a time series, and which they call a *periodically extended* EOF technique. It works by dividing a series of length $n = n'\tau$ into n' blocks of length τ. A covariance matrix, $\mathbf{S}^{EX}$, is then computed as

$$\begin{bmatrix} \mathbf{S}_{11}^{EX} & \mathbf{S}_{12}^{EX} & \cdots & \mathbf{S}_{1n'}^{EX} \\ \mathbf{S}_{21}^{EX} & \mathbf{S}_{22}^{EX} & \cdots & \mathbf{S}_{2n'}^{EX} \\ \vdots & \vdots & & \vdots \\ \mathbf{S}_{n'1}^{EX} & \mathbf{S}_{n'2}^{EX} & \cdots & \mathbf{S}_{n'n'}^{EX} \end{bmatrix},$$

in which the (i, j)th element of $\mathbf{S}_{kl}^{EX}$ is the sample covariance between the measurements at the ith location at time k in each block and at the jth location at time l in the same block. The description of the technique in Kim and Wu (1999) is sketchy, but it appears that whereas in ordinary PCA, covariances are calculated by averaging over *all* time points, here the averaging is done over times at the same point within each block *across* blocks. A similar averaging is implicit in cyclo-stationary POP analysis. Examples of periodically extended EOFs are given by Kim and Wu (1999).

12.2.6 Examples and Comparisons

Relatively few examples have been given in this section, due in part to their complexity and their rather specialized nature. Several of the techniques described claim to be able to detect stationary and propagating waves in the standard type of spatio-temporal meteorological data. Each of the methods has proponents who give nice examples, both real and artificial, in which their techniques appear to work well. However, as with many other aspects of PCA and related procedures, caution is needed, lest enthusiasm leads to 'over-interpretation.' With this caveat firmly in mind, examples of SSA can be found in Elsner and Tsonis (1996), Vautard (1995); MSSA in Plaut and Vautard (1994), Mote et al. (2000); POPs in Kooperberg and O'Sullivan (1996) with cyclo-stationary POPs in addition in von Storch and Zwiers (1999, Chapter 15); Hilbert EOFs in Horel (1984), Cai and Baines (2001); MTM-SVD in Mann and Park (1999); cyclo-stationary and periodically extended EOFs in Kim and Wu (1999). This last paper compares eight techniques (PCA, PCA plus rotation, extended EOFs (MSSA), Hilbert EOFs, cyclo-stationary EOFs, periodically extended EOFs, POPs and cyclo-stationary POPs) on artificial data sets with stationary patterns, with patterns that are stationary in space but which have amplitudes changing in time, and with patterns that have periodic oscillations in space. The results largely confirm what might be expected. The procedures designed to find oscillatory patterns do not perform well for stationary data, the converse holds when oscillations are present, and those techniques devised for cyclo-stationary data do best on such data.

12.3 Functional PCA

There are many circumstances when the data are curves. A field in which such data are common is that of chemical spectroscopy (see, for example, Krzanowski et al. (1995), Mertens (1998)). Other examples include the trace on a continuously recording meteorological instrument such as a barograph, or the trajectory of a diving seal. Other data, although measured at discrete intervals, have an underlying continuous functional form. Examples include the height of a child at various ages, the annual cycle of temperature recorded as monthly means, or the speed of an athlete during a race.

The basic ideas of PCA carry over to this continuous (functional) case, but the details are different. In Section 12.3.1 we describe the general set-up in functional PCA (FPCA), and discuss methods for estimating functional PCs in Section 12.3.2. Section 12.3.3 presents an example. Finally, Section 12.3.4 briefly covers some additional topics including curve registration, bivariate FPCA, smoothing, principal differential analysis, prediction, discrimination, rotation, density estimation and robust FPCA.

A key reference for functional PCA is the book by Ramsay and Silverman (1997), written by two researchers in the field who, together with Besse (see, for example, Besse and Ramsay, 1986), have been largely responsible for bringing these ideas to the attention of statisticians. We draw heavily on this book in the present section. However, the ideas of PCA in a continuous domain have also been developed in other fields, such as signal processing, and the topic is an active research area. The terminology 'Karhunen-Loève expansion' is in common use in some disciplines to denote PCA in a continuous domain. Diamantaras and Kung (1996, Section 3.2) extend the terminology to cover the case where the data are discrete time series with a theoretically infinite number of time points.

In atmospheric science the Karhunen-Loève expansion has been used in the case where the continuum is spatial, and the different observations correspond to different discrete times. Preisendorfer and Mobley (1988, Section 2d) give a thorough discussion of this case and cite a number of earlier references dating back to Obukhov (1947). Bouhaddou et al. (1987) independently consider a spatial context for what they refer to as 'principal component analysis of a stochastic process,' but which is PCA for a (two-dimensional spatial) continuum of variables. They use their approach to approximate both a spatial covariance function and the underlying spatial stochastic process, and compare it with what they regard as the less flexible alternative of kriging. Guttorp and Sampson (1994) discuss similar ideas in a wider review of methods for estimating spatial covariance matrices. Durbin and Knott (1972) derive a special case of functional PCA in the context of goodness-of-fit testing (see Section 14.6.2).

12.3.1 The Basics of Functional PCA (FPCA)

When data are functions, our usual data structure x_{ij}, $i = 1, 2, \ldots, n$; $j = 1, 2, \ldots, p$ is replaced by $x_i(t)$, $i = 1, 2, \ldots, n$ where t is continuous in some interval. We assume, as elsewhere in the book, that the data are centred, so that a mean curve $\bar{x} = \frac{1}{n} \sum_{i=1}^{n} \tilde{x}_i(t)$ has been subtracted from each of the original curves $\tilde{x}_i(t)$. Linear functions of the curves are now integrals instead of sums, that is $z_i = \int a(t)x_i(t)dt$ rather than $z_i = \sum_{j=1}^{p} a_j x_{ij}$. In 'ordinary' PCA the first PC has weights $a_{11}, a_{21}, \ldots, a_{p1}$, which maximize the sample variance $\text{var}(z_i)$, subject to $\sum_{j=1}^{p} a_{j1}^2 = 1$. Because the data are centred,

$$\text{var}(z_{i1}) = \frac{1}{n-1} \sum_{i=1}^{n} z_{i1}^2 = \frac{1}{n-1} \sum_{i=1}^{n} \left[\sum_{j=1}^{p} a_{j1} x_{ij} \right]^2.$$

Analogously for curves, we find $a_1(t)$ which maximizes

$$\frac{1}{n-1} \sum_{i=1}^{n} z_{i1}^2 = \frac{1}{n-1} \sum_{i=1}^{n} \left[\int a_1(t)x_i(t)\, dt \right]^2,$$

subject to $\int a_1^2(t)dt = 1$. Here, and elsewhere, the integral is over the range of values of t for which the data are observed. Subsequent FPCs are defined successively, as with ordinary PCA, to maximise

$$\text{var}(z_{ik}) = \frac{1}{n-1} \sum_{i=1}^{n} \left[\int a_k(t)x_i(t)\, dt \right]^2,$$

subject to $\int a_k^2(t)dt = 1$; $\int a_k(t)a_h(t)dt = 0$, $k = 2, 3, \ldots$; $h = 1, 2, \ldots$; $h < k$.

The sample covariance between $x(s)$ and $x(t)$ can be defined as $S(s,t) = \frac{1}{(n-1)} \sum_{i=1}^{n} x_i(s)x_i(t)$, with a corresponding definition for correlation, and to find the functional PCs an eigenequation is solved involving this covariance function. Specifically, we solve

$$\int S(s,t)a(t)dt = la(s). \tag{12.3.1}$$

Comparing (12.3.1) to the eigenequation $\mathbf{Sa} = l\mathbf{a}$ for ordinary PCA, the pre-multiplication of a vector of weights by a matrix of covariances on the left-hand side is replaced by an integral operator in which the covariance function is multiplied by a weight *function* $a(t)$ and then integrated. In ordinary PCA the number of solutions to the eigenequation is usually p, the number of variables. Here the p variables are replaced by a infinite number of values for t but the number of solutions is still finite, because the number of curves n is finite. The number of non-zero eigenvalues $l_1, l_2, \ldots$, and corresponding functions $a_1(t), a_2(t), \ldots$ cannot exceed $(n-1)$.

12.3.2   Calculating Functional PCs (FPCs)

Unless the curves are all fairly simple functional forms, it is not possible to solve (12.3.1) exactly. Ramsay and Silverman (1997, Section 6.4) give three computational methods for FPCA, but in most circumstances they represent *approximate* solutions to (12.3.1). In the first of the three, the data are discretized. The values $x_i(t_1), x_i(t_2), \ldots, x_i(t_p)$ form the ith row of an $(n \times p)$ data matrix which is then analysed using standard PCA. The times $t_1, t_2, \ldots, t_p$ are usually chosen to be equally spaced in the range of continuous values for t. To convert the eigenvectors found from this PCA into functional form, it is necessary to renormalize the eigenvectors and then interpolate them with a suitable smoother (Ramsay and Silverman, 1997, Section 6.4.1).

A second approach assumes that the curves $x_i(t)$ can be expressed linearly in terms of a set of G basis functions, where G is typically less than n. If $\phi_1(t), \phi_2(t), \ldots, \phi_G(t)$ are the basis functions, then $x_i(t) = \sum_{g=1}^{G} c_{ig}\phi_g(t)$, $i = 1, 2, \ldots, n$ or, in matrix form, $\mathbf{x}(t) = \mathbf{C}\phi(t)$, where $\mathbf{x}'(t) = (x_1(t), x_2(t), \ldots, x_n(t))$, $\phi'(t) = (\phi_1(t), \phi_2(t), \ldots, \phi_G(t))$ and $\mathbf{C}$ is an $(n \times G)$ matrix with (i, g)th element c_{ig}.

The sample covariance between $x(s)$ and $x(t)$ can be written

$$\frac{1}{n-1}\mathbf{x}'(s)\mathbf{x}(t) = \frac{1}{n-1}\phi'(s)\mathbf{C}'\mathbf{C}\phi(t).$$

Any eigenfunction $a(t)$ can be expressed in terms of the basis functions as $a(t) = \sum_{g=1}^{G} b_g\phi_g(t) = \phi'(t)\mathbf{b}$ for some vector of coefficients $\mathbf{b}' = (b_1, b_2, \ldots, b_G)$. The left-hand-side of equation (12.3.1) is then

$$\int S(s,t)a(t)\, dt = \int \frac{1}{n-1}\phi'(s)\mathbf{C}'\mathbf{C}\phi(t)\phi'(t)\mathbf{b}\, dt$$

$$= \frac{1}{n-1}\phi'(s)\mathbf{C}'\mathbf{C}\left[\int \phi(t)\phi'(t)\, dt\right]\mathbf{b}.$$

The integral is a $(G \times G)$ matrix $\mathbf{W}$ whose (g, h)th element is $\int \phi_g(t)\phi_h(t)dt$. If the basis is orthogonal, $\mathbf{W}$ is simply the identity matrix $\mathbf{I}_G$. Hence choosing an orthogonal basis in circumstances where such a choice makes sense, as with a Fourier basis for periodic data, gives simplified calculations. In general (12.3.1) becomes

$$\frac{1}{n-1}\phi'(s)\mathbf{C}'\mathbf{C}\mathbf{W}\mathbf{b} = l\phi'(s)\mathbf{b}$$

but, because this equation must hold for all available values of s, it reduces to

$$\frac{1}{n-1}\mathbf{C}'\mathbf{C}\mathbf{W}\mathbf{b} = l\mathbf{b}. \tag{12.3.2}$$

When $\int a^2(t)dt = 1$ it follows that

$$1 = \int a^2(t)\, dt = \int \mathbf{b}'\phi(t)\phi'(t)\mathbf{b}\, dt = \mathbf{b}'\mathbf{W}\mathbf{b}.$$

If $a_k(t)$ is written in terms of the basis functions as $a_k(t) = \sum_{g=1}^{G} b_{kg}\phi_g(t)$, with a similar expression for $a_l(t)$, then $a_k(t)$ is orthogonal to $a_l(t)$ if $\mathbf{b}'_k\mathbf{W}\mathbf{b}_l = 0$, where $\mathbf{b}'_k = (b_{k1}, b_{k2}, \ldots, b_{kG})$, and $\mathbf{b}'_l$ is defined similarly.

In an eigenequation, the eigenvector is usually normalized to have unit length (norm). To convert (12.3.2) into this form, define $\mathbf{u} = \mathbf{W}^{\frac{1}{2}}\mathbf{b}$. Then $\mathbf{u}'\mathbf{u} = 1$ and (12.3.2) can be written

$$\frac{1}{n-1}\mathbf{W}^{\frac{1}{2}}\mathbf{C}'\mathbf{C}\mathbf{W}^{\frac{1}{2}}\mathbf{u} = l\mathbf{u}. \tag{12.3.3}$$

Equation (12.3.3) is solved for l and $\mathbf{u}$, $\mathbf{b}$ is obtained as $\mathbf{W}^{-\frac{1}{2}}\mathbf{u}$, and finally $a(t) = \phi'(t)\mathbf{b} = \phi'(t)\mathbf{W}^{-\frac{1}{2}}\mathbf{u}$.

The special case where the basis is orthogonal has already been mentioned. Here $\mathbf{W} = \mathbf{I}_G$, so $\mathbf{b} = \mathbf{u}$ is an eigenvector of $\frac{1}{n-1}\mathbf{C}'\mathbf{C}$. Another special case, noted by Ramsay and Silverman (1997), occurs when the data curves themselves are taken as the basis. Then $\mathbf{C} = \mathbf{I}_n$ and $\mathbf{u}$ is an eigenvector of $\frac{1}{n-1}\mathbf{W}$.

The third computational method described by Ramsay and Silverman (1997) involves applying numerical quadrature schemes to the integral on the left-hand side of (12.3.1). Castro et al. (1986) used this approach in an early example of functional PCA. Quadrature methods can be adapted to cope with irregularly spaced data (Ratcliffe and Solo, 1998). Aguilera et al. (1995) compare a method using a trigonometric basis with one based on a trapezoidal scheme, and find that the behaviour of the two algorithms is similar except at the extremes of the time interval studied, where the trapezoidal method is superior.

Preisendorfer and Mobley (1988, Section 2d) have two interesting approaches to finding functional PCs in the case where t represents spatial position and different observations correspond to different discrete times. In the first the eigenequation (12.3.1) is replaced by a dual eigenequation, obtained by using a relationship similar to that between $\mathbf{X}'\mathbf{X}$ and $\mathbf{X}\mathbf{X}'$, which was noted in the proof of Property G4 in Section 3.2. This dual problem is discrete, rather than continuous, and so is easier to solve. Its eigenvectors are the PC scores for the continuous problem and an equation exists for calculating the eigenvectors of the original continuous eigenequation from these scores.

The second approach is similar to Ramsay and Silverman's (1997) use of basis functions, but Preisendorfer and Mobley (1988) also compare the basis functions and the derived eigenvectors (EOFs) in order to explore the physical meaning of the latter. Bouhaddou et al. (1987) independently proposed the use of interpolating basis functions in the implementation of a continuous version of PCA, given what is necessarily a discrete-valued data set.

12.3.3   Example - 100 km Running Data

Here we revisit the data that were first introduced in Section 5.3, but in a slightly different format. Recall that they consist of times taken for ten 10 km sections by 80 competitors in a 100 km race. Here we convert the data into speeds over each section for each runner. Ignoring 'pitstops,' it seems reasonable to model the speed of each competitor through the race as a continuous curve. In this example the horizontal axis represents position in (one-dimensional) space rather than time. Figure 12.9 shows the speed for each competitor, with the ten discrete points joined by straight lines. Despite the congestion of lines on the figure, the general pattern of slowing down is apparent. Figure 12.10 shows the coefficients for the first three ordinary PCs of these speed data. In Figure 12.11 the piecewise linear plots of Figure 12.10 are smoothed using a spline basis. Finally, Figure 12.12 gives the eigenfunctions from a FPCA of the data, using spline basis functions, implemented in S-Plus.

There are strong similarities between Figures 12.10–12.12, though some differences exist in the details. The first PC, as with the 'time taken' version

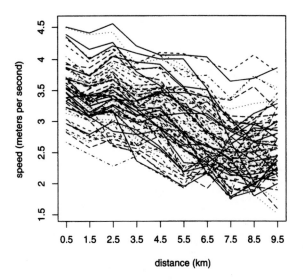

Figure 12.9. Plots of speed for 80 competitors in a 100 km race.

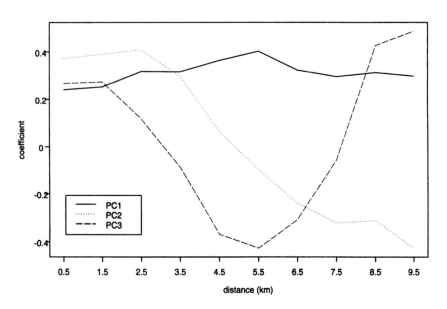

Figure 12.10. Coefficients for first three PCs from the 100 km speed data.

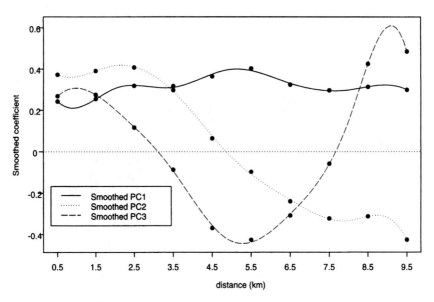

Figure 12.11. Smoothed version of Figure 12.10 using a spline basis; dots are coefficients from Figure 12.10.

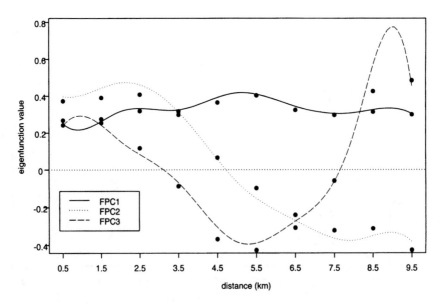

Figure 12.12. Coefficients (eigenfunctions) for the first three components in a functional PCA of the 100 km speed data using a spline basis; dots are coefficients from Figure 12.10.

of the data in Section 5.3, is a measure of overall speed throughout the race. The second component in all cases represents the degree to which runners slow down during the race. The pattern in Figure 12.10 is very much like that in Table 5.2. Figures 12.11 and 12.12 have a similar shape for this component, although the absolute values of the eigenfunction in Figure 12.12 are slightly larger than the coefficients in Figure 12.11. The reason for this is that the speed for the first 10 km section is plotted at distance equal to 0.5 km, the speed for the second section at 1.5 km and so on. The eigenfunctions for the FPCs are then calculated over the interval 0.5 to 9.5, a range of 9, whereas there are 10 elements in the ordinary PC eigenvectors, leading to different normalizations. Calculating the FPCs over the range 0 to 10 leads to erratic behaviour at the ends of the interval. The pattern for the third PC is again similar in all three figures, except at the extreme right-hand end.

12.3.4   Further Topics in FPCA

This subsection collects together briefly a number of topics that are relevant to FPCA. As in the rest of this section, the main reference is Ramsay and Silverman (1997).

Curve Registration

In some examples where the data are functions, the individual curves $x_i(t)$ may be observed over different ranges for t. Most procedures for analysing functional data assume the same range of t for all curves. Furthermore, it is often the case that the horizontal extent of the curve is of less importance than its shape. For example, suppose that the trajectory of underwater dives made by seals is of interest (see Schreer et al. (1998)). Although the lengths of the dives may be informative, so are their shapes, and to compare these the curves should be aligned to start and finish at the same respective times. This process is known as *curve registration*. Another example arises when outlines of sliced carrots are examined in order to determine to which variety of carrot an individual specimen belongs (Horgan et al., 2001). More complicated situations are discussed by Ramsay and Silverman (1997, Chapter 5), in which several landmarks along the curves must be aligned, leading to differential stretching and expansion of 'time' for different parts of the curves. Capra and Müller (1997) use an accelerated time model to align mortality curves for medflies. They refer to their accelerated time as 'eigenzeit.'

Sometimes alignment is desirable in the vertical, as well the horizontal, direction. Consider again the dive trajectories of seals. If the shapes of the dives are more important than their depths, then the dives can be aligned vertically at their deepest points.

To analyse only the registered curves, without taking into account of what was done in the registration process, is to throw away information.

Keeping one or more parameters for each curve, defining how registration was done, leads to what Ramsay and Silverman (1997) called 'mixed data.' Each observation for such data consists of a curve, together with p other 'ordinary' variables. Ramsay and Silverman (1997, Chapter 8) discuss the analysis of such data.

Bivariate FPCA

In other cases, the data are not 'mixed' but there is more than one curve associated with each individual. An example involving changes of angles in both hip and knee during a gait cycle is described by Ramsay and Silverman (1997, Section 6.5). They discuss the analysis of bivariate curves from this example using bivariate FPCA. Suppose that the two sets of curves are $x_1(t), x_2(t), \ldots, x_n(t)$; $y_1(t), y_2(t), \ldots, y_n(t)$. Define a bivariate covariance function $S(s,t)$ as

$$\left[\begin{array}{cc} S_{XX}(s,t) & S_{XY}(s,t) \\ S_{YX}(s,t) & S_{YY}(s,t) \end{array} \right],$$

where $\mathbf{S}_{XX}(s,t)$, $\mathbf{S}_{YY}(s,t)$ are covariance functions defined, as earlier, for $(x(s), x(t))$ and $(y(s), y(t))$, respectively, and $\mathbf{S}_{XY}(s,t)$ has elements that are covariances between $x(s)$ and $y(t)$. Suppose that

$$z_{Xi} = \int a_X(t) x_i(t)\, dt, \quad z_{Yi} = \int a_Y(t) y_i(t) dt.$$

Finding $a_X(t)$, $a_Y(t)$ to maximize $\frac{1}{n-1}\sum_{i=1}^{n}(z_{Xi}^2 + z_{Yi}^2)$ leads to the eigenequations

$$\int S_{XX}(s,t) a_X(t)\, dt + \int S_{XY}(s,t) a_Y(t)\, dt = l a_X(t)$$

$$\int S_{YX}(s,t) a_X(t)\, dt + \int S_{YY}(s,t) a_Y(t)\, dt = l a_Y(t).$$

This analysis can be extended to the case of more than two curves per individual.

Smoothing

If the data are not smooth, the weighting functions $a(t)$ in FPCA may not be smooth either. With most curves, an underlying smoothness is expected, with the superimposed roughness being due to noise that should ideally be removed. Ramsay and Silverman (1997, Chapter 7) tackle this problem. Their main approach incorporates a roughness penalty into FPCA's variance-maximizing problem. The second derivative of a curve is often taken as a measure of its roughness and, if $D^2 a(t)$ represents the second derivative of $a(t)$, a smooth curve requires a small value of $D^2 a(t)$. Ramsay and Silverman's approach is to maximize

$$\frac{\frac{1}{n-1}\sum_{i=1}^{n}[\int a(t) x_i(t) dt]^2}{\int a^2(t) dt + \lambda \int (D^2 a(t))^2 dt} \tag{12.3.4}$$

subject to $\int a^2(t)dt = 1$, where λ is a tuning parameter. Taking $\lambda = 0$ simply gives unsmoothed FPCA; as λ increases, so does the smoothness of the optimal $a(t)$. Ramsay and Silverman (1997, Section 7.3) show that solving the optimization problem reduces to solving the eigenequation

$$\int S(s,t)a(t)dt = l(1 + \lambda D^4)a(s), \qquad (12.3.5)$$

where $D^4 a(s)$ is the fourth derivative of $a(s)$. In their Section 7.4 they give some computational methods for (approximately) solving equation (12.3.5), with the choice of λ discussed in Section 7.5. Minimizing (12.3.4) by solving (12.3.5) implies a form of orthogonality between successive $a_k(t)$ that is different from the usual one. Ocaña et al. (1999) interpret this as a different choice of norm and inner product, and discuss the choice of norm in FPCA from a theoretical point of view. If (12.3.5) is solved by using basis functions for the data, the number of basis functions and the smoothing parameter λ both need to be chosen. Ratcliffe and Solo (1998) address the problem of choosing both simultaneously and propose an improved cross-validation procedure for doing so.

An alternative to incorporating smoothing directly into FPCA is to smooth the data first and then conduct unsmoothed FPCA on the smoothed curves. The smoothing step uses the well-known idea of minimizing the sum of squares between fitted and observed curves, supplemented by a penalty function based on lack of smoothness (a roughness penalty). The quantity we wish to minimize can be written succicntly as

$$\|x(t) - \hat{x}(t)\|^2 + \lambda\|\hat{x}(t)\|^2, \qquad (12.3.6)$$

where $x(t)$, $\hat{x}(t)$ denote observed and fitted curves, respectively, $\|\,.\,\|$ denotes an appropriately chosen norm, not necessarily the same in the two parts of the expression, and λ is a tuning parameter. The second norm is often related to the second derivative, $D^2[\hat{x}(t)]$, which is a commonly used measure of roughness.

Besse et al. (1997) use a similar approach in which they optimize a criterion which, like (12.3.6), is a sum of two terms, one a measure of fit and the other a roughness penalty with a tuning parameter λ. Besse et al.'s model assumes that, apart from noise, the curves lie in a q-dimensional space. A version $\hat{R}_q$ of the criterion R_q, suggested by Besse and de Falguerolles (1993) and discussed in Section 6.1.5, is minimized with respect to q and λ simultaneously. If smoothing of the data is done using splines, then the PCA on the resulting interpolated data is equivalent to a PCA on the original data, but with a different metric (Besse, 1988). We return to this topic in Section 14.2.2.

A more complicated approach to smoothing followed by PCA, involving different smoothing parameters for different FPCs, is discussed by Ramsay and Silverman (1997, Section 7.8). Grambsch et al. (1995) also use a smoothing process—though a different one—for the data, followed by PCA on the smoothed data. Kneip (1994) independently suggested smoothing the data, followed by PCA on the smoothed data, in the context of fitting a model in which a small number of functions is assumed to underlie a set of regression curves. Theoretical properties of Kneip's (1994) procedure are examined in detail in his paper. Champely and Doledec (1997) use lo(w)ess (locally weighted scatterplot smoothing—Cleveland, 1979, 1981) to fit smooth trend and periodic curves to water quality data, and then apply FPCA separately to the trend and periodic curves.

Principal Differential Analysis

Principal differential analysis (PDA), a term coined by Ramsay (1996) and discussed in Ramsay and Silverman (1997, Chapter 14) is another method of approximating a set of curves by a smaller number of functions. Although PDA has some similarities to FPCA, which we note below, it concentrates on finding functions that have a certain type of smoothness, rather than maximizing variance. Define a linear differential operator

$$L = w_0 I + w_1 D + \ldots + w_{m-1} D^{m-1} + D^m,$$

where D^i, as before, denotes the ith derivative operator and I is the identity operator. PDA finds weights $w_0, w_1, \ldots, w_{m-1}$ for which $[Lx_i(t)]^2$ is small for each observed curve $x_i(t)$. Formally, we minimize $\sum_{i=1}^{n} \int [Lx_i(t)]^2 \, dt$ with respect to $w_0, w_1, \ldots, w_{m-1}$.

Once $w_0, w_1, \ldots, w_{m-1}$ and hence L are found, any curve satisfying $Lx(t) = 0$ can be expressed as a linear combination of m linearly independent functions spanning the null space of the operator L. Any observed curve $x_i(t)$ can be approximated by expanding it in terms of these m functions. This is similar to PCA, where the original data can be approximated by expanding them in terms of the first few (say m) PCs. The difference is that PCA finds an m-dimensional space with a least squares fit to the original data, whereas PDA finds a space which penalizes roughness. This last interpretation follows because $Lx_i(t)$ is typically rougher than $x_i(t)$, and PDA aims to make $Lx_i(t)$, or rather $[Lx_i(t)]^2$, as small as possible when averaged over i and t. An application of PDA to the study of variations in handwriting is presented by Ramsay (2000).

Prediction and Discrimination

Aguilera et al. (1997, 1999a) discuss the idea of predicting a continuous time series by regressing functional PCs for the future on functional PCs from the past. To implement this methodology, Aguilera et al. (1999b) propose cutting the series into a number of segments n of equal length, which are then treated as n realizations of the same underlying process. Each segment is in turn divided into two parts, with the second, shorter, part to be predicted from the first. In calculating means and covariance functions, less weight is given to the segments from the early part of the series than to those closer to the present time from where the future is to be predicted. Besse et al. (2000) also use functional PCA to predict time series, but in the context of a smoothed first order autoregressive model. The different replications of the function correspond to different years, while the function ranges over the months within years. A 'local' version of the technique is developed in which the assumption of stationarity is relaxed.

Hall et al. (2001) advocate the use of functional PCA as a dimension-reducing step in the context of discriminating between different types of radar signal. Although this differs from the usual set-up for PCA in discriminant analysis (see Section 9.1) because it notionally has an infinite number of variables in a continuum, there is still the possibility that some of the later discarded components may contain non-trivial discriminatory power.

Rotation

As with ordinary PCA, the interpretation of FPCA may be improved by rotation. In addition to the conventional rotation of coefficients in a subset of PCs (see Section 11.1), Ramsay and Silverman (1997, Section 6.3.3) suggest that the coefficients $b_1, b_2, \ldots, b_m$ of the first m eigenfunctions with respect to a chosen set of basis functions, as defined in Section 12.3.2, could also be rotated to help interpretation. Arbuckle and Friendly (1977) propose a variation on the usual rotation criteria of Section 11.1 for variables that are measurements at discrete points on a continuous curve. They suggest rotating the results of an ordinary PCA towards smoothness rather than towards simplicity as usually defined (see Section 11.1).

Density Estimation

Kneip and Utikal (2001) discuss functional PCA as a means of examining common structure and differences in a set of probability density functions. The densities are first estimated from a number of data sets using kernel density estimators, and these estimates are then subjected to functional PCA. As well as a specific application, which examines how densities evolve in data sets collected over a number of years, Kneip and Utikal (2001) introduce some new methodology for estimating functional PCs and for deciding how many components should be retained to represent the density estimates adequately. Their paper is followed by published discussion from four sets of discussants.

Robustness

Locantore et al. (1999) consider robust estimation of PCs for functional data (see Section 10.4).

12.4   PCA and Non-Independent Data—Some Additional Topics

In this section we collect together a number of topics from time series, and other contexts in which non-independent data arise, where PCA or related techniques are used. Section 12.4.1 describes PCA in the frequency domain, and Section 12.4.2 covers growth curves and longitudinal data. In Section 12.4.3 a summary is given of another idea (optimal fingerprints) from climatology, though one that is rather different in character from those presented in Section 12.2. Section 12.4.4 discusses spatial data, and the final subsection provides brief coverage of a few other topics, including non-independence induced by survey designs.

12.4.1   PCA in the Frequency Domain

The idea of PCA in the frequency domain clearly has no counterpart for data sets consisting of independent observations. Brillinger (1981, Chapter 9) devotes a whole chapter to the subject (see also Priestley et al. (1974)). To see how frequency domain PCs are derived, note that PCs for a p-variate random vector $\mathbf{x}$, with zero mean, can be obtained by finding $(p \times q)$ matrices $\mathbf{B}$, $\mathbf{C}$ such that

$$E[(\mathbf{x} - \mathbf{Cz})'(\mathbf{x} - \mathbf{Cz})]$$

is minimized, where $\mathbf{z} = \mathbf{B}'\mathbf{x}$. This is equivalent to the criterion that defines Property A5 in Section 2.1 It turns out that $\mathbf{B} = \mathbf{C}$ and that the columns of $\mathbf{B}$ are the first q eigenvectors of $\boldsymbol{\Sigma}$, the covariance matrix of $\mathbf{x}$, so that the elements of $\mathbf{z}$ are the first q PCs for $\mathbf{x}$. This argument can be extended to a time series of p variables as follows. Suppose that our series is $\ldots \mathbf{x}_{-1}$, $\mathbf{x}_0, \mathbf{x}_1, \mathbf{x}_2, \ldots$ and that $E[\mathbf{x}_t] = \mathbf{0}$ for all t. Define

$$\mathbf{z}_t = \sum_{u=-\infty}^{\infty} \mathbf{B}'_{t-u}\mathbf{x}_u,$$

and estimate $\mathbf{x}_t$ by $\sum_{u=-\infty}^{\infty} \mathbf{C}_{t-u}\mathbf{z}_u$, where

$$\ldots \mathbf{B}_{t-1}, \mathbf{B}_t, \mathbf{B}_{t+1}, \mathbf{B}_{t+2}, \ldots, \mathbf{C}_{t-1}, \mathbf{C}_t, \mathbf{C}_{t+1}, \mathbf{C}_{t+2}, \ldots$$

are $(p \times q)$ matrices that minimize

$$E\left[\left(\mathbf{x}_t - \sum_{u=-\infty}^{\infty} \mathbf{C}_{t-u}\mathbf{z}_u\right)^{*}\left(\mathbf{x}_t - \sum_{u=-\infty}^{\infty} \mathbf{C}_{t-u}\mathbf{z}_u\right)\right],$$

where $*$ denotes conjugate transpose. The difference between this formulation and that for ordinary PCs above is that the relationships between $\mathbf{z}$ and $\mathbf{x}$ are in terms of all values of $\mathbf{x}_t$ and $\mathbf{z}_t$, at different times, rather than between a single $\mathbf{x}$ and $\mathbf{z}$. Also, the derivation is in terms of general complex series, rather than being restricted to real series. It turns out (Brillinger, 1981, p. 344) that

$$\mathbf{B}'_u = \frac{1}{2\pi} \int_0^{2\pi} \tilde{\mathbf{B}}(\lambda) e^{iu\lambda} d\lambda$$

$$\mathbf{C}_u = \frac{1}{2\pi} \int_0^{2\pi} \tilde{\mathbf{C}}(\lambda) e^{iu\lambda} d\lambda,$$

where $\tilde{\mathbf{C}}(\lambda)$ is a $(p \times q)$ matrix whose columns are the first q eigenvectors of the matrix $\mathbf{F}(\lambda)$ given in (12.1.4), and $\tilde{\mathbf{B}}(\lambda)$ is the conjugate transpose of $\tilde{\mathbf{C}}(\lambda)$.

The q series that form the elements of $\mathbf{z}_t$ are called the first q PC series of $\mathbf{x}_t$. Brillinger (1981, Sections 9.3, 9.4) discusses various properties and estimates of these PC series, and gives an example in Section 9.6 on monthly temperature measurements at 14 meteorological stations. Principal component analysis in the frequency domain has also been used on economic time series, for example on Dutch provincial unemployment data (Bartels, 1977, Section 7.7).

There is a connection between frequency domain PCs and PCs defined in the time domain (Brillinger, 1981, Section 9.5). The connection involves Hilbert transforms and hence, as noted in Section 12.2.3, frequency domain PCA has links to HEOF analysis. Define the vector of variables $\mathbf{y}_t^H(\lambda) = (\mathbf{x}'_t(\lambda), \mathbf{x}'^H_t(\lambda))'$, where $\mathbf{x}_t(\lambda)$ is the contribution to $\mathbf{x}_t$ at frequency λ (Brillinger, 1981, Section 4.6), and $\mathbf{x}_t^H(\lambda)$ is its Hilbert transform. Then the covariance matrix of $\mathbf{y}_t^H(\lambda)$ is proportional to

$$\left[\begin{array}{cc} \mathrm{Re}(\mathbf{F}(\lambda)) & \mathrm{Im}(\mathbf{F}(\lambda)) \\ -\mathrm{Im}(\mathbf{F}(\lambda)) & \mathrm{Re}(\mathbf{F}(\lambda)) \end{array} \right],$$

where the functions $\mathrm{Re}(.)$, $\mathrm{Im}(.)$ denote the real and imaginary parts, respectively, of their argument. A PCA of $\mathbf{y}_t^H$ gives eigenvalues that are the eigenvalues of $\mathbf{F}(\lambda)$ with a corresponding pair of eigenvectors

$$\left[\begin{array}{c} \mathrm{Re}(\tilde{C}_j(\lambda)) \\ \mathrm{Im}(\tilde{C}_j(\lambda)) \end{array} \right], \left[\begin{array}{c} -\mathrm{Im}(\tilde{C}_j(\lambda)) \\ \mathrm{Re}(\tilde{C}_j(\lambda)) \end{array} \right],$$

where $\tilde{C}_j(\lambda)$ is the jth column of $\tilde{\mathbf{C}}(\lambda)$.

Horel (1984) interprets HEOF analysis as frequency domain PCA averaged over all frequency bands. When a single frequency of oscillation dominates the variation in a time series, the two techniques become the same. The averaging over frequencies of HEOF analysis is presumably the reason behind Plaut and Vautard's (1994) claim that it is less good than MSSA at distinguishing propagating patterns with different frequencies.

Preisendorfer and Mobley (1988) describe a number of ways in which PCA is combined with a frequency domain approach. Their Section 4e discusses the use of PCA after a vector field has been transformed into the frequency domain using Fourier analysis, and for scalar-valued fields their Chapter 12 examines various combinations of real and complex-valued harmonic analysis with PCA.

Stoffer (1999) describes a different type of frequency domain PCA, which he calls the *spectral envelope*. Here a PCA is done on the spectral matrix $\mathbf{F}(\lambda)$ relative to the time domain covariance matrix $\mathbf{\Gamma}_0$. This is a form of generalized PCA for $\mathbf{F}(\lambda)$ with $\mathbf{\Gamma}_0$ as a metric (see Section 14.2.2), and leads to solving the eigenequation $[\mathbf{F}(\lambda) - l(\lambda)\mathbf{\Gamma}_0]\mathbf{a}(\lambda) = 0$ for varying angular frequency λ. Stoffer (1999) advocates the method as a way of discovering whether the p series $x_1(t), x_2(t), \ldots x_p(t)$ share common signals and illustrates its use on two data sets involving pain perception and blood pressure.

The idea of cointegration is important in econometrics. It has a technical definition, but can essentially be described as follows. Suppose that the elements of the p-variate time series $\mathbf{x}_t$ are stationary after, but not before, differencing. If there are one or more vectors $\boldsymbol{\alpha}$ such that $\boldsymbol{\alpha}'\mathbf{x}_t$ is stationary without differencing, the p series are cointegrated. Tests for cointegration based on the variances of frequency domain PCs have been put forward by a number of authors. For example, Cubadda (1995) points out problems with previously defined tests and suggests a new one.

12.4.2   Growth Curves and Longitudinal Data

A common type of data that take the form of curves, even if they are not necessarily recorded as such, consists of measurements of growth for animals or children. Some curves such as heights are monotonically increasing, but others such as weights need not be. The idea of using principal components to summarize the major sources of variation in a set of growth curves dates back to Rao (1958), and several of the examples in Ramsay and Silverman (1997) are of this type. Analyses of growth curves are often concerned with predicting future growth, and one way of doing this is to use principal components as predictors. A form of generalized PC regression developed for this purpose is described by Rao (1987).

Caussinus and Ferré (1992) use PCA in a different type of analysis of growth curves. They consider a 7-parameter model for a set of curves, and estimate the parameters of the model separately for each curve. These 7-parameter estimates are then taken as values of 7 variables to be analyzed by PCA. A two-dimensional plot in the space of the first two PCs gives a representation of the relative similarities between members of the set of curves. Because the parameters are not estimated with equal precision, a weighted version of PCA is used, based on the fixed effects model described in Section 3.9.

Growth curves constitute a special case of longitudinal data, also known as 'repeated measures,' where measurements are taken on a number of individuals at several different points of time. Berkey et al. (1991) use PCA to model such data, calling their model a 'longitudinal principal component model.' As they are in Rao's (1987) generalized principal component regression, the PCs are used for prediction.

Growth curves are also the subject of James et al. (2000). Their objective is to model growth curves when the points at which the curves are measured are possibly irregular, different for different individuals (observations), and sparse. Various models are proposed for such data and the differences and connections between models are discussed. One of their models is the reduced rank model

$$x_i(t) = \mu(t) + \sum_{k=1}^{q} a_k(t) z_{ik} + \epsilon_i(t), \ i = 1, 2, \dots, n, \quad (12.4.1)$$

where $x_i(t)$ represents the growth curve for the ith individual, $\mu(t)$ is a mean curve, $\epsilon_i(t)$ is an error term for the ith individual and $a_k(t), z_{ik}$ are curves defining the principal components and PC scores, respectively, as in Section 12.3.1. James et al. (2000) consider a restricted form of this model in which $\mu(t)$ and $a_k(t)$ are expressed in terms of a spline basis, leading to a model

$$\mathbf{x}_i = \mathbf{\Phi}_i \mathbf{b}_0 + \mathbf{\Phi}_i \mathbf{B} \mathbf{z} + \epsilon_i, \ i = 1, 2, \dots, n. \quad (12.4.2)$$

Here $\mathbf{x}_i, \epsilon_i$ are vectors of values $x_i(t), \epsilon_i(t)$ at the times for which measurements are made on the ith individual; $\mathbf{b}_0, \mathbf{B}$ contain coefficients in the expansions of $\mu(t), a_k(t)$, respectively in terms of the spline basis; and $\mathbf{\Phi}_i$ consists of values of that spline basis at the times measured for the ith individual. When all individuals are measured at the same time, the subscript i disappears from $\mathbf{\Phi}_i$ in (12.4.2) and the error term has covariance matrix $\sigma^2 \mathbf{I}_p$, where p is the (common) number of times that measurements are made. James et al. (2000) note that the approach is then equivalent to a PCA of the spline coefficients in $\mathbf{B}$. More generally, when the times of measurement are different for different individuals, the analysis is equivalent to a PCA with respect to the metric $\mathbf{\Phi}_i' \mathbf{\Phi}_i$. This extends the idea of metric-based PCA described in Section 14.2.2 in allowing different (non-diagonal) metrics for different observations. James et al. (2000) discuss how to choose the number of knots in the spline basis, the choice of q in equation (12.4.1), and how to construct bootstrap-based confidence intervals for the mean function, the curves defining the principal components, and the individual curves.

As noted in Section 9.3.4, redundancy analysis can be formulated as PCA on the predicted responses in a multivariate regression. Van den Brink and ter Braak (1999) extend this idea so that some of the predictor variables in the regression are not included among the predictors on which the PCA is done. The context in which they implement their technique is where species abundances depend on time and on which 'treatment' is applied. The results of this analysis are called *principal response curves*.

12.4.3 Climate Change—Fingerprint Techniques

In climate change detection, the objective is not only to discover whether change is taking place, but also to explain any changes found. If certain causes are suspected, for example increases in greenhouse gases, variations in solar output, or volcanic activity, then changes in these quantities can be built into complex atmospheric models, and the models are run to discover what happens to parameters of interest such as the global pattern of temperature. Usually changes in one (or more) of the potential causal variables will manifest themselves, according to the model, in different ways in different geographical regions and for different climatic variables. The predicted patterns of change are sometimes known as 'fingerprints' associated with changes in the causal variable(s). In the detection of climate change it is usually more productive to attempt to detect changes that resemble such fingerprints than to search broadly over a wide range of possible changes. The paper by Hasselmann (1979) is usually cited as the start of interest in this type of climate change detection, and much research has been done subsequently. Very similar techniques have been derived via a number of different approaches. Zwiers (1999) gives a good, though not exhaustive, summary of several of these techniques, together with a number of applications. North and Wu (2001) describe a number of recent developments, again with applications.

The basic idea is that the observed data, which are often values of some climatic variable x_{tj}, where t indexes time and j indexes spatial position, can be written

$$x_{tj} = s_{tj} + e_{tj}.$$

Here s_{tj} is the deterministic response to changes in the potential causal variables (the signal), and e_{tj} represents the stochastic noise associated with 'normal' climate variation.

Suppose that an optimal detection variable A_t at time t is constructed as a linear combination of the observed data x_{sj} for $s = t, (t-1), \ldots, (t-l+1)$ and $j = 1, 2, \ldots, m$, where m is the number of spatial locations for which data are available. The variable A_t can be written as $A_t = \mathbf{w}'\mathbf{x}$, where $\mathbf{x}$ is an ml-vector of observed measurements at times $t, (t-1), \ldots, (t-l+1)$, and all m spatial locations, and $\mathbf{w}$ is a vector of weights. Then $\mathbf{w}$ is chosen to maximize the signal to noise ratio

$$\frac{[E(A_t)]^2}{\mathrm{var}(A_t)} = \frac{[\mathbf{w}'\mathbf{s}_t]^2}{\mathbf{w}'\boldsymbol{\Sigma}_e\mathbf{w}},$$

where $\mathbf{s}_t$ is an ml-vector of known signal at time t and $\boldsymbol{\Sigma}_e$ is the spatio-temporal covariance matrix of the noise term. It is straightforward to show that the optimal detector, sometimes known as the *optimal fingerprint*, has the form $\hat{\mathbf{w}} = \boldsymbol{\Sigma}_e^{-1}\mathbf{s}_t$. The question then arises: Where does PCA fit into this methodology?

Underlying the answer to this question is the fact that Σ_e needs to be estimated, usually from a 'control run' of the atmospheric model in which no changes are made to the potential causal variables. Because the dimension ml of Σ_e is large, it may be nearly singular, and estimation of Σ_e^{-1}, which is required in calculating $\hat{\mathbf{w}}$, will be unstable. To avoid this instability, the estimate of Σ_e is replaced by the first few terms in its spectral decomposition (Property A3 of Section 2.1), that is by using its first few PCs. Allen and Tett (1999) discuss the choice of how many PCs to retain in this context. 'First few' is perhaps not quite the right phrase in some of these very large problems; North and Wu (2001) suggest keeping up to 500 out of 3600 in one of their studies.

Expressing the optimal fingerprint in terms of the PCs of the estimate of Σ_e also has advantages in interpretation, as in principal component regression (Chapter 8), because of the uncorrelatedness of the PCs (Zwiers, 1999, equations (13), (14)).

12.4.4 Spatial Data

Many of the techniques discussed earlier in the chapter have a spatial dimension. In Section 12.2, different points in space mostly correspond to different 'variables.' For MSSA, variables correspond to combinations of time lag and spatial position, and Plaut and Vautard (1994) refer to the output of MSSA as 'space-time EOFs.' North and Wu (2001) use the same term in a different situation where variables also correspond to space-time combinations, but with time in separated rather than overlapping blocks.

The possibility was also raised in Section 12.3 that the continuum of variables underlying the curves could be in space rather than time. The example of Section 12.3.3 is a special case of this in which space is one-dimensional, while Bouhaddou et al. (1987) and Guttorp and Sampson (1994) discuss estimation of continuous (two-dimensional) spatial covariances. Extended EOF analysis (see Sections 12.2.1, 14.5) has several variables measured at each spatial location, but each *space × variable* combination is treated as a separate 'variable' in this type of analysis. Another example in which variables correspond to location is described by Boyles (1996). Here a quality control regime includes measurements taken on a sample of n parts from a production line. The measurements are made at p locations forming a lattice on each manufactured part. We return to this example in Section 13.7.

In some situations where several variables are measured at each of n spatial locations, the different spatial locations correspond to different *observations* rather than variables, but the observations are typically not independent. Suppose that a vector $\mathbf{x}$ of p variables is measured at each of n spatial locations. Let the covariance between the elements x_j and x_k of $\mathbf{x}$ for two observations which are separated in space by a vector $\mathbf{h}$ be the (j, k)th element of a matrix $\Sigma_{\mathbf{h}}$. This expression assumes second order stationarity in the sense that the covariances do not depend on the location of the two observations, only on the vector joining them. On the other hand, it does not require isotropy—the covariance may depend on the direction of $\mathbf{h}$ as well as its length. The *intrinsic correlation model* (Wackernagel, 1995, Chapter 22) assumes that $\Sigma_{\mathbf{h}} = \rho_{\mathbf{h}}\Sigma$. Because all terms in Σ are multiplied by the same spatial factor, this factor cancels when correlations are calculated from the covariances and the correlation between x_j and x_k does not depend on $\mathbf{h}$. Wackernagel (1995, Chapter 22) suggests testing whether or not this model holds by finding principal components based on sample covariance matrices for the p variables. Cross-covariances between the resulting PCs are then found at different separations $\mathbf{h}$. For $k \neq l$, the kth and lth PCs should be uncorrelated for different values of $\mathbf{h}$ under the intrinsic correlation model, because $\Sigma_{\mathbf{h}}$ has the same eigenvectors for all $\mathbf{h}$.

An extension of the intrinsic correlation model to a 'linear model of coregionalization' is noted by Wackernagel (1995, Chapter 24). In this model the variables are expressed as a sum of $(S + 1)$ spatially uncorrelated components, and the covariance matrix now takes the form

$$\Sigma_{\mathbf{h}} = \sum_{u=0}^{S} \rho_{u\mathbf{h}}\Sigma_u.$$

Wackernagel (1995, Chapter 25) suggests that separate PCAs of the estimates of the matrices $\Sigma_0, \Sigma_1, \ldots, \Sigma_S$ may be informative, but it is not clear how these matrices are estimated, except that they each represent different spatial scales.

As well as dependence on $\mathbf{h}$, the covariance or correlation between x_j and x_k may depend on the nature of the measurements made (point measurements, averages over an area where the area might be varied) and on the size of the domain. Vargas-Guzmán et al. (1999) discuss each of these aspects, but concentrate on the last. They describe a procedure they name *growing scale PCA*, in which the nature of the measurements is fixed and averaging (integration) takes place with respect to $\mathbf{h}$, but the size of the domain is allowed to vary continuously. As it does, the PCs and their variances also evolve continuously. Vargas-Guzmán et al. illustrate this technique and also the linear model of coregionalization with a three-variable example. The basic form of the PCs is similar at all domain sizes and stabilizes as the size increases, but there are visible changes for small domain sizes. The example also shows the changes that occur in the PCs for four different areal extents of the individual measurements. Buell (1975) also discussed the dependence of PCA on size and shape of spatial domains (see Section 11.4), but his emphasis was on the shape of the domain and, unlike Vargas-Guzmán et al. (1999), he made no attempt to produce a continuum of PCs dependent on size.

Kaplan et al. (2001) consider optimal interpolation and smoothing of spatial field data that evolve in time. There is a basic first-order autoregressive structure relating the vector of true values of the field at time $(t+1)$ to that at time t, as in equation (12.2.1), but in addition the values are measured with error, so that there is a second equation, relating the observed field $\mathbf{y}_t$ to the true field $\mathbf{x}_t$, which is

$$\mathbf{y}_t = \mathbf{\Xi}\mathbf{x}_t + \boldsymbol{\xi}_t, \qquad (12.4.3)$$

where $\mathbf{\Xi}$ is a $(p \times p)$ matrix and $\boldsymbol{\xi}_t$ is a vector of observational errors. In the most general case the matrices $\mathbf{\Upsilon}$ and $\mathbf{\Xi}$ in equations (12.2.1), (12.4.1) may depend on t, and the covariance matrices of the error terms $\boldsymbol{\epsilon}_t$ and $\boldsymbol{\xi}_t$ need not be proportional to the identity matrix, or even diagonal. There are standard procedures for optimal interpolation and smoothing, but these become computationally prohibitive for the large data sets considered by Kaplan and researchers. They therefore suggest projecting the data onto the first few principal components of the covariance matrix of the anomaly field, that is, the field constructed by subtracting the long-term climatic mean from the measurements at each spatial location in the field. Kaplan et al. (2001) describe a number of subtleties associated with the approach. The main problem is the need to estimate the covariance matrices associated with $\boldsymbol{\epsilon}_t$ and $\boldsymbol{\xi}_t$. Difficulties arise because of possible non-stationarity in both means and covariance, and because of changing spatial coverage over time, among other things, but Kaplan and his coworkers propose solutions to overcome the difficulties.

Among the procedures considered computationally 'extremely expensive' by Kaplan et al. (2001) is the Kalman filter. Wikle and Cressie (1999) propose a form of the Kalman filter in which there is an additional non-dynamic term that captures small-scale spatial variability. Apart from this term, their model is similar to that of Kaplan et al. and they, too, suggest dimension reduction using principal components. Wikle and Cressie's model has space defined continuously but the principal components that give their dimension reduction are derived from predicted values of the spatial process on a regular grid.

In a spatial discriminant analysis example, Storvik (1993) suggests linearly transforming the data, but not to PCs. Instead, he finds linear functions $\mathbf{a}_k'\mathbf{x}$ of $\mathbf{x}$ that successively minimize autocorrelation between $\mathbf{a}_k'\mathbf{x}(s)$ and $\mathbf{a}_k'\mathbf{x}(s + \Delta)$, where the argument for $\mathbf{x}$ denotes spatial position and Δ is a fixed distance apart in space. It turns out that the functions are derived via an eigenanalysis of $\mathbf{S}^{-1}\mathbf{S}_\Delta$, where $\mathbf{S}$ is the usual sample covariance matrix for $\mathbf{x}$ and $\mathbf{S}_\Delta$ is the sample covariance matrix for $\mathbf{x}(s)-\mathbf{x}(s+\Delta)$. This analysis has some resemblance to the procedure for POP analysis (Section 12.2.2), but in a spatial rather than time series context.

12.4.5   Other Aspects of Non-Independent Data and PCA

A different type of non-independence between observations is induced in sample surveys for which the survey design is more complex than simple random sampling. Skinner et al. (1986) consider what happens to PCA when the selection of a sample of observations on p variables $\mathbf{x}$ depends on a vector of covariates $\mathbf{z}$. They use Taylor expansions to approximate the changes (compared to simple random sampling) in the eigenvalues and eigenvectors of the sample covariance matrix when samples are chosen in this way. Skinner et al. present a simulation study in which stratified samples are taken, based on the value of a single covariate z, whose correlations with 6 measured variables range from 0.48 to 0.60. Substantial biases arise in the estimates of the eigenvalues and eigenvectors for some of the stratification schemes examined, and these biases are well-approximated using the Taylor expansions. By assuming multivariate normality, Skinner and coworkers show that a maximum likelihood estimate of the covariance matrix can be obtained, given knowledge of the survey design, and they show that using the eigenvalues and eigenvectors of this estimate corrects for the biases found for the usual covariance matrix. Tortora (1980) presents an example illustrating the effect of a disproportionate stratified survey design on some of the rules for variable selection using PCA that are described in Section 6.3.

Similar structure to that found in sample surveys can arise in observational as well as design-based data sets. For example, Konishi and Rao (1992) consider multivariate data, which could be genetic or epidemiological, for samples of families. There are correlations between members of the same family, and families may be of unequal sizes. Konishi and Rao (1992) propose a model for the correlation structure in such familial data, which they use to suggest estimates of the principal components underlying the data.

Solow (1994) notes that when a trend is present in a set of time series, it is often a major source of variation and will be found by an early PC. However, he argues that to identify trends it is better to look for a linear combination of the p series that has maximum autocorrelation, rather than maximum variance. Solow (1994) solves this maximization problem by finding the smallest eigenvalues and corresponding eigenvectors of $\mathbf{S}^{-1}\mathbf{S}_D$, where $\mathbf{S}_D$ is the covariance matrix of first differences of the series. This is the same eigenproblem noted in the Section 12.4.4, which Storvik (1993) solved in a spatial setting, but its objectives are different here so that autocorrelation is maximized rather than minimized. Solow (1994) also modifies his procedure to deal with compositional data, based on the approach suggested by Aitchison (1983) for PCA (see Section 13.3).

Peña and Box (1987) consider a factor analysis model for time series in which p time series are assumed to be derived linearly from a smaller number m of underlying factors. Each factor follows an autoregressive moving-average (ARMA) model. Peña and Box (1987) use the eigenvalues and eigenvectors of covariance matrices between the series measured simultaneously (essentially a PCA) and at various lags to deduce the structure of their factor model. They claim that considerable simplification is possible compared to a multivariate ARMA model involving all p of the original series.

Wold (1994) suggests exponentially weighted moving principal components in the context of process control (see Section 13.7), and Diamantaras and Kung (1996, Section 3.5) advocate PCs based on weighted covariance matrices for multivariate time series, with weights decreasing exponentially for less recent observations.

Yet another rôle for PCs in the analysis of time series data is presented by Doran (1976). In his paper, PCs are used to estimate the coefficients in a regression analysis of one time series variable on several others. The idea is similar to that of PC regression (see Section 8.1), but is more complicated as it involves the frequency domain. Consider the distributed lag model, which is a time series version of the standard regression model $\mathbf{y} = \mathbf{X}\boldsymbol{\beta} + \boldsymbol{\epsilon}$ of equation (8.1.1), with the time series structure leading to correlation between elements of $\boldsymbol{\epsilon}$. There is a decomposition of the least squares estimator of the regression coefficients $\boldsymbol{\beta}$ rather like equation (8.1.8) from PC regression, except that the eigenvalues are replaced by ratios of spectral density estimates for the predictors (signal) and error (noise). Doran (1976) suggests using an estimate in which the terms corresponding to the smallest values of this signal to noise ratio are omitted from the least squares decomposition.

13
Principal Component Analysis for Special Types of Data

The viewpoint taken in much of this text is that PCA is mainly a descriptive tool with no need for rigorous distributional or model assumptions. This implies that it can be used on a wide range of data, which can diverge considerably from the 'ideal' of multivariate normality. There are, however, certain types of data where some modification or special care is desirable when performing PCA. Some instances of this have been encountered already, for example in Chapter 9 where the data are grouped either by observations or by variables, and in Chapter 12 where observations are non-independent. The present chapter describes a number of other special types of data for which standard PCA should be modified in some way, or where related techniques may be relevant.

Section 13.1 looks at a number of ideas involving PCA for discrete data. In particular, correspondence analysis, which was introduced as a graphical technique in Section 5.4, is discussed further, and procedures for dealing with data given as ranks are also described.

When data consist of measurements on animals or plants it is sometimes of interest to identify 'components' of variation that quantify size and various aspects of shape. Section 13.2 examines modifications of PCA that attempt to find such components.

In Section 13.3, compositional data in which the p elements of $\mathbf{x}$ are constrained to sum to the same constant (usually 1 or 100) for all observations are discussed, and in Section 13.4 the rôle of PCA in analysing data from designed experiments is described.

Section 13.5 looks at a number of ways of defining 'common' PCs, or common subspaces, when the observations come from several distinct populations, and also examines how PCs from the different populations can be compared.

Section 13.6 discusses possible ways of dealing with missing data in a PCA, and Section 13.7 describes the use of PCs in statistical process control. Finally, Section 13.8 covers a number of other types of data rather briefly. These include vector or directional data, data presented as intervals, species abundance data and large data sets.

13.1 Principal Component Analysis for Discrete Data

When PCA is used as a descriptive technique, there is no reason for the variables in the analysis to be of any particular type. At one extreme, **x** may have a multivariate normal distribution, in which case all the relevant inferential results mentioned in Section 3.7 can be used. At the opposite extreme, the variables could be a mixture of continuous, ordinal or even binary (0/1) variables. It is true that variances, covariances and correlations have especial relevance for multivariate normal **x**, and that linear functions of binary variables are less readily interpretable than linear functions of continuous variables. However, the basic objective of PCA—to summarize most of the 'variation' that is present in the original set of p variables using a smaller number of derived variables—can be achieved regardless of the nature of the original variables.

For data in which all variables are binary, Gower (1966) points out that using PCA *does* provide a plausible low-dimensional representation. This follows because PCA is equivalent to a principal coordinate analysis based on the commonly used definition of similarity between two individuals (observations) as the proportion of the p variables for which the two individuals take the same value (see Section 5.2). Cox (1972), however, suggests an alternative to PCA for binary data. His idea, which he calls 'permutational principal components,' is based on the fact that a set of data consisting of p binary variables can be expressed in a number of different but equivalent ways. As an example, consider the following two variables from a cloud-seeding experiment:

$$x_1 = \begin{cases} 1 & \text{if rain falls in seeded area,} \\ 0 & \text{if no rain falls in seeded area} \end{cases}$$

$$x_2 = \begin{cases} 1 & \text{if rain falls in control area,} \\ 0 & \text{if no rain falls in control area.} \end{cases}$$

Instead of x_1, x_2 we could define

$$x_1' = \begin{cases} 0 & \text{if both areas have rain or both areas dry} \\ 1 & \text{if one area has rain, the other is dry} \end{cases}$$

and $x_2' = x_1$. There is also a third possibility namely $x_1'' = x_1'$, $x_2'' = x_2$. For $p > 2$ variables there are many more possible permutations of this type, and Cox (1972) suggests that an alternative to PCA might be to transform to *independent* binary variables using such permutations. Bloomfield (1974) investigates Cox's suggestion in more detail, and presents an example having four variables. In examples involving two variables we can write

$$x_1' = x_1 + x_2 \quad \text{(modulo 2), using the notation above.}$$

For more than two variables, not all permutations can be written in this way, but Bloomfield restricts his attention to those permutations which can. Thus, for a set of p binary variables $x_1, x_2, \ldots, x_p$, he considers transformations to $z_1, z_2, \ldots, z_p$ such that, for $k = 1, 2, \ldots, p$, we have either

$$z_k = x_j \quad \text{for some } j,$$

or

$$z_k = x_i + x_j \quad \text{(modulo 2)} \quad \text{for some } i, j, \quad i \neq j.$$

He is thus restricting attention to *linear* transformations of the variables (as in PCA) and the objective in this case is to choose a transformation that simplifies the structure between variables. The data can be viewed as a contingency table, and Bloomfield (1974) interprets a simpler structure as one that reduces high order interaction terms between variables. This idea is illustrated on a 4-variable example, and several transformations are examined, but (unlike PCA) there is no algorithm for finding a unique 'best' transformation.

A second special type of discrete data occurs when, for each observation, only ranks and not actual values are given for each variable. For such data, all the columns of the data matrix $\mathbf{X}$ have the same sum, so that the data are constrained in a way similar to that which holds for compositional data (see Section 13.3). Gower (1967) discusses some geometric implications that follow from the constraints imposed by this type of ranked data and by compositional data.

Another possible adaptation of PCA to discrete data is to replace variances and covariances by measures of dispersion and association that are more relevant to discrete variables. For the particular case of contingency table data, many different measures of association have been suggested (Bishop et al., 1975, Chapter 11). It is also possible to define measures of variation other than variance for such data, for example Gini's measure (see Bishop et al., 1975, Section 11.3.4). An approach of this sort is proposed by Korhonen and Siljamäki (1998) for ordinal data. The objective they tackle is to find an 'optimal' ordering or ranking of the multivariate data so that, instead of assigning a score on a continuum to each of n observations, what is required is a rank between 1 and n. They note that the

first PC maximizes the (weighted) sum of squared correlations between the
PC and each of the p variables. The weights are unity for correlation-based
PCA and equal to the sample variances for covariance-based PCA. These
results follow from the sample version of Property A6 in Section 2.3 and
the discussion that follows the property.

Korhonen and Siljamäki (1998) define the *first ordinal principal component* as the ranking of the n observations for which the sum of squared
rank correlation coefficients between the ordinal PC and each of the p variables is maximized. They suggest that either Spearman's rank correlation
or Kendall's τ could be used as the 'rank correlation' in this definition. The
first ordinal PC can be computed when the data themselves are ordinal,
but it can also be obtained for continuous data, and may be useful if an
optimal ordering of the data, rather than a continuous derived variable that
maximizes variance, is of primary interest. Computationally the first ordinal principal component may be found by an exhaustive search for small
examples, but the optimization problem is non-trivial for moderate or large
data sets. Korhonen and Siljamäki (1998) note that the development of a
fast algorithm for the procedure is a topic for further research, as is the
question of whether it is useful and feasible to look for 2nd, 3rd, ... ordinal
PCs.

Baba (1995) uses the simpler procedure for ranked data of calculating
rank correlations and conducting an ordinary PCA on the resulting (rank)
correlation matrix. It is shown that useful results may be obtained for some
types of data.

Returning to contingency tables, the usual 'adaptation' of PCA to such
data is correspondence analysis. This technique is the subject of the remainder of this section. Correspondence analysis was introduced in Section 5.4
as a graphical technique, but is appropriate to discuss it in a little more
detail here because, as well as being used as a graphical means of displaying
contingency table data (see Section 5.4), the technique has been described
by some authors (for example, De Leeuw and van Rijckevorsel, 1980) as a
form of PCA for nominal data. To see how this description is valid, consider, as in Section 5.4, a data set of n observations arranged in a two-way
contingency table, with n_{ij} denoting the number of observations that take
the ith value for the first (row) variable and the jth value for the second
(column) variable, $i = 1, 2, \ldots, r$; $j = 1, 2, \ldots, c$. Let $\mathbf{N}$ be the $(r \times c)$ ma-
trix with (i, j)th element n_{ij}, and define $\mathbf{P} = \frac{1}{n}\mathbf{N}, \mathbf{r} = \mathbf{Pl}_c, \mathbf{c} = \mathbf{P'l}_r$, and
$\mathbf{X} = \mathbf{P} - \mathbf{rc'}$, where $\mathbf{l}_c, \mathbf{l}_r$, are vectors of c and r elements, respectively,
with all elements unity. If the variable defining the rows of the contingency
table is independent of the variable defining the columns, then the matrix
of 'expected counts' is given by $n\mathbf{rc'}$. Thus, $\mathbf{X}$ is a matrix of the residuals
that remain when the 'independence' model is fitted to $\mathbf{P}$.

The generalized singular value decomposition (SVD) of $\mathbf{X}$, is defined by

$$\mathbf{X} = \mathbf{VMB'}, \tag{13.1.1}$$

where $\mathbf{V}'\boldsymbol{\Omega}\mathbf{V} = \mathbf{I}$, $\mathbf{B}'\boldsymbol{\Phi}\mathbf{B} = \mathbf{I}$, $\boldsymbol{\Omega}$ and $\boldsymbol{\Phi}$ are $(r \times r)$ and $(c \times c)$ matrices, respectively, and $\mathbf{M}$, $\mathbf{I}$ are diagonal and identity matrices whose dimensions equal the rank of $\mathbf{X}$ (see Section 14.2.1). If $\boldsymbol{\Omega} = \mathbf{D}_r^{-1}$, $\boldsymbol{\Phi} = \mathbf{D}_c^{-1}$, where $\mathbf{D}_r$, $\mathbf{D}_c$ are diagonal matrices whose diagonal entries are the elements of $\mathbf{r}$, $\mathbf{c}$, respectively, then the columns of $\mathbf{B}$ define principal axes for the set of r 'observations' given by the rows of $\mathbf{X}$. Similarly, the columns of $\mathbf{V}$ define principal axes for the set of c 'observations' given by the columns of $\mathbf{X}$, and from the first q columns of $\mathbf{B}$ and $\mathbf{V}$, respectively, we can derive the 'coordinates' of the row and column profiles of $\mathbf{N}$ in q-dimensional space (see Greenacre, 1984, p. 87) which are the end products of a correspondence analysis.

A correspondence analysis is therefore based on a generalized SVD of $\mathbf{X}$, and, as will be shown in Section 14.2.1, this is equivalent to an 'ordinary' SVD of

$$\tilde{\mathbf{X}} = \boldsymbol{\Omega}^{1/2}\mathbf{X}\boldsymbol{\Phi}^{1/2}$$
$$= \mathbf{D}_r^{-1/2}\mathbf{X}\mathbf{D}_c^{-1/2}$$
$$= \mathbf{D}_r^{-1/2}\left(\frac{1}{n}\mathbf{N} - \mathbf{rc}'\right)\mathbf{D}_c^{-1/2}.$$

The SVD of $\tilde{\mathbf{X}}$ can be written

$$\tilde{\mathbf{X}} = \mathbf{WKC}' \tag{13.1.2}$$

with $\mathbf{V}$, $\mathbf{M}$, $\mathbf{B}$ of (13.1.1) defined in terms of $\mathbf{W}$, $\mathbf{K}$, $\mathbf{C}$ of (13.1.2) as

$$\mathbf{V} = \boldsymbol{\Omega}^{-1/2}\mathbf{W}, \quad \mathbf{M} = \mathbf{K}, \quad \mathbf{B} = \boldsymbol{\Phi}^{-1/2}\mathbf{C}.$$

If we consider $\tilde{\mathbf{X}}$ as a matrix of r observations on c variables, then the coefficients of the PCs for $\tilde{\mathbf{X}}$ are given in the columns of $\mathbf{C}$, and the coordinates (scores) of the observations with respect to the PCs are given by the elements of $\mathbf{WK}$ (see the discussion of the biplot with $\alpha = 1$ in Section 5.3). Thus, the positions of the row points given by correspondence analysis are rescaled versions of the values of the PCs for the matrix $\tilde{\mathbf{X}}$. Similarly, the column positions given by correspondence analysis are rescaled versions of values of PCs for the matrix $\tilde{\mathbf{X}}'$, a matrix of c observations on r variables. In this sense, correspondence analysis can be thought of as a form of PCA for a transformation $\tilde{\mathbf{X}}$ of the original contingency table $\mathbf{N}$ (or a generalized PCA for $\mathbf{X}$; see Section 14.2.1).

Because of the various optimality properties of PCs discussed in Chapters 2 and 3, and also the fact that the SVD provides a sequence of 'best-fitting' approximations to $\tilde{\mathbf{X}}$ of rank $1, 2, \ldots$ as defined by equation (3.5.4), it follows that correspondence analysis provides coordinates for rows and columns of $\mathbf{N}$ that give the best fit in a small number of dimensions (usually two) to a transformed version $\tilde{\mathbf{X}}$ of $\mathbf{N}$. This rather convoluted definition of correspondence analysis demonstrates its connection with PCA, but there are a number of other definitions that turn out to be equivalent, as shown in Greenacre (1984, Chapter 4). In particular, the techniques of reciprocal averaging and dual (or optimal) scaling are widely used in ecology and psychology, respectively. The rationale behind each technique is different, and differs in turn from that given above for correspondence analysis, but numerically all three techniques provide the same results for a given table of data.

The ideas of correspondence analysis can be extended to contingency tables involving more than two variables (Greenacre, 1984, Chapter 5), and links with PCA remain in this case, as will now be discussed very briefly. Instead of doing a correspondence analysis using the $(r \times c)$ matrix $\mathbf{N}$, it is possible to carry out the same type of analysis on the $[n.. \times (r+c)]$ indicator matrix $\mathbf{Z} = (\mathbf{Z}_1 \ \mathbf{Z}_2)$. Here $\mathbf{Z}_1$ is $(n.. \times r)$ and has (i,j)th element equal to 1 if the ith observation takes the jth value for the first (row) variable, and zero otherwise. Similarly, $\mathbf{Z}_2$ is $(n.. \times c)$, with (i,j)th element equal to 1 if the ith observation takes the jth value for the second (column) variable and zero otherwise. If we have a contingency table with more than two variables, we can extend the correspondence analysis based on $\mathbf{Z}$ by adding further indicator matrices $\mathbf{Z}_3, \mathbf{Z}_4, \ldots$, to $\mathbf{Z}$, one matrix for each additional variable, leading to 'multiple correspondence analysis' (see also Section 14.1.1). Another alternative to carrying out the analysis on $\mathbf{Z} = (\mathbf{Z}_1 \ \mathbf{Z}_2 \ \mathbf{Z}_3 \ldots)$ is to base the correspondence analysis on the so-called Burt matrix $\mathbf{Z}'\mathbf{Z}$ (Greenacre, 1984, p. 140).

In the case where each variable can take only two values, Greenacre (1984, p. 145) notes two relationships between (multiple) correspondence analysis and PCA. He states that correspondence analysis of $\mathbf{Z}$ is closely related to PCA of a matrix $\mathbf{Y}$ whose ith column is one of the two columns of $\mathbf{Z}_i$, standardized to have unit variance. Furthermore, the correspondence analysis of the Burt matrix $\mathbf{Z}'\mathbf{Z}$ is equivalent to a PCA of the correlation matrix $\frac{1}{n..}\mathbf{Y}'\mathbf{Y}$. Thus, the idea of correspondence analysis as a form of PCA for nominal data is valid for any number of binary variables. A final relationship between correspondence analysis and PCA (Greenacre, 1984, p. 183) occurs when correspondence analysis is done for a special type of 'doubled' data matrix, in which each variable is repeated twice, once in its original form, and the second time in a complementary form (for details, see Greenacre (1984, Chapter 6)).

We conclude this section by noting one major omission, namely the nonlinear principal component analyses of Gifi (1990, Chapter 4). These are most relevant to discrete data, but we defer detailed discussion of them to Section 14.1.1.

13.2 Analysis of Size and Shape

In a number of examples throughout the book the first PC has all its coefficients of the same sign and is a measure of 'size.' The orthogonality constraint in PCA then demands that subsequent PCs are contrasts between variables or measures of 'shape.' When the variables are physical measurements on animals or plants the terms 'size' and 'shape' take on real meaning. The student anatomical measurements that were analysed in Sections 1.1, 4.1, 5.1 and 10.1 are of this type, and there is a large literature on the study of size and shape for non-human animals. Various approaches have been adopted for quantifying size and shape, only some of which involve PCA. We shall concentrate on the latter, though other ideas will be noted briefly.

The study of relationships between size and shape during the growth of organisms is sometimes known as *allometry* (Hills, 1982). The idea of using the first PC as a measure of size, with subsequent PCs defining various aspects of shape, dates back at least to Jolicoeur (1963). Sprent (1972) gives a good review of early work in the area from a mathematical/statistical point of view, and Blackith and Reyment (1971, Chapter 12) provide references to a range of early examples. It is fairly conventional, for reasons explained in Jolicoeur (1963), to take logarithms of the data, with PCA then conducted on the covariance matrix of the log-transformed data.

In circumstances where all the measured variables are thought to be of equal importance, it seems plausible that size should be an weighted average of the (log-transformed) variables with all weights equal. This is known as *isometric size*. While the first PC may have roughly equal weights (coefficients), sampling variability ensures that they are never exactly equal. Somers (1989) argues that the first PC contains a mixture of size and shape information, and that in order to examine 'shape,' an isometric component rather than the first PC should be removed. A number of ways of 'removing' isometric size and then quantifying different aspects of shape have been suggested and are now discussed.

Recall that the covariance matrix (of the log-transformed data) can be written using the spectral decomposition in equation (3.1.4) as

$$\mathbf{S} = l_1\mathbf{a}_1\mathbf{a}_1' + l_2\mathbf{a}_2\mathbf{a}_2' + \ldots + l_p\mathbf{a}_p\mathbf{a}_p'.$$

Removal of the first PC is achieved by removing the first term in this decomposition. The first, second, ..., PCs of the reduced matrix are then the second, third, ...PCs of $\mathbf{S}$. Somers (1986) suggests removing $l_0\mathbf{a}_0\mathbf{a}_0'$ from $\mathbf{S}$, where $\mathbf{a}_0 = \frac{1}{\sqrt{p}}(1, 1, \ldots, 1)$ is the isometric vector and l_0 is the sample variance of $\mathbf{a}_0'\mathbf{x}$, and then carrying out a 'PCA' on the reduced matrix. This procedure has a number of drawbacks (Somers, 1989; Sundberg, 1989), including the fact that, unlike PCs, the shape components found in this way are correlated and have vectors of coefficients that are not orthogonal.

One alternative suggested by Somers (1989) is to find 'shape' components by doing a 'PCA' on a doubly-centred version of the log-transformed data. The double-centering is considered to remove size (see also Section 14.2.3) because the isometric vector is one of the eigenvectors of its 'covariance' matrix, with zero eigenvalue. Hence the vectors of coefficients of the shape components are orthogonal to the isometric vector, but the shape components themselves are correlated with the isometric component. Cadima and Jolliffe (1996) quote an example in which these correlations are as large as 0.92.

Ranatunga (1989) introduced a method for which the shape components are uncorrelated with the isometric component, but her technique sacrifices orthogonality of the vectors of coefficients. A similar problem, namely losing either uncorrelatedness or orthogonality when searching for simple alternatives to PCA, was observed in Chapter 11. In the present context, however, Cadima and Jolliffe (1996) derived a procedure combining aspects of double-centering and Ranatunga's approach and gives shape components that are both uncorrelated with the isometric component and have vectors of coefficients orthogonal to $\mathbf{a}_0$. Unfortunately, introducing one desirable property leads to the loss of another. As pointed out by Mardia et al. (1996), if $\mathbf{x}_h = c\mathbf{x}_i$ where $\mathbf{x}_h$, $\mathbf{x}_i$ are two observations and c is a constant, then in Cadima and Jolliffe's (1996) method the scores of the two observations are different on the shape components. Most definitions of shape consider two observations related in this manner to have the same shape.

Decomposition into size and shape of the variation in measurements made on organisms is a complex problem. None of the terms 'size,' 'shape,' 'isometric' or 'allometry' is uniquely defined, which leaves plenty of scope for vigorous debate on the merits or otherwise of various procedures (see, for example, Bookstein (1989); Jungers et al. (1995)).

One of the other approaches to the analysis of size and shape is to define a scalar measure of size, and then calculate a shape vector as the original vector $\mathbf{x}$ of p measurements divided by the size. This is intuitively reasonable, but needs a definition of size. Darroch and Mosimann (1985) list a number of possibilities, but home in on $g_{\mathbf{a}}(\mathbf{x}) = \prod_{k=1}^{p} x_k^{a_k}$, where $\mathbf{a}' = (a_1, a_2, \ldots, a_p)$ and $\sum_{k=1}^{p} a_k = 1$. The size is thus a generalization of the geometric mean. Darroch and Mosimann (1985) discuss a number of properties of the shape vector $\mathbf{x}/g_{\mathbf{a}}(\mathbf{x})$ and its logarithm, and advocate the use of PCA on the log-transformed shape vector, leading to shape components. The log shape vector generalizes the vector $\mathbf{v}$ used by Aitchison (1983) in the analysis of compositional data (see Section 13.3), but the PCs are invariant with respect to the choice of $\mathbf{a}$. As with Aitchison's (1983) analysis, the covariance matrix of the log shape data has the isometric vector $\mathbf{a}_0$ as an eigenvector, with zero eigenvalue. Hence all the shape components are contrasts between log-transformed variables. Darroch and Mosimann (1985) give an example in which both the first and last shape components are of interest.

The analysis of shapes goes well beyond the size and shape of organisms (see, for example, Dryden and Mardia (1998) and Bookstein (1991)). A completely different approach to the analysis of shape is based on 'landmarks.' These are well-defined points on an object whose coordinates define the shape of the object, after the effects of location, scale and rotation have been removed. Landmarks can be used to examine the shapes of animals and plants, but they are also relevant to the analysis of shapes of many other types of object. Here, too, PCA can be useful and it may be implemented in a variety of forms. Kent (1994) distinguishes four versions depending on the choice of coordinates and on the choice of whether to use real or complex PCA (see also Section 13.8).

Kent (1994) describes two coordinate systems due to Kendall (1984) and to Bookstein (1991). The two systems arise because of the different possible ways of removing location and scale. If there are p landmarks, then the end result in either system of coordinates is that each object is represented by a set of $(p-1)$ two-dimensional vectors. There is now a choice of whether to treat the data as measurements on $2(p-1)$ real variables, or as measurements on $(p-1)$ complex variables where the two coordinates at each landmark point give the real and imaginary parts. Kent (1994) discusses the properties of the four varieties of PCA thus produced, and comments that complex PCA is rather uninformative. He gives an example of real PCA for both coordinate systems.

Horgan (2000) describes an application of PCA to the comparison of shapes of carrots. After bringing the carrots into the closest possible alignment, distances are calculated between each pair of carrots based on the amount of non-overlap of their shapes. A principal coordinate analysis (Section 5.2) is done on these distances, and Horgan (2000) notes that this is equivalent to a principal component analysis on binary variables representing the presence or absence of each carrot at a grid of points in two-dimensional space. Horgan (2000) also notes the similarity between his technique and a PCA on the grey scale levels of aligned images, giving so-called eigenimages. This latter procedure has been used to analyse faces (see, for example Craw and Cameron, 1992) as well as carrots (Horgan, 2001).

13.3 Principal Component Analysis for Compositional Data

Compositional data consist of observations $\mathbf{x}_1, \mathbf{x}_2, \ldots, \mathbf{x}_n$ for which each element of $\mathbf{x}_i$ is a proportion, and the elements of $\mathbf{x}_i$ are constrained to sum to unity. Such data occur, for example, when a number of chemical compounds or geological specimens or blood samples are analysed, and the proportion in each of a number of chemical elements is recorded. As noted in Section 13.1, Gower (1967) discusses some geometric implications that follow from the constraints on the elements of $\mathbf{x}$, but the major reference for PCA on compositional data is Aitchison (1983). Because of the constraints on the elements of $\mathbf{x}$, and also because compositional data apparently often exhibit non-linear rather than linear structure among their variables, Aitchison (1983) proposes that PCA be modified for such data.

At first sight, it might seem that no real difficulty is implied by the condition

$$x_{i1} + x_{i2} + \cdots + x_{ip} = 1, \tag{13.3.1}$$

which holds for each observation. If a PCA is done on $\mathbf{x}$, there will be a PC with zero eigenvalue identifying the constraint. This PC can be ignored because it is entirely predictable from the form of the data, and the remaining PCs can be interpreted as usual. A counter to this argument is that correlations and covariances, and hence PCs, cannot be interpreted in the usual way when the constraint is present. In particular, the constraint (13.3.1) introduces a bias towards negative values among the correlations, so that a set of compositional variables that are 'as independent as possible' will not all have zero correlations between them.

One way of overcoming this problem is to do the PCA on a subset of $(p-1)$ of the p compositional variables, but this idea has the unsatisfactory feature that the choice of which variable to leave out is arbitrary, and different choices will lead to different PCs. For example, suppose that two variables have much larger variances than the other $(p-2)$ variables. If a PCA is based on the covariance matrix for $(p-1)$ of the variables, then the result will vary considerably, depending on whether the omitted variable has a large or small variance. Furthermore, there remains the restriction that the $(p-1)$ chosen variables must sum to no more than unity, so that the interpretation of correlations and covariances is still not straightforward.

The alternative that is suggested by Aitchison (1983) is to replace $\mathbf{x}$ by $\mathbf{v} = \log[\mathbf{x}/g(\mathbf{x})]$, where $g(\mathbf{x}) = (\prod_{i=1}^{p} x_i)^{\frac{1}{p}}$ is the geometric mean of the elements of $\mathbf{x}$. Thus, the jth element of $\mathbf{v}$ is

$$v_j = \log x_j - \frac{1}{p} \sum_{i=1}^{p} \log x_i, \quad j = 1, 2, \ldots, p. \tag{13.3.2}$$

A PCA is then done for $\mathbf{v}$ rather than $\mathbf{x}$. There is one zero eigenvalue whose eigenvector is the isometric vector with all elements equal; the remaining eigenvalues are positive and, because the corresponding eigenvectors are orthogonal to the final eigenvector, they define *contrasts* (that is, linear functions whose coefficients sum to zero) for the $\log x_j$.

Aitchison (1983) also shows that these same functions can equivalently be found by basing a PCA on the non-symmetric set of variables $\mathbf{v}^{(j)}$, where

$$\mathbf{v}^{(j)} = \log[\mathbf{x}^{(j)}/x_j] \tag{13.3.3}$$

and $\mathbf{x}^{(j)}$ is the $(p-1)$-vector obtained by deleting the jth element x_j from $\mathbf{x}$. The idea of transforming to logarithms before doing the PCA can, of course, be used for data other than compositional data (see also Section 13.2). However, there are a number of particular advantages of the log-ratio transformation (13.3.2), or equivalently (13.3.3), for compositional data. These include the following, which are discussed further by Aitchison (1983)

(i) It was noted above that the constraint (13.3.1) introduces a negative bias to the correlations between the elements of $\mathbf{x}$, so that any notion of 'independence' between variables will not imply zero correlations. A number of ideas have been put forward concerning what should constitute 'independence,' and what 'null correlations' are implied, for compositional data. Aitchison (1982) presents arguments in favour of a definition of independence in terms of the structure of the covariance matrix of $\mathbf{v}^{(j)}$ (see his equations (4.1) and (5.1)). With this definition, the PCs based on $\mathbf{v}$ (or $\mathbf{v}^{(j)}$) for a set of 'independent' variables are simply the elements of $\mathbf{v}$ (or $\mathbf{v}^{(j)}$) arranged in descending size of their variances. This is equivalent to what happens in PCA for 'ordinary' data with independent variables.

(ii) There is a tractable class of probability distributions for $\mathbf{v}^{(j)}$ and for linear contrasts of the elements of $\mathbf{v}^{(j)}$, but there is no such tractable class for linear contrasts of the elements of $\mathbf{x}$ when $\mathbf{x}$ is restricted by the constraint (13.3.1).

(iii) Because the log-ratio transformation removes the effect of the constraint on the interpretation of covariance, it is possible to define distances between separate observations of $\mathbf{v}$ in a way that is not possible with $\mathbf{x}$.

(iv) It is easier to examine the variability of subcompositions (subsets of $\mathbf{x}$ renormalized to sum to unity) compared to that of the whole composition, if the comparison is done in terms of $\mathbf{v}$ rather than $\mathbf{x}$.

Aitchison (1983) provides examples in which the proposed PCA of $\mathbf{v}$ is considerably superior to a PCA of $\mathbf{x}$. This seems to be chiefly because there is curvature inherent in many compositional data sets; the proposed analysis is very successful in uncovering correct curved axes of maximum variation, whereas the usual PCA, which is restricted to linear functions of $\mathbf{x}$, is not. However, Aitchison's (1983) proposal does not *necessarily* make much difference to the results of a PCA, as is illustrated in the example given below in Section 13.3.1. Aitchison (1986, Chapter 8) covers similar material to Aitchison (1983), although more detail is given, including examples, of the analysis of subdecompositions.

A disadvantage of Aitchison's (1983) approach is that it cannot handle zeros for any of the x_j (see equation (13.3.2)). One possibility is to omit from the analysis any variables which have zeros, though discarding information in this way is undesirable. Alternatively, any zeros can be replaced by a small positive number, but the results are sensitive to the choice of that number. Bacon-Shone (1992) proposes an approach to compositional data based on ranks, which allows zeros to be present. The values of x_{ij}, $i = 1, 2, \ldots, n$; $j = 1, 2, \ldots, p$ are ranked either within rows or within columns or across the whole data matrix, and the data values are then replaced by their ranks, which range from 1 to p, from 1 to n, or from 1 to np, respectively, depending on the type of ranking. These ranks are then scaled within each row so that each row sum equals 1, as is true for the original data.

Bacon-Shone (1992) does not use PCA on these rank-transformed data, but Baxter (1993) does. He looks at several approaches for a number of compositional examples from archaeology, and demonstrates that for typical archaeological data, which often include zeros, Bacon-Shone's (1992) procedure is unsatisfactory because it is too sensitive to the ranked zeros. Baxter (1993) also shows that both Aitchison's and Bacon-Shone's approaches can be misleading when there are small but non-zero elements in the data. He claims that simply ignoring the compositional nature of the data and performing PCA on the original data is often a more informative alternative in archaeology than these approaches.

Kaciak and Sheahan (1988) advocate the use of uncentred PCA (see Section 14.2.3), apparently without a log transformation, for the analysis of compositional data, and use it in a market segmentation example.

13.3.1   Example: 100 km Running Data

In Sections 5.3 and 12.3.3, a data set was discussed which consisted of times taken for each of ten 10 km sections by 80 competitors in a 100 km race. If, instead of recording the actual time taken in each section, we look at the proportion of the total time taken for each section, the data then become compositional in nature. A PCA was carried out on these compositional data, and so was a modified analysis as proposed by Aitchison (1983). The coefficients and variances for the first two PCs are given for the unmodified and modified analyses in Tables 13.1, 13.2, respectively. It can be seen that the PCs defined in Tables 13.1 and 13.2 have very similar coefficients, with angles between corresponding vectors of coefficients equal to 8° for both first and second PCs. This similarity continues with later PCs. The first PC is essentially a linear contrast between times early and late in the race, whereas the second PC is a 'quadratic' contrast with times early and late in the race contrasted with those in the middle.

Comparison of Tables 13.1 and 13.2 with Table 5.2 shows that converting the data to compositional form has removed the first (overall time) component, but the coefficients for the second PC in Table 5.2 are very similar to those of the first PC in Tables 13.1 and 13.2. This correspondence continues to later PCs, with the third, fourth, ... PCs for the 'raw' data having similar coefficients to those of the second, third,... PCs for the compositional data.

Table 13.1. First two PCs: 100 km compositional data.

	Coefficients Component 1	Coefficients Component 2
First 10 km	0.42	0.19
Second 10 km	0.44	0.18
Third 10 km	0.44	0.00
Fourth 10 km	0.40	−0.23
Fifth 10 km	0.05	−0.56
Sixth 10 km	−0.18	−0.53
Seventh 10 km	−0.20	−0.15
Eighth 10 km	−0.27	−0.07
Ninth 10 km	−0.24	0.30
Tenth 10 km	−0.27	0.41
Eigenvalue	4.30	2.31
Cumulative percentage of total variation	43.0	66.1

Table 13.2. First two PCs: Aitchison's (1983) technique for 100 km compositional data.

	Coefficients Component 1	Coefficients Component 2
First 10 km	0.41	0.19
Second 10 km	0.44	0.17
Third 10 km	0.42	−0.06
Fourth 10 km	0.36	−0.31
Fifth 10 km	−0.04	−0.57
Sixth 10 km	−0.25	−0.48
Seventh 10 km	−0.24	−0.08
Eighth 10 km	−0.30	−0.01
Ninth 10 km	−0.24	0.30
Tenth 10 km	−0.25	0.43
Eigenvalue	4.38	2.29
Cumulative percentage of total variation	43.8	66.6

13.4 Principal Component Analysis in Designed Experiments

In Chapters 8 and 9 we discussed ways in which PCA could be used as a preliminary to, or in conjunction with, other standard statistical techniques. The present section gives another example of the same type of application; here we consider the situation where p variables are measured in the course of a designed experiment. The standard analysis would be either a set of separate analyses of variance (ANOVAs) for each variable or, if the variables are correlated, a multivariate analysis of variance (MANOVA—Rencher, 1995, Chapter 6) could be done.

As an illustration, consider a two-way model of the form

$$\mathbf{x}_{ijk} = \boldsymbol{\mu} + \boldsymbol{\tau}_j + \boldsymbol{\beta}_k + \boldsymbol{\epsilon}_{ijk}, \ i = 1, 2, \ldots, n_{jk}; \ j = 1, 2, \ldots, t; \ k = 1, 2, \ldots, b,$$

where $\mathbf{x}_{ijk}$ is the ith observation for treatment j in block k of a p-variate vector $\mathbf{x}$. The vector $\mathbf{x}_{ijk}$ is therefore the sum of an overall mean $\boldsymbol{\mu}$, a treatment effect $\boldsymbol{\tau}_j$, a block effect $\boldsymbol{\beta}_k$ and an error term $\boldsymbol{\epsilon}_{ijk}$.

The most obvious way in which PCA can be used in such analyses is simply to replace the original p variables by their PCs. Then either separate ANOVAs can be done on each PC, or the PCs can be analysed using MANOVA. Jackson (1991, Sections 13.5–13.7) discusses the use of separate ANOVAs for each PC in some detail. In the context of analysing growth curves (see Section 12.4.2) Rao (1958) suggests that 'methods of multivariate analysis for testing the differences between treatments' can be implemented on the first few PCs, and Rencher (1995, Section 12.2) advocates PCA as a first step in MANOVA when p is large. However, as noted by Rao (1964), for most types of designed experiment this simple analysis is often not particularly useful. This is because the overall covariance matrix represents a *mixture* of contributions from within treatments and blocks, between treatments, between blocks, and so on, whereas we usually wish to *separate* these various types of covariance. Although the PCs are uncorrelated overall, they are not necessarily so, even approximately, with respect to between-group or within-group variation. This is a more complicated manifestation of what occurs in discriminant analysis (Section 9.1), where a PCA based on the covariance matrix of the raw data may prove confusing, as it inextricably mixes up variation between and within populations. Instead of a PCA of all the $\mathbf{x}_{ijk}$, a number of other PCAs have been suggested and found to be useful in some circumstances.

Jeffers (1962) looks at a PCA of the (treatment × block) means $\bar{\mathbf{x}}_{jk}$, $j = 1, 2, \ldots, t$; $k = 1, 2, \ldots, b$, where

$$\bar{\mathbf{x}}_{jk} = \frac{1}{n_{jk}} \sum_{i=1}^{n_{jk}} \mathbf{x}_{ijk},$$

that is, a PCA of a data set with tb observations on a p-variate random vector. In an example on tree seedlings, he finds that ANOVAs carried out on the first five PCs, which account for over 97% of the variation in the original eight variables, give significant differences between treatment means (averaged over blocks) for the first and fifth PCs. This result contrasts with ANOVAs for the original variables, where there were no significant differences. The first and fifth PCs can be readily interpreted in Jeffers' (1962) example, so that transforming to PCs produces a clear advantage in terms of detecting interpretable treatment differences. However, PCs will not always be interpretable and, as in regression (Section 8.2), there is no reason to expect that treatment differences will necessarily manifest themselves in high variance, rather than low variance, PCs. For example, while Jeffer's first component accounts for over 50% of total variation, his fifth component accounts for less than 5%.

Jeffers (1962) looked at 'between' treatments and blocks PCs, but the PCs of the 'within' treatments or blocks covariance matrices can also provide useful information. Pearce and Holland (1960) give an example having four variables, in which different treatments correspond to different rootstocks, but which has no block structure. They carry out separate PCAs for within- and between-rootstock variation. The first PC is similar in the two cases, measuring general size. Later PCs are, however, different for the two analyses, but they are readily interpretable in both cases so that the two analyses each provide useful but separate information.

Another use of 'within-treatments' PCs occurs in the case where there are several populations, as in discriminant analysis (see Section 9.1), and 'treatments' are defined to correspond to different populations. If each population has the same covariance matrix Σ, and within-population PCs based on Σ are of interest, then the 'within-treatments' covariance matrix provides an estimate of Σ. Yet another way in which 'error covariance matrix PCs' can contribute is if the analysis looks for potential outliers as suggested for multivariate regression in Section 10.1.

A different way of using PCA in a designed experiment is described by Mandel (1971, 1972). He considers a situation where there is only one variable, which follows the two-way model

$$x_{jk} = \mu + \tau_j + \beta_k + \varepsilon_{jk}, \ j = 1, 2, \ldots, t; \ k = 1, 2, \ldots, b, \qquad (13.4.1)$$

that is, there is only a single observation on the variable x at each combination of treatments and blocks. In Mandel's analysis, estimates $\hat{\mu}$, $\hat{\tau}_j$, $\hat{\beta}_k$ are found for μ, τ_j, β_k, respectively, and residuals are calculated as $e_{jk} = x_{jk} - \hat{\mu} - \hat{\tau}_j - \hat{\beta}_k$. The main interest is then in using e_{jk} to estimate the non-additive part ε_{jk} of the model (13.4.1). This non-additive part is assumed to take the form

$$\varepsilon_{jk} = \sum_{h=1}^{m} u_{jh} l_h a_{kh}, \qquad (13.4.2)$$

where m, l_h, u_{jh}, a_{kh}, are suitably chosen constants. Apart from slight changes in notation, the right-hand side of (13.4.2) is the same as that of the singular value decomposition (SVD) in (3.5.3). Thus, the model (13.4.2) is fitted by finding the SVD of the matrix $\mathbf{E}$ whose (j, k)th element is e_{jk}, or equivalently finding the PCs of the covariance matrix based on the data matrix $\mathbf{E}$. This analysis also has links with correspondence analysis (see Section 13.1). In both cases we find an SVD of a two-way table of residuals, the difference being that in the present case the elements of the table are residuals from an additive model for a quantitative variable, rather than residuals from an independence (multiplicative) model for counts. As noted in Section 1.2, R.A. Fisher used the SVD in a two-way analysis of an agricultural trial, leading to an eigenanalysis of a multiple of a covariance matrix as long ago as 1925.

A substantial amount of work has been done on the model defined by (13.4.1) and (13.4.2). Freeman (1975) showed that Mandel's approach can be used for incomplete as well as complete two-way tables, and a number of authors have constructed tests for the rank m of the interaction term. For example, Boik (1986) develops likelihood ratio and union-intersection tests, and Milliken and Johnson (1989) provide tables of critical points for likelihood ratio statistics. Boik (1986) also points out that the model is a reduced-rank regression model (see Section 9.3.4).

Shafii and Price (1998) give an example of a more complex design in which seed yields of 6 rapeseed cultivars are measured in 27 environments spread over three separate years, with 4 replicates at each of the (6×27) combinations. There are additional terms in the model compared to (13.4.1), but the non-additive part is still represented as in (13.4.2). The first two terms in (13.4.2) are deemed significant using Milliken and Johnson's (1989) tables; they account for 80% of the variability that would be explained by taking m of full rank 5. The results of the analysis of the non-additive term are interpreted using biplots (see Section 5.3).

Gower and Krzanowski (1999) consider the situation in which the data have a MANOVA structure but where the assumptions that underlie formal MANOVA procedures are clearly invalid. They suggest a number of graphical displays to represent and interpret such data; one is based on weighted PCA. Goldstein (1995, Section 4.5) notes the possibility of using PCA to explore the structure of various residual matrices in a multilevel model.

Planned surveys are another type of designed experiment, and one particular type of survey design is based on stratified sampling. Pla (1991) suggests that when the data from a survey are multivariate, the first PC can be used to define the strata for a stratified sampling scheme. She shows that stratification in this manner leads to reduced sampling variability compared to stratification based on only one or two variables. Skinner et al. (1986) and Tortora (1980) demonstrate the effect of the non-independence induced by other methods of stratification on subsequently calculated PCs (see Section 12.4.5).

Observations of the atmosphere certainly do not constitute a designed experiment, but a technique proposed by Zheng et al. (2001) is included here because of its connections to analysis of variance. Measurements of meteorological fields can be thought of as the sum of long-term variability caused by external forcing and the slowly varying internal dynamics of the atmosphere, and short-term day-to-day weather variability. The first term is potentially predictable over seasonal or longer time scales, whereas the second term is not. It is therefore of interest to separate out the potentially predictable component and examine its major sources of variation. Zheng et al. (2001) do this by estimating the covariance matrix of the day-to-day variation and subtracting it from the 'overall' covariance matrix, which assumes independence of short- and long-term variation. A PCA is then done on the resulting estimate of the covariance matrix for the potentially predictable variation, in order to find potentially predictable patterns.

Returning to designed experiments, in optimal design it is desirable to know the effect of changing a design by deleting design points or augmenting it with additional design points. Jensen (1998) advocates the use of principal components of the covariance matrix of predicted values at a chosen set of design points to investigate the effects of such augmentation or deletion. He calls these components *principal predictors*, though they are quite different from the entities with the same name defined by Thacker (1999) and discussed in Section 9.3.3. Jensen (1998) illustrates the use of his principal predictors for a variety of designs.

13.5   Common Principal Components and Comparisons of Principal Components

Suppose that observations on a p-variate random vector $\mathbf{x}$ may have come from any one of G distinct populations, and that the mean and covariance matrix for the gth population are, respectively, $\boldsymbol{\mu}_g, \boldsymbol{\Sigma}_g, g = 1, 2, \ldots, G$. This is the situation found in discriminant analysis (see Section 9.1) although in discriminant analysis it is often assumed that all the $\boldsymbol{\Sigma}_g$ are the same, so that the populations differ only in their means. If the $\boldsymbol{\Sigma}_g$ *are* all the same, then the 'within-population PCs' are the same for all G populations, though, as pointed out in Section 9.1, within-population PCs are often different from PCs found by pooling data from all populations together.

If the $\boldsymbol{\Sigma}_g$ are different, then there is no uniquely defined set of within-population PCs that is common to all populations. However, a number of authors have examined 'common principal components,' which can usefully be defined in some circumstances where the $\boldsymbol{\Sigma}_g$ are not all equal. The idea of 'common' PCs arises if we suspect that the same components underlie the covariance matrices of each group, but that they have different weights in different groups. For example, if anatomical measurements are made on different but closely related species of animals, then the same general 'size' and 'shape' components (see Section 13.2) may be present for each species, but with varying importance. Similarly, if the same variables are measured on the same individuals but at different times, so that 'groups' correspond to different times as in longitudinal studies (see Section 12.4.2), then the components may remain the same but their relative importance may vary with time.

One way of formally expressing the presence of 'common PCs' as just defined is by the hypothesis that there is an orthogonal matrix $\mathbf{A}$ that simultaneously diagonalizes all the $\boldsymbol{\Sigma}_g$ so that

$$\mathbf{A}'\boldsymbol{\Sigma}_g\mathbf{A} = \boldsymbol{\Lambda}_g, \tag{13.5.1}$$

where $\boldsymbol{\Lambda}_g$, $g = 1, 2, \ldots, G$ are all diagonal. The kth column of $\mathbf{A}$ gives the coefficients of the kth common PC, and the (diagonal) elements of $\boldsymbol{\Lambda}_g$ give the variances of these PCs for the gth population. Note that the order of these variances need not be the same for all g, so that different PCs may have the largest variance in different populations.

In a series of papers in the 1980s Flury developed ways of estimating and testing the model implied by (13.5.1). Much of this work later appeared in a book (Flury, 1988), in which (13.5.1) is the middle level of a 5-level hierarchy of models for a set of G covariance matrices. The levels are these:

- $\boldsymbol{\Sigma}_1 = \boldsymbol{\Sigma}_2 = \ldots = \boldsymbol{\Sigma}_G$ (equality).

- $\boldsymbol{\Sigma}_g = \rho_g\boldsymbol{\Sigma}_1$ for some positive constants $\rho_2, \rho_3, \ldots, \rho_g$ (proportionality).

- $\mathbf{A}'\boldsymbol{\Sigma}_g\mathbf{A} = \boldsymbol{\Lambda}_g$ (the common PC model).

- Equation (13.5.1) can also be written, using spectral decompositions (2.1.10) of each covariance matrix, as

$$\boldsymbol{\Sigma}_g = \lambda_{g1}\boldsymbol{\alpha}_1\boldsymbol{\alpha}_1' + \lambda_{g2}\boldsymbol{\alpha}_2\boldsymbol{\alpha}_2' + \ldots + \lambda_{gp}\boldsymbol{\alpha}_p\boldsymbol{\alpha}_p'.$$

Level 4 (the partial common PC model) replaces this by

$$\boldsymbol{\Sigma}_g = \lambda_{g1}\boldsymbol{\alpha}_1\boldsymbol{\alpha}_1' + \ldots + \lambda_{gq}\boldsymbol{\alpha}_q\boldsymbol{\alpha}_q' + \lambda_{g(q+1)}\boldsymbol{\alpha}_{q+1}^{(g)}\boldsymbol{\alpha}_{q+1}^{'(g)} + \ldots + \lambda_{gp}\boldsymbol{\alpha}_p^{(g)}\boldsymbol{\alpha}_p^{'(g)} \tag{13.5.2}$$

Thus, q of the p PCs have common eigenvectors in the G groups, whereas the other $(p - q)$ do not. The ordering of the components in (13.5.2) need not in general reflect the size of the eigenvalues. Any subset of q components can be 'common.'

- No restriction on $\boldsymbol{\Sigma}_1, \boldsymbol{\Sigma}_2, \ldots, \boldsymbol{\Sigma}_g$.

The first and last levels of the hierarchy are trivial, but Flury (1988) devotes a chapter to each of the three intermediate ones, covering maximum likelihood estimation, asymptotic inference and applications. The partial common PC model is modified to give an additional level of the hierarchy in which q of the components span a common subspace, but there is no requirement that the individual components should be the same. Likelihood ratio tests are described by Flury (1988) for comparing the fit of models at different levels of the hierarchy. Model selection can also be made on the basis of Akaike's information criterion (AIC) (Akaike, 1974).

One weakness of the theory described in Flury (1988) is that it is only really applicable to covariance-based PCA and not to the more frequently encountered correlation-based analysis.

Lefkovitch (1993) notes that Flury's (1988) procedure for fitting a common PC model can be time-consuming for moderate or large data sets. He proposes a technique that produces what he calls *consensus components*, which are much quicker to find. They are based on the so-called *polar decomposition* of a data matrix, and approximately diagonalize two or more covariance matrices simultaneously. In the examples that Lefkovitch (1993) presents the consensus and common PCs are similar and, if the common PCs are what is really wanted, the consensus components provide a good starting point for the iterative process that leads to common PCs.

A number of topics that were described briefly by Flury (1988) in a chapter on miscellanea were subsequently developed further. Schott (1988) derives an approximate test of the partial common PC model for $G = 2$ when the common subspace is restricted to be that spanned by the *first q* PCs. He argues that in dimension-reducing problems this, rather than *any* q-dimensional subspace, is the subspace of interest. His test is extended to $G > 2$ groups in Schott (1991), where further extensions to correlation-based analyses and to robust PCA are also considered. Yuan and Bentler (1994) provide a test for linear trend in the last few eigenvalues under the common PC model.

Flury (1988, Section 8.5) notes the possibility of using models within his hierarchy in the multivariate Behrens-Fisher problem of testing equality between the means of two p-variate groups when their covariance matrices cannot be assumed equal. Nel and Pienaar (1998) develop this idea, which also extends Takemura's (1985) decomposition of Hotelling's T^2 statistic with respect to principal components when equality of covariances is assumed (see also Section 9.1). Flury et al. (1995) consider the same two-group set-up, but test the hypothesis that a subset of the p means is the same in the two groups, while simultaneously estimating the covariance matrices under the common PC model. Bartoletti et al. (1999) consider tests for the so-called *allometric extension* model defined by Hills (1982). In this model, the size and shape of organisms (see Section 13.2) are such that for two groups of organisms not only is the first PC common to both groups, but the difference in means between the two groups also lies in the same direction as this common PC.

In the context of discriminant analysis, Bensmail and Celeux (1996) use a similar hierarchy of models for the covariance matrices of G groups to that of Flury (1988), though the hierarchy is augmented to include special cases of interest such as identity or diagonal covariance matrices. Krzanowski (1990) and Flury (1995) also discuss the use of the common principal component model in discriminant analysis (see Section 9.1).

Flury and Neuenschwander (1995) look at the situation in which the assumption that the G groups of variables are independent is violated. This can occur, for example, if the same variables are measured for the *same individuals* at G different times. They argue that, in such circumstances, when $G = 2$ the common PC model can provide a useful alternative to canonical correlation analysis (CCA) (see Section 9.3) for examining relationships between two groups and, unlike CCA, it is easily extended to the case of $G > 2$ groups. Neuenschwander and Flury (2000) discuss in detail the theory underlying the common PC model for dependent groups.

Krzanowski (1984a) describes a simpler method of obtaining estimates of $\mathbf{A}$ and $\mathbf{\Lambda}_g$ based on the fact that if (13.5.1) is true then the columns of $\mathbf{A}$ contain the eigenvectors not only of $\mathbf{\Sigma}_1, \mathbf{\Sigma}_2, \dots, \mathbf{\Sigma}_G$ individually but of any linear combination of $\mathbf{\Sigma}_1, \mathbf{\Sigma}_2, \dots, \mathbf{\Sigma}_G$. He therefore uses the eigenvectors of $\mathbf{S}_1 + \mathbf{S}_2 + \dots + \mathbf{S}_G$, where $\mathbf{S}_g$ is the sample covariance matrix for the gth population, to estimate $\mathbf{A}$, and then substitutes this estimate and $\mathbf{S}_g$ for $\mathbf{A}$ and $\mathbf{\Sigma}_g$, respectively, in (13.5.1) to obtain estimates of $\mathbf{\Lambda}_g$, $g = 1, 2, \dots, G$.

To assess whether or not (13.5.1) is true, the estimated eigenvectors of $\mathbf{S}_1 + \mathbf{S}_2 + \dots + \mathbf{S}_G$ can be compared with those estimated for some other weighted sum of $\mathbf{S}_1, \mathbf{S}_2, \dots, \mathbf{S}_G$ chosen to have different eigenvectors from $\mathbf{S}_1 + \mathbf{S}_2 + \dots + \mathbf{S}_G$ if (13.5.1) does not hold. The comparison between eigenvectors can be made either informally or using methodology developed by Krzanowski (1979b), which is now described.

Suppose that sample covariance matrices $\mathbf{S}_1$, $\mathbf{S}_2$ are available for two groups of individuals, and that we wish to compare the two sets of PCs found from $\mathbf{S}_1$ and $\mathbf{S}_2$. Let $\mathbf{A}_{1q}, \mathbf{A}_{2q}$ be $(p \times q)$ matrices whose columns contain the coefficients of the first q PCs based on $\mathbf{S}_1, \mathbf{S}_2$, respectively. Krzanowski's (1979b) idea is to find the minimum angle δ between the subspaces defined by the q columns of $\mathbf{A}_{1q}$ and $\mathbf{A}_{2q}$, together with associated vectors in these two subspaces that subtend this minimum angle. This suggestion is based on an analogy with the *congruence coefficient*, which has been widely used in factor analysis to compare two sets of factor loadings (Korth and Tucker, 1975) and which, for two vectors, can be interpreted as the cosine of the angle between those vectors. In Krzanowski's (1979b) set-up, it turns out that δ is given by

$$\delta = \cos^{-1} \left(\nu_1^{1/2} \right),$$

where ν_1 is the first (largest) eigenvalue of $\mathbf{A}'_{1q} \mathbf{A}_{2q} \mathbf{A}'_{2q} \mathbf{A}_{1q}$, and the vectors that subtend the minimum angle are related, in a simple way, to the corresponding eigenvector.

The analysis can be extended by looking at the second, third,..., eigenvalues and corresponding eigenvectors of $\mathbf{A}'_{1q} \mathbf{A}_{2q} \mathbf{A}'_{2q} \mathbf{A}_{1q}$; from these can be found pairs of vectors, one each in the two subspaces spanned by $\mathbf{A}_{1q}, \mathbf{A}_{2q}$, that have minimum angles between them, subject to being orthogonal to previous pairs of vectors. Krzanowski (1979b) notes that the sum of eigenvalues $\text{tr}(\mathbf{A}'_{1q}\mathbf{A}_{2q}\mathbf{A}'_{2q}\mathbf{A}_{1q})$ can be used as an overall measure of the similarity between the two subspaces. Crone and Crosby (1995) define a transformed version of this trace as an appropriate measure of subspace similarity, examine some of its properties, and apply it to an example from satellite meteorology.

Another extension is to $G > 2$ groups of individuals with covariance matrices $\mathbf{S}_1, \mathbf{S}_2, \ldots, \mathbf{S}_G$ and matrices $\mathbf{A}_{1q}, \mathbf{A}_{2q}, \ldots, \mathbf{A}_{Gq}$ containing the first q eigenvectors (PC coefficients) for each group. We can then look for a vector that minimizes

$$\Delta = \sum_{g=1}^{G} \cos^2 \delta_g,$$

where δ_g is the angle that the vector makes with the subspace defined by the columns of $\mathbf{A}_{gq}$. This objective is achieved by finding eigenvalues and eigenvectors of

$$\sum_{g=1}^{G} \mathbf{A}_{gq}\mathbf{A}'_{gq},$$

and Krzanowski (1979b) shows that for $g = 2$ the analysis reduces to that given above.

In Krzanowski (1979b) the technique is suggested as a descriptive tool— if δ (for $G = 2$) or Δ (for $G > 2$) is 'small enough' then the subsets of q PCs for the G groups are similar, but there is no formal definition of 'small enough.' In a later paper, Krzanowski (1982) investigates the behaviour of δ using simulation, both when all the individuals come from populations with the same covariance matrix, and when the covariance matrices are different for the two groups of individuals. The simulation encompasses several different values for p, q and for the sample sizes, and it also includes several different structures for the covariance matrices. Krzanowski (1982) is therefore able to offer some limited guidance on what constitutes a 'small enough' value of δ, based on the results from his simulations.

As an example, consider anatomical data similar to those discussed in Sections 1.1, 4.1, 5.1, 10.1 and 10.2 that were collected for different groups of students in different years. Comparing the first three PCs found for the 1982 and 1983 groups of students gives a value of $2.02°$ for δ; the corresponding value for 1982 and 1984 is $1.25°$, and that for 1983 and 1984 is $0.52°$. Krzanowski (1982) does not have a table of simulated critical angles for the sample sizes and number of variables relevant to this example. In addition, his tables are for covariance matrices whereas the student data PCAs are for correlation matrices. However, for illustration we note that the three values quoted above are well below the critical angles corresponding to the values of p, q and sample size in his tables closest to those of the present example. Hence, if Krzanowski's tables are at all relevant for correlation matrices, the sets of the first three PCs are not significantly different for the three years 1982, 1983, 1984 as might be expected from such small angles.

If all three years are compared simultaneously, then the angles between the subspaces formed by the first three PCs and the nearest vector to all three subspaces are

1982	1983	1984
1.17°	1.25°	0.67°

Again, the angles are very small; although no tables are available for assessing the significance of these angles, they seem to confirm the impression given by looking at the years two at a time that the sets of the first three PCs are not significantly different for the three years.

Two points should be noted with respect to Krzanowski's technique. First, it can only be used to compare *subsets* of PCs—if $q = p$, then $\mathbf{A}_{1p}, \mathbf{A}_{2p}$ will usually span p-dimensional space (unless either $\mathbf{S}_1$ or $\mathbf{S}_2$ has zero eigenvalues), so that δ is necessarily zero. It seems likely that the technique will be most valuable for values of q that are small compared to p. The second point is that while δ is clearly a useful measure of the closeness of two subsets of PCs, the vectors and angles found from the second, third, ..., eigenvalues and eigenvectors of $\mathbf{A}'_{1q}\mathbf{A}_{2q}\mathbf{A}'_{2q}\mathbf{A}_{1q}$ are successively less valuable. The first two or three angles give an idea of the overall difference between the two subspaces, provided that q is not too small. However, if we reverse the analysis and look at the *smallest* eigenvalue and corresponding eigenvector of $\mathbf{A}'_{1q}\mathbf{A}_{2q}\mathbf{A}'_{2q}\mathbf{A}_{1q}$, then we find the *maximum* angle between vectors in the two subspaces (which will often be 90°, unless q is small). Thus, the last few angles and corresponding vectors need to be interpreted in a rather different way from that of the first few. The general problem of interpreting angles other than the first can be illustrated by again considering the first three PCs for the student anatomical data from 1982 and 1983. We saw above that $\delta = 2.02°$, which is clearly very small; the second and third angles for these data are 25.2° and 83.0°, respectively. These angles are fairly close to the 5% critical values given in Krzanowski (1982) for the second and third angles when $p = 8$, $q = 3$ and the sample sizes are each 50 (our data have $p = 7$, $q = 3$ and sample sizes around 30), but it is difficult to see what this result implies. In particular, the fact that the third angle is close to 90° might intuitively suggest that the first three PCs are significantly different for 1982 and 1983. Intuition is, however, contradicted by Krzanowski's Table I, which shows that for sample sizes as small as 50 (and, hence, certainly for samples of size 30), the 5% critical value for the third angle is nearly 90°. For $q = 3$ this is not particularly surprising—the dimension of $\mathbf{A}'_{1q}\mathbf{A}_{2q}\mathbf{A}'_{2q}\mathbf{A}_{1q}$ is (3×3) so the third angle is the *maximum* angle between subspaces.

Cohn (1999) considers four test statistics for deciding the equivalence or otherwise of subspaces defined by sets of q PCs derived from each of two covariance matrices corresponding to two groups of observations. One of the statistics is the likelihood ratio test used by Flury (1988) and two others are functions of the eigenvalues, or corresponding cosines, derived by Krzanowski (1979b). The fourth statistic is based on a sequence of two-dimensional rotations from one subspace towards the other, but simulations show it to be less reliable than the other three. There are a number of novel aspects to Cohn's (1999) study. The first is that the observations within the two groups are not independent; in his motivating example the data are serially correlated time series. To derive critical values for the test statistics, a bootstrap procedure is used, with resampling in blocks because of the serial correlation. The test statistics are compared in a simulation study and on the motivating example.

Keramidas et al. (1987) suggest a graphical procedure for comparing eigenvectors of several covariance matrices $S_1, S_2, \ldots, S_G$. Much of the paper is concerned with the comparison of a *single* eigenvector from each matrix, either with a common predetermined vector or with a 'typical' vector that maximizes the sum of squared cosines between itself and the G eigenvectors to be compared. If a_{gk} is the kth eigenvector for the gth sample covariance matrix, $g = 1, 2, \ldots, G$, and a_{0k} is the predetermined or typical vector, then distances

$$_k\delta_g^2 = \min[(a_{gk} - a_{0k})'(a_{gk} - a_{0k}), (a_{gk} + a_{0k})'(a_{gk} + a_{0k})]$$

are calculated. If the sample covariance matrices are drawn from the same population, then $_k\delta_g^2$ has an approximate gamma distribution, so Keramidas et al. (1987) suggest constructing gamma Q-Q plots to detect differences from this null situation. Simulations are given for both the null and non-null cases. Such plots are likely to be more useful when G is large than when there is only a handful of covariance matrices to be compared.

Keramidas et al. (1987) extend their idea to compare subspaces spanned by two or more eigenvectors. For two subspaces, their overall measure of similarity, which reduces to $_k\delta_g^2$ when single eigenvectors are compared, is the sum of the square roots $\nu_k^{1/2}$ of eigenvalues of $A_{1q}' A_{2q} A_{2q}' A_{1q}$. Recall that Krzanowski (1979b) uses the sum of these eigenvalues, not their square roots as his measure of overall similarity. Keramidas et al. (1987) stress that individual eigenvectors or subspaces should only be compared when their eigenvalues are well-separated from adjacent eigenvalues so that the eigenvectors or subspaces are well-defined.

Ten Berge and Kiers (1996) take a different and more complex view of common principal components than Flury (1988) or Krzanowski (1979b). They refer to *generalizations* of PCA to G (≥ 2) groups of individuals, with different generalizations being appropriate depending on what is taken as the defining property of PCA. They give three different defining criteria and draw distinctions between the properties of principal components found by each. Although the different criteria lead to a number of different generalizations, it is arguable just how great a distinction should be drawn between the three ungeneralized analyses (see Cadima and Jolliffe (1997); ten Berge and Kiers (1997)).

The first property considered by ten Berge and Kiers (1996) corresponds to Property A1 of Section 2.1, in which $\text{tr}(\mathbf{B}'\boldsymbol{\Sigma}\mathbf{B})$ is minimized. For G groups of individuals treated separately this leads to minimization of $\sum_{g=1}^{G} \text{tr}(\mathbf{B}'_g\boldsymbol{\Sigma}_g\mathbf{B}_g)$, but taking $\mathbf{B}_g$ the same for each group gives *simultaneous components* that minimize

$$\sum_{g=1}^{G} \text{tr}(\mathbf{B}'\boldsymbol{\Sigma}_g\mathbf{B}) = \text{tr}[\mathbf{B}'(\sum_{g=1}^{G}\boldsymbol{\Sigma}_g)\mathbf{B}]$$

$$= G \ \text{tr}(\mathbf{B}'\bar{\boldsymbol{\Sigma}}\mathbf{B}),$$

where $\bar{\boldsymbol{\Sigma}}$ is the average of $\boldsymbol{\Sigma}_1, \boldsymbol{\Sigma}_2, \ldots, \boldsymbol{\Sigma}_G$.

Ten Berge and Kiers' (1996) second property is a sample version of Property A5 in Section 2.1. They express this property as minimizing $\|\mathbf{X} - \mathbf{XBC}'\|^2$. For G groups treated separately, the quantity

$$\sum_{g=1}^{G} \|\mathbf{X}_g - \mathbf{X}_g\mathbf{B}_g\mathbf{C}'_g\|^2 \tag{13.5.3}$$

is minimized. Ten Berge and Kiers (1996) distinguish three different ways of adapting this formulation to find *simultaneous* components.

- Minimize $\sum_{g=1}^{G} \|\mathbf{X}_g - \mathbf{X}_g\mathbf{BC}'_g\|^2$.

- Minimize $\sum_{g=1}^{G} \|\mathbf{X}_g - \mathbf{X}_g\mathbf{B}_g\mathbf{C}'\|^2$.

- Minimize (13.5.3) subject to $\boldsymbol{\Sigma}_g\mathbf{B}_g = \mathbf{SD}_g$, where $\mathbf{D}_g$ is diagonal and $\mathbf{S}$ is a 'common component structure.'

The third optimality criterion considered by Ten Berge and Kiers (1996) is that noted at the end of Section 2.1, and expressed by Rao (1964) as minimizing $\|\boldsymbol{\Sigma} - \boldsymbol{\Sigma}\mathbf{B}(\mathbf{B}'\boldsymbol{\Sigma}\mathbf{B})^{-1}\mathbf{B}'\boldsymbol{\Sigma}\|$. Ten Berge and Kiers (1996) write this as minimizing $\|\boldsymbol{\Sigma} - \mathbf{FF}'\|^2$, which extends to G groups by minimizing $\sum_{g=1}^{G} \|\boldsymbol{\Sigma}_g - \mathbf{F}_g\mathbf{F}'_g\|^2$. This can then be modified to give simultaneous components by minimizing $\sum_{g=1}^{G} \|\boldsymbol{\Sigma}_g - \mathbf{FF}'\|^2$. They show that this criterion and the criterion based on Property A1 are both equivalent to the second of their generalizations derived from Property A5.

Ten Berge and Kiers (1996) compare properties of the three generalizations of Property A5, but do not reach any firm conclusions as to which is preferred. They are, however, somewhat dismissive of Flury's (1988) approach on the grounds that it has at its heart the simultaneous diagonalization of G covariance matrices and 'it is by no means granted that components which diagonalize the covariance or correlation matrices reflect the most important sources of variation in the data.'

Muller (1982) suggests that canonical correlation analysis of PCs (see Section 9.3) provides a way of comparing the PCs based on two sets of variables, and cites some earlier references. When the two sets of variables are, in fact, the same variables measured for two groups of observations, Muller's analysis is equivalent to that of Krzanowski (1979b); the latter paper notes the links between canonical correlation analysis and its own technique.

In a series of five technical reports, Preisendorfer and Mobley (1982) examine various ways of comparing data sets measured on the same variables at different times, and part of their work involves comparison of PCs from different sets (see, in particular, their third report, which concentrates on comparing the singular value decompositions (SVDs, Section 3.5) of two data matrices $\mathbf{X}_1, \mathbf{X}_2$). Suppose that the SVDs are written

$$\mathbf{X}_1 = \mathbf{U}_1 \mathbf{L}_1 \mathbf{A}_1'$$
$$\mathbf{X}_2 = \mathbf{U}_2 \mathbf{L}_2 \mathbf{A}_2'.$$

Then Preisendorfer and Mobley (1982) define a number of statistics that compare $\mathbf{U}_1$ with $\mathbf{U}_2$, $\mathbf{A}_1$ with $\mathbf{A}_2$, $\mathbf{L}_1$ with $\mathbf{L}_2$ or compare any two of the three factors in the SVD for $\mathbf{X}_1$ with the corresponding factors in the SVD for $\mathbf{X}_2$. All of these comparisons are relevant to comparing PCs, since $\mathbf{A}$ contains the coefficients of the PCs, $\mathbf{L}$ provides the standard deviations of the PCs, and the elements of $\mathbf{U}$ are proportional to the PC scores (see Section 3.5). The 'significance' of an observed value of any one of Preisendorfer and Mobley's statistics is assessed by comparing the value to a 'reference distribution', which is obtained by simulation. Preisendorfer and Mobley's (1982) research is in the context of atmospheric science. A more recent application in this area is that of Sengupta and Boyle (1998), who illustrate the use of Flury's (1988) common principal component model to compare different members of an ensemble of forecasts from a general circulation model (GCM) and to compare outputs from different GCMs. Applications in other fields of the common PC model and its variants can be found in Flury (1988, 1997).

When the same variables are measured on the same n individuals in the different data sets, it may be of interest to compare the configurations of the points defined by the n individuals in the subspaces of the first few PCs in each data set. In this case, Procrustes analysis (or generalized Procrustes analysis) provides one possible way of doing this for two (more than two) data sets (see Krzanowski and Marriott (1994, Chapter 5)). The technique in general involves the SVD of the product of one data matrix and the transpose of the other, and because of this Davison (1983, Chapter 8) links it to PCA.

13.6 Principal Component Analysis in the Presence of Missing Data

In all the examples given in this text, the data sets are complete. However, it is not uncommon, especially for large data sets, for some of the values of some of the variables to be missing. The most usual way of dealing with such a situation is to delete, entirely, any observation for which at least one of the variables has a missing value. This is satisfactory if missing values are few, but clearly wasteful of information if a high proportion of observations have missing values for just one or two variables. To meet this problem, a number of alternatives have been suggested.

The first step in a PCA is usually to compute the covariance or correlation matrix, so interest often centres on estimating these matrices in the presence of missing data. There are a number of what Little and Rubin (1987, Chapter 3) call 'quick' methods. One option is to compute the (j,k)th correlation or covariance element-wise, using all observations for which the values of both x_j and x_k are available. Unfortunately, this leads to covariance or correlation matrices that are not necessarily positive semidefinite. Beale and Little (1975) note a modification of this option. When computing the summation $\sum_i (x_{ij} - \bar{x}_j)(x_{ik} - \bar{x}_k)$ in the covariance or correlation matrix, $\bar{x}_j$, $\bar{x}_k$ are calculated from all available values of x_j, x_k, respectively, instead of only for observations for which both x_j and x_k have values present, They state that, at least in the regression context, the results can be unsatisfactory. However, Mehrota (1995), in discussing robust estimation of covariance matrices (see Section 10.4), argues that the problem of a possible lack of positive semi-definiteness is less important than making efficient use of as many data as possible. He therefore advocates element-wise estimation of the variances and covariances in a covariance matrix, with possible adjustment if positive semi-definiteness is lost.

Another quick method is to replace missing values for variable x_j by the mean value $\bar{x}_j$, calculated from the observations for which the value of x_j is available. This is a simple way of 'imputing' rather than ignoring missing values. A more sophisticated method of imputation is to use regression of the missing variables on the available variables case-by-case. An extension to the idea of imputing missing values is *multiple imputation*. Each missing value is replaced by a value drawn from a probability distribution, and this procedure is repeated M times (Little and Rubin, 1987, Section 12.4; Schafer, 1997, Section 4.3). The analysis, in our case PCA, is then done M times, corresponding to each of the M different sets of imputed values. The variability in the results of the analyses gives an indication of the uncertainty associated with the presence of missing values.

A different class of procedures is based on maximum likelihood estimation (Little and Rubin, 1987, Section 8.2). The well-known EM algorithm (Dempster et al., 1977) can easily cope with maximum likelihood estimation of the mean and covariance matrix in the presence of missing values, under the assumption of multivariate normality. Little and Rubin (1987, Section 8.2) describe three versions of the EM algorithm for solving this problem; a number of other authors, for example, Anderson (1957), De Ligny et al. (1981), tackled the same problem earlier by less efficient means.

The multivariate normal assumption is a restrictive one, and Little (1988) relaxes it by adapting the EM algorithm to find MLEs when the data are from a multivariate t-distribution or from a mixture of two multivariate normals with different covariance matrices. He calls these 'robust' methods for dealing with missing data because they assume longer-tailed distributions than multivariate normal. Little (1988) conducts a simulation study, the results of which demonstrate that his robust MLEs cope well with missing data, compared to other methods discussed earlier in this section. However, the simulation study is limited to multivariate normal data, and to data from distributions that are similar to those assumed by the robust MLEs. It is not clear that the good performance of the robust MLEs would be repeated for other distributions. Little and Rubin (1987, Section 8.3) also extend their multivariate normal procedures to deal with covariance matrices on which some structure is imposed. Whilst this may be appropriate for factor analysis it is less relevant for PCA.

Another adaptation of the EM algorithm for estimation of covariance matrices, the *regularized EM algorithm*, is given by Schneider (2001). It is particularly useful when the number of variables exceeds the number of observations. Schneider (2001) adds a diagonal matrix to the current estimate of the covariance matrix before inverting the matrix, a similar idea to that used in ridge regression.

Tipping and Bishop (1999a) take the idea of maximum likelihood estimation using the EM algorithm further. They suggest an iterative algorithm in which their EM procedure for estimating the probabilistic PCA model (Section 3.9) is combined with Little and Rubin's (1987) methodology for estimating the parameters of a multivariate normal distribution in the presence of missing data. The PCs are estimated directly, rather than by going through the intermediate step of estimating the covariance or correlation matrix. An example in which data are randomly deleted from a data set is used by Tipping and Bishop (1999a) to illustrate their procedure.

In the context of satellite-derived sea surface temperature measurements with missing data caused by cloud cover, Houseago-Stokes and Challenor (2001) compare Tipping and Bishop's procedure with a standard interpolation technique followed by PCA on the interpolated data. The two procedures give similar results but the new method is computationally much more efficent. This is partly due to the fact that only the first few PCs are found and that they are calculated directly, without the intermediate step of estimating the covariance matrix. Houseago-Stokes and Challenor note that the quality of interpolated data using probabilistic PCA depends on the number of components q in the model. In the absence of prior knowledge about q, there is, at present, no procedure for choosing its value without repeating the analysis for a range of values.

Most published work, including Little and Rubin (1987), does not explicitly deal with PCA, but with the estimation of covariance matrices in general. Tipping and Bishop (1999a) is one of relatively few papers that focus specifically on PCA when discussing missing data. Another is Wiberg (1976). His approach is via the singular value decomposition (SVD), which gives a least squares approximation of rank m to the data matrix $\mathbf{X}$. In other words, the approximation $_m\tilde{x}_{ij}$ minimizes

$$\sum_{i=1}^{n}\sum_{j=1}^{p}(_mx_{ij} - x_{ij})^2,$$

where $_mx_{ij}$ is any rank m approximation to x_{ij} (see Section 3.5). Principal components can be computed from the SVD (see Section 3.5 and Appendix A1). With missing data, Wiberg (1976) suggests minimizing the same quantity, but with the summation only over values of (i, j) for which x_{ij} is not missing; PCs can then be estimated from the modified SVD. The same idea is implicitly suggested by Gabriel and Zamir (1979). Wiberg (1976) reports that for simulated multivariate normal data his method is slightly worse than the method based on maximum likelihood estimation. However, his method has the virtue that it can be used regardless of whether or not the data come from a multivariate normal distribution.

For the specialized use of PCA in analysing residuals from an additive model for data from designed experiments (see Section 13.4), Freeman (1975) shows that incomplete data can be easily handled, although modifications to procedures for deciding the rank of the model are needed. Michailidis and de Leeuw (1998) note three ways of dealing with missing data in non-linear multivariate analysis, including non-linear PCA (Section 14.1).

A special type of 'missing' data occurs when observations or variables correspond to different times or different spatial locations, but with irregular spacing between them. In the common atmospheric science set-up, where variables correspond to spatial locations, Karl et al. (1982) examine differences between PCAs when locations are on a regularly spaced grid, and when they are irregularly spaced. Unsurprisingly, for the irregular data the locations in areas with the highest density of measurements tend to increase their loadings on the leading PCs, compared to the regularly spaced data. This is because of the larger correlations observed in the high-density regions. Kaplan et al. (2001) discuss methodology based on PCA for interpolating spatial fields (see Section 12.4.4). Such interpolation is, in effect, imputing missing data.

Another special type of data in which some values are missing occurs when candidates choose to take a subset of p' out of p examinations, with different candidates choosing different subsets. Scores on examinations not taken by a candidate are therefore 'missing.' Shibayama (1990) devises a method for producing a linear combination of the examination scores that represents the overall performance of each candidate. When $p' = p$ the method is equivalent to PCA.

Anderson et al. (1983) report a method that they attribute to Dear (1959), which is not for dealing with missing values in a PCA, but which uses PCA to impute missing data in a more general context. The idea seems to be to first substitute zeros for any missing cells in the data matrix, and then find the SVD of this matrix. Finally, the leading term in the SVD, corresponding to the first PC, is used to approximate the missing values. If the data matrix is column-centred, this is a variation on using means of variables in place of missing values. Here there is the extra SVD step that adjusts the mean values using information from other entries in the data matrix.

Finally, note that there is a similarity of purpose in robust estimation of PCs (see Section 10.4) to that present in handling missing data. In both cases we identify particular observations which we cannot use in unadjusted form, either because they are suspiciously extreme (in robust estimation), or because they are not given at all (missing values). To completely ignore such observations may throw away valuable information, so we attempt to estimate 'correct' values for the observations in question. Similar techniques may be relevant in each case. For example, we noted above the possibility of imputing missing values for a particular observation by regressing the missing variables on the variables present for that observation, an idea that dates back at least to Beale and Little (1975), Frane (1976) and Gleason and Staelin (1975) (see Jackson (1991, Section 14.1.5)). A similar idea, namely robust regression of the variables on each other, is included in Devlin et al.'s (1981) study of robust estimation of PCs (see Section 10.4).

13.7   PCA in Statistical Process Control

The topic of this section, finding outliers, is closely linked to that of Section 10.1, and many of the techniques used are based on those described in that section. However, the literature on using PCA in multivariate statistical process control (SPC) is sufficiently extensive to warrant its own section. In various manufacturing processes improved technology means that greater numbers of variables are now measured in order to monitor whether or not a process is 'in control.' It has therefore become increasingly relevant to use multivariate techniques for control purposes, rather than simply to monitor each variable separately.

The main ways in which PCA is used in this context are (Martin et al., 1999):

- One- or two-dimensional plots of PC scores. It was noted in Section 10.1 that both the first few and the last few PCs may be useful for detecting (different types of) outliers, and plots of both are used in process control. In the published discussion of Roes and Does (1995), Sullivan et al. (1995) argue that the last few PCs are perhaps more useful in SPC than the first few, but in their reply to the discussion Roes and Does disagree. If p is not too large, such arguments can be overcome by using a scatterplot matrix to display all two-dimensional plots of PC scores simultaneously. Plots can be enhanced by including equal-probability contours, assuming approximate multivariate normality, corresponding to warning and action limits for those points that fall outside them (Jackson, 1991, Section 1.7; Martin et al., 1999).

- Hotelling's T^2. It was seen in Section 10.1 that this is a special case for $q = p$ of the statistic d_{2i}^2 in equation (10.1.2). If multivariate normality is assumed, the distribution of T^2 is known, and control limits can be set based on that distribution (Jackson, 1991, Section 1.7).

- The *squared prediction error* (SPE). This is none other than the statistic d_{1i}^2 in equation (10.1.1). It was proposed by Jackson and Mudholkar (1979), who constructed control limits based on an approximation to its distribution. They prefer d_{1i}^2 to d_{2i}^2 for computational reasons and because of its intuitive appeal as a sum of squared residuals from the $(p - q)$-dimensional space defined by the first $(p - q)$ PCs. However, Jackson and Hearne (1979) indicate that the complement of d_{2i}^2, in which the sum of squares of the first few rather than the last few renormalized PCs is calculated, may be useful in process control when the objective is to look for *groups* of 'out-of-control' or outlying observations, rather than single outliers. Their basic statistic is decomposed to give separate information about variation *within* the sample (group) of potentially outlying observations, and about the difference between the sample mean and some known standard value. In addition, they propose an alternative statistic based on absolute, rather than squared, values of PCs. Jackson and Mudholkar (1979) also extend their proposed control procedure, based on d_{1i}^2, to the multiple-outlier case, and Jackson (1991, Figure 6.2) gives a sequence of significance tests for examining subgroups of observations in which each test is based on PCs in some way.

Eggett and Pulsipher (1989) compare T^2, SPE, and the complement of d_{2i}^2 suggested by Jackson and Hearne (1979), in a simulation study and find the third of these statistics to be inferior to the other two. On the basis of their simulations, they recommend Hotelling's T^2 for large samples, with SPE or univariate control charts preferred for small samples. They also discussed the possibility of constructing CUSUM charts based on the three statistics.

The control limits described so far are all based on the assumption of approximate multivariate normality. Martin and Morris (1996) introduce a non-parametric procedure that provides warning and action contours on plots of PCs. These contours can be very different from the normal-based ellipses. The idea of the procedure is to generate bootstrap samples from the data set and from each of these calculate the value of a (possibly vector-valued) statistic of interest. A smooth approximation to the probability density of this statistic is then constructed using kernel density estimation, and the required contours are derived from this distribution. Coleman (1985) suggests that when using PCs in quality control, the PCs should be estimated robustly (see Section 10.4). Sullivan et al. (1995) do this by omitting some probable outliers, identified from an initial scan of the data, before carrying out a PCA.

When a variable is used to monitor a process over time, its successive values are likely to be correlated unless the spacing between observations is large. One possibility for taking into account this autocorrelation is to plot an exponentially weighted moving average of the observed values. Wold (1994) suggests that similar ideas should be used when the monitoring variables are PC scores, and he describes an algorithm for implementing 'exponentially weighted moving principal components analysis.'

Data often arise in SPC for which, as well as different variables and different times of measurement, there is a third 'mode,' namely different batches. So-called multiway, or three-mode, PCA can then be used (see Section 14.5 and Nomikos and MacGregor (1995)). Grimshaw et al. (1998) note the possible use of multiway PCA simultaneously on both the variables monitoring the process and the variables measuring inputs or initial conditions, though they prefer a regression-based approach involving modifications of Hotelling's T^2 and the SPE statistic.

Boyles (1996) addresses the situation in which the number of variables exceeds the number of observations. The sample covariance matrix $\mathbf{S}$ is then singular and Hotelling's T^2 cannot be calculated. One possibility is to replace $\mathbf{S}^{-1}$ by $\sum_{k=1}^{r} l_k^{-1} \mathbf{a}_k \mathbf{a}_k'$ for $r < n$, based on the first r terms in the spectral decomposition of $\mathbf{S}$ (the sample version of Property A3 in Section 2.1). However, the data of interest to Boyles (1996) have variables measured at points of a regular lattice on the manufactured product. This structure implies that a simple pattern exists in the population covariance matrix $\mathbf{\Sigma}$. Using knowledge of this pattern, a positive definite estimate of $\mathbf{\Sigma}$ can be calculated and used in T^2 in place of $\mathbf{S}$. Boyles finds appropriate estimates for three different regular lattices.

Lane et al. (2001) consider the case where a several products or processes are monitored simultaneously. They apply Flury's common PC subspace model (Section 13.5) to this situation. McCabe (1986) suggests the use of principal variables (see Section 6.3) to replace principal components in quality control.

Apley and Shi (2001) assume that a vector of p measured features from a process or product can be modelled as in the probabilistic PCA model of Tipping and Bishop (1999a), described in Section 3.9. The vector therefore has covariance matrix $\mathbf{BB'} + \sigma^2\mathbf{I}_p$, where in the present context the q columns of $\mathbf{B}$ are taken to represent the effects of q uncorrelated faults on the p measurements. The vectors of principal component coefficients (loadings) that constitute the columns of $\mathbf{B}$ thus provide information about the nature of the faults. To allow for the fact that the faults may not be uncorrelated, Apley and Shi suggest that interpreting the faults may be easier if the principal component loadings are rotated towards simple structure (see Section 11.1).

13.8 Some Other Types of Data

In this section we discuss briefly some additional types of data with special features.

Vector-valued or Directional Data—Complex PCA

Section 12.2.3 discussed a special type of complex PCA in which the series $\mathbf{x}_t + i\mathbf{x}_t^H$ is analysed, where $\mathbf{x}_t$ is a p-variate time series, $\mathbf{x}_t^H$ is its Hilbert transform and $i = \sqrt{-1}$. More generally, if $\mathbf{x}_t$, $\mathbf{y}_t$ are two real-valued p-variate series, PCA can be done on the complex series $\mathbf{x}_t + i\mathbf{y}_t$, and this general form of complex PCA is relevant not just in a time series context, but whenever two variables are recorded in each cell of the $(n \times p)$ data matrix. This is then a special case of three-mode data (Section 14.5) for which the index for the third mode takes only two values.

One situation in which such data arise is for landmark data (see Section 13.2). Another is when the data consist of vectors in two dimensions, as with directional data. A specific example is the measurement of wind, which involves both strength and direction, and can be expressed as a vector whose elements are the zonal (x or easterly) and meridional (y or northerly) components.

Suppose that $\mathbf{X}$ is an $(n \times p)$ data matrix whose (h,j)th element is $x_{hj} + iy_{hj}$. A complex covariance matrix is defined as

$$\mathbf{S} = \frac{1}{n-1}\mathbf{X}^\dagger\mathbf{X},$$

where $\mathbf{X}^\dagger$ is the conjugate transpose of $\mathbf{X}$. Complex PCA is then done by finding the eigenvalues and eigenvectors of $\mathbf{S}$. Because $\mathbf{S}$ is Hermitian the eigenvalues are real and can still be interpreted as proportions of total variance accounted for by each complex PC. However, the eigenvectors are complex, and the PC scores, which are obtained as in the real case by multiplying the data matrix by the matrix of eigenvectors, are also complex.

Hence both the PC scores and their vectors of loadings have real and imaginary parts that can be examined separately. Alternatively, they can be expressed in polar coordinates, and displayed as arrows whose lengths and directions are defined by the polar coordinates. Such displays for loadings are particularly useful when the variables correspond to spatial locations, as in the example of wind measurements noted above, so that a map of the arrows can be constructed for the eigenvectors. For such data, the 'observations' usually correspond to different times, and a different kind of plot is needed for the PC scores. For example, Klink and Willmott (1989) use two-dimensional contour plots in which the horizontal axis corresponds to time (different observations), the vertical axis gives the angular coordinate of the complex score, and contours represent the amplitudes of the scores.

The use of complex PCA for wind data dates back to at least Walton and Hardy (1978). An example is given by Klink and Willmott (1989) in which two versions of complex PCA are compared. In one, the real and imaginary parts of the complex data are zonal (west-east) and meridional (south-north) wind velocity components, while wind speed is ignored in the other with real and imaginary parts corresponding to sines and cosines of the wind direction. A third analysis performs separate PCAs on the zonal and meridional wind components, and then recombines the results of these scalar analyses into vector form. Some similarities are found between the results of the three analyses, but there are non-trivial differences. Klink and Willmott (1989) suggest that the velocity-based complex PCA is most appropriate for their data. Von Storch and Zwiers (1999, Section 16.3.3) have an example in which ocean currents, as well as wind stresses, are considered.

One complication in complex PCA is that the resulting complex eigenvectors can each be arbitrarily rotated in the complex plane. This is different in nature from rotation of (real) PCs, as described in Section 11.1, because the variance explained by each component is unchanged by rotation. Klink and Willmott (1989) discuss how to produce solutions whose mean direction is not arbitrary, so as to aid interpretation.

Preisendorfer and Mobley (1988, Section 2c) discuss the theory of complex-valued PCA in some detail, and extend the ideas to quaternion-valued and matrix-valued data sets. In their Section 4e they suggest that it may sometimes be appropriate with vector-valued data to take Fourier transforms of each element in the vector, and conduct PCA in the frequency domain. There are, in any case, connections between complex PCA and PCA in the frequency domain (see Section 12.4.1 and Brillinger (1981, Chapter 9)).

PCA for Data Given as Intervals

Sometimes, because the values of the measured variables are imprecise or because of other reasons, an interval of values is given for a variable rather than a single number. An element of the $(n \times p)$ data matrix is then an interval $(\underline{x_{ij}}, \overline{x_{ij}})$ instead of the single value x_{ij}. Chouakria et al. (2000)

discuss two adaptations of PCA for such data. In the first, called the VER-
TICES method, the ith row of the $(n \times p)$ data matrix is replaced by the
2^p distinct rows whose elements have either $\underline{x_{ij}}$ or $\overline{x_{ij}}$ in their jth column.
A PCA is then done on the resulting $(n2^p \times p)$ matrix. The value or score
of a PC from this analysis can be calculated for each of the $n2^p$ rows of the
new data matrix. For the ith observation there are 2^p such scores and an
interval can be constructed for the observation, bounded by the smallest
and largest of these scores. In plotting the observations, either with respect
to the original variables or with respect to PCs, each observation is represented by a rectangle or hyperrectangle in two or higher-dimensional space.
The boundaries of the (hyper)rectangle are determined by the intervals for
the variables or PC scores. Chouakria et al. (2000) examine a number of
indices measuring the quality of representation of an interval data set by
a small number of 'interval PCs' and the contributions of each observation
to individual PCs.

For large values of p, the VERTICES method produces very large matrices. As an alternative, Chouakria et al. suggest the CENTERS procedure,
in which a PCA is done on the $(n \times p)$ matrix whose (i, j)th element is
$(\underline{x_{ij}} + \overline{x_{ij}})/2$. The immediate results give a single score for each observation
on each PC, but Chouakria and coworkers use the intervals of possible val-
ues for the variables to construct intervals for the PC scores. This is done
by finding the combinations of allowable values for the variables, which,
when inserted in the expression for a PC in terms of the variables, give the
maximum and minimum scores for the PC. An example is given to compare
the VERTICES and CENTERS approaches.

Ichino and Yaguchi (1994) describe a generalization of PCA that can be
used on a wide variety of data types, including discrete variables in which a
measurement is a subset of more than one of the possible values for a variable; continuous variables recorded as intervals are also included. To carry
out PCA, the measurement on each variable is converted to a single value.
This is done by first calculating a 'distance' between any two observations
on each variable, constructed from a formula that involves the union and
intersection of the values of the variable taken by the two observations.
From these distances a 'reference event' is found, defined as the observation whose sum of distances from all other observations is minimized, where
distance here refers to the sum of 'distances' for each of the p variables.
The coordinate of each observation for a particular variable is then taken
as the distance on that variable from the reference event, with a suitably
assigned sign. The coordinates of the n observations on the p variables thus
defined form a data set, which is then subjected to PCA.

Species Abundance Data

These data are common in ecology—an example was given in Section 5.4.1.
When the study area has diverse habitats and many species are included,

there may be a large number of zeros in the data. If two variables x_j and x_k simultaneously record zero for a non-trivial number of sites, the calculation of covariance or correlation between this pair of variables is likely to be distorted. Legendre and Legendre (1983, p. 285) argue that data are better analysed by nonmetric multidimensional scaling (Cox and Cox, 2001) or with correspondence analysis (as in Section 5.4.1), rather than by PCA, when there are many such 'double zeros' present. Even when such zeros are not a problem, species abundance data often have highly skewed distributions and a transformation; for example, taking logarithms, may be advisable before PCA is contemplated.

Another unique aspect of species abundance data is an interest in the diversity of species at the various sites. It has been argued that to examine diversity, it is more appropriate to use uncentred than column-centred PCA. This is discussed further in Section 14.2.3, together with doubly centred PCA which has also found applications to species abundance data.

Large Data Sets

The problems of large data sets are different depending on whether the number of observations n or the number of variables p is large, with the latter typically causing greater difficulties than the former. With large n there may be problems in viewing graphs because of superimposed observations, but it is the size of the covariance or correlation matrix that usually determines computational limitations. However, if $p > n$ it should be remembered (Property G4 of Section 3.2) that the eigenvectors of $\mathbf{X'X}$ corresponding to non-zero eigenvalues can be found from those of the smaller matrix $\mathbf{XX'}$.

For very large values of p, Preisendorfer and Mobley (1988, Chapter 11) suggest splitting the variables into subsets of manageable size, performing PCA on each subset, and then using the separate eigenanalyses to approximate the eigenstructure of the original large data matrix. Developments in computer architecture may soon allow very large problems to be tackled much faster using neural network algorithms for PCA (see Appendix A1 and Diamantaras and Kung (1996, Chapter 8)).

14
Generalizations and Adaptations of Principal Component Analysis

The basic technique of PCA has been generalized or adapted in many ways, and some have already been discussed, in particular in Chapter 13 where adaptations for special types of data were described. This final chapter discusses a number of additional generalizations and modifications; for several of them the discussion is very brief in comparison to the large amount of material that has appeared in the literature.

Sections 14.1 and 14.2 present some definitions of 'non-linear PCA' and 'generalized PCA,' respectively. In both cases there are connections with correspondence analysis, which was discussed at somewhat greater length in Section 13.1. Non-linear extensions of PCA (Section 14.1) include the Gifi approach, principal curves, and some types of neural network, while the generalizations of Section 14.2 cover many varieties of weights, metrics, transformations and centerings.

Section 14.3 describes modifications of PCA that may be useful when secondary or 'instrumental' variables are present, and in Section 14.4 some possible alternatives to PCA for data that are are non-normal are discussed. These include independent component analysis (ICA).

Section 14.5 introduces the ideas of three-mode and multiway PCA. These analyses are appropriate when the data matrix, as well as having two dimensions corresponding to individuals and variables, respectively, has one or more extra dimensions corresponding, for example, to time.

The penultimate miscellaneous section (14.6) collects together some ideas from neural networks and goodness-of-fit, and presents some other modifications of PCA. The chapter ends with a few concluding remarks.

14.1 Non-Linear Extensions of Principal Component Analysis

One way of introducing non-linearity into PCA is what Gnanadesikan (1977) calls 'generalized PCA.' This extends the vector of p variables $\mathbf{x}$ to include *functions* of the elements of $\mathbf{x}$. For example, if $p = 2$, so $\mathbf{x}' = (x_1, x_2)$, we could consider linear functions of $\mathbf{x}'_+ = (x_1, x_2, x_1^2, x_2^2, x_1 x_2)$ that have maximum variance, rather than restricting attention to linear functions of $\mathbf{x}'$. In theory, any functions $g_1(x_1, x_2, \ldots, x_p), g_2(x_1, x_2, \ldots, x_p), \ldots, g_h(x_1, x_2, \ldots, x_p)$ of $x_1, x_2, \ldots, x_p$ could be added to the original vector $\mathbf{x}$, in order to construct an extended vector $\mathbf{x}_+$ whose PCs are then found. In practice, however, Gnanadesikan (1977) concentrates on quadratic functions, so that the analysis is a procedure for finding quadratic rather than linear functions of $\mathbf{x}$ that maximize variance.

An obvious alternative to Gnanadesikan's (1977) proposal is to *replace* $\mathbf{x}$ by a function of $\mathbf{x}$, rather than *add* to $\mathbf{x}$ as in Gnanadesikan's analysis. Transforming $\mathbf{x}$ in this way might be appropriate, for example, if we are interested in products of powers of the elements of $\mathbf{x}$. In this case, taking logarithms of the elements and doing a PCA on the transformed data provides a suitable analysis. Another possible use of transforming to non-linear PCs is to detect near-constant, non-linear relationships between the variables. If an appropriate transformation is made, such relationships will be detected by the last few PCs of the transformed data. Transforming the data is suggested before doing a PCA for allometric data (see Section 13.2) and for compositional data (Section 13.3). Kazmierczak (1985) also advocates logarithmic transformation followed by double-centering (see Section 14.2.3) for data in which it is important for a PCA to be invariant to changes in the units of measurement and to the choice of which measurement is used as a 'reference.' However, as noted in the introduction to Chapter 4, transformation of variables should only be undertaken, in general, after careful thought about whether it is appropriate for the data set at hand.

14.1.1 Non-Linear Multivariate Data Analysis—Gifi and Related Approaches

The most extensively developed form of non-linear multivariate data analysis in general, and non-linear PCA in particular, is probably the Gifi (1990) approach. 'Albert Gifi' is the *nom de plume* of the members of the Department of Data Theory at the University of Leiden. As well as the 1990 book, the Gifi contributors have published widely on their system of multivariate analysis since the 1970s, mostly under their own names. Much of it is not easy reading. Here we attempt only to outline the approach. A rather longer, accessible, description is provided by Krzanowski and Marriott (1994, Chapter 8), and Michailidis and de Leeuw (1998) give a review.

Gifi's (1990) form of non-linear PCA is based on a generalization of the result that if, for an $(n \times p)$ data matrix $\mathbf{X}$, we minimize

$$\text{tr}\{(\mathbf{X} - \mathbf{YB}')'(\mathbf{X} - \mathbf{YB}')\}, \qquad (14.1.1)$$

with respect to the $(n \times q)$ matrix $\mathbf{Y}$ whose columns are linear functions of columns of $\mathbf{X}$, and with respect to the $(q \times p)$ matrix $\mathbf{B}'$ where the columns of $\mathbf{B}$ are orthogonal, then the optimal $\mathbf{Y}$ consists of the values (scores) of the first q PCs for the n observations, and the optimal matrix $\mathbf{B}$ consists of the coefficients of the first q PCs. The criterion (14.1.1) corresponds to that used in the sample version of Property A5 (see Section 2.1), and can be rewritten as

$$\text{tr}\left\{ \sum_{j=1}^{p} (\mathbf{x}_j - \mathbf{Yb}_j)'(\mathbf{x}_j - \mathbf{Yb}_j) \right\}, \qquad (14.1.2)$$

where $\mathbf{x}_j$, $\mathbf{b}_j$ are the jth columns of $\mathbf{X}$, $\mathbf{B}'$, respectively.

Gifi's (1990) version of non-linear PCA is designed for categorical variables so that there are no immediate values of $\mathbf{x}_j$ to insert in (14.1.2). Any variables that are continuous are first converted to categories; then values need to be derived for each category of every variable. We can express this algebraically as the process minimizing

$$\text{tr}\left\{ \sum_{j=1}^{p} (\mathbf{G}_j \mathbf{c}_j - \mathbf{Yb}_j)'(\mathbf{G}_j \mathbf{c}_j - \mathbf{Yb}_j) \right\}, \qquad (14.1.3)$$

where $\mathbf{G}_j$ is an $(n \times g_j)$ indicator matrix whose (h, i)th value is unity if the hth observation is in the ith category of the jth variable and is zero otherwise, and $\mathbf{c}_j$ is a vector of length g_j containing the values assigned to the g_j categories of the jth variable. The minimization takes place with respect to both $\mathbf{c}_j$ and $\mathbf{Yb}_j$, so that the difference from (linear) PCA is that there is optimization over the values of the variables in addition to optimization of the scores on the q components. The solution is found by an alternating least squares (ALS) algorithm which alternately fixes the $\mathbf{c}_j$ and minimizes with respect to the $\mathbf{Yb}_j$, then fixes the $\mathbf{Yb}_j$ at the new values and minimizes with respect to the $\mathbf{c}_j$, fixes the $\mathbf{c}_j$ at the new values and minimizes over $\mathbf{Yb}_j$, and so on until convergence. This is implemented by the Gifi-written PRINCALS computer program (Gifi, 1990, Section 4.6) which is incorporated in the SPSS software.

A version of non-linear PCA also appears in another guise within the Gifi system. For two categorical variables we have a contingency table that can be analysed by correspondence analysis (Section 13.1). For more than two categorical variables there is an extension of correspondence analysis, called multiple correspondence analysis (see Section 13.1 and Greenacre,

1984, Chapter 5). This technique has at its core the idea of assigning scores to each category of each variable. It can be shown that if a PCA is done on the correlation matrix of these scores, the first PC is equivalent to the first non-trivial multiple correspondence analysis dimension (Bekker and de Leeuw, 1988). These authors give further discussion of the relationships between the different varieties of non-linear PCA.

Mori et al. (1998) combine the Gifi approach with the procedure described by Tanaka and Mori (1997) for selecting a subset of variables (see Section 6.3). Using the optimal values of the variables c_j found by minimizing (14.1.3), variables are selected in the same way as in Tanaka and Mori (1997). The results can be thought of as either an extension of Tanaka and Mori's method to qualitative data, or as a simplication of Gifi's non-linear PCA by using only a subset of variables.

An approach that overlaps with—but differs from—the main Gifi ideas underlying non-linear PCA is described by Meulman (1986). Categorical data are again transformed to give optimal scores or values for each category of each variable, and simultaneously a small number of optimal dimensions is found within which to represent these scores. The 'non-linearity' of the technique becomes more obvious when a continuous variable is fitted into this framework by first dividing its range of values into a finite number of categories and then assigning a value to each category. The non-linear transformation is thus a step function. Meulman's (1986) proposal, which is known as the distance approach to nonlinear multivariate data analysis, differs from the main Gifi (1990) framework by using different optimality criteria (loss functions) instead of (14.1.3). Gifi's (1990) algorithms concentrate on the representation of the *variables* in the analysis, so that representation of the *objects* (observations) can be suboptimal. The distance approach directly approximates distances between objects. Krzanowski and Marriott (1994, Chapter 8) give a readable introduction to, and an example of, the distance approach.

An example of Gifi non-linear PCA applied in an agricultural context and involving a mixture of categorical and numerical variables is given by Kroonenberg et al. (1997). Michailidis and de Leeuw (1998) discuss various aspects of stability for Gifi-based methods, and Verboon (1993) describes a robust version of a Gifi-like procedure.

A sophisticated way of replacing the variables by functions of the variables, and hence incorporating non-linearity, is described by Besse and Ferraty (1995). It is based on an adaptation of the fixed effects model which was introduced in Section 3.9. The adaptation is that, whereas before we had $E(\mathbf{x}_i) = \mathbf{z}_i$, now $E[\mathbf{f}(\mathbf{x}_i)] = \mathbf{z}_i$, where $\mathbf{f}(\mathbf{x}_i)$ is a p-dimensional vector of functions of $\mathbf{x}_i$. As before, $\mathbf{z}_i$ lies in a q-dimensional subspace F_q, but var($\mathbf{e}_i$) is restricted to be $\sigma^2 \mathbf{I}_p$. The quantity to be minimized is similar to (3.9.1) with $\mathbf{x}_i$ replaced by $\mathbf{f}(\mathbf{x}_i)$. In the current problem it is necessary to choose q and then optimize with respect to the q-dimensional subspace F_q *and* with respect to the functions $\mathbf{f}(.)$. The functions must be restricted in some way to make the optimization problem tractable. One choice is to use step functions, which leads back towards Gifi's (1990) system of non-linear PCA. Besse and Ferraty (1995) favour an approach based on splines. They contrast their proposal, in which flexibility of the functional transformation is controlled by the choice of smoothing parameters, with earlier spline-based procedures controlled by the number and positioning of knots (see, for example, van Rijckevorsel (1988) and Winsberg (1988)). Using splines as Besse and Ferraty do is equivalent to adding a roughness penalty function to the quantity to be minimized. This is similar to Besse et al.'s (1997) approach to analysing functional data described in Section 12.3.4 using equation (12.3.6).

As with Gifi's (1990) non-linear PCA, Besse and Ferraty's (1995) proposal is implemented by means of an alternating least squares algorithm and, as in Besse and de Falgerolles (1993) for the linear case (see Section 6.1.5), bootstrapping of residuals from a q-dimensional model is used to decide on the best fit. Here, instead of simply using the bootstrap to choose q, simultaneous optimization with respect q and with respect to the smoothing parameters which determine the function $\mathbf{f}(\mathbf{x})$ is needed. At this stage it might be asked 'where is the PCA in all this?' The name 'PCA' is still appropriate because the q-dimensional subspace is determined by an optimal set of q linear functions of the vector of transformed random variables $\mathbf{f}(\mathbf{x})$, and it is these linear functions that are the non-linear PCs.

14.1.2   Additive Principal Components and Principal Curves

Fowlkes and Kettenring (1985) note that one possible objective for transforming data before performing a PCA is to find near-singularities in the transformed data. In other words, $\mathbf{x}' = (x_1, x_2, \ldots, x_p)$ is transformed to $\mathbf{f}'(\mathbf{x}) = (f_1(x_1), f_2(x_2), \ldots, f_p(x_p))$, and we are interested in finding linear functions $\mathbf{a}'\mathbf{f}(\mathbf{x})$ of $\mathbf{f}(\mathbf{x})$ for which $\mathrm{var}[\mathbf{a}'\mathbf{f}(\mathbf{x})] \approx 0$. Fowlkes and Kettenring (1985) suggest looking for a transformation that minimizes the determinant of the correlation matrix of the transformed variables. The last few PCs derived from this correlation matrix should then identify the required near-constant relationships, if any exist.

A similar idea underlies *additive principal components*, which are discussed in detail by Donnell et al. (1994). The additive principal components take the form $\sum_{j=1}^{p} \phi_j(x_j)$ instead of $\sum_{j=1}^{p} a_j x_j$ in standard PCA, and, as with Fowlkes and Kettenring (1985), interest centres on components for which $\mathrm{var}[\sum_{j=1}^{p} \phi_j(x_j)]$ is small. To define a non-linear analogue of PCA there is a choice of either an algebraic definition that minimizes variance, or a geometric definition that optimizes expected squared distance from the additive manifold $\sum_{j=1}^{p} \phi_j(x_j) = \mathrm{const}$. Once we move away from linear PCA, the two definitions lead to different solutions, and Donnell et al. (1994) choose to minimize variance. The optimization problem to be solved is then to successively find p-variate vectors $\boldsymbol{\phi}^{(k)}, k = 1, 2, \ldots$, whose elements are $\phi_j^{(k)}(x_j)$, which minimize

$$\operatorname{var}\left[\sum_{j=1}^{p} \phi_j^{(k)}(x_j)\right]$$

subject to $\sum_{j=1}^{p} \operatorname{var}[\phi_j^{(k)}(x_j)] = 1$, and for $k > 1, k > l$,

$$\sum_{j=1}^{p} \operatorname{cov}[\phi_j^{(k)}(x_j)\phi_j^{(l)}(x_j)] = 0.$$

As with linear PCA, this reduces to an eigenvalue problem. The main choice to be made is the set of functions $\phi(.)$ over which optimization is to take place. In an example Donnell et al. (1994) use splines, but their theoretical results are quite general and they discuss other, more sophisticated, smoothers. They identify two main uses for low-variance additive principal components, namely to fit additive implicit equations to data and to identify the presence of 'concurvities,' which play the same rôle and cause the same problems in additive regression as do collinearities in linear regression.

Principal curves are included in the same section as additive principal components despite the insistence by Donnell and coworkers in a response to discussion of their paper by Flury that they are very different. One difference is that although the range of functions allowed in additive principal components is wide, an equation *is* found relating the variables via the functions $\phi_j(x_j)$, whereas a principal curve is just that, a smooth curve with no necessity for a parametric equation. A second difference is that additive principal components concentrate on low-variance relationships, while principal curves minimize variation *orthogonal* to the curve.

There is nevertheless a similarity between the two techniques, in that both replace an optimum line or plane produced by linear PCA by an optimal non-linear curve or surface. In the case of principal curves, a smooth one-dimensional curve is sought that passes through the 'middle' of the data set. With an appropriate definition of 'middle,' the first PC gives the best straight line through the middle of the data, and principal curves generalize this using the idea of self-consistency, which was introduced at the end of Section 2.2. We saw there that, for p-variate random vectors $\mathbf{x}, \mathbf{y}$, the vector of random variables $\mathbf{y}$ is self-consistent for $\mathbf{x}$ if $E[\mathbf{x}|\mathbf{y}] = \mathbf{y}$. Consider a smooth curve in the p-dimensional space defined by $\mathbf{x}$. The curve can be written $\mathbf{f}(\lambda)$, where λ defines the position along the curve, and the vector $\mathbf{f}(\lambda)$ contains the values of the elements of $\mathbf{x}$ for a given value of λ. A curve $\mathbf{f}(\lambda)$ is self-consistent, that is, a *principal curve*, if $E[\mathbf{x} \mid f^{-1}(\mathbf{x}) = \lambda] = \mathbf{f}(\lambda)$, where $f^{-1}(\mathbf{x})$ is the value of λ for which $\|\mathbf{x} - \mathbf{f}(\lambda)\|$ is minimized. What this means intuitively is that, for any given value of λ, say λ_0, the average of all values of $\mathbf{x}$ that have $\mathbf{f}(\lambda_0)$ as their closest point on the curve is precisely $\mathbf{f}(\lambda_0)$.

It follows from the discussion in Section 2.2 that for multivariate normal and elliptical distributions the first principal component defines a principal curve, though there may also be other principal curves which are different from the first PC. Hence, non-linear principal curves may be thought of as a generalization of the first PC for other probability distributions. The discussion so far has been in terms of probability distributions, but a similar idea can be defined for samples. In this case a curve is fitted iteratively, alternating between 'projection' and 'conditional-expectation' steps. In a projection step, the closest point on the current curve is found for each observation in the sample, and the conditional-expectation step then calculates the average of observations closest to each point on the curve. These averages form a new curve to be used in the next projection step. In a finite data set there will usually be at most one observation corresponding to a given point on the curve, so some sort of smoothing is required to form the averages. Hastie and Stuetzle (1989) provide details of some possible smoothing schemes, together with examples. They also discuss the possibility of extension from curves to higher-dimensional surfaces.

Tarpey (1999) describes a 'lack-of-fit' test that can be used to decide whether or not a principal curve is simply the first PC. The test involves the idea of *principal points* which, for populations, are defined as follows. Suppose that $\mathbf{x}$ is a p-variate random vector and $\mathbf{y}$ is a discrete p-variate random vector, taking only the k values $\mathbf{y}_1, \mathbf{y}_2, \ldots, \mathbf{y}_k$. If $\mathbf{y}$ is such that $E[\|\mathbf{x} - \mathbf{y}\|^2]$ is minimized over all possible choices of the k values for $\mathbf{y}$, then $\mathbf{y}_1, \mathbf{y}_2, \ldots, \mathbf{y}_k$ are the k principal points for the distribution of $\mathbf{x}$. There is a connection with self-consistency, as $\mathbf{y}$ is self-consistent for $\mathbf{x}$ in this case. Flury (1993) discusses several methods for finding principal points in a sample.

There is another link between principal points and principal components, namely that if $\mathbf{x}$ has a multivariate normal or elliptical distribution, and the principal points $\mathbf{y}_1, \mathbf{y}_2, \ldots, \mathbf{y}_k$ for the distribution lie in a q ($< p$) subspace, then the subspace is identical to that spanned by the vectors of coefficients defining the first q PCs of $\mathbf{x}$ (Flury, 1995, Theorem 2.3). Tarpey (2000) introduces the idea of parallel principal axes, which are parallel hyperplanes orthogonal to the axis defined by the first PC that intersect that axis at the principal points of the marginal distribution of $\mathbf{x}$ along the axis. He shows that self-consistency of parallel principal axes characterizes multivariate normal distributions.

14.1.3   Non-Linearity Using Neural Networks

A considerable amount of work has been done on PCA in the context of neural networks. Indeed, there is a book on the subject (Diamantaras and Kung, 1996) which gives a good overview. Here we describe only those developments that provide non-linear extensions of PCA. Computational matters are discussed in Appendix A1, and other aspects of the PCA/neural networks interface are covered briefly in Section 14.6.1.

Diamantaras and Kung (1996, Section 6.6) give a general definition of non-linear PCA as minimizing

$$E[\|\mathbf{x} - \mathbf{g}(\mathbf{h}(\mathbf{x}))\|^2], \qquad (14.1.4)$$

where $\mathbf{y} = \mathbf{h}(\mathbf{x})$ is a $q(< p)$-dimensional function of $\mathbf{x}$ and $\mathbf{g}(\mathbf{y})$ is a p-dimensional function of $\mathbf{y}$. The functions $\mathbf{g}(.)$, $\mathbf{h}(.)$ are chosen from some given sets of non-linear functions so as to minimize (14.1.4). When $\mathbf{g}(.)$ and $\mathbf{h}(.)$ are restricted to be linear functions, it follows from Property A5 of Section 2.1 that minimizing (14.1.4) gives the usual (linear) PCs.

Diamantaras and Kung (1996, Section 6.6.1) note that for some types of network allowing non-linear functions leads to no improvement in minimizing (14.1.4) compared to the linear case. Kramer (1991) describes a network for which improvement does occur. There are two parts to the network, one that creates the components z_k from the p variables x_j, and a second that approximates the p variables given a reduced set of m ($< p$) components. The components are constructed from the variables by means of the formula

$$z_k = \sum_{l=1}^{N} w_{lk2}\sigma\left[\sum_{j=1}^{p} w_{jl1}x_j + \theta_l\right],$$

where

$$\sigma\left[\sum_{j=1}^{p} w_{jl1}x_j + \theta_l\right] = \left[1 + \exp\left(-\sum_{j=1}^{p} w_{jl1}x_j - \theta_l\right)\right]^{-1}, \qquad (14.1.5)$$

in which $w_{lk2}, w_{jl1}, \theta_l, \ j = 1, 2, \ldots, p; \ k = 1, 2, \ldots, m; \ l = 1, 2, \ldots, N$ are constants to be chosen, and N is the number of nodes in the hidden layer. A similar equation relates the estimated variables $\hat{x}_j$ to the components z_k, and Kramer (1991) combines both relationships into a single network. The objective is find the values of all the unknown constants so as to minimize the Euclidean norm of the matrix of residuals formed by estimating n values of each x_j by the corresponding values of $\hat{x}_j$. This is therefore a special case of Diamantaras and Kung's general formulation with $\mathbf{g}(.), \mathbf{h}(.)$ both restricted to the class of non-linear functions defined by (14.1.5).

For Kramer's network, m and N need to chosen, and he discusses various strategies for doing this, including the use of information criteria such as AIC (Akaike, 1974) and the comparison of errors in training and test sets to avoid overfitting. In the approach just described, m components are calculated simultaneously, but Kramer (1991) also discusses a sequential version in which one component at a time is extracted. Two examples are given of very different sizes. One is a two-variable artificial example in which non-linear PCA finds a built-in non-linearity. The second is from chemical engineering with 100 variables, and again non-linear PCA appears to be superior to its linear counterpart.

Since Kramer's paper appeared, a number of authors in the neural network literature have noted limitations to his procedure and have suggested alternatives or modifications (see, for example, Jia et al., 2000), although it is now used in a range of disciplines including climatology (Monahan, 2001). Dong and McAvoy (1996) propose an algorithm that combines the principal curves of Section 14.1.2 (Hastie and Stuetzle, 1989) with the auto-associative neural network set-up of Kramer (1991). Principal curves alone do not allow the calculation of 'scores' with respect to the curves for new observations, but their combination with a neural network enables such quantities to be computed.

An alternative approach, based on a so-called input-training net, is suggested by Tan and Mavrovouniotis (1995). In such networks, the inputs are not fixed, but are trained along with the other parameters of the network. With a single input the results of the algorithm are equivalent to principal curves, but with a larger number of inputs there is increased flexibility to go beyond the additive model underlying principal curves.

Jia et al. (2000) use Tan and Mavrovouniotis's (1995) input-training net, but have an ordinary linear PCA as a preliminary step. The non-linear algorithm is then conducted on the first m linear PCs, where m is chosen to be sufficiently large, ensuring that only PCs with very small variances are excluded. Jia and coworkers suggest that around 97% of the total variance should be retained to avoid discarding dimensions that might include important non-linear variation. The non-linear components are used in process control (see Section 13.7), and in an example they give improved fault detection compared to linear PCs (Jia et al., 2000). The preliminary step reduces the dimensionality of the data from 37 variables to 12 linear PCs, whilst retaining 98% of the variation.

Kambhatla and Leen (1997) introduce non-linearity in a different way, using a piecewise-linear or 'local' approach. The p-dimensional space defined by the possible values of $\mathbf{x}$ is partitioned into Q regions, and linear PCs are then found separately for each region. Kambhatla and Leen (1997) note that this local PCA provides a faster algorithm than a global non-linear neural network. A clustering algorithm is used to define the Q regions. Roweis and Saul (2000) describe a locally linear embedding algorithm that also generates local linear reconstructions of observations, this time based on a set of 'neighbours' of each observation. Tarpey (2000) implements a similar but more restricted idea. He looks separately at the first PCs within two regions defined by the sign of the first PC for the whole data set, as a means of determining the presence of non-linear structure in the data.

14.1.4 Other Aspects of Non-Linearity

We saw in Section 5.3 that biplots can provide an informative way of displaying the results of a PCA. Modifications of these 'classical' biplots to become non-linear are discussed in detail by Gower and Hand (1996, Chapter 6), and a shorter description is given by Krzanowski and Marriott (1994, Chapter 8). The link between non-linear biplots and PCA is somewhat tenuous, so we introduce them only briefly. Classical biplots are based on the singular value decomposition of the data matrix $\mathbf{X}$, and provide a best possible rank 2 approximation to $\mathbf{X}$ in a least squares sense (Section 3.5). The distances between observations in the 2-dimensional space of the biplot with $\alpha = 1$ (see Section 5.3) give optimal approximations to the corresponding Euclidean distances in p-dimensional space (Krzanowski and Marriott, 1994). Non-linear biplots replace Euclidean distance by other distance functions. In plots thus produced the straight lines or arrows representing variables in the classical biplot are replaced by curved trajectories. Different trajectories are used to interpolate positions of observations on the plots and to predict values of the variables given the plotting position of an observation. Gower and Hand (1996) give examples of interpolation biplot trajectories but state that they 'do not yet have an example of prediction nonlinear biplots.'

Tenenbaum et al. (2000) describe an algorithm in which, as with non-linear biplots, distances between observations other than Euclidean distance are used in a PCA-related procedure. Here so-called geodesic distances are approximated by finding the shortest paths in a graph connecting the observations to be analysed. These distances are then used as input to what seems to be principal coordinate analysis, a technique which is related to PCA (see Section 5.2).

14.2 Weights, Metrics, Transformations and Centerings

Various authors have suggested 'generalizations' of PCA. We have met examples of this in the direction of non-linearity in the previous section. A number of generalizations introduce weights or metrics on either observations or variables or both. The related topics of weights and metrics make up two of the three parts of the present section; the third is concerned with different ways of transforming or centering the data.

14.2.1 Weights

We start with a definition of generalized PCA which was given by Greenacre (1984, Appendix A). It can viewed as introducing either weights or metrics into the definition of PCA. Recall the singular value decomposition (SVD) of the $(n \times p)$ data matrix $\mathbf{X}$ defined in equation (3.5.1), namely

$$\mathbf{X} = \mathbf{ULA}'. \tag{14.2.1}$$

The matrices $\mathbf{A}, \mathbf{L}$ give, respectively, the eigenvectors and the square roots of the eigenvalues of $\mathbf{X}'\mathbf{X}$, from which the coefficients and variances of the PCs for the sample covariance matrix $\mathbf{S}$ are easily found.

In equation (14.2.1) we have $\mathbf{U}'\mathbf{U} = \mathbf{I}_r, \mathbf{A}'\mathbf{A} = \mathbf{I}_r$, where r is the rank of $\mathbf{X}$, and $\mathbf{I}_r$ is the identity matrix of order r. Suppose now that Ω and Φ are specified positive-definite symmetric matrices and that we replace (14.2.1) by a generalized SVD

$$\mathbf{X} = \mathbf{VMB}', \qquad (14.2.2)$$

where $\mathbf{V}, \mathbf{B}$ are $(n \times r)$, $(p \times r)$ matrices, respectively satisfying $\mathbf{V}'\Omega\mathbf{V} = \mathbf{I}_r$, $\mathbf{B}'\Phi\mathbf{B} = \mathbf{I}_r$, and $\mathbf{M}$ is a $(r \times r)$ diagonal matrix.

This representation follows by finding the ordinary SVD of $\tilde{\mathbf{X}} = \Omega^{1/2}\mathbf{X}\Phi^{1/2}$. If we write the usual SVD of $\tilde{\mathbf{X}}$ as

$$\tilde{\mathbf{X}} = \mathbf{WKC}', \qquad (14.2.3)$$

where $\mathbf{K}$ is diagonal, $\mathbf{W}'\mathbf{W} = \mathbf{I}_r, \mathbf{C}'\mathbf{C} = \mathbf{I}_r$, then

$$\mathbf{X} = \Omega^{-1/2}\tilde{\mathbf{X}}\Phi^{-1/2}$$

$$= \Omega^{-1/2}\mathbf{WKC}'\Phi^{-1/2}.$$

Putting $\mathbf{V} = \Omega^{-1/2}\mathbf{W}, \mathbf{M} = \mathbf{K}, \mathbf{B} = \Phi^{-1/2}\mathbf{C}$ gives (14.2.2), where $\mathbf{M}$ is diagonal, $\mathbf{V}'\Omega\mathbf{V} = \mathbf{I}_r$, and $\mathbf{B}'\Phi\mathbf{B} = \mathbf{I}_r$, as required. With this representation, Greenacre (1984) defines generalized PCs as having coefficients given by the columns of $\mathbf{B}$, in the case where Ω is diagonal. Rao (1964) suggested a similar modification of PCA, to be used when oblique rather than orthogonal axes are desired. His idea is to use the transformation $\mathbf{Z} = \mathbf{XB}$, where $\mathbf{B}'\Phi\mathbf{B} = \mathbf{I}$, for some specified positive-definite matrix, Φ; this idea clearly has links with generalized PCA, as just defined.

It was noted in Section 3.5 that, if we take the usual SVD and retain only the first m PCs so that x_{ij} is approximated by

$$_m\tilde{x}_{ij} = \sum_{k=1}^{m} u_{ik}l_k^{1/2}a_{jk} \quad \text{(with notation as in Section 3.5)},$$

then $_m\tilde{x}_{ij}$ provides a best possible rank m approximation to x_{ij} in the sense of minimizing $\sum_{i=1}^{n}\sum_{j=1}^{p}(_m x_{ij} - x_{ij})^2$ among all possible rank m approximations $_m x_{ij}$. It can also be shown (Greenacre, 1984, p. 39) that if Ω, Φ are both diagonal matrices, with elements ω_i, $i = 1, 2, \ldots, n$; ϕ_j, $j = 1, 2, \ldots, p$, respectively, and if $_m\tilde{x}_{ij} = \sum_{k=1}^{m} v_{ik}m_k b_{jk}$, where v_{ik}, m_k, b_{jk} are elements of $\mathbf{V}, \mathbf{M}, \mathbf{B}$ defined in (14.2.2), then $_m\tilde{x}_{ij}$ minimizes

$$\sum_{i=1}^{n}\sum_{j=1}^{p}\omega_i\phi_j(_m x_{ij} - x_{ij})^2 \qquad (14.2.4)$$

among all possible rank m approximations $_m x_{ij}$ to x_{ij}. Thus, the special case of generalized PCA, in which Φ as well as Ω is diagonal, is

a form of 'weighted PCA,' where different variables can have different
weights, $\phi_1, \phi_2, \ldots, \phi_p$, and different observations can have different weights
$\omega_1, \omega_2, \ldots, \omega_n$. Cochran and Horne (1977) discuss the use of this type of
weighted PCA in a chemistry context.

It is possible that one set of weights, but not the other, is present. For
example, if $\omega_1 = \omega_2 = \cdots = \omega_n$ but the ϕ_j are different, then only the
variables have different weights and the observations are treated identi-
cally. Using the correlation matrix rather than the covariance matrix is
a special case in which $\phi_j = 1/s_{jj}$ where s_{jj} is the sample variance of
the jth variable. Deville and Malinvaud (1983) argue that the choice of
$\phi_j = 1/s_{jj}$ is somewhat arbitrary and that other weights may be appropriate in some circumstances, and Rao (1964) also suggested the possibility of
using weights for variables. Gower (1966) notes the possibility of dividing
the variables by their ranges rather than standard deviations, or by their
means (for positive random variables, leading to an analysis of coefficients
of variation), or 'even [by] the cube root of the sample third moment.' In
other circumstances, there may be reasons to allow different observations
to have different weights, although the variables remain equally weighted.

In practice, it must be rare that an obvious uniquely appropriate set of
the ω_i or ϕ_j is available, though a general pattern may suggest itself. For
example, when data are time series Diamantaras and Kung (1996, Section
3.5) suggest basing PCA on a weighted estimate of the covariance matrix,
where weights of observations decrease geometrically as the distance in time
from the end of the series increases. In forecasting functional data, Aguilera et al. (1999b) cut a time series into segments, which are then treated
as different realizations of a single series (see Section 12.3.4). Those segments corresponding to more recent parts of the original series are given
greater weight in forming functional PCs for forecasting than are segments
further in the past. Both linearly decreasing and exponentially decreasing
weights are tried. These are examples of weighted observations. An example
of weighted variables, also from functional PCA, is presented by Ramsay
and Abrahamowicz (1989). Here the functions are varying binomial parameters, so that at different parts of the curves, weights are assigned that are
inversely proportional to the binomial standard deviation for that part of
the curve.

An even more general set of weights than that given in (14.2.4) is
proposed by Gabriel and Zamir (1979). Here $\mathbf{X}$ is approximated by
minimizing

$$\sum_{i=1}^{n} \sum_{j=1}^{p} w_{ij} (_m\hat{x}_{ij} - x_{ij})^2, \qquad (14.2.5)$$

where the rank m approximation to $\mathbf{X}$ has elements $_m\hat{x}_{ij}$ of the form $_m\hat{x}_{ij} = \sum_{k=1}^{m} g_{ik} h_{jk}$ for suitably chosen constants $g_{ik}, h_{jk}, i = 1, 2, \ldots, n, j = 1, 2, \ldots, p, k = 1, 2, \ldots, m$. This does not readily fit into the generalized

PC framework above unless the w_{ij} can be written as products $w_{ij} = \omega_i \phi_j$, $i = 1, 2, \ldots, n$; $j = 1, 2, \ldots, p$, although this method involves similar ideas. The examples given by Gabriel and Zamir (1979) can be expressed as contingency tables, so that correspondence analysis rather than PCA may be more appropriate, and Greenacre (1984), too, develops generalized PCA as an offshoot of correspondence analysis (he shows that another special case of the generalized SVD (14.2.2) produces correspondence analysis, a result which was discussed further in Section 13.1). The idea of weighting could, however, be used in PCA for any type of data, provided that suitable weights can be defined.

Gabriel and Zamir (1979) suggest a number of ways in which special cases of their weighted analysis may be used. As noted in Section 13.6, it can accommodate missing data by giving zero weight to missing elements of $\mathbf{X}$. Alternatively, the analysis can be used to look for 'outlying cells' in a data matrix. This can be achieved by using similar ideas to those introduced in Section 6.1.5 in the context of choosing how many PCs to retain. Any particular element x_{ij} of $\mathbf{X}$ is estimated by least squares based on a subset of the data that does not include x_{ij}. This (rank m) estimate $_m\hat{x}_{ij}$ is readily found by equating to zero a subset of weights in (14.2.5), including w_{ij}, The difference between x_{ij} and $_m\hat{x}_{ij}$ provides a better measure of the 'outlyingness' of x_{ij} compared to the remaining elements of $\mathbf{X}$, than does the difference between x_{ij} and a rank m estimate, $_m\tilde{x}_{ij}$, based on the SVD for the entire matrix $\mathbf{X}$. This result follows because $_m\hat{x}_{ij}$ is not affected by x_{ij}, whereas x_{ij} contributes to the estimate $_m\tilde{x}_{ij}$.

Commandeur et al. (1999) describe how to introduce weights for both variables and observations into Meulman's (1986) distance approach to nonlinear multivariate data analysis (see Section 14.1.1).

In the standard atmospheric science set-up, in which variables correspond to spatial locations, weights may be introduced to take account of uneven spacing between the locations where measurements are taken. The weights reflect the size of the area for which a particular location (variable) is the closest point. This type of weighting may also be necessary when the locations are *regularly* spaced on a latitude/longitude grid. The areas of the corresponding grid cells decrease towards the poles, and allowance should be made for this if the latitudinal spread of the data is moderate or large. An obvious strategy is to assign to the grid cells weights that are proportional to their areas. However, if there is a strong positive correlation within cells, it can be argued that doubling the area, for example, does not double the amount of independent information and that weights should reflect this. Folland (1988) implies that weights should be proportional to $(Area)^c$, where c is between $\frac{1}{2}$ and 1. Hannachi and O'Neill (2001) weight their data by the cosine of latitude.

Buell (1978) and North et al. (1982) derive weights for irregularly spaced atmospheric data by approximating a continuous version of PCA, based on an equation similar to (12.3.1).

14.2.2 Metrics

The idea of defining PCA with respect to a metric or an inner-product dates back at least to Dempster (1969, Section 7.6). Following the publication of Cailliez and Pagès (1976) it became, together with an associated 'duality diagram,' a popular view of PCA in France in the 1980s (see, for example, Caussinus, 1986; Escoufier, 1987). In this framework, PCA is defined in terms of a triple $(\mathbf{X}, \mathbf{Q}, \mathbf{D})$, the three elements of which are:

- the matrix $\mathbf{X}$ is the $(n \times p)$ data matrix, which is usually but not necessarily column-centred;

- the $(p \times p)$ matrix $\mathbf{Q}$ defines a metric on the p variables, so that the distance between two observations $\mathbf{x}_j$ and $\mathbf{x}_k$ is $(\mathbf{x}_j - \mathbf{x}_k)'\mathbf{Q}(\mathbf{x}_j - \mathbf{x}_k)$;

- the $(n \times n)$ matrix $\mathbf{D}$ is usually diagonal, and its diagonal elements consist of a set of weights for the n observations. It can, however, be more general, for example when the observations are not independent, as in time series (Caussinus, 1986; Escoufier, 1987).

The usual definition of covariance-based PCA has $\mathbf{Q} = \mathbf{I}_p$ the identity matrix, and $\mathbf{D} = \frac{1}{n}\mathbf{I}_n$, though to get the sample covariance matrix with divisor $(n-1)$ it is necessary to replace n by $(n-1)$ in the definition of $\mathbf{D}$, leading to a set of 'weights' which do not sum to unity. Correlation-based PCA is achieved either by standardizing $\mathbf{X}$, or by taking $\mathbf{Q}$ to be the diagonal matrix whose jth diagonal element is the reciprocal of the standard deviation of the jth variable, $j = 1, 2, \ldots, p$.

Implementation of PCA with a general triple $(\mathbf{X}, \mathbf{Q}, \mathbf{D})$ is readily achieved by means of the generalized SVD, described in Section 14.2.1, with $\mathbf{\Phi}$ and $\mathbf{\Omega}$ from that section equal to $\mathbf{Q}$ and $\mathbf{D}$ from this section. The coefficients of the generalized PCs are given in the columns of the matrix $\mathbf{B}$ defined by equation (14.2.2). Alternatively, they can be found from an eigenanalysis of $\mathbf{X}'\mathbf{D}\mathbf{X}\mathbf{Q}$ or $\mathbf{X}\mathbf{Q}\mathbf{X}'\mathbf{D}$ (Escoufier, 1987).

A number of particular generalizations of the standard form of PCA fit within this framework. For example, Escoufier (1987) shows that, in addition to the cases already noted, it can be used to: transform variables; to remove the effect of an observation by putting it at the origin; to look at subspaces orthogonal to a subset of variables; to compare sample and theoretical covariance matrices; and to derive correspondence and discriminant analyses. Maurin (1987) examines how the eigenvalues and eigenvectors of a generalized PCA change when the matrix $\mathbf{Q}$ in the triple is changed.

The framework also has connections with the fixed effects model of Section 3.9. In that model, the observations $\mathbf{x}_i$ are such that $\mathbf{x}_i = \mathbf{z}_i + \mathbf{e}_i$, where $\mathbf{z}_i$ lies in a q-dimensional subspace and $\mathbf{e}_i$ is an error term with zero mean and covariance matrix $\frac{\sigma^2}{w_i}\mathbf{\Gamma}$. Maximum likelihood estimation of the model, assuming a multivariate normal distribution for $\mathbf{e}$, leads to a generalized PCA, where $\mathbf{D}$ is diagonal with elements w_i and $\mathbf{Q}$ (which is denoted

$\mathbf{M}$ in (3.9.1)) is equal to $\mathbf{\Gamma}^{-1}$ (see Besse (1994b)). Futhermore, it can be shown (Besse, 1994b) that $\mathbf{\Gamma}^{-1}$ is approximately optimal even without the assumption of multivariate normality. Optimality is defined here as finding $\mathbf{Q}$ for which $E[\frac{1}{n}\sum_{i=1}^{n}\|\mathbf{z}_i - \hat{\mathbf{z}}_i\|_{\mathbf{A}}^2]$ is minimized, where $\mathbf{A}$ is any given Euclidean metric. The matrix $\mathbf{Q}$ enters this expression because $\hat{\mathbf{z}}_i$ is the $\mathbf{Q}$-orthogonal projection of $\mathbf{x}_i$ onto the optimal q-dimensional subspace.

Of course, the model is often a fiction, and even when it might be believed, $\mathbf{\Gamma}$ will typically not be known. There are, however, certain types of data where plausible estimators exist for $\mathbf{\Gamma}$. One is the case where the data fall into groups or clusters. If the groups are known, then within-group variation can be used to estimate $\mathbf{\Gamma}$, and generalized PCA is equivalent to a form of discriminant analysis (Besse, 1994b). In the case of unknown clusters, Caussinus and Ruiz (1990) use a form of generalized PCA as a projection pursuit technique to find such clusters (see Section 9.2.2). Another form of generalized PCA is used by the same authors to look for outliers in a data set (Section 10.1).

Besse (1988) searches for an 'optimal' metric in a less formal manner. In the context of fitting splines to functional data, he suggests several families of metric that combine elements of closeness between vectors with closeness between their smoothness. A family is indexed by a parameter playing a similar rôle to λ in equation (12.3.6), which governs smoothness. The optimal value of λ, and hence the optimal metric, is chosen to give the most clear-cut decision on how many PCs to retain.

Thacker (1996) independently came up with a similar approach, which he refers to as *metric-based PCA*. He assumes that associated with a set of p variables $\mathbf{x}$ is a covariance matrix $\mathbf{E}$ for errors or uncertainties. If $\mathbf{S}$ is the covariance matrix of $\mathbf{x}$, then rather than finding $\mathbf{a}'\mathbf{x}$ that maximizes $\mathbf{a}'\mathbf{S}\mathbf{a}$, it may be more relevant to maximize $\frac{\mathbf{a}'\mathbf{S}\mathbf{a}}{\mathbf{a}'\mathbf{E}\mathbf{a}}$. This reduces to solving the eigenproblem

$$\mathbf{S}\mathbf{a}_k = l_k\mathbf{E}\mathbf{a}_k \qquad (14.2.6)$$

for $k = 1, 2, \ldots, p$.

Second, third, and subsequent $\mathbf{a}_k$ are subject to the constraints $\mathbf{a}_h'\mathbf{E}\mathbf{a}_k = 0$ for $h < k$. In other words, $\mathbf{a}_1'\mathbf{x}, \mathbf{a}_2'\mathbf{x}, \ldots$ are uncorrelated with respect to the *error* covariance matrix. The eigenvalue l_k corresponding to the eigenvector $\mathbf{a}_k$ is equal to the ratio of the variances $\mathbf{a}_k'\mathbf{S}\mathbf{a}_k$, $\mathbf{a}_k'\mathbf{E}\mathbf{a}_k$ of $\mathbf{a}_k'\mathbf{x}$ calculated using the overall covariance matrix $\mathbf{S}$ and the error covariance matrix $\mathbf{E}$, respectively. To implement the technique it is necessary to know $\mathbf{E}$, a similar difficulty to requiring knowledge of $\mathbf{\Gamma}$ to choose an optimal metric for the fixed effects model. Another way of viewing the optimization problem is seeing that we maximize the variance $\mathbf{a}_k'\mathbf{S}\mathbf{a}_k$ of $\mathbf{a}_k'\mathbf{x}$ subject to the normalization constraint $\mathbf{a}_k'\mathbf{E}\mathbf{a}_k = 1$, so that the normalization is in terms of the error variance of $\mathbf{a}_k'\mathbf{x}$ rather than in terms of the length of $\mathbf{a}$, as in ordinary PCA.

Metric-based PCA, as defined by Thacker (1996), corresponds to the triple $(\mathbf{X}, \mathbf{E}^{-1}, \frac{1}{n}\mathbf{I}_n)$, and $\mathbf{E}$ plays the same rôle as does $\mathbf{\Gamma}$ in the fixed effects model. Tipping and Bishop's (1999a) model (Section 3.9) can be fitted as a special case with $\mathbf{E} = \sigma^2\mathbf{I}_p$. In this case the $\mathbf{a}_k$ are simply eigenvectors of $\mathbf{S}$.

Consider a model for $\mathbf{x}$ in which $\mathbf{x}$ is the sum of a signal and an independent noise term, so that the overall covariance matrix can be decomposed as $\mathbf{S} = \mathbf{S}_S + \mathbf{S}_N$, where $\mathbf{S}_S$, $\mathbf{S}_N$ are constructed from signal and noise, respectively. If $\mathbf{S}_N = \mathbf{E}$, then $\mathbf{S}_S = \mathbf{S} - \mathbf{E}$ and

$$\mathbf{S}_S\mathbf{a}_k = \mathbf{S}\mathbf{a}_k - \mathbf{E}\mathbf{a}_k = l_k\mathbf{E}\mathbf{a}_k - \mathbf{E}\mathbf{a}_k = (l_k - 1)\mathbf{E}\mathbf{a}_k,$$

so the $\mathbf{a}_k$ are also eigenvectors of the signal covariance matrix, using the metric defined by $\mathbf{E}^{-1}$. Hannachi (2000) demonstrates equivalences between

- Thacker's technique;

- a method that finds a linear function of $\mathbf{x}$ that minimizes the probability density of noise for a fixed value of the probability density of the data, assuming both densities are multivariate normal;

- maximization of signal to noise ratio as defined by Allen and Smith (1997).

Diamantaras and Kung (1996, Section 7.2) discuss maximization of signal to noise ratio in a neural network context using what they call 'oriented PCA.' Their optimization problem is again equivalent to that of Thacker (1996). The fingerprint techniques in Section 12.4.3 also analyse signal to noise ratios, but in that case the signal is defined as a squared expectation, rather than in terms of a signal covariance matrix.

Because any linear transformation of $\mathbf{x}$ affects both the numerator and denominator of the ratio in the same way, Thacker's (1996) technique shares with canonical variate analysis and CCA an invariance to the units of measurement. In particular, unlike PCA, the results from covariance and correlation matrices are equivalent.

14.2.3   Transformations and Centering

Data may be transformed in a variety of ways before PCA is carried out, and we have seen a number of instances of this elsewhere in the book. Transformations are often used as a way of producing non-linearity (Section 14.1) and are a frequent preprocessing step in the analysis of special types of data. For example, discrete data may be ranked (Section 13.1) and size/shape data, compositional data and species abundance data (Sections 13.2, 13.3, 13.8) may each be log-transformed before a PCA is done. The log transformation is particularly common and its properties, with and without standardization, are illustrated by Baxter (1995) using a number of examples from archaeology.

Standardization, in the sense of dividing each column of the data matrix by its standard deviation, leads to PCA based on the correlation matrix, and its pros and cons are discussed in Sections 2.3 and 3.3. This can be thought of a version of weighted PCA (Section 14.2.1). So, also, can dividing each column by its range or its mean (Gower, 1966), in the latter case giving a matrix of coefficients of variation. Underhill (1990) suggests a biplot based on this matrix (see Section 5.3.2). Such plots are only relevant when variables are non-negative, as with species abundance data.

Principal components are linear functions of x whose coefficients are given by the eigenvectors of a covariance or correlation matrix or, equivalently, the eigenvectors of a matrix $\mathbf{X'X}$. Here $\mathbf{X}$ is a $(n \times p)$ matrix whose (i, j)th element is the value for the ith observation of the jth variable, *measured about the mean* for that variable. Thus, the columns of $\mathbf{X}$ have been centred, so that the sum of each column is zero, though Holmes-Junca (1985) notes that centering by either medians or modes has been suggested as an alternative to centering by means.

Two alternatives to 'column-centering' are:

(i) the columns of $\mathbf{X}$ are left uncentred, that is x_{ij} is now the value for the ith observation of the jth variable, *as originally measured;*

(ii) *both* rows and columns of $\mathbf{X}$ are centred, so that sums of rows, as well as sums of columns, are zero.

In either (i) or (ii) the analysis now proceeds by looking at linear functions of x whose coefficients are the eigenvectors of $\mathbf{X'X}$, with $\mathbf{X}$ now non-centred or doubly centred. Of course, these linear functions no longer maximize variance, and so are not PCs according to the usual definition, but it is convenient to refer to them as non-centred and doubly centred PCs, respectively.

Non-centred PCA is a fairly well-established technique in ecology (Ter Braak,1983). It has also been used in chemistry (Jackson, 1991, Section 3.4; Cochran and Horne, 1977) and geology (Reyment and Jöreskog, 1993). As noted by Ter Braak (1983), the technique projects observations onto the best fitting plane (or flat) *through the origin*, rather than through the centroid of the data set. If the data are such that the origin is an important point of reference, then this type of analysis can be relevant. However, if the centre of the observations is a long way from the origin, then the first 'PC' will dominate the analysis, and will simply reflect the position of the centroid. For data that consist of counts of a number of biological species (the variables) at various sites (the observations), Ter Braak (1983) claims that non-centred PCA is better than standard (centred) PCA at simultaneously representing within-site diversity and between-site diversity of species (see also Digby and Kempton (1987, Section 3.5.5)). Centred PCA is better at representing between-site species diversity than non-centred PCA, but it is more difficult to deduce within-site diversity from a centred PCA.

Reyment and Jöreskog (1993, Section 8.7) discuss an application of the method (which they refer to as Imbrie's Q-mode method) in a similar context concerning the abundance of various marine micro-organisms in cores taken at a number of sites on the seabed. The same authors also suggest that this type of analysis is relevant for data where the p variables are amounts of p chemical constituents in n soil or rock samples. If the degree to which two samples have the same *proportions* of each constituent is considered to be an important index of similarity between samples, then the similarity measure implied by non-centred PCA is appropriate (Reyment and Jöreskog, 1993, Section 5.4). An alternative approach if proportions are of interest is to reduce the data to compositional form (see Section 13.3).

The technique of empirical orthogonal teleconnections (van den Dool et al., 2000), described in Section 11.2.3, operates on uncentred data. Here matters are confused by referring to *uncentred* sums of squares and cross-products as 'variances' and 'correlations.' Devijver and Kittler (1982, Section 9.3) use similar misleading terminology in a population derivation and discussion of uncentred PCA.

Doubly centred PCA was proposed by Buckland and Anderson (1985) as another method of analysis for data that consist of species counts at various sites. They argue that centred PCA of such data may be dominated by a 'size' component, which measures the relative abundance of the various species. It is possible to simply ignore the first PC, and concentrate on later PCs, but an alternative is provided by double centering, which 'removes' the 'size' PC. The same idea has been suggested in the analysis of size/shape data (see Section 13.2). Double centering introduces a component with zero eigenvalue, because the constraint $x_{i1} + x_{i2} + \ldots + x_{ip} = 0$ now holds for all i. A further alternative for removing the 'size' effect of different abundances of different species is, for some such data sets, to record only whether a species is present or absent at each site, rather than the actual counts for each species.

In fact, what is being done in double centering is the same as Mandel's (1971, 1972) approach to data in a two-way analysis of variance (see Section 13.4). It removes main effects due to rows/observations/sites, and due to columns/variables/species, and concentrates on the interaction between species and sites. In the regression context, Hoerl et al. (1985) suggest that double centering can remove 'non-essential ill-conditioning,' which is caused by the presence of a row (observation) effect in the original data. Kazmierczak (1985) advocates a logarithmic transformation of data, followed by double centering. This gives a procedure that is invariant to pre- and post-multiplication of the data matrix by diagonal matrices. Hence it is invariant to different weightings of observations and to different scalings of the variables.

One reason for the suggestion of both non-centred and doubly-centred PCA for counts of species at various sites is perhaps that it is not entirely clear which of 'sites' and 'species' should be treated as 'variables' and which as 'observations.' Another possibility is to centre with respect to sites, but not species, in other words, carrying out an analysis with sites rather than species as the variables. Buckland and Anderson (1985) analyse their data in this way.

Yet another technique which has been suggested for analysing some types of site-species data is correspondence analysis (see, for example, Section 5.4.1 and Gauch, 1982). As pointed out in Section 13.4, correspondence analysis has some similarity to Mandel's approach, and hence to doubly centred PCA. In doubly centred PCA we analyse the residuals from an additive model for row and column (site and species) effects, whereas in correspondence analysis the residuals from a multiplicative (independence) model are considered.

Both uncentred and doubly centred PCA perform eigenanalyses on matrices whose elements are not covariances or correlations, but which can still be viewed as measures of similarity or association between pairs of variables. Another technique in the same vein is proposed by Elmore and Richman (2001). Their idea is to find 'distances' between variables which can then be converted into similarities and an eigenanalysis done on the resulting similarity matrix. Although Elmore and Richman (2001) note a number of possible distance measures, they concentrate on Euclidean distance, so that the distance d_{jk} between variables j and k is

$$\left[\sum_{i=1}^{n}(x_{ij} - x_{ik})^2\right]^{\frac{1}{2}}.$$

If D is largest of the p^2 d_{jk}, the corresponding similarity matrix is defined to have elements

$$s_{jk} = 1 - \frac{d_{jk}}{D}.$$

The procedure is referred to as *PCA based on ES (Euclidean similarity)*. There is an apparent connection with principal coordinate analysis (Section 5.2) but for ES-based PCA it is distances between variables, rather than between observations, that are analysed.

The technique is only appropriate if the variables are all measured in the same units—it makes no sense to compute a distance between a vector of temperatures and a vector of heights, for example. Elmore and Richman (2001) report that the method does better at finding known 'modes' in a data set than PCA based on either a covariance or a correlation matrix. However, as with uncentred and doubly centred PCA, it is much less clear than it is for PCA what is optimized by the technique, and hence it is more difficult to know how to interpret its results.

14.3   Principal Components in the Presence of Secondary or Instrumental Variables

Rao (1964) describes two modifications of PCA that involve what he calls 'instrumental variables.' These are variables which are of secondary importance, but which may be useful in various ways in examining the variables that are of primary concern. The term 'instrumental variable' is in widespread use in econometrics, but in a rather more restricted context (see, for example, Darnell (1994, pp. 197–200)).

Suppose that $\mathbf{x}$ is, as usual, a p-element vector of primary variables, and that $\mathbf{w}$ is a vector of s secondary, or instrumental, variables. Rao (1964) considers the following two problems, described respectively as 'principal components *of* instrumental variables' and 'principal components ... *uncorrelated with* instrumental variables':

(i) Find linear functions $\gamma_1'\mathbf{w}, \gamma_2'\mathbf{w}, \ldots,$ of $\mathbf{w}$ that best predict $\mathbf{x}$.

(ii) Find linear functions $\alpha_1'\mathbf{x}, \alpha_2'\mathbf{x}, \ldots$ with maximum variances that, as well as being uncorrelated with each other, are also uncorrelated with $\mathbf{w}$.

For (i), Rao (1964) notes that $\mathbf{w}$ may contain some or all of the elements of $\mathbf{x}$, and gives two possible measures of predictive ability, corresponding to the trace and Euclidean norm criteria discussed with respect to Property A5 in Section 2.1. He also mentions the possibility of introducing weights into the analysis. The two criteria lead to different solutions to (i), one of which is more straightforward to derive than the other. There is a superficial resemblance between the current problem and that of canonical correlation analysis, where relationships between two sets of variables are also investigated (see Section 9.3), but the two situations are easily seen to be different. However, as noted in Sections 6.3 and 9.3.4, the methodology of Rao's (1964) PCA of instrumental variables has reappeared under other names. In particular, it is equivalent to redundancy analysis (van den Wollenberg, 1977) and to one way of fitting a reduced rank regression model (Davies and Tso, 1982).

The same technique is derived by Esposito (1998). He projects the matrix $\mathbf{X}$ onto the space spanned by $\mathbf{W}$, where $\mathbf{X}, \mathbf{W}$ are data matrices associated with $\mathbf{x}, \mathbf{w}$, and then finds principal components of the projected data. This leads to an eigenequation

$$\mathbf{S}_{XW}\mathbf{S}_{WW}^{-1}\mathbf{S}_{WX}\mathbf{a}_k = l_k\mathbf{a}_k,$$

which is the same as equation (9.3.5). Solving that equation leads to redundancy analysis. Kazi-Aoual et al. (1995) provide a permutation test, using the test statistic $\text{tr}(\mathbf{S}_{WX}\mathbf{S}_{XX}^{-1}\mathbf{S}_{XW})$ to decide whether there is any relationship between the $\mathbf{x}$ and $\mathbf{w}$ variables.

Inevitably, the technique has been generalized. For example, Sabatier et al. (1989) do so using the generalization of PCA described in Section 14.2.2, with triples $(\mathbf{X}, \mathbf{Q}_1, \mathbf{D})$, $(\mathbf{W}, \mathbf{Q}_2, \mathbf{D})$. They note that Rao's (1964) unweighted version of PCA of instrumental variables results from doing a generalized PCA on $\mathbf{W}$, with $\mathbf{D} = \frac{1}{n}\mathbf{I}_n$, and $\mathbf{Q}_2$ chosen to minimize $\|\mathbf{XX'} - \mathbf{WQ}_2\mathbf{W'}\|$, where $\|.\|$ denotes Euclidean norm. Sabatier et al. (1989) extend this to minimize $\|\mathbf{XQ}_1\mathbf{X'D} - \mathbf{WQ}_2\mathbf{W'D}\|$ with respect to $\mathbf{Q}_2$. They show that for various choices of $\mathbf{Q}_1$ and $\mathbf{D}$, a number of other statistical techniques arise as special cases. Another generalization is given by Takane and Shibayama (1991). For an $(n_1 \times p_1)$ data matrix $\mathbf{X}$, consider the prediction of $\mathbf{X}$ not only from an $(n_1 \times p_2)$ matrix of additional variables measured on the same individuals, but also from an $(n_2 \times p_1)$ matrix of observations on a different set of n_2 individuals for the same variables as in $\mathbf{X}$. PCA of instrumental variables occurs as a special case when only the first predictor matrix is present. Takane et al. (1995) note that redundancy analysis, and Takane and Shibayama's (1991) extension of it, amount to projecting the data matrix $\mathbf{X}$ onto a subspace that depends on the external information $\mathbf{W}$ and then conducting a PCA on the projected data. This projection is equivalent to putting constraints on the PCA, with the same constraints imposed in all dimensions. Takane et al. (1995) propose a further generalization in which different constraints are possible in different dimensions. The principal response curves of van den Brink and ter Braak (1999) (see Section 12.4.2) represent another extension.

One situation mentioned by Rao (1964) in which problem type (ii) (principal components uncorrelated with instrumental variables) might be relevant is when the data $\mathbf{x}_1, \mathbf{x}_2, \ldots, \mathbf{x}_n$ form a multiple time series with p variables and n time points, and it is required to identify linear functions of $\mathbf{x}$ that have large variances, but which are uncorrelated with 'trend' in the time series (see Section 4.5 for an example where the first PC is dominated by trend). Rao (1964) argues that such functions can be found by defining instrumental variables which represent trend, and then solving the problem posed in (ii), but he gives no example to illustrate this idea. A similar idea is employed in some of the techniques discussed in Section 13.2 that attempt to find components that are uncorrelated with an isometric component in the analysis of size and shape data. In the context of neural networks, Diamantaras and Kung (1996, Section 7.1) describe a form of 'constrained PCA' in which the requirement of uncorrelatedness in Rao's method is replaced by orthogonality of the vectors of coefficients in the constrained PCs to the subspace spanned by a set of constraints (see Section 14.6.1).

Kloek and Mennes (1960) also discussed the use of PCs as 'instrumental variables,' but in an econometric context. In their analysis, a number of dependent variables $\mathbf{y}$ are to be predicted from a set of predictor variables $\mathbf{x}$. Information is also available concerning another set of variables $\mathbf{w}$ (the instrumental variables) not used directly in predicting $\mathbf{y}$, but which can be used to obtain improved estimates of the coefficients **B** in the equation predicting **y** from **x**. Kloek and Mennes (1960) examine a number of ways in which PCs of **w** or PCs of the residuals obtained from regressing **w** on **x** or PCs of the combined vector containing all elements of **w** *and* **x**, can be used as 'instrumental variables' in order to obtain improved estimates of the coefficients **B**.

14.4   Alternatives to Principal Component Analysis for Non-Normal Distributions

We have noted several times that for many purposes it is not necessary to assume any particular distribution for the variables **x** in a PCA, although some of the properties of Chapters 2 and 3 rely on the assumption of multivariate normality.

One way of handling possible non-normality, especially if the distribution has heavy tails, is to use robust estimation of the covariance or correlation matrix, or of the PCs themselves. The estimates may be designed to allow for the presence of aberrant observations in general, or may be based on a specific non-normal distribution with heavier tails, as in Baccini et al. (1996) (see Section 10.4). In inference, confidence intervals or tests of hypothesis may be constructed without any need for distributional assumptions using the bootstrap or jackknife (Section 3.7.2). The paper by Dudziński et al. (1995), which was discussed in Section 10.3, investigates the effect of non-normality on repeatability of PCA, albeit in a small simulation study.

Another possibility is to assume that the vector **x** of random variables has a known distribution other than the multivariate normal. A number of authors have investigated the case of elliptical distributions, of which the multivariate normal is a special case. For example, Waternaux (1984) considers the usual test statistic for the null hypothesis H_{0q}, as defined in Section 6.1.4, of equality of the last $(p-q)$ eigenvalues of the covariance matrix. She shows that, with an adjustment for kurtosis, the same asymptotic distribution for the test statistic is valid for all elliptical distributions with finite fourth moments. Jensen (1986) takes this further by demonstrating that for a range of hypotheses relevant to PCA, tests based on a multivariate normal assumption have identical level and power for all distributions with ellipsoidal contours, even those without second moments. Things get more complicated outside the class of elliptical distributions, as shown by Waternaux (1984) for H_{0q}.

Jensen (1987) calls the linear functions of **x** that successively maximize 'scatter' of a conditional distribution, where conditioning is on previously derived linear functions, *principal variables*. Unlike McCabe's (1984) usage of the same phrase, these 'principal variables' are not a subset of the original variables, but linear functions of them. Jensen (1997) shows that when 'scatter' is defined as variance and **x** has a multivariate normal distribution, then his principal variables turn out to be the principal components. This result is discussed following the spectral decomposition of the covariance matrix (Property A3) in Section 2.1. Jensen (1997) greatly extends the result by showing that for a family of elliptical distributions and for a wide class of definitions of scatter, his principal variables are the same as principal components.

An idea which may be considered an extension of PCA to non-normal data is described by Qian et al. (1994). They investigate linear transformations of the p-variable vector **x** to q ($< p$) derived variables **y** that minimize what they call an index of predictive power. This index is based on minimum description length or stochastic complexity (see, for example, Rissanen and Yu (2000)) and measures the difference in stochastic complexity between **x** and **y**. The criterion is such that the optimal choice of **y** depends on the probability distribution of **x**, and Qian and coworkers (1994) show that for multivariate normal **x**, the derived variables **y** are the first q PCs. This can be viewed as an additional property of PCA, but confusingly they take it as a *definition* of principal components. This leads to their 'principal components' being different from the usual principal components when the distribution of **x** is nonnormal. They discuss various properties of their components and include a series of tests of hypotheses for deciding how many components are needed to adequately represent all the original variables.

Another possible extension of PCA to non-normal data is hinted at by O'Hagan (1994, Section 2.15). For a multivariate normal distribution, the covariance matrix is given by the negative of the inverse 'curvature' of the log-probability density function, where 'curvature' is defined as the matrix of second derivatives with respect to the elements of **x**. In the Bayesian setup where **x** is replaced by a vector of parameters $\boldsymbol{\theta}$, O'Hagan (1994) refers to the curvature evaluated at the modal value of $\boldsymbol{\theta}$ as the *modal dispersion matrix*. He suggests finding eigenvectors, and hence principal axes, based on this matrix, which is typically not the covariance matrix for non-normal distributions.

14.4.1 Independent Component Analysis

The technique, or family of techniques, known as *independent component analysis* (ICA) has been the subject of a large amount of research, starting in the late 1980s, especially in signal processing. It has been applied to various biomedical and imaging problems, and is beginning to be used in other fields such as atmospheric science. By the end of the 1990s it had its own annual workshops and at least one book (Lee, 1998). Although it is sometimes presented as a competitor to PCA, the links are not particularly strong-as we see below it seems closer to factor analysis—so the description here is brief. Stone and Porrill (2001) provide a more detailed introduction.

PCA has as its main objective the successive maximization of variance, and the orthogonality and uncorrelatedness constraints are extras, which are included to ensure that the different components are measuring separate things. By contrast, independent component analysis (ICA) takes the 'separation' of components as its main aim. ICA starts from the view that uncorrelatedness is rather limited as it only considers a lack of *linear* relationship, and that ideally components should be statistically *independent*. This is a stronger requirement than uncorrelatedness, with the two only equivalent for normal (Gaussian) random variables. ICA can thus be viewed as a generalization of PCA to non-normal data, which is the reason for including it in the present section. However this may lead to the mistaken belief, as implied by Aires et al. (2000), that PCA assumes normality, which it does not. Aires and coworkers also describe PCA as assuming a model in which the variables are linearly related to a set of underlying components, apart from an error term. This is much closer to the set-up for factor analysis, and it is this 'model' that ICA generalizes.

ICA assumes, instead of the factor analysis model $\mathbf{x} = \mathbf{\Lambda f} + \mathbf{e}$ given in equation (7.1.1), that $\mathbf{x} = \mathbf{\Lambda(f)}$, where $\mathbf{\Lambda}$ is some, not necessarily linear, function and the elements of $\mathbf{f}$ are independent. The components (factors) $\mathbf{f}$ are estimated by $\hat{\mathbf{f}}$, which is a function of $\mathbf{x}$. The family of functions from which $\mathbf{\Lambda}$ can be chosen must be defined. As in much of the ICA literature so far, Aires et al. (2000) and Stone and Porrill (2001) concentrate on the special case where $\mathbf{\Lambda}$ is restricted to linear functions. Within the chosen family, functions are found that minimize an 'objective cost function,' based on information or entropy, which measures how far are the elements of $\hat{\mathbf{f}}$ from independence. This differs from factor analysis in that the latter has the objective of *explaining correlations*. Some details of a 'standard' ICA method, including its entropy criterion and an algorithm for implementation, are given by Stone and Porrill (2001).

Typically, an iterative method is used to find the optimal $\hat{\mathbf{f}}$, and like projection pursuit (see Section 9.2.2), a technique with which Stone and Porrill (2001) draw parallels, it is computationally expensive. As with projection pursuit, PCA can be used to reduce dimensionality (use the first m, rather than all p) before starting the ICA algorithm, in order to reduce the computational burden (Aires et al., 2000; Stone and Porrill, 2001). It is also suggested by Aires and coworkers that the PCs form a good starting point for the iterative algorithm, as they are uncorrelated. These authors give an example involving sea surface temperature, in which they claim that the ICs are physically more meaningful than PCs. The idea that physically meaningful signals underlying a data set should be independent is a major motivation for ICA. This is very different from the view taken in some applications of factor analysis or rotated PCA, where it is believed that underlying factors will very often be correlated, and that it is too restrictive to force them to be uncorrelated, let alone independent (see, for example Cattell (1978, p. 128); Richman (1986)).

14.5 Three-Mode, Multiway and Multiple Group Principal Component Analysis

Principal component analysis is usually done on a single $(n \times p)$ data matrix $\mathbf{X}$, but there are extensions to many other data types. In this section we discuss briefly the case where there are additional 'modes' in the data. As well as rows (individuals) and columns (variables) there are other layers, such as different time periods.

The ideas for three-mode methods were first published by Tucker in the mid-1960s (see, for example, Tucker, 1966) and by the early 1980s the topic of *three-mode principal component analysis* was, on its own, the subject of a 398-page book (Kroonenberg, 1983a). A 33-page annotated bibliography (Kroonenberg, 1983b) gave a comprehensive list of references for the slightly wider topic of three-mode factor analysis. The term 'three-mode' refers to data sets that have three modes by which the data may be classified. For example, when PCs are obtained for several groups of individuals as in Section 13.5, there are three modes corresponding to variables, groups and individuals. Alternatively, we might have n individuals, p variables and t time points, so that 'individuals,' 'variables' and 'time points' define the three modes. In this particular case we have effectively n time series of p variables, or a single time series of np variables. However, the time points need not be equally spaced, nor is the time-order of the t repetitions necessarily relevant in the sort of data for which three-mode PCA is used, in the same way that neither individuals nor variables usually have any particular a priori ordering.

Let x_{ijk} be the observed value of the jth variable for the ith individual measured on the kth occasion. The basic idea in three-mode analysis is to approximate x_{ijk} by the model

$$\tilde{x}_{ijk} = \sum_{h=1}^{m}\sum_{l=1}^{q}\sum_{r=1}^{s} a_{ih} b_{jl} c_{kr} g_{hlr}.$$

The values m, q, s are less, and if possible very much less, than n, p, t, respectively, and the parameters $a_{ih}, b_{jl}, c_{kr}, g_{hlr}$, $i = 1, 2, \ldots, n$, $h = 1, 2, \ldots, m$, $j = 1, 2, \ldots, p$, $l = 1, 2, \ldots, q$, $k = 1, 2, \ldots, t$, $r = 1, 2, \ldots, s$ are chosen to give a good fit of $\tilde{x}_{ijk}$ to x_{ijk} for all i, j, k. There are a number of methods for solving this problem and, like ordinary PCA, they involve finding eigenvalues and eigenvectors of cross-product or covariance matrices, in this case by combining two of the modes (for example, combine individuals and observations to give a mode with np categories) before finding cross-products. Details will not be given here (see Tucker (1966) or Kroonenberg (1983a), where examples may also be found).

The substantial literature on the subject that existed at the time of Kroonenberg's (1983a) book has continued to grow. A key reference, collecting together material from many of those working in the area in the late 1980s, is Coppi and Bolasco (1989). Research is still being done on various extensions, special cases and properties of the three-mode model (see, for example, Timmerman and Kiers (2000)). One particular extension is to the case where more than three modes are present. Such data are usually called 'multiway' rather than 'multimode' data.

Although multiway analysis has its roots in the psychometric literature, it has more recently been adopted enthusiastically by the chemometrics community. Volume 14, Issue 3 of the *Journal of Chemometrics*, published in 2000, is a special issue on multiway analysis. The issue contains relatively little on multiway PCA itself, but there is no shortage of articles on it in the chemometrics literature and in the overlapping field of process control (see, for example, Dahl et al. (1999)). In process control the three most commonly encountered modes are different control variables, different time intervals and different batch runs (Nomikos and MacGregor, 1995).

Another context in which three-mode data arise is in atmospheric science, where one mode is spatial location, a second is time and a third is a set of different meteorological variables. It was noted in Section 12.2.1 that the analysis of such data, which amalgamates the p locations and n different meteorological variables into a combined set of np variables, is sometimes known as extended EOF analysis.

An alternative strategy for analysing data of this type is to consider pairs of two modes, fixing the third, and then perform some form of PCA on each chosen pair of modes. There are six possible pairs, leading to six possible analyses. These are known as O-, P-, Q-, R-, S- and T-mode analyses (Richman, 1986), a terminology that has its roots in psychology (Cattell, 1978, Chapter 12). In atmospheric science the most frequently used mode is S-mode (locations = variables; times = observations; meteorological variable fixed), but T-mode (times = variables; locations = observations; meteorological variable fixed) is not uncommon (see, for example, Salles et al. (2001)). Richman (1986) discusses the other four possibilities. Weare (1990) describes a tensor-based variation of PCA for data having four 'dimensions,' three in space together with time. He notes a similarity between his technique and three-mode factor analysis.

Some types of multiway data convert naturally into other forms. In some cases one of the modes corresponds to different groups of individuals measured on the same variables, so that the analyses of Section 13.5 may be relevant. In other circumstances, different modes may correspond to different groups of variables. For two such groups, Section 9.3 describes a number of techniques with some connection to PCA, and many of these procedures can be extended to more than two groups. For example, Casin (2001) reviews a number of techniques for dealing with K sets of variables, most of which involve a PCA of the data arranged in one way or another. He briefly compares these various methods with his own 'generalization' of PCA, which is now described.

Suppose that $\mathbf{X}_k$ is an $(n \times p_k)$ data matrix consisting of measurements of p_k variables on n individuals, $k = 1, 2, \ldots, K$. The same individuals are observed for each k. The first step in Casin's (2001) procedure is a PCA based on the correlation matrix obtained from the $(n \times p)$ supermatrix

$$\mathbf{X} = (\mathbf{X}_1 \ \ \mathbf{X}_2 \ \ \ldots \ \ \mathbf{X}_K),$$

where $p = \sum_{k=1}^{K} p_k$. The first PC, $z^{(1)}$, thus derived is then projected onto the subspaces spanned by the columns of $\mathbf{X}_k$, $k = 1, 2, \ldots, K$, to give a 'first component' $z_k^{(1)}$ for each $\mathbf{X}_k$. To obtain a second component, residual matrices $\mathbf{X}_k^{(2)}$ are calculated. The jth column of $\mathbf{X}_k^{(2)}$ consists of residuals from a regression of the jth column of $\mathbf{X}_k$ on $z_k^{(1)}$. A covariance matrix PCA is then performed for the supermatrix

$$\mathbf{X}^{(2)} = (\mathbf{X}_1^{(2)} \ \ \mathbf{X}_2^{(2)} \ \ \ldots \ \ \mathbf{X}_K^{(2)}).$$

The first PC from this analysis is next projected onto the subspaces spanned by the columns of $\mathbf{X}_k^{(2)}$, $k = 1, 2, \ldots, K$ to give a second component $z_k^{(2)}$ for $\mathbf{X}_k$. This is called a 'second auxiliary' by Casin (2001). Residuals from regressions of the columns of $\mathbf{X}_k^{(2)}$ on $z_k^{(2)}$ give matrices $\mathbf{X}_k^{(3)}$, and a covariance matrix PCA is carried out on the supermatrix formed from these matrices. From this, third auxiliaries $z_k^{(3)}$ are calculated, and so on. Unlike an ordinary PCA of $\mathbf{X}$, which produces p PCs, the number of auxiliaries for the kth group of variables is only p_k. Casin (2001) claims that this procedure is a sensible compromise between separate PCAs for each $\mathbf{X}_k$, which concentrate on within-group relationships, and extensions of canonical correlation analysis, which emphasize relationships between groups.

Van de Geer (1984) reviews the possible ways in which linear relationships between two groups of variables can be quantified, and then discusses how each might be generalized to more than two groups (see also van de Geer (1986)). One of the properties considered by van de Geer (1984) in his review is the extent to which within-group, as well as between-group, structure is considered. When within-group variability is taken into account there are links to PCA, and one of van de Geer's (1984) generalizations is equivalent to a PCA of all the variables in the K groups, as in extended EOF analysis. Lafosse and Hanafi (1987) extend Tucker's inter-battery model, which was discussed in Section 9.3.3, to more than two groups.

14.6   Miscellanea

This penultimate section discusses briefly some topics involving PCA that do not fit very naturally into any of the other sections of the book.

14.6.1   Principal Components and Neural Networks

This subject is sufficiently large to have a book devoted to it (Diamantaras and Kung, 1996). The use of neural networks to provide non-linear extensions of PCA is discussed in Section 14.1.3 and computational aspects are revisited in Appendix A1. A few other related topics are noted here, drawing mainly on Diamantaras and Kung (1996), to which the interested reader is referred for further details. Much of the work in this area is concerned with constructing efficient algorithms, based on neural networks, for deriving PCs. There are variations depending on whether a single PC or several PCs are required, whether the first or last PCs are of interest, and whether the chosen PCs are found simultaneously or sequentially. The advantage of neural network algorithms is greatest when data arrive sequentially, so that the PCs need to be continually updated. In some algorithms the transformation to PCs is treated as deterministic; in others noise is introduced (Diamantaras and Kung, 1996, Chapter 5). In this latter case, the components are written as

$$\mathbf{y} = \mathbf{B}'\mathbf{x} + \mathbf{e},$$

and the original variables are approximated by

$$\hat{\mathbf{x}} = \mathbf{Cy} = \mathbf{CB}'\mathbf{x} + \mathbf{Ce},$$

where $\mathbf{B}, \mathbf{C}$ are $(p \times q)$ matrices and $\mathbf{e}$ is a noise term. When $\mathbf{e} = \mathbf{0}$, minimizing $E[(\hat{\mathbf{x}} - \mathbf{x})'(\hat{\mathbf{x}} - \mathbf{x})]$ with respect to $\mathbf{B}$ and $\mathbf{C}$ leads to PCA (this follows from Property A5 of Section 2.1), but the problem is complicated by the presence of the term $\mathbf{Ce}$ in the expression for $\hat{\mathbf{x}}$. Diamantaras and Kung (1996, Chapter 5) describe solutions to a number of formulations of the problem of finding optimal $\mathbf{B}$ and $\mathbf{C}$. Some constraints on $\mathbf{B}$ and/or $\mathbf{C}$ are necessary to make the problem well-defined, and the different formulations correspond to different constraints. All solutions have the common feature that they involve combinations of the eigenvectors of the covariance matrix of $\mathbf{x}$ with the eigenvectors of the covariance matrix of $\mathbf{e}$. As with other signal/noise problems noted in Sections 12.4.3 and 14.2.2, there is the necessity either to know the covariance matrix of $\mathbf{e}$ or to be able to estimate it separately from that of $\mathbf{x}$.

Networks that implement extensions of PCA are described in Diamantaras and Kung (1996, Chapters 6 and 7). Most have links to techniques developed independently in other disciplines. As well as non-linear extensions, the following analysis methods are discussed:

- Linear Approximation Asymmetric PCA. This leads to an equation that is equivalent to (9.3.2). Hence the technique is the same as redundancy analysis, one form of reduced rank regression and PCA of instrumental variables (Sections 9.3.3, 9.3.4, 14.3).

- Cross-correlation Asymmetric PCA. This reduces to finding the SVD of the matrix of covariances between two sets of variables, and so is equivalent to maximum covariance analysis (Section 9.3.3).

- Constrained PCA. This technique finds 'principal components' that are constrained to be orthogonal to a space defined by a set of constraint vectors. It is therefore closely related to the idea of projecting orthogonally to the isometric vector for size and shape data (Section 13.2) and is similar to Rao's (1964) PCA uncorrelated with instrumental variables (Section 14.3). A soft-constraint version of this technique, giving a compromise between constrained PCA and ordinary PCA, is discussed in Diamantaras and Kung (1996, Section 7.3).

- Oriented PCA. In general terms, the objective is to find $\mathbf{a}_1, \mathbf{a}_2, \ldots,$ $\mathbf{a}_k, \ldots$ that successively maximize $\frac{\mathbf{a}_k' \mathbf{S}_1 \mathbf{a}_k}{\mathbf{a}_k' \mathbf{S}_2 \mathbf{a}_k}$, where $\mathbf{S}_1, \mathbf{S}_2$ are two covariance matrices. Diamantaras and Kung (1996, Section 7.2) note that special cases include canonical discriminant analysis (Section 9.1) and maximization of a signal to noise ratio (Sections 12.4.3, 14.2.2).

Xu and Yuille (1992) describe a neural network approach based on statistical physics that gives a robust version of PCA (see Section 10.4). Fancourt and Principe (1998) propose a network that is tailored to find PCs for locally stationary time series.

As well as using neural networks to find PCs, the PCs can also be used as inputs to networks designed for other purposes. Diamantaras and Kung (1996, Section 4.6) give examples in which PCs are used as inputs to discriminant analysis (Section 9.1) and image processing. McGinnis (2000) uses them in a neural network approach to predicting snowpack accumulation from 700 mb geopotential heights.

14.6.2 Principal Components for Goodness-of-Fit Statistics

The context of this application of PCA is testing whether or not a (univariate) set of data $y_1, y_2, \ldots, y_n$ could have arisen from a given probability distribution with cumulative distribution function $G(y)$; that is, we want a goodness-of-fit test. If the transformation

$$x_i = G(y_i), \quad i = 1, 2, \ldots, n$$

is made, then we can equivalently test whether or not $x_1, x_2, \ldots, x_n$ are from a uniform distribution on the range $(0, 1)$. Assume, without loss of generality, that $x_1 \leq x_2 \leq \ldots \leq x_n$, and define the sample distribution function as $F_n(x) = i/n$ for $x_i \leq x < x_{i+1}$, $i = 0, 1, \ldots, n$, where x_0, x_{n+1} are defined as $0, 1$ respectively. Then a well-known test statistic is the Cramér-von Mises statistic:

$$W_n^2 = n \int_0^1 (F_n(x) - x)^2 \, dx.$$

Like most all-purpose goodness-of-fit statistics, W_n^2 can detect many different types of discrepancy between the observations and $G(y)$; a large value of W_n^2 on its own gives no information about what type has occurred. For this reason a number of authors, for example Durbin and Knott (1972), Durbin et al. (1975), have looked at decompositions of W_n^2 into a number of separate 'components,' each of which measures the degree to which a different type of discrepancy is present.

It turns out that a 'natural' way of partitioning W_n^2 is (Durbin and Knott, 1972)

$$W_n^2 = \sum_{k=1}^{\infty} z_{nk}^2,$$

where

$$z_{nk} = (2n)^{1/2} \int_0^1 (F_n(x) - x) \, \sin \, (k\pi x) \, dx, \quad k = 1, 2, \ldots,$$

are the PCs of $\sqrt{n}(F_n(x) - x)$. The phrase 'PCs of $\sqrt{n}(F_n(x) - x)$' needs further explanation, since $\sqrt{n}(F_n(x) - x)$ is not, as is usual when defining PCs, a p-variable vector. Instead, it is an infinite-dimensional random variable corresponding to the continuum of values for x between zero and one. Durbin and Knott (1972) solve an equation of the form (12.3.1) to obtain eigenfunctions $a_k(x)$, and hence corresponding PCs

$$z_{nk} = \sqrt{n} \int_0^1 a_k(x)(F_n(x) - x) \, dx,$$

where $a_k(x) = \sqrt{2} \sin(k\pi x)$.

The components z_{nk}, $k = 1, 2, \ldots$ are discussed in considerable detail, from both theoretical and practical viewpoints, by Durbin and Knott (1972), and Durbin et al. (1975), who also give several additional references for the topic.

Another use of PCA in goodness-of-fit testing is noted by Jackson (1991, Section 14.3), namely using an extension to the multivariate case of the Shapiro-Wilk test for normality, based on PCs rather than on the original variables. Kaigh (1999) also discusses something described as 'principal components' in the context of goodness-of-fit, but these appear to be related to Legendre polynomials, rather than being the usual variance-maximizing PCs.

14.6.3   Regression Components, Sweep-out Components and Extended Components

Ottestad (1975) proposed an alternative to PCA that he called *regression components*. He developed this for standardized variables, and hence it is a correlation-based analysis. The new variables, or regression components, $y_1, y_2, \ldots, y_p$ are defined in terms of the original (standardized) variables $x_1, x_2, \ldots, x_p$ as $y_1 = x_1$, $y_2 = x_2 - b_{21}x_1$, $y_3 = x_3 - b_{31}x_1 - b_{32}x_2$, $\ldots$,

$$y_p = x_p - b_{p1}x_1 - b_{p2}x_2 - \ldots b_{p(p-1)}x_{(p-1)},$$

where b_{jk} is the regression coefficient of x_k in a regression of x_j on all other variables on the right hand-side of the equation defining y_j. It should be stressed that the labelling in these defining equations has been chosen for simplicity to correspond to the order in which the y variables are defined. It will usually be different from the labelling of the data as originally recorded. The x variables can be selected in any order to define the y variables and the objective of the technique is to choose a best order from the $p!$ possibilities. This is done by starting with y_p, for which x_p is chosen to be the original variable that has maximum multiple correlation with the other $(p - 1)$ variables. The next variable $x_{(p-1)}$, from which $y_{(p-1)}$ is defined, minimizes $(1 + b_{p(p-1)})^2(1 - R^2)$, where R^2 denotes the multiple correlation of $x_{(p-1)}$ with $x_{(p-2)}, x_{(p-3)}, \ldots, x_1$, and so on until only x_1 is left. The reasoning behind the method, which gives uncorrelated components, is that it provides results that are simpler to interpret than PCA in the examples that Ottestad (1975) studies. However, orthogonality of vectors of coefficients and successive variance maximization are both lost. Unlike the techniques described in Chapter 11, no explicit form of simplicity is targeted and neither is there any overt attempt to limit variance loss, so the method is quite different from PCA.

A variation on the same theme is proposed by Atiqullah and Uddin (1993). They also produce new variables $y_1, y_2, \ldots, y_p$ from a set of measured variables $x_1, x_2, \ldots, x_p$ in a sequence $y_1 = x_1$, $y_2 = x_2 - b_{21}x_1$, $y_3 = x_3 - b_{31}x_1 - b_{32}x_2$, $\ldots$,

$$y_p = x_p - b_{p1}x_1 - b_{p2}x_2 - \ldots b_{p(p-1)}x_{(p-1)},$$

but for a different set of b_{kj}. Although the details are not entirely clear it appears that, unlike Ottestad's (1975) method, the ordering in the sequence is not determined by statistical criteria, but simply corresponds to the labels on the original x variables. Atiqullah and Uddin (1993) transform the covariance matrix for the x variables into upper triangular form, with diagonal elements equal to unity. The elements of this matrix above the diagonal are then the b_{kj}. As with Ottestad's method, the new variables, called *sweep-out components*, are uncorrelated.

Rather than compare variances of $y_1, y_2, \ldots, y_p$, which do not sum to $\sum_{j=1}^{p} \text{var}(x_i)$, both Ottestad (1975) and Atiqullah and Uddin (1993) decompose var($\sum_{j=1}^{p} x_j$) into parts due to each component. In this respect there are similarities with a method proposed by Vermeiren et al. (2001), which they call *extended principal component analysis*. This also decomposes var($\sum_{j=1}^{p} x_j$), but does so with rescaled PCs. Denote the kth such component by $z_k^E = \mathbf{a}_k^{E'} \mathbf{x}$, where $\mathbf{a}_k^E = c_k \mathbf{a}_k$, $\mathbf{a}_k$ is the usual vector of coefficients for the kth PC with $\mathbf{a}_k' \mathbf{a}_k = 1$, and c_k is a rescaling constant. Vermeiren et al. (2001) stipulate that $\sum_{k=1}^{p} z_k^E = \sum_{j=1}^{p} x_j$ and then show that this condition is satisfied by $\mathbf{c} = \mathbf{A}' \mathbf{1_p}$, where the kth column of $\mathbf{A}$ is $\mathbf{a}_k$ and the kth element of $\mathbf{c}$ is c_k. Thus c_k is the sum of the coefficients in $\mathbf{a}_k$ and will be large when all coefficients in $\mathbf{a}_k$ are of the same sign, or when a PC is dominated by a single variable. The importance of such PCs is enhanced by the rescaling. Conversely, c_k is small for PCs that are contrasts between groups of variables, and rescaling makes these components less important. The rescaled or 'extended' components are, like the unscaled PCs z_k, uncorrelated, so that

$$\text{var}\left[\sum_{j=1}^{p} x_j\right] = \text{var}\left[\sum_{k=1}^{p} z_k^E\right] = \sum_{k=1}^{p} \text{var}(z_k^E) = \sum_{k=1}^{p} c_k^2 \, \text{var}(z_k) = \sum_{k=1}^{p} c_k^2 l_k.$$

Hence var$[\sum_{j=1}^{p} x_j]$ may be decomposed into contributions $c_k^2 l_k$, $k = 1, 2, \ldots, p$ from each rescaled component. Vermeiren et al. (2001) suggest that such a decomposition is relevant when the variables are constituents of a financial portfolio.

14.6.4 Subjective Principal Components

Korhonen (1984) proposes a technique in which a user has input into the form of the 'components.' The slightly tenuous link with PCA is that it is assumed that the user wishes to maximize correlation between the chosen component and one or more of the original variables. The remarks following the spectral decomposition (Property A3) in Section 2.1, Property A6 in Section 2.3, and the discussion of different normalization constraints at the end of that section, together imply that the first few PCs tend to have large correlations with the variables, especially in a correlation matrix-based PCA. Korhonen's (1984) procedure starts by presenting the user with the correlations between the elements of $\mathbf{x}$ and the 'component' $\mathbf{a}_0' \mathbf{x}$, where $\mathbf{a}_0$ is the isometric vector $\frac{1}{\sqrt{p}}(1, 1, \ldots, 1)$ (see Section 13.2). The user is then invited to choose a variable for which the correlation is desired to be larger. The implications for other correlations of modifying $\mathbf{a}_0$ so as to increase the selected correlation are displayed graphically. On the basis of this information, the user then chooses by how much to increase the correlation and hence change $\mathbf{a}_0$, giving the first *subjective principal component*.

If second, third, $\ldots$, subjective components are desired, emphasizing correlations with different variables, a similar procedure is repeated in the space orthogonal to the components found so far. Clearly, this technique loses the variance maximization property of PCA but, like the techniques of Section 11.2, it can be thought of as an alternative that simplifies interpretation. In the present case simplification is in the direction of the user's expectations.

14.7 Concluding Remarks

It has been seen in this book that PCA can be used in a wide variety of different ways. Many of the topics covered, especially in the last four chapters, are of recent origin and it is likely that there will be further advances in the near future that will help to clarify the usefulness, in practice, of some of the newer techniques. Developments range from an increasing interest in model-based approaches on the one hand to the mainly algorithmic ideas of neural networks on the other. Additional uses and adaptations of PCA are certain to be proposed and, given the large number of fields of application in which PCA is employed, it is inevitable that there are already some uses and modifications of which the present author is unaware.

In conclusion, it should be emphasized again that, far from being an old and narrow technique, PCA is the subject of much recent research and has great versatility, both in the ways in which it can be applied, and in the fields of application for which it is useful.

Appendix A
Computation of Principal Components

This Appendix is the only part of the book that has shrunk compared to the first edition, where it consisted of two sections. The first described efficient methods for deriving PCs, that is efficient techniques from numerical analysis for calculating eigenvectors and eigenvalues of positive semi-definite matrices; the second section discussed the facilities for computing PCs, and performing related analyses, which were then available in five of the best known statistical computer packages.

The first topic has been updated in this edition and some general comments on the second topic are included. However much of the detail on the latter topic has been removed from this edition, mainly because such material rapidly becomes out of date. This is readily illustrated by two quotations from the first edition.

> Despite the likelihood that personal computers will become the main tool for ... users of PCA ... [it] is still usually carried out on mainframe computers ... [T]he author has no experience yet of PCA on personal computers.

> MINITAB does not have any direct instructions for finding PCs.

Five packages were described in the first edition—BMDP, GENSTAT, MINITAB, SAS and SPSS^X. Since then a number a new packages or languages have appeared. Perhaps the biggest change is the greatly expanded use by statisticians of S-PLUS and its 'open source' relative R. The MATLAB software should also be mentioned. Although it is not primarily a statistical package, it has found increasing favour among statisticians as a programming environment within which new statistical techniques can be implemented. PCA is also included in some neural network software.

All the main statistical software packages incorporate procedures for finding the basic results of a PCA. There are some variations in output, such as the choice of normalization constraints used for the vectors of loadings or coefficients. This can cause confusion for the unwary user (see Section 11.1), who may also be confused by the way in which some software erroneously treats PCA as a special case of factor analysis (see Chapter 7). However, even with this misleading approach, numerically correct answers are produced by all the major statistical software packages, and provided that the user is careful to ensure that he or she understands the details of how the output is presented, it is usually adequate to use whichever software is most readily available.

Most statistical packages produce the basic results of a PCA satisfactorily, but few provide much in the way of extensions or add-ons as standard features. Some allow (or even encourage) rotation, though not necessarily in a sufficiently flexible manner to be useful, and some will display biplots. With most it is fairly straightforward to use the output from PCA in another part of the package so that PC regression, or discriminant or cluster analysis using PCs instead of the measured variables (see Chapters 8 and 9) are easily done. Beyond that, there are two possibilities for many of the extensions to PCA. Either software is available from the originator of the technique, or extra functions or code can be added to the more flexible software, such as S-PLUS or R.

A.1 Numerical Calculation of Principal Components

Most users of PCA, whether statisticians or non-statisticians, have little desire to know about efficient algorithms for computing PCs. Typically, a statistical program or package can be accessed that performs the analysis automatically. Thus, the user does not need to write his or her own programs; often the user has little or no interest in whether or not the software available performs its analyses efficiently. As long as the results emerge, the user is satisfied.

However, the type of algorithm used can be important, in particular if some of the last few PCs are of interest or if the data set is very large. Many programs for PCA are geared to looking mainly at the first few PCs, especially if PCA is included only as part of a factor analysis routine. In this case, several algorithms can be used successfully, although some will encounter problems if any pairs of the eigenvalues are very close together. When the last few or all of the PCs are to be calculated, difficulties are more likely to arise for some algorithms, particularly if some of the eigenvalues are very small.

Finding PCs reduces to finding the eigenvalues and eigenvectors of a positive-semidefinite matrix. We now look briefly at some of the possible algorithms that can be used to solve such an eigenproblem.

The Power Method

A form of the power method was described by Hotelling (1933) in his original paper on PCA, and an accelerated version of the technique was presented in Hotelling (1936). In its simplest form, the power method is a technique for finding the largest eigenvalue and the corresponding eigenvector of a $(p \times p)$ matrix $\mathbf{T}$. The idea is to choose an initial p-element vector $\mathbf{u}_0$, and then form the sequence

$$\mathbf{u}_1 = \mathbf{T}\mathbf{u}_0$$
$$\mathbf{u}_2 = \mathbf{T}\mathbf{u}_1 \quad = \mathbf{T}^2\mathbf{u}_0$$
$$\vdots \qquad \vdots$$
$$\mathbf{u}_r = \mathbf{T}\mathbf{u}_{r-1} = \mathbf{T}^r\mathbf{u}_0$$
$$\vdots \qquad \vdots$$

If $\boldsymbol{\alpha}_1, \boldsymbol{\alpha}_2, \ldots, \boldsymbol{\alpha}_p$ are the eigenvectors of $\mathbf{T}$, then they form a basis for p-dimensional space, and we can write, for arbitrary $\mathbf{u}_0$,

$$\mathbf{u}_0 = \sum_{k=1}^{p} \kappa_k \boldsymbol{\alpha}_k$$

for some set of constants $\kappa_1, \kappa_2, \ldots, \kappa_p$. Then

$$\mathbf{u}_1 = \mathbf{T}\mathbf{u}_0 = \sum_{k=1}^{p} \kappa_k \mathbf{T}\boldsymbol{\alpha}_k = \sum_{k=1}^{p} \kappa_k \lambda_k \boldsymbol{\alpha}_k,$$

where $\lambda_1, \lambda_2, \ldots, \lambda_p$ are the eigenvalues of $\mathbf{T}$. Continuing, we get for $r = 2, 3, \ldots$

$$\mathbf{u}_r = \sum_{k=1}^{p} \kappa_k \lambda_k^r \boldsymbol{\alpha}_k$$

and

$$\frac{\mathbf{u}_r}{(\kappa_1 \lambda_1^r)} = \left(\boldsymbol{\alpha}_1 + \frac{\kappa_2}{\kappa_1} \left(\frac{\lambda_2}{\lambda_1} \right)^r \boldsymbol{\alpha}_2 + \cdots + \frac{\kappa_p}{\kappa_1} \left(\frac{\lambda_p}{\lambda_1} \right)^r \boldsymbol{\alpha}_p \right).$$

Assuming that the first eigenvalue of $\mathbf{T}$ is distinct from the remaining eigenvalues, so that $\lambda_1 > \lambda_2 \geq \cdots \geq \lambda_p$, it follows that a suitably normalized version of $\mathbf{u}_r \to \boldsymbol{\alpha}_1$ as $r \to \infty$. It also follows that the ratios of corresponding elements of $\mathbf{u}_r$ and $\mathbf{u}_{r-1} \to \lambda_1$ as $r \to \infty$.

The power method thus gives a simple algorithm for finding the first (largest) eigenvalue of a covariance or correlation matrix and its corresponding eigenvector, from which the first PC and its variance can be derived. It works well if $\lambda_1 \gg \lambda_2$, but converges only slowly if λ_1 is not well separated from λ_2. Speed of convergence also depends on the choice of the initial vector $\mathbf{u}_0$; convergence is most rapid if $\mathbf{u}_0$ is close to $\boldsymbol{\alpha}_1$,

If $\lambda_1 = \lambda_2 > \lambda_3$, a similar argument to that given above shows that a suitably normalized version of $\mathbf{u}_r \rightarrow \boldsymbol{\alpha}_1 + (\kappa_2/\kappa_1)\boldsymbol{\alpha}_2$ as $r \rightarrow \infty$. Thus, the method does not lead to $\boldsymbol{\alpha}_1$, but it still provides information about the space spanned by $\boldsymbol{\alpha}_1$, $\boldsymbol{\alpha}_2$. Exact equality of eigenvalues is extremely unlikely for *sample* covariance or correlation matrices, so we need not worry too much about this case.

Rather than looking at all $\mathbf{u}_r$, $r = 1, 2, 3, \ldots$, attention can be restricted to $\mathbf{u}_1, \mathbf{u}_2, \mathbf{u}_4, \mathbf{u}_8, \ldots$ (that is $\mathbf{T}\mathbf{u}_0, \mathbf{T}^2\mathbf{u}_0, \mathbf{T}^4\mathbf{u}_0, \mathbf{T}^8\mathbf{u}_0, \ldots$) by simply squaring each successive power of $\mathbf{T}$. This accelerated version of the power method was suggested by Hotelling (1936). The power method can be adapted to find the second, third, ... PCs, or the last few PCs (see Morrison, 1976, p. 281), but it is likely to encounter convergence problems if eigenvalues are close together, and accuracy diminishes if several PCs are found by the method. Simple worked examples for the first and later components can be found in Hotelling (1936) and Morrison (1976, Section 8.4) .

There are various adaptations to the power method that partially overcome some of the problems just mentioned. A large number of such adaptations are discussed by Wilkinson (1965, Chapter 9), although some are not directly relevant to positive-semidefinite matrices such as covariance or correlation matrices. Two ideas that *are* of use for such matrices will be mentioned here. First, the origin can be shifted, that is the matrix $\mathbf{T}$ is replaced by $\mathbf{T} - \rho\mathbf{I}_p$, where $\mathbf{I}_p$ is the identity matrix, and ρ is chosen to make the ratio of the first two eigenvalues of $\mathbf{T} - \rho\mathbf{I}_p$ much larger than the corresponding ratio for $\mathbf{T}$, hence speeding up convergence.

A second modification is to use inverse iteration (with shifts), in which case the iterations of the power method are used but with $(\mathbf{T} - \rho\mathbf{I}_p)^{-1}$ replacing $\mathbf{T}$. This modification has the advantage over the basic power method with shifts that, by using appropriate choices of ρ (different for different eigenvectors), convergence to *any* of the eigenvectors of $\mathbf{T}$ can be achieved. (For the basic method it is only possible to converge in the first instance to $\boldsymbol{\alpha}_1$ or to $\boldsymbol{\alpha}_p$.) Furthermore, it is not necessary to explicitly calculate the inverse of $\mathbf{T} - \rho\mathbf{I}_p$, because the equation $\mathbf{u}_r = (\mathbf{T} - \rho\mathbf{I}_p)^{-1}\mathbf{u}_{r-1}$ can be replaced by $(\mathbf{T} - \rho\mathbf{I}_p)\mathbf{u}_r = \mathbf{u}_{r-1}$. The latter equation can then be solved using an efficient method for the solution of systems of linear equations (see Wilkinson, 1965, Chapter 4). Overall, computational savings with inverse iteration can be large compared to the basic power method (with or without shifts), especially for matrices with special structure, such as tridiagonal matrices. It turns out that an efficient way of computing PCs is to first transform the covariance or correlation matrix to tridiagonal form using, for example, either the Givens or Householder transformations (Wilkinson, 1965, pp. 282, 290), and then to implement inverse iteration with shifts on this tridiagonal form.

There is one problem with shifting the origin that has not yet been mentioned. This is the fact that to choose efficiently the values of ρ that determine the shifts, we need some preliminary idea of the eigenvalues of $\mathbf{T}$. This preliminary estimation can be achieved by using the method of bisection, which in turn is based on the Sturm sequence property of tridiagonal matrices. Details will not be given here (see Wilkinson, 1965, pp. 300–302), but the method provides a quick way of finding approximate values of the eigenvalues of a tridiagonal matrix. In fact, bisection could be used to find the eigenvalues to any required degree of accuracy, and inverse iteration implemented solely to find the eigenvectors.

Two major collections of subroutines for finding eigenvalues and eigenvectors for a wide variety of classes of matrix are the EISPACK package (Smith et al., 1976), which is distributed by IMSL, and parts of the NAG library of subroutines. In both of these collections, there are recommendations as to which subroutines are most appropriate for various types of eigenproblem. In the case where only a few of the eigenvalues and eigenvectors of a real symmetric matrix are required (corresponding to finding just a few of the PCs for a covariance or correlation matrix) both EISPACK and NAG recommend transforming to tridiagonal form using Householder transformations, and then finding eigenvalues and eigenvectors using bisection and inverse iteration respectively. NAG and EISPACK both base their subroutines on algorithms published in Wilkinson and Reinsch (1971), as do the 'Numerical Recipes' for eigensystems given by Press et al. (1992, Chapter 11).

The QL Algorithm

If all of the PCs are required, then methods other than those just described may be more efficient. For example, both EISPACK and NAG recommend that we should still transform the covariance or correlation matrix to tridiagonal form, but at the second stage the so-called QL algorithm should now be used, instead of bisection and inverse iteration. Chapter 8 of Wilkinson (1965) spends over 80 pages describing the QR and LR algorithms (which are closely related to the QL algorithm), but only a very brief outline will be given here.

The basic idea behind the QL algorithm is that any non-singular matrix $\mathbf{T}$ can be written as $\mathbf{T} = \mathbf{QL}$, where $\mathbf{Q}$ is orthogonal and $\mathbf{L}$ is lower triangular. (The $\mathbf{QR}$ algorithm is similar, except that $\mathbf{T}$ is written instead as $\mathbf{T} = \mathbf{QR}$, where $\mathbf{R}$ is upper triangular, rather than lower triangular.) If $\mathbf{T}_1 = \mathbf{T}$ and we write $\mathbf{T}_1 = \mathbf{Q}_1\mathbf{L}_1$, then $\mathbf{T}_2$ is defined as $\mathbf{T}_2 = \mathbf{L}_1\mathbf{Q}_1$. This is the first step in an iterative procedure. At the next step, $\mathbf{T}_2$ is written as $\mathbf{T}_2 = \mathbf{Q}_2\mathbf{L}_2$ and $\mathbf{T}_3$ is defined as $\mathbf{T}_3 = \mathbf{L}_2\mathbf{Q}_2$. In general, $\mathbf{T}_r$ is written as $\mathbf{Q}_r\mathbf{L}_r$ and $\mathbf{T}_{r+1}$ is then defined as $\mathbf{L}_r\mathbf{Q}_r$, $r = 1, 2, 3, \ldots$, where $\mathbf{Q}_1$, $\mathbf{Q}_2$, $\mathbf{Q}_3, \ldots$ are orthogonal matrices, and $\mathbf{L}_1$, $\mathbf{L}_2$, $\mathbf{L}_3, \ldots$ are lower triangular. It can be shown that $\mathbf{T}_r$ converges to a diagonal matrix, with the eigenvalues of **T** in decreasing absolute size down the diagonal. Eigenvectors can be found by accumulating the transformations in the QL algorithm (Smith et al., 1976, p. 468).

As with the power method, the speed of convergence of the QL algorithm depends on the ratios of consecutive eigenvalues. The idea of incorporating shifts can again be implemented to improve the algorithm and, unlike the power method, efficient strategies exist for finding appropriate shifts that do not rely on prior information about the eigenvalues (see, for example, Lawson and Hanson (1974, p. 109)). The QL algorithm can also cope with equality between eigenvalues.

It is probably fair to say that the algorithms described in detail by Wilkinson (1965) and Wilkinson and Reinsch (1971), and implemented in various IMSL and NAG routines, have stood the test of time. They still provide efficient ways of computing PCs in many circumstances. However, we conclude the Appendix by discussing two alternatives. The first is implementation *via* the singular value decomposition (SVD) of the data matrix, and the second consists of the various algorithms for PCA that have been suggested in the neural networks literature. The latter is a large topic and will be summarized only briefly.

One other type of algorithm that has been used recently to find PCs is the EM algorithm (Dempster et al., 1977). This is advocated by Tipping and Bishop (1999a,b) and Roweis (1997), and has its greatest value in cases where some of the data are missing (see Section 13.6).

Singular Value Decomposition

The suggestion that PCs may best be computed using the SVD of the data matrix (see Section 3.5) is not new. For example, Chambers (1977, p. 111) talks about the SVD providing the best approach to computation of principal components and Gnanadesikan (1977, p. 10) states that '... the recommended algorithm for ... obtaining the principal components is either the ... QR method ... or the singular value decomposition.' In constructing the SVD, it turns out that similar algorithms to those given above can be used. Lawson and Hanson (1974, p. 110) describe an algorithm (see also Wilkinson and Reinsch (1971)) for finding the SVD, which has two stages; the first uses Householder transformations to transform to an upper bidiagonal matrix, and the second applies an adapted QR algorithm. The method is therefore not radically different from that described earlier.

As noted at the end of Section 8.1, the SVD can also be useful in computations for regression (Mandel, 1982; Nelder, 1985), so the SVD has further advantages if PCA is used in conjunction with regression. Nash and Lefkovitch (1976) describe an algorithm that uses the SVD to provide a variety of results for regression, as well as PCs.

Another point concerning the SVD is that it provides simultaneously not only the coefficients and variances for the PCs, but also the scores of each observation on each PC, and hence all the information that is required to construct a biplot (see Section 5.3). The PC scores would otherwise need to be derived as an extra step after calculating the eigenvalues and eigenvectors of the covariance or correlation matrix $\mathbf{S} = \frac{1}{n-1}\mathbf{X}'\mathbf{X}$.

The values of the PC scores are related to the eigenvectors of $\mathbf{XX}'$, which can be derived from the eigenvectors of $\mathbf{X}'\mathbf{X}$ (see the proof of Property G4 in Section 3.2); conversely, the eigenvectors of $\mathbf{X}'\mathbf{X}$ can be found from those of $\mathbf{XX}'$. In circumstances where the sample size n is smaller than the number of variables p, $\mathbf{XX}'$ has smaller dimensions than $\mathbf{X}'\mathbf{X}$, so that it can be advantageous to use the algorithms described above, based on the power method or QL method, on a multiple of $\mathbf{XX}'$ rather than $\mathbf{X}'\mathbf{X}$ in such cases. Large computational savings are possible when $n \ll p$, as in chemical spectroscopy or in the genetic example of Hastie et al. (2000), which is described in Section 9.2 and which has $n = 48$, $p = 4673$. Algorithms also exist for updating the SVD if data arrive sequentially (see for example Berry et al. (1995)).

Neural Network Algorithms

Neural networks provide ways of extending PCA, including some non-linear generalizations (see Sections 14.1.3, 14.6.1). They also give alternative algorithms for estimating 'ordinary' PCs. The main difference between these algorithms and the techniques described earlier in the Appendix is that most are 'adaptive' rather than 'batch' methods. If the whole of a data set is collected before PCA is done and parallel processing is not possible, then batch methods such as the QR algorithm are hard to beat (see Diamantaras and Kung, 1996 (hereafter DK96), Sections 3.5.3, 4.4.1). On the other hand, if data arrive sequentially and PCs are re-estimated when new data become available, then adaptive neural network algorithms come into their own. DK96, Section 4.2.7 note that 'there is a plethora of alternative [neural network] techniques that perform PCA.' They describe a selection of single-layer techniques in their Section 4.2, with an overview of these in their Table 4.1. Different algorithms arise depending on

- whether the first or last few PCs are of interest;

- whether one or more than one PC is required;

- whether individual PCs are wanted or whether subspaces spanned by several PCs will suffice;

- whether the network is required to be biologically plausible.

DK96, Section 4.2.7 treat finding the last few PCs as a different technique, calling it *minor component analysis.*

In their Section 4.4, DK96 compare the properties, including speed, of seven algorithms using simulated data. In Section 4.5 they discuss multilayer networks.

Neural network algorithms are feasible for larger data sets than batch methods because they are better able to take advantage of developments in computer architecture. DK96, Chapter 8, discuss the potential for exploiting parallel VSLI (very large scale integration) systems, where the most appropriate algorithms may be different from those for non-parallel systems (DK96, Section 3.5.5). They discuss both digital and analogue implementations and their pros and cons (DK96, Section 8.3). Classical eigenvector-based algorithms are not easily parallelizable, whereas neural network algorithms are (DK96 pp. 205–207).

References

Aguilera, A.M., Gutiérrez, R., Ocaña, F.A. and Valderrama, M.J. (1995). Computational approaches to estimation in the principal component analysis of a stochastic process. *Appl. Stoch. Models Data Anal.*, **11**, 279–299.

Aguilera, A.M., Ocaña, F.A. and Valderrama, M.J. (1997). An approximated principal component prediction model for continuous time stochastic processes. *Appl. Stoch. Models Data Anal.*, **13**, 61–72.

Aguilera, A.M., Ocaña, F.A. and Valderrama, M.J. (1999a). Forecasting with unequally spaced data by a functional principal component analysis. *Test*, **8**, 233–253.

Aguilera, A.M., Ocaña, F.A. and Valderrama, M.J. (1999b). Forecasting time series by functional PCA. Discussion of several weighted approaches. *Computat. Statist.*, **14**, 443–467.

Ahamad, B. (1967). An analysis of crimes by the method of principal components. *Appl. Statist.*, **16**, 17–35.

Aires, F., Chedin, A. and Nadal, J.P. (2000). Independent component analysis of multivariate time series: Application to tropical SST variability. *J. Geophys. Res.—Atmos.*, **105 (D13)**, 17,437–17,455.

Aitchison, J. (1982). The statistical analysis of compositional data (with discussion). *J. R. Statist. Soc. B*, **44**, 139–177.

Aitchison, J. (1983). Principal component analysis of compositional data. *Biometrika*, **70**, 57–65.

Aitchison, J. (1986). *The Statistical Analysis of Compositional Data*. London: Chapman and Hall.

Akaike, H. (1974). A new look at the statistical model identification. *IEEE Trans. Autom. Cont.*, **19**, 716–723.

Aldenderfer, M.S. and Blashfield, R.K. (1984). *Cluster Analysis*. Beverly Hills: Sage.

Aldrin, M. (2000). Multivariate prediction using softly shrunk reduced-rank regression. *Amer. Statistician*, **54**, 29–34.

Ali, A., Clarke, G.M. and Trustrum, K. (1985). Principal component analysis applied to some data from fruit nutrition experiments. *Statistician*, **34**, 365–369.

Al-Kandari, N. (1998). *Variable Selection and Interpretation in Principal Component Analysis*. Unpublished Ph.D. thesis, University of Aberdeen.

Al-Kandari, N.M. and Jolliffe, I.T. (2001). Variable selection and interpretation of covariance principal components. *Commun. Statist.—Simul. Computat.*, **30**, 339-354.

Allan, R., Chambers, D., Drosdowsky, W., Hendon, H., Latif, M., Nicholls, N., Smith, I., Stone, R. and Tourre, Y. (2001). Is there an Indian Ocean dipole, and is it independent of the El Niño—Southern Oscillation? *CLIVAR Exchanges*, **6**, 18–22.

Allen, D.M. (1974). The relationship between variable selection and data augmentation and a method for prediction. *Technometrics*, **16**, 125–127.

Allen M.R. and Robertson, A.W. (1996). Distinguishing modulated oscillations from coloured noise in multivariate datasets. *Climate Dynam.*, **12**, 775–784.

Allen M.R. and Smith, L.A. (1996). Monte Carlo SSA: Detecting irregular oscillations in the presence of colored noise. *J. Climate*, **9**, 3373–3404.

Allen M.R. and Smith, L.A. (1997). Optimal filtering in singular spectrum analysis. *Phys. Lett. A*, **234**, 419–428.

Allen, M.R. and Tett, S.F.B. (1999). Checking for model consistency in optimal fingerprinting. *Climate Dynam.*, **15**, 419–434.

Ambaum, M.H.P., Hoskins, B.J. and Stephenson, D.B. (2001). Arctic oscillation or North Atlantic Oscillation. *J. Climate*, **14**, 3495–3507.

Anderson, A.B., Basilevsky, A. and Hum, D.P.J. (1983). Missing data: A review of the literature. In *Handbook of Survey Research*, eds. P.H. Rossi, J.D. Wright and A.B. Anderson, 415–494.

Anderson, T.W. (1957). Maximum likelihood estimates for a multivariate normal distribution when some observations are missing. *J. Amer. Statist. Assoc.*, **52**, 200–203.

Anderson, T.W. (1963). Asymptotic theory for principal component analysis. *Ann. Math. Statist.*, **34**, 122–148.

Anderson, T.W. (1984). Estimating linear statistical relationships. *Ann. Statist.*, **12**, 1–45.

Andrews, D.F. (1972). Plots of high-dimensional data. *Biometrics*, **28**, 125–136.

Apley, D.W. and Shi, J. (2001). A factor-analysis method for diagnosing variability in multivariate manufacturing processes. *Technometrics*, **43**, 84–95.

Arbuckle, J. and Friendly, M.L. (1977). On rotating to smooth functions. *Psychometrika*, **42**, 127–140.

Asselin de Beauville, J.-P. (1995). Non parametric discrimination by the nearest principal axis method (NPA)—preliminary study. In *Data Science and Its Application*, eds. Y. Escoufier, B. Fichet, E. Diday, L. Lebart, C. Hayashi, N. Ohsumi and Y. Baba, 145–154. Tokyo: Academic Press.

Atiqullah, M. and Uddin, M. (1993). Sweep-out components analysis. *J. Appl. Statist. Sci.*, **1**, 67–79.

Baba, Y. (1995). Scaling methods for ranked data. In *Data Science and Its Application*, eds. Y. Escoufier, B. Fichet, E. Diday, L. Lebart, C. Hayashi, N. Ohsumi and Y. Baba, 133–144. Tokyo: Academic Press.

Baccini, A., Besse, P. and de Falguerolles, A. (1996). A L_1-norm PCA and a heuristic approach. In *Ordinal and Symbolic Data Analysis*, eds. E. Diday, Y. Lechevalier and O. Opitz, 359–368. Berlin: Springer-Verlag.

Bacon-Shone, J. (1992). Ranking methods for compositional data. *Appl. Statist.*, **41**, 533–537.

Bargmann, R.E. and Baker, F.D. (1977). A minimax approach to component analysis. In *Applications of Statistics*, ed. P.R. Krishnaiah, 55–69. Amsterdam: North-Holland.

Barnett, T.P. and Hasselmann, K. (1979). Techniques of linear prediction, with application to oceanic and atmospheric fields in the tropical Pacific. *Rev. Geophys. Space Phys.*, **17**, 949–968.

Barnett, V. (1981). *Interpreting Multivariate Data*. Chichester: Wiley.

Barnett, V. and Lewis, T. (1994). *Outliers in Statistical Data*, 3rd edition. Chichester: Wiley.

Bärring, L. (1987). Spatial patterns of daily rainfall in central Kenya: Application of principal component analysis, common factor analysis and spatial correlation. *J. Climatol.*, **7**, 267–289.

Bartels, C.P.A. (1977). *Economic Aspects of Regional Welfare, Income Distribution and Unemployment*. Leiden: Martinus Nijhoff.

Bartholomew, D.J. and Knott, M. (1999). *Latent Variable Models and Factor Analysis*, 2nd edition. London: Arnold.

Bartkowiak, A. (1982). The choice of representative variables by stepwise regression. *Zastosowania Matematyki*, **17**, 527–538.

Bartkowiak, A. (1991). How to reveal dimensionality of the data? In *Applied Stochastic Models and Data Analysis*, eds. R. Gutiérrez and M.J. Valderrama, 55–64. Singapore: World Scientific.

Bartkowiak, A., Lukasik, S., Chwistecki, K., Mrukowicz, M. and Morgenstern, W. (1988). Transformation of data and identification of outliers—as experienced in an epidemiological study. *EDV in Medizin und Biologie*, **19**, 64–69.

Bartkowiak, A. and Szustalewicz, A. (1996). Some issues connected with a 3D representation of multivariate data points. *Machine Graph. Vision*, **5**, 563–577.

Bartlett, M.S. (1950). Tests of significance in factor analysis. *Brit. J. Psychol. Statist. Section*, **3**, 77–85.

Bartoletti, S., Flury, B.D. and Nel, D.G. (1999). Allometric extension. *Biometrics*, **55**, 1210–1214.

Bartzokas, A., Metaxas, D.A. and Ganas, I.S. (1994). Spatial and temporal sea-surface temperatures in the Mediterranean. *Int. J. Climatol.*, **14**, 201–213.

Basilevsky, A. and Hum, D.P.J. (1979). Karhunen-Loève analysis of historical time series with an application to plantation births in Jamaica. *J. Amer. Statist. Assoc.*, **74**, 284–290.

Baskerville, J.C. and Toogood, J.H. (1982). Guided regression modeling for prediction and exploration of structure with many explanatory variables. *Technometrics*, **24**, 9–17.

Bassett, E.E., Clewer, A., Gilbert, P. and Morgan, B.J.T. (1980). Forecasting numbers of households: The value of social and economic information. Unpublished report. University of Kent.

Baxter, M.J. (1993). Principal component analysis of ranked compositional data: An empirical study. Research Report 17/93, Department of Mathematics, Statistics and Operational Research, Nottingham Trent University.

Baxter, M.J. (1995). Standardization and transformation in principal component analysis, with applications to archaeometry. *Appl. Statist.*, **44**, 513–527.

Beale, E.M.L. and Little, R.J.A. (1975). Missing values in multivariate analysis. *J. R. Statist. Soc. B*, **37**, 129–145.

Bekker, P. and de Leeuw, J. (1988). Relations between variants of nonlinear principal component analysis. In *Component and Correspondence Analysis. Dimension Reduction by Functional Approximation*, eds. J.L.A. van Rijckevorsel and J. de Leeuw, 1–31. Chichester: Wiley.

Belsley, D.A. (1984). Demeaning conditioning diagnostics through centering (with comments). *Amer. Statistician*, **38**, 73–93.

Beltrando, G. (1990). Space-time variability of rainfall in April and October–November over East Africa during the period 1932–1983. *Int. J. Climatol.*, **10**, 691–702.

Beltrami, E. (1873). Sulle funzioni bilineari. *Giornale di Mathematiche di Battaglini*, **11**, 98–106.

Benasseni, J. (1986a). Une amélioration d'un résultat concernant l'influence d'une unité statistique sur les valeurs propres en analyse en composantes principales. *Statistique et Analyse des Données*, **11**, 42–63.

Benasseni, J. (1986b). Stabilité en A.C.P. par rapport aux incertitudes de mesure. In *Data Analysis and Informatics 4*, eds. E. Diday, Y. Escoufier, L. Lebart, J.P. Pagès, Y. Schektman, R. Tomassone, 523–533. Amsterdam: North Holland.

Benasseni, J. (1987a). Perturbation des poids des unités statistique et approximation en analyse en composantes principales. *RAIRO—Recherche Operationelle—Operations Research*, **21**, 175–198.

Benasseni, J. (1987b). Sensitivity of principal component analysis to data perturbation. *Fifth International Symposium: Data Analysis and Informatics*, Tome 1, 249–256.

Benasseni, J. (1990). Sensitivity coefficients for the subspaces spanned by principal components. *Commun. Statist.—Theor. Meth.* **19**, 2021–2034.

Bensmail, H. and Celeux, G. (1996). Regularized Gaussian discriminant analysis through eigenvalue decomposition. *J. Amer. Statist. Assoc.*, **91**, 1743–1748.

Bentler, P.M. and Yuan, K.-H. (1996). Test of linear trend in eigenvalues of a covariance matrix with application to data analysis. *Brit. J. Math. Statist. Psychol.*, 49, 299–312.

Bentler, P.M. and Yuan, K.-H. (1998). Tests for linear trend in the smallest eigenvalues of the correlation matrix. *Psychometrika*, **63**, 131–144.

Benzécri, J.-P. (1980). *L'Analyse des Données*. Tome (Vol.) 2: *L'Analyse des Correspondances*, 3rd edition. Paris: Dunod.

Benzécri, J.-P. (1992). *Correspondence Analysis Handbook*. New York: Marcel Dekker.

Benzi, R., Deidda, R. and Marrocu, M. (1997). Characterization of temperature and precipitation fields over Sardinia with principal component analysis and singular spectrum analysis. *Int. J. Climatol.*, **17**, 1231–1262.

Beran, R. and Srivastava, M.S. (1985). Bootstrap tests and confidence regions for functions of a covariance matrix. *Ann. Statist.*, **13**, 95–115.

Berkey, C.S., Laird, N.M., Valadian, I. and Gardner, J. (1991). Modeling adolescent blood pressure patterns and their prediction of adult pressures. *Biometrics*, **47**, 1005–1018.

Berry, M.W., Dumais, S.T. and Letsche, T.A. (1995). Computational methods for intelligent information access. Paper presented at Supercomputing '95, San Diego, December 1995.

Bertrand, D., Qannari, E.M. and Vigneau, E. (2001). Latent root regression analysis: An alternative method to PLS. *Chemometrics Intell. Lab. Syst.*, **58**, 227–234.

Berk, K.N. (1984). Validating regression procedures with new data. *Technometrics*, **26**, 331–338.

Besse, P. (1988). Spline functions and optimal metric in linear principal component analysis. In *Component and Correspondence Analysis. Dimension Reduction by Functional Approximation*, eds. J.L.A. van Rijckevorsel and J. de Leeuw, 81–101. Chichester: Wiley.

Besse, P. (1992). PCA stability and choice of dimensionality. *Stat. Prob. Lett.*, **13**, 405–410.

Besse, P.C. (1994a). Insight of a dreamed PCA. In *SEUGI/CLUB SAS Proceedings*, 744–759.

Besse, P.C. (1994b). Models for multivariate data analysis. In *COMPSTAT 94*, eds. R. Dutter and W. Grossmann, 271–285. Heidelberg: Physica-Verlag.

Besse, P.C., ⌐   t, H. and Ferraty, F. (1997). Simultaneous non-parame⁺   :ons of unbalanced longitudinal data. *Computat. St⌐   .nal.*, **24**, 255–270.

 , Cardot, H. and Stephenson, D.B. (2000). Autoregressive fore-
Cing of some functional climatic variations. *Scand. J. Statist.*, **27**, ы/3–687.

Besse, P. and de Falguerolles, A. (1993). Application of resampling methods to the choice of dimension in principal component analysis. In *Computer Intensive Methods in Statistics*, eds. W. Härdle and L. Simar, 167–176. Heidelberg: Physica-Verlag.

Besse, P.C. and Ferraty, F. (1995). A fixed effect curvilinear model. *Computat. Statist.*, **10**, 339–351.

Besse, P. and Ferre, L. (1993). Sur l'usage de la validation croisée en analyse en composantes principales. *Rev. Statistique Appliquée*, **41**, 71–76.

Besse, P. and Ramsay, J.O. (1986). Principal components analysis of sampled functions. *Psychometrika*, **51**, 285–311.

Bhargava, R.P. and Ishizuka, T. (1981). Selection of a subset of variables from the viewpoint of variation—an alternative to principal component analysis. In *Proc. Indian Statist. Inst. Golden Jubilee Int. Conf. on Statistics: Applications and New Directions*, 33–44.

Bibby, J. (1980). Some effects of rounding optimal estimates. *Sankhya B*, **42**, 165–178.

Bishop, C.M. (1995) *Neural Networks for Pattern Recognition*. Oxford: Clarendon Press.

Bishop, C.M. (1999). Bayesian PCA. In *Advances in Neural Information Processing Systems, 11*, eds. S.A. Solla, M.S. Kearns and D.A. Cohn, 382–388. Cambridge: MIT Press.

Bishop, Y.M.M., Fienberg, S.E. and Holland, P.W. (1975). *Discrete Multivariate Analysis: Theory and Practice*. Cambridge: MIT Press.

Blackith, R.E. and Reyment, R.A. (1971). *Multivariate Morphometrics*. London: Academic Press.

Bloomfield, P. (1974). Linear transformations for multivariate binary data. *Biometrics*, **30**, 609–617.

Bloomfield, P. and Davis, J.M. (1994). Orthogonal rotation of complex principal components. *Int. J. Climatol.*, **14**, 759–775.

Böhning, D. (1999). *Computer-Assisted Analysis of Mixtures and Applications Meta-analysis, Disease Mapping and Others*. Boca Raton: Chapman and Hall/CRC.

Boik, R.J. (1986). Testing the rank of a matrix with applications to the analysis of interaction in ANOVA. *J. Amer. Statist. Assoc.*, **81**, 243–248.

Bolton, R.J. and Krzanowski, W.J. (1999). A characterization of principal components for projection pursuit. *Amer. Statistician*, **53**, 108–109.

Boneh, S. and Mendieta, G.R. (1994). Variable selection in regression models using principal components. *Commun. Statist.—Theor. Meth.*, **23**, 197–213.

Bookstein, F.L. (1989). 'Size and shape': A comment on semantics. *Syst. Zool.*, **38**, 173–180.

Bookstein, F.L. (1991). *Morphometric Tools for Landmark Data: Geometry and Biology*. Cambridge: Cambridge University Press.

Bouhaddou, O., Obled, C.H. and Dinh, T.P. (1987). Principal component analysis and interpolation of stochastic processes: Methods and simulation. *J. Appl. Statist.*, **14**, 251–267.

Boyles, R.A. (1996). Multivariate process analysis with lattice data. *Technometrics*, **38**, 37–49.

Bretherton, C.S., Smith, C. and Wallace, J.M. (1992). An intercomparison of methods for finding coupled patterns in climate data. *J. Climate*, **5**, 541–560.

Briffa, K.R., Jones, P.D., Wigley, T.M.L., Pilcher, J.R. and Baillie, M.G.L. (1986). Climate reconstruction from tree rings: Part 2, spatial reconstruction of summer mean sea-level pressure patterns over Great Britain. *J. Climatol.*, **6**, 1–15.

Brillinger, D.R. (1981). *Time Series: Data Analysis and Theory*. Expanded edition. San Francisco: Holden-Day.

Brockwell, P.J. and Davis, R.A. (1996). *Introduction to Time Series and Forecasting*. New York: Springer.

Brooks, S. (1992). *Constrained Principal Components*. Unpublished M.Sc. project report. University of Kent at Canterbury.

Brooks, S.P. (1994). Diagnostics for principal components: Influence functions as diagnostic tools. *J. Appl. Statist.*, **43**, 483–494.

Browne, M.W. (1979). The maximum-likelihood solution in inter-battery factor analysis. *Brit. J. Math. Stat. Psychol.*, **32**, 75–86.

Bryant, E.H. and Atchley, W.R. (1975). *Multivariate Statistical Methods: Within Group Covariation*. Stroudsberg: Halsted Press.

Buckland, S.T. and Anderson, A.J.B. (1985). Multivariate analysis of Atlas data. In *Statistics in Ornithology*, eds. B.J.T. Morgan and P.M. North, 93–112. Berlin: Springer-Verlag.

Buell, C.E. (1975). The topography of the empirical orthogonal functions. *Fourth Conference on Probability and Statistics in Atmospheric Science*, 188–193. American Meteorological Society.

Buell, C.E. (1978). The number of significant proper functions of two-dimensional fields. *J. Appl. Meteorol.*, **17**, 717–722.

Buell, C.E. (1979). On the physical interpretation of empirical orthogonal functions. *Sixth Conference on Probability and Statistics in Atmospheric Science*, 112–117. American Meteorological Society.

Burnham, A.J., MacGregor, J.F. and Viveros, R. (1999). Latent variable multivariate regression modelling. *Chemometrics Intell. Lab. Syst.*, **48**, 167–180.

Butler, N.A. and Denham, M.C. (2000). The peculiar shrinkage properties of partial least squares regression. *J. R. Statist. Soc. B*, **62**, 585–593.

Cadima, J. (2000). A scale-invariant component analysis: MCCA. Technical report 4, Departamento de Matemática, Instituto Superior de Agronomia, Universidade Técnica de Lisboa.

Cadima, J., Cerdeira, J.O. and Minhoto, M. (2002). A computational study of algorithms for variable selection in multivariate statistics. Submitted for publication.

Cadima, J. and Jolliffe, I.T. (1995). Loadings and correlations in the interpretation of principal components. *J. Appl. Statist.*, **22**, 203–214.

Cadima, J.F.C.L. and Jolliffe, I.T. (1996). Size- and shape-related principal component analysis. *Biometrics*, **52**, 710–716.

Cadima, J. and Jolliffe, I. (1997). Some comments on ten Berge, J.M.F. and Kiers, H.A.L. (1996). Optimality criteria for principal component analysis and generalizations. *Brit. J. Math. Stat. Psychol.*, **50**, 365–366.

Cadima, J.F.C.L. and Jolliffe, I.T. (2001). Variable selection and the interpretation of principal subspaces. *J. Agri. Biol. Environ. Statist.*, **6**, 62–79.

Cahalan, R.F. (1983). EOF spectral estimation in climate analysis. *Second International Meeting on Statistical Climatology*, Preprints volume, 4.5.1–4.5.7.

Cai, W. and Baines, P. (2001). Forcing of the Antarctic Circumpolar Wave by ENSO teleconnections. *J. Geophys. Res.—Oceans*, **106**, 9019–9038.

Cailliez, F. and Pagès, J.-P. (1976). *Introduction à l'Analyse des Données*. Paris: SMASH.

Calder, P. (1986). *Influence Functions in Multivariate Analysis*. Unpublished Ph.D. thesis, University of Kent at Canterbury.

Campbell, N.A. (1980). Robust procedures in multivariate analysis 1: Robust covariance estimation. *Appl. Statist.*, **29**, 231–237.

Campbell, N.A. and Atchley, W.R. (1981). The geometry of canonical variate analysis. *Syst. Zool.*, **30**, 268–280.

Capra, W.B. and Müller, H.-G. (1997). An accelerated-time model for response curves. *J. Amer. Statist. Assoc.*, **92**, 72–83.

Carr, D.B. (1998). Multivariate graphics. In *Encyclopedia of Biostatistics*, eds. P. Armitage and T. Colton, 2864–2886. Chichester: Wiley.

Casin, Ph. (2001). A generalization of principal component analysis to K sets of variables. *Computat. Statist. Data Anal.*, **35**, 417–428.

Castro, P.E., Lawton, W.H., and Sylvestre, E.A. (1986). Principal modes of variation for processes with continuous sample curves. *Technometrics*, **28**, 329–337.

Cattell, R.B. (1966). The scree test for the number of factors. *Multiv. Behav. Res.*, **1**, 245–276.

Cattell, R.B. (1978). *The Scientific Use of Factor Analysis in Behavioral and Life Sciences*. New York: Plenum Press.

Cattell, R.B. and Vogelmann, S. (1977). A comprehensive trial of the scree and KG criteria for determining the number of factors. *Mult. Behav. Res.*, **12**, 289–325.

Caussinus, H. (1986). Models and uses of principal component analysis: A comparison emphasizing graphical displays and metric choices. In *Multidimensional Data Analysis*, eds. J. de Leeuw, W. Heiser, J. Meulman and F. Critchley, 149–178. Leiden: DSWO Press.

Caussinus, H. (1987). Discussion of 'What is projection pursuit?' by Jones and Sibson. *J. R. Statist. Soc. A*, **150**, 26.

Caussinus, H. and Ferré, L. (1992). Comparing the parameters of a model for several units by means of principal component analysis. *Computat. Statist. Data Anal.*, **13**, 269–280.

Caussinus, H., Hakam, S. and Ruiz-Gazen, A. (2001). Projections révélatrices contrôlées. Recherche d'individus atypiques. To appear in *Rev. Statistique Appliquée*.

Caussinus, H. and Ruiz, A. (1990) Interesting projections of multidimensional data by means of generalized principal component analysis. In *COMPSTAT 90*, eds. K. Momirovic and V. Mildner, 121–126. Heidelberg: Physica-Verlag.

Caussinus, H. and Ruiz-Gazen, A. (1993). Projection pursuit and generalized principal component analysis. In *New Directions in Statistical Data Analysis and Robustness*, eds. S. Morgenthaler, E. Ronchetti and W.A. Stahel, 35–46. Basel: Birkhäuser Verlag.

Caussinus, H. and Ruiz-Gazen, A. (1995). Metrics for finding typical structures by means of principal component analysis. In *Data Science and Its Application*, eds. Y. Escoufier, B. Fichet, E. Diday, L. Lebart, C. Hayashi, N. Ohsumi and Y. Baba, 177–192. Tokyo: Academic Press.

Chambers, J.M. (1977). *Computational Methods for Data Analysis*. New York: Wiley.

Chambers, J.M., Cleveland, W.S., Kleiner, B. and Tukey, P.A. (1983). *Graphical Methods for Data Analysis*. Belmont: Wadsworth.

Champely, S. and Doledec, S. (1997). How to separate long-term trends from periodic variation in water quality monitoring. *Water Res.*, **11**, 2849–2857.

Chang, W.-C. (1983). On using principal components before separating a mixture of two multivariate normal distributions. *Appl. Statist.*, **32**, 267–275.

Chatfield, C. and Collins, A.J. (1989). *Introduction to Multivariate Analysis*. London: Chapman and Hall.

Cheng, C.-L. and van Ness, J.W. (1999). *Statistical Regression with Measurement Error*. London: Arnold.

Chernoff, H. (1973). The use of faces to represent points in k-dimensional space graphically. *J. Amer. Statist. Assoc.*, **68**, 361–368.

Cherry, S. (1997). Some comments on singular value decomposition analysis. *J. Climate*, **10**, 1759–1761.

Chipman, H.A. and Gu, H. (2002). Interpretable dimension reduction. To appear in *J. Appl. Statist.*

Chouakria, A., Cazes, P. and Diday, E. (2000). Symbolic principal component analysis. In *Analysis of Symbolic Data. Exploratory Methods for Extracting Statistical Information from Complex Data*, eds. H.-H. Bock and E. Diday, 200–212. Berlin: Springer-Verlag.

Clausen, S.-E. (1998). *Applied Correspondence Analysis: An Introduction*. Thousand Oaks: Sage.

Cleveland, W.S. (1979). Robust locally weighted regression and smoothing scatterplots. *J. Amer. Statist. Assoc.*, **74**, 829–836.

Cleveland, W.S. (1981). LOWESS: A program for smoothing scatterplots by robust locally weighted regression. *Amer. Statistician*, **35**, 54.

Cleveland, W.S. and Guarino, R. (1976). Some robust statistical procedures and their application to air pollution data. *Technometrics*, **18**, 401–409.

Cochran, R.N. and Horne, F.H., (1977). Statistically weighted principal component analysis of rapid scanning wavelength kinetics experiments. *Anal. Chem.*, **49**, 846–853.

Cohen, S.J. (1983). Classification of 500 mb height anomalies using obliquely rotated principal components. *J. Climate Appl. Meteorol.*, **22**, 1975–1988.

Cohn, R.D. (1999). Comparisons of multivariate relational structures in serially correlated data. *J. Agri. Biol. Environ. Statist.*, **4**, 238–257.

Coleman, D. (1985). Hotelling's T^2, robust principal components, and graphics for SPC. Paper presented at the 1985 Annual Meeting of the American Statistical Association.

Commandeur, J.J.F, Groenen, P.J.F and Meulman, J.J. (1999). A distance-based variety of nonlinear multivariate data analysis, including weights for objects and variables. *Psychometrika*, **64**, 169–186.

Compagnucci, R.H., Araneo, D. and Canziani, P.O. (2001). Principal sequence pattern analysis: A new approach to classifying the evolution of atmospheric systems. *Int. J. Climatol.*, **21**, 197–217.

Compagnucci, R.H. and Salles, M.A. (1997). Surface pressure patterns during the year over Southern South America. *Int. J. Climatol.*, **17**, 635–653.

Cook, R.D. and Weisberg, S. (1982). *Residuals and Influence in Regression*. New York: Chapman and Hall.

Cook, R.D. (1986). Assessment of local influence. *J. R. Statist. Soc. B*, **48**, 133–169 (including discussion).

Coppi, R. and Bolasco, S. (eds.) (1989). *Multiway Data Analysis*. Amsterdam: North-Holland.

Corbitt, B. and Ganesalingam, S. (2001). Comparison of two leading multivariate techniques in terms of variable selection for linear discriminant analysis. *J. Statist. Manag. Syst.*, **4**, 93–108.

Corsten, L.C.A. and Gabriel, K.R. (1976). Graphical exploration in comparing variance matrices. *Biometrics*, **32**, 851–863.

Cox, D.R. (1972). The analysis of multivariate binary data. *Appl. Statist.*, **21**, 113–120.

Cox, T.F. and Cox, M.A.A. (2001). *Multidimensional Scaling*, 2nd edition. Boca Raton: Chapman and Hall.

Craddock, J.M. (1965). A meteorological application of principal component analysis. *Statistician*, **15**, 143–156.

Craddock, J.M. and Flintoff, S. (1970). Eigenvector representations of Northern Hemispheric fields. *Q.J.R. Met. Soc.*, **96**, 124–129.

Craddock, J.M. and Flood, C.R. (1969). Eigenvectors for representing the 500 mb. geopotential surface over the Northern Hemisphere. *Q.J.R. Met. Soc.*, **95**, 576–593.

Craw, I. and Cameron, P. (1992). Face recognition by computer. *Proc. Br. Machine Vision Conf.*, 489–507. Berlin: Springer-Verlag.

Critchley, F. (1985). Influence in principal components analysis. *Biometrika*, **72**, 627–636.

Crone, L.J. and Crosby, D.S. (1995). Statistical applications of a metric on subspaces to satellite meteorology. *Technometrics*, **37**, 324–328.

Croux, C. and Haesbroeck, G. (2000). Principal component analysis based on robust estimators of the covariance or correlation matrix: Influence functions and efficiencies. *Biometrika*, **87**, 603–618.

Croux, C. and Ruiz-Gazen, A. (1995). A fast algorithm for robust principal components based on projection pursuit. In *COMPSTAT 96*, ed. A. Prat, 211–216.

Croux, C. and Ruiz-Gazen, A. (2000). High breakdown estimators for principal components: the projection-pursuit approach revisited. Preprint 2000/149. Institut de Statistique et de Recherche Opérationelle, Université Libre de Bruxelles.

Cuadras, C.M. (1998). Comment on 'Some cautionary notes on the use of principal components regression'. *Amer. Statistician*, **52**, 371.

Cubadda, G. (1995). A note on testing for seasonal co-integration using principal components in the frequency domain. *J. Time Series Anal.*, **16**, 499–508.

Dahl, K.S., Piovoso, M.J. and Kosanovich, K.A. (1999). Translating third-order data analysis methods to chemical batch processes. *Chemometrics Intell. Lab. Syst.*, **46**, 161–180.

Daigle, G. and Rivest, L.-P. (1992). A robust biplot. *Canad. J. Statist.*, **20**, 241–255.

Daling, J.R. and Tamura, H. (1970). Use of orthogonal factors for selection of variables in a regression equation—an illustration. *Appl. Statist.*, **19**, 260–268.

Darnell, A.C. (1994). *A Dictionary of Econometrics*. Aldershot: Edward Elgar.

Darroch, J.N. and Mosimann, J.E. (1985). Canonical and principal components of shape. *Biometrika*, **72**, 241–252.

Daudin, J.J., Duby, C. and Trécourt, P. (1988). Stability of principal component analysis studied by the bootstrap method. *Statistics*, **19**, 241–258.

Daudin, J.J., Duby, C. and Trécourt, P. (1988). PCA stability studied by the bootstrap and the infinitesimal jackknife method. *Statistics*, **20**, 255–270.

Daultrey, S. (1976). *Principal Components Analysis*. Norwich: Geo Abstracts.

Davenport, M. and Studdert-Kennedy, G. (1972). The statistical analysis of aesthetic judgment: An exploration. *Appl. Statist.*, **21**, 324–333.

Davies, P.T. and Tso, M.K.-S. (1982). Procedures for reduced-rank regression. *Appl. Statist.*, **31**, 244–255.

Davison, M.L. (1983). *Multidimensional Scaling*. New York: Wiley.

Dawkins, B. (1990). Reply to Comment on Dawkins (1989) by W.F. Kuhfeld. *Amer. Statistician*, **44**, 58–60.

Dear, R.E. (1959). *A Principal Components Missing Data Method for Multiple Regression Models*. SP-86. Santa Monica: Systems Development Corporation.

de Falguerolles, A. (2000). GBMs: GLMs with bilinear terms. In *COMPSTAT 2000*, eds. J.G. Bethlehem and P.G.M. van der Heijden, 53–64. Heidelberg: Physica-Verlag.

de Falguerolles, A. and Jmel, S. (1993). Un critère de choix de variables en analyses en composantes principales fondé sur des modèles graphiques gaussiens particuliers. *Canad. J. Statist.*, **21**, 239–256.

de Leeuw, J. (1986). Comment on Caussinus. In *Multidimensional Data Analysis*, eds. J. de Leeuw, W. Heiser, J. Meulman and F. Critchley, 171–176. Leiden: DSWO Press.

de Leeuw, J. and van Rijckevorsel, J. (1980). Homals and Princals. Some generalizations of principal components analysis. In *Data Analysis and Informatics*, eds. E. Diday, L. Lebart, J.P. Pagès and R. Tomassone, 231–242. Amsterdam: North-Holland.

de Ligny, C.L., Nieuwdorp, G.H.E., Brederode, W.K., Hammers, W.E. and van Houwelingen, J.C. (1981). An application of factor analysis with missing data. *Technometrics*, **23**, 91–95.

Dempster, A.P. (1969). *Elements of Continuous Multivariate Analysis*. Reading, Massachusetts: Addison-Wesley.

Dempster, A.P., Laird, N.M. and Rubin, D.B. (1977). Maximum likelihood from incomplete data via the EM algorithm. *J. R. Statist. Soc. B*, **39**, 1–38 (including discussion).

Denham, M.C. (1985). Unpublished postgraduate diploma project report. University of Kent at Canterbury.

DeSarbo, W. and Jedidi, K. (1987). Redundancy analysis. In *Encyclopedia of Statistical Science, Vol 7*, eds. S. Kotz and N.L. Johnson, 662–666. New York: Wiley.

Devijver, P.A. and Kittler, J. (1982). *Pattern Recognition: A Statistical Approach*. Englewood Cliffs: Prentice Hall.

Deville, J.C. and Malinvaud, E. (1983). Data analysis in official socio-economic statistics (with discussion). *J. R. Statist. Soc. A*, **146**, 335–361.

Devlin, S.J., Gnanadesikan, R. and Kettenring, J.R. (1975). Robust estimation and outlier detection with correlation coefficients. *Biometrika*, **62**, 531–545.

Devlin, S.J., Gnanadesikan, R. and Kettenring, J.R. (1981). Robust estimation of dispersion matrices and principal components. *J. Amer. Statist. Assoc.*, **76**, 354–362.

Diaconis, P. and Efron, B. (1983). Computer-intensive methods in statistics. *Scientific Amer.*, **248**, 96–108.

Diamantaras, K.I. and Kung, S.Y. (1996). *Principal Component Neural Networks Theory and Applications*. New York: Wiley.

Digby, P.G.N. and Kempton, R.A. (1987). *Multivariate Analysis of Ecological Communities*. London: Chapman and Hall.

Dillon, W.R., Mulani, N. and Frederick, D.G. (1989). On the use of component scores in the presence of group structures. *J. Consum. Res.*, **16**, 106–112.

Dong, D. and McAvoy, T.J. (1996). Non-linear principal component analysis based on principal curve and neural networks. *Computers Chem. Engng.*, **20**, 65–78.

Donnell, D.J., Buja, A. and Stuetzle, W. (1994). Analysis of additive dependencies and concurvities using smallest additive principal components. *Ann. Statist.*, **22**, 1635–1673.

Doran, H.E. (1976). A spectral principal components estimator of the distributed lag model. *Int. Econ. Rev.*, **17**, 8–25.

Draper, N.R. and Smith, H. (1998). *Applied Regression Analysis*, 3rd edition. New York: Wiley.

Dryden, I.L. and Mardia, K.V. (1998). *Statistical Shape Analysis*. Chichester: Wiley.

Dudziński, M.L., Norris, J.M., Chmura, J.T. and Edwards, C.B.H. (1975). Repeatability of principal components in samples: Normal and non-normal data sets compared. *Multiv. Behav. Res.*, **10**, 109–117.

Dunn, J.E. and Duncan, L. (2000). Partitioning Mahalanobis D^2 to sharpen GIS classification. *University of Arkansas Statistical Laboratory Technical Report No. 29.*

Dunteman, G.H. (1989). *Principal Components Analysis.* Beverly Hills: Sage.

Durbin, J. (1984). Time series analysis. Present position and potential developments: Some personal views. *J. R. Statist. Soc. A*, **147**, 161–173.

Durbin, J. and Knott, M. (1972). Components of Cramér–von Mises statistics I. *J. R. Statist. Soc. B*, **34**, 290–307 (correction, **37**, 237).

Durbin, J., Knott, M. and Taylor, C.C. (1975). Components of Cramér–von Mises statistics II. *J. R. Statist. Soc. B*, **37**, 216–237.

Eastment, H.T. and Krzanowski, W.J. (1982). Cross-validatory choice of the number of components from a principal component analysis. *Technometrics*, **24**, 73–77.

Efron, B. and Tibshirani, R.J. (1993). *An Introduction to the Bootstrap.* New York: Chapman and Hall.

Eggett, D.L. and Pulsipher, B.A. (1989). Principal components in multivariate control charts. Paper presented at the American Statistical Association Annual Meeeting, August 1989, Washington, D.C.

Elmore, K.L. and Richman, M.B. (2001). Euclidean distance as a similarity metric for principal component analysis. *Mon. Weather Rev.*, **129**, 540–549.

Elsner, J.B. and Tsonis, A.A. (1996). *Singular Spectrum Analysis: A New Tool in Time Series Analyis.* New York: Plenum Press.

Escoufier, Y. (1986). A propos du choix des variables en analyse des données. *Metron*, **44**, 31–47.

Escoufier, Y. (1987). The duality diagram: a means for better practical application. In *Developments in Numerical Ecology*, eds. P. Legendre and L. Legendre, 139-156. Berlin: Springer-Verlag.

Esposito, V. (1998). Deterministic and probabilistic models for symmetrical and non symmetrical principal component analysis. *Metron*, **56**, 139–154.

Everitt, B.S. (1978). *Graphical Techniques for Multivariate Data.* London: Heinemann Educational Books.

Everitt, B.S. and Dunn, G. (2001). *Applied Multivariate Data Analysis*, 2nd edition. London: Arnold.

Everitt, B.S., Landau, S. and Leese, M. (2001). *Cluster Analysis*, 4th edition. London: Arnold.

Fancourt, C.L. and Principe, J.C. (1998). Competitive principal component analysis for locally stationary time series. *IEEE Trans. Signal Proc.*, **11**, 3068–3081.

Farmer, S.A. (1971). An investigation into the results of principal component analysis of data derived from random numbers. *Statistician*, **20**, 63–72.

Feeney, G.J. and Hester, D.D. (1967). Stock market indices: A principal components analysis. In *Risk Aversion and Portfolio Choice*, eds. D.D. Hester and J. Tobin, 110–138. New York: Wiley.

Fellegi, I.P. (1975). Automatic editing and imputation of quantitative data. *Bull. Int. Statist. Inst.*, **46**, (3), 249–253.

Ferré, L. (1990). A mean square error criterion to determine the number of components in a principal component analysis with known error structure. Preprint. Laboratoire de Statistique et Probabilités, University Paul Sabatier, Toulouse.

Ferré, L. (1995a). Improvement of some multidimensional estimates by reduction of dimensionality. *J. Mult. Anal.*, **54**, 147–162.

Ferré, L. (1995b). Selection of components in principal component analysis: A comparison of methods. *Computat. Statist. Data Anal.*, **19**, 669–682.

Filzmoser, P. (2000). Orthogonal principal planes. *Psychometrika*, **65**, 363–376.

Fisher, R.A. and Mackenzie, W.A. (1923). Studies in crop variation II. The manurial response of different potato varieties. *J. Agri. Sci.*, **13**, 311–320.

Flury, B. (1988). *Common Principal Components and Related Models*. New York: Wiley.

Flury, B.D. (1993). Estimation of principal points. *Appl. Statist.*, **42**, 139–151.

Flury B.D. (1995). Developments in principal component analysis. In *Recent Advances in Descriptive Multivariate Analysis*, ed. W.J. Krzanowski, 14–33. Oxford: Clarendon Press.

Flury, B.D. (1997). *A First Course in Multivariate Statistics*. New York: Springer.

Flury, B.D., Nel, D.G. and Pienaar, I. (1995). Simultaneous detection of shift in means and variances. *J. Amer. Statist. Assoc.*, **90**, 1474–1481.

Flury, B.D. and Neuenschwander, B.E. (1995). Principal component models for patterned covariance matrices with applications to canonical correlation analysis of several sets of variables. In *Recent Advances in Descriptive Multivariate Analysis*, ed. W.J. Krzanowski, 90–112. Oxford: Clarendon Press.

Flury, B. and Riedwyl, H. (1981). Graphical representation of multivariate data by means of asymmetrical faces. *J. Amer. Statist. Assoc.*, **76**, 757–765.

Flury, B. and Riedwyl, H. (1988). *Multivariate Statistics. A Practical Approach*. London: Chapman and Hall.

Folland, C. (1988). The weighting of data in an EOF analysis. *Met 0 13 Discussion Note 113*. UK Meteorological Office.

Folland, C.K., Parker, D.E. and Newman, M. (1985). Worldwide marine temperature variations on the season to century time scale. *Proceedings of the Ninth Annual Climate Diagnostics Workshop*, 70–85.

Fomby, T.B., Hill, R.C. and Johnson, S.R. (1978). An optimal property of principal components in the context of restricted least squares. *J. Amer. Statist. Assoc.*, **73**, 191–193.

Foster, P. (1998). Exploring multivariate data using directions of high density. *Statist. Computing*, **8**, 347–355.

Fowlkes, E.B. and Kettenring, J.R. (1985). Comment on 'Estimating optimal transformations for multiple regression and correlation' by L. Breiman and J.H. Friedman. *J. Amer. Statist. Assoc.*, **80**, 607–613.

Frane, J.W. (1976). Some simple procedures for handling missing data in multivariate analysis. *Psychometrika*, **41**, 409–415.

Frank, I.E. and Friedman, J.H. (1989). Classification: Oldtimers and newcomers. *J. Chemometrics*, **3**, 463–475.

Frank, I.E. and Friedman, J.H. (1993). A statistical view of some chemometrics tools. *Technometrics*, **35**, 109–148 (including discussion).

Franklin, S.B., Gibson, D.J., Robertson, P.A., Pohlmann, J.T. and Fralish, J.S. (1995). Parallel analysis: A method for determining significant principal components. *J. Vegetat. Sci.*, **6**, 99–106.

Freeman, G.H. (1975). Analysis of interactions in incomplete two-way tables. *Appl. Statist.*, **24**, 46–55.

Friedman, D.J. and Montgomery, D.C. (1985). Evaluation of the predictive performance of biased regression estimators. *J. Forecasting*, **4**, 153-163.

Friedman, J.H. (1987). Exploratory projection pursuit. *J. Amer. Statist. Assoc.*, **82**, 249–266.

Friedman, J.H. (1989). Regularized discriminant analysis. *J. Amer. Statist. Assoc.*, **84**, 165–175.

Friedman, J.H. and Tukey, J.W. (1974). A projection pursuit algorithm for exploratory data analysis. *IEEE Trans. Computers C*, **23**, 881–889.

Friedman, S. and Weisberg, H.F. (1981). Interpreting the first eigenvalue of a correlation matrix. *Educ. Psychol. Meas.*, **41**, 11–21.

Frisch, R. (1929). Correlation and scatter in statistical variables. *Nordic Statist. J.*, **8**, 36–102.

Fujikoshi, Y., Krishnaiah, P.R. and Schmidhammer, J. (1985). Effect of additional variables in principal component analysis, discriminant analysis and canonical correlation analysis. *Tech. Report 85-31*, Center for Multivariate Analysis, University of Pittsburgh.

Gabriel, K.R. (1971). The biplot graphic display of matrices with application to principal component analysis. *Biometrika*, **58**, 453–467.

Gabriel, K.R. (1978). Least squares approximation of matrices by additive and multiplicative models. *J. R. Statist. Soc. B*, **40**, 186–196.

Gabriel, K.R. (1981). Biplot display of multivariate matrices for inspection of data and diagnosis. In *Interpreting Multivariate Data*, ed. V. Barnett, 147–173. Chichester: Wiley.

Gabriel K.R. (1995a). Biplot display of multivariate categorical data, with comments on multiple correspondence analysis. In *Recent Advances in Descriptive Multivariate Analysis*, ed. W.J. Krzanowski, 190–226. Oxford: Clarendon Press.

Gabriel K.R. (1995b). MANOVA biplots for two-way contingency tables. In *Recent Advances in Descriptive Multivariate Analysis*, ed. W.J. Krzanowski, 227–268. Oxford: Clarendon Press.

Gabriel, K.R. (2002). Goodness of fit of biplots and correspondence analysis. To appear in *Biometrika*.

Gabriel, K.R. and Odoroff C.L. (1983). Resistant lower rank approximation of matrices. Technical report 83/02, Department of Statistics, University of Rochester, New York.

Gabriel, K.R. and Odoroff, C.L. (1990). Biplots in biomedical research. *Statist. Med.*, **9**, 469–485.

Gabriel, K.R. and Zamir, S. (1979). Lower rank approximation of matrices by least squares with any choice of weights. *Technometrics*, **21**, 489–498.

Garnham, N. (1979). Some aspects of the use of principal components in multiple regression. Unpublished M.Sc. dissertation. University of Kent at Canterbury.

Garthwaite, P.H. (1994). An interpretation of partial least squares. *J. Amer. Statist. Assoc.*, **89**, 122–127.

Gauch, H.G. (1982). *Multivariate Analysis in Community Ecology.* Cambridge: Cambridge University Press.

Geladi, P. (1988). Notes on the history and nature of partial least squares (PLS) modelling. *J. Chemometrics*, **2**, 231–246.

Gifi, A. (1990). *Nonlinear Multivariate Analysis.* Chichester: Wiley.

Girshick, M.A. (1936). Principal components. *J. Amer. Statist. Assoc.*, **31**, 519–528.

Girshick, M.A. (1939). On the sampling theory of roots of determinantal equations. *Ann. Math. Statist.*, **10**, 203–224.

Gittins, R. (1969). The application of ordination techniques. In *Ecological Aspects of the Mineral Nutrition of Plants*, ed. I. H. Rorison, 37–66. Oxford: Blackwell Scientific Publications.

Gittins, R. (1985). *Canonical Analysis. A Review with Applications in Ecology.* Berlin: Springer.

Gleason, T.C. and Staelin, R. (1975). A proposal for handling missing data. *Psychometrika*, **40**, 229–252.

Gnanadesikan, R. (1977). *Methods for Statistical Data Analysis of Multivariate Observations.* New York: Wiley.

Gnanadesikan, R. and Kettenring, J.R. (1972). Robust estimates, residuals, and outlier detection with multiresponse data. *Biometrics*, **28**, 81–124.

Goldstein, H. (1995). *Multilevel Statistical Models*, 2nd edition. London: Arnold.

Goldstein, M. and Dillon, W.R. (1978). *Discrete Discriminant Analysis.* New York: Wiley.

Golyandina, N.E., Nekrutin, V.V. and Zhigljavsky, A.A. (2001). *Analysis of Time Series Structure. SSA and Related Techniques.* Boca Raton: Chapman and Hall.

Gonzalez, P.L., Evry, R., Cléroux, R. and Rioux, B. (1990). Selecting the best subset of variables in principal component analysis. In *COMP-STAT 90*, eds. K. Momirovic and V. Mildner, 115–120. Heidelberg: Physica-Verlag.

Good, I.J. (1969). Some applications of the singular value decomposition of a matrix. *Technometrics*, **11**, 823–831.

Gordon, A.D. (1999). *Classification*, 2nd edition. Boca Raton: Chapman and Hall/CRC.

Gower, J.C. (1966). Some distance properties of latent root and vector methods used in multivariate analysis. *Biometrika*, **53**, 325–338.

Gower, J.C. (1967). Multivariate analysis and multidimensional geometry. *Statistician*, **17**, 13–28.

Gower, J.C. and Hand, D.J. (1996). *Biplots.* London: Chapman and Hall.

Gower, J.C. and Krzanowski, W.J. (1999). Analysis of distance for structured multivariate data and extensions to multivariate analysis of variance. *Appl. Statist.*, **48**, 505–519.

Grambsch, P.M., Randall, B.L., Bostick, R.M., Potter, J.D. and Louis, T.A. (1995). Modeling the labeling index distribution: An application of functional data analysis. *J. Amer. Statist. Assoc.*, **90**, 813–821.

Green, B.F. (1977). Parameter sensitivity in multivariate methods. *J. Multiv. Behav. Res.*, **12**, 263–287.

Greenacre, M.J. (1984). *Theory and Applications of Correspondence Analysis.* London: Academic Press.

Greenacre, M.J. (1993). *Correspondence Analysis in Practice.* London: Academic Press.

Greenacre, M. and Hastie, T. (1987). The geometric interpretation of correspondence analysis. *J. Amer. Statist. Assoc.*, **82**, 437–447.

Grimshaw, S.D., Shellman, S.D. and Hurwitz, A.M. (1998). Real-time process monitoring for changing inputs. *Technometrics*, **40**, 283–296.

Grossman, G.D., Nickerson, D.M. and Freeman, M.C. (1991). Principal component analyses of assemblage structure data: Utility of tests based on eigenvalues. *Ecology*, **72**, 341–347.

Guiot, J. (1981). *Analyse Mathématique de Données Geophysique. Applications à la Dendroclimatologie.* Unpublished Ph.D. dissertation, Université Catholique de Louvain.

Gunst, R.F. (1983). Regression analysis with multicollinear predictor variables: Definition, detection and effects. *Commun. Statist. —Theor. Meth.*, **12**, 2217–2260.

Gunst, R.F. and Mason, R.L. (1977a). Biased estimation in regression: An evaluation using mean squared error. *J. Amer. Statist. Assoc.*, **72**, 616–628.

Gunst, R.F. and Mason, R.L. (1977b). Advantages of examining multi-collinearities in regression analysis. *Biometrics*, **33**, 249–260.

Gunst, R.F. and Mason, R.L. (1979). Some considerations in the evaluation of alternative prediction equations. *Technometrics*, **21**, 55–63.

Gunst, R.F. and Mason, R.L. (1980). *Regression Analysis and Its Applications: A Data-Oriented Approach.* New York: Dekker.

Gunst, R.F., Webster, J.T. and Mason, R.L. (1976). A comparison of least squares and latent root regression estimators. *Technometrics*, **18**, 75–83.

Guttorp, P. and Sampson, P.D. (1994). Methods for estimating heterogeneous spatial covariance functions with environmental applications. In *Handbook of Statistics, Vol. 12*, eds. G.P. Patil and C.R. Rao, 661–689. Amsterdam: Elsevier.

Hadi, A.S. and Nyquist, H. (1993). Further theoretical results and a comparison between two methods for approximating eigenvalues of perturbed covariance matrices. *Statist. Computing*, **3**, 113–123.

Hadi, A.S. and Ling, R.F. (1998). Some cautionary notes on the use of principal components regression. *Amer. Statistician*, **52**, 15–19.

Hall, P., Poskitt, D.S., and Presnell, B. (2001). A functional data-analytic approach to signal discrimination. *Technometrics*, **43**, 1–9.

Hamilton, J.D. (1994). *Time Series Analysis.* Princeton: Princeton University Press.

Hampel, F.R. (1974). The influence curve and its role in robust estimation. *J. Amer. Statist. Assoc.*, **69**, 383–393.

Hampel, F.R., Ronchetti, E.M., Rousseeuw, P.J. and Stahel, W.A. (1986). *Robust Statistics: The Approach Based on Influence Functions.* New York: Wiley.

Hand, D.J. (1982). *Kernel Discriminant Analysis.* Chichester: Research Studies Press.

Hand, D.J. (1998). Data mining: Statistics and more? *Amer. Statistician*, **52**, 112–118.

Hand, D, Mannila, H. and Smyth, P. (2001). *Principles of Data Mining.* Cambridge: MIT Press.

Hannachi, A. (2000). Probabilistic-based approach to optimal filtering. *Phys. Rev. E*, **61**, 3610–3619.

Hannachi, A. and O'Neill, A. (2001). Atmospheric multiple equilibria and non-Gaussian behaviour in model simulations. *Q.J.R. Meteorol. Soc. 127*, 939-958.

Hansch, C., Leo, A., Unger, S.H., Kim, K.H., Nikaitani, D. and Lien, E.J. (1973). 'Aromatic' substituent constants for structure–activity correlations. *J. Medicinal Chem.*, **16**, 1207–1216.

Hasselmann, K. (1979). On the signal-to-noise problem in atmospheric response studies. In *Meteorology Over the Tropical Oceans*, ed. B. D. Shaw, 251–259. Bracknell: Royal Meteorological Society.

Hasselmann, K. (1988). PIPs and POPs: The reduction of complex dynamical systems using principal interaction and oscillation patterns. *J. Geophys. Res.*, **93**, 11,015–11,021.

Hastie, T. and Stuetzle, W. (1989). Principal curves. *J. Amer. Statist. Assoc.*, **84**, 502–516.

Hastie, T., Tibshirani, R., Eisen, M.B., Alizadeh, A., Levy, R., Staudt, L., Chan, W.C., Botstein, D. and Brown, P. (2000). 'Gene shaving' as a method for identifying distinct sets of genes with similar expression patterns. *Genome Biol.*, **1**, research 0003.1–003.21.

Hausmann, R. (1982). Constrained multivariate analysis. In *Optimisation in Statistics*, eds. S.H. Zanckis and J.S. Rustagi, 137–151. Amsterdam: North-Holland.

Hawkins, D.M. (1973). On the investigation of alternative regressions by principal component analysis. *Appl. Statist.*, **22**, 275–286.

Hawkins, D.M. (1974). The detection of errors in multivariate data using principal components. *J. Amer. Statist. Assoc.*, **69**, 340–344.

Hawkins, D.M. (1980). *Identification of Outliers*. London: Chapman and Hall.

Hawkins, D.M. and Eplett, W.J.R. (1982). The Cholesky factorization of the inverse correlation or covariance matrix in multiple regression. *Technometrics*, **24**, 191–198.

Hawkins, D.M. and Fatti, L.P. (1984). Exploring multivariate data using the minor principal components. *Statistician*, **33**, 325–338.

Helland, I.S. (1988). On the structure of partial least squares regression. *Commun. Statist.—Simul.*, **17**, 581–607.

Helland, I.S. (1990). Partial least squares regression and statistical models. *Scand. J. Statist.*, **17**, 97–114.

Helland, I.S. and Almøy, T. (1994). Comparison of prediction methods when only a few components are relevant. *J. Amer. Statist. Assoc.*, **89**, 583–591 (correction **90**, 399).

Heo, M. and Gabriel, K.R. (2001). The fit of graphical displays to patterns of expectations. *Computat. Statist. Data Anal.*, **36**, 47-67.

Hill, R.C., Fomby, T.B. and Johnson, S.R. (1977). Component selection norms for principal components regression. *Commun. Statist.*, **A6**, 309–334.

Hills, M. (1982). Allometry. In *Encyclopedia of Statistical Sciences Vol 1*, eds. S. Kotz and N.L. Johnson, 48–54. New York: Wiley.

Hoaglin, D.C., Mosteller, F. and Tukey, J.W. (1983). *Understanding Robust and Exploratory Data Analysis*. New York: Wiley.

Hocking, R.R. (1976). The analysis and selection of variables in linear regression. *Biometrics*, **32**, 1–49.

Hocking, R.R. (1984). Discussion of 'K-clustering as a detection tool for influential subsets in regression' by J.B. Gray and R.F. Ling. *Technometrics*, **26**, 321–323.

Hocking, R.R., Speed, F.M. and Lynn, M.J. (1976). A class of biased estimators in linear regression. *Technometrics*, **18**, 425–437.

Hoerl, A.E. and Kennard, R.W. (1970a). Ridge regression: Biased estimation for nonorthogonal problems. *Technometrics*, **12**, 55–67.

Hoerl, A.E. and Kennard, R.W. (1970b). Ridge regression: Applications to nonorthogonal problems. *Technometrics*, **12**, 69–82.

Hoerl, A.E., Kennard, R.W. and Hoerl, R.W. (1985). Practical use of ridge regression: A challenge met. *Appl. Statist.*, **34**, 114–120.

Hoerl, R.W., Schuenemeyer, J.H. and Hoerl, A.E. (1986). A simulation of biased estimation and subset selection regression techniques. *Technometrics*, **28**, 369–380.

Holmes-Junca, S. (1985). *Outils Informatiques pour l'evaluation de la Pertinence d'un Resultat en Analyse des Données.* Unpublished Ph.D. thesis, Université des Sciences et Techniques du Languedoc.

Horel, J.D. (1984). Complex principal component analysis: Theory and examples. *J. Climate Appl. Meteorol.*, **23**, 1660–1673.

Horgan, G.W. (2000). Principal component analysis of random particles. *J. Math. Imaging Vision*, **12**, 169–175.

Horgan, G.W. (2001). The statistical analysis of plant part appearance—A review. *Computers Electronics Agri.*, **31**, 169–190.

Horgan, G.W., Talbot, M. and Davey, J.C. (2001). Use of statistical image analysis to discriminate carrot cultivars. *Computers Electronics Agri.*, **31**, 191–199.

Horn, J.L. (1965). A rationale and test for the number of factors in a factor analysis. *Psychometrika*, **30**, 179–185.

Hotelling, H. (1933). Analysis of a complex of statistical variables into principal components. *J. Educ. Psychol.*, **24**, 417–441, 498–520.

Hotelling, H. (1936). Simplified calculation of principal components. *Psychometrika*, **1**, 27–35.

Hotelling, H. (1957). The relations of the newer multivariate statistical methods to factor analysis. *Brit. J. Statist. Psychol.*, **10**, 69–79.

Houseago-Stokes, R. and Challenor, P. (2001). Using PPCA to estimate EOFs in the presence of missing data. Submitted for publication.

Householder, A.S. and Young, G. (1938). Matrix approximation and latent roots. *Amer. Math. Mon.*, **45**, 165–171.

Hsuan, F.C. (1981). Ridge regression from principal component point of view. *Commun. Statist.*, **A10**, 1981–1995.

Hu, Q. (1997). On the uniqueness of the singular value decomposition in meteorological applications. *J. Climate*, **10**, 1762–1766.

Huang, D-Y. and Tseng, S-T. (1992). A decision procedure for determining the number of components in principal component analysis. *J. Statist. Plan. Inf.*, **30**, 63–71.

Huber, P.J. (1964). Robust estimation for a location parameter. *Ann. Math. Stat.*, **35**, 73–101.

Huber, P.J. (1981). *Robust Statistics.* New York: Wiley.

Huber, P.J. (1985). Projection pursuit. *Ann. Statist.*, **13**, 435–475 (including discussion).

Hudlet, R. and Johnson, R.A. (1982). An extension of some optimal properties of principal components. *Ann. Inst. Statist. Math.*, **34**, 105–110.

Huettmann, F. and Diamond, A.W. (2001). Using PCA scores to classify species communities: An example for pelagic seabird distribution. *J. Appl. Statist.*, **28**, 843–853.

Hunt, A. (1978). The elderly at home. *OPCS Social Survey Division*, Publication SS 1078. London: HMSO.

Ibazizen, M. (1986). *Contribution de l'étude d'une Analyse en Composantes Principales Robuste.* Unpublished Ph.D. thesis. Université Paul Sabatier de Toulouse.

Ichino, M. and Yaguchi, H. (1994). Generalized Minkowski matrices for mixed feature-type data analysis. *IEEE Trans. Syst. Man Cybernet.*, **24**, 698–708.

Iglarsh, H.J. and Cheng, D.C. (1980). Weighted estimators in regression with multicollinearity. *J. Statist. Computat. Simul.*, **10**, 103–112.

Imber, V. (1977). A classification of the English personal social services authorities. *DHSS Statistical and Research Report Series.* No. 16. London: HMSO.

Jackson, D.A. (1993). Stopping rules in principal components analysis: A comparison of heuristical and statistical approaches. *Ecology*, **74**, 2204–2214.

Jackson, J.E. (1981). Principal components and factor analysis: Part III—What is factor analysis? *J. Qual. Tech.*, **13**, 125–130.

Jackson, J.E. (1991). *A User's Guide to Principal Components.* New York: Wiley.

Jackson, J.E. and Hearne, F.T. (1973). Relationships among coefficients of vectors used in principal components. *Technometrics*, **15**, 601–610.

Jackson, J.E. and Hearne, F.T. (1979). Hotelling's T_M^2 for principal components—What about absolute values? *Technometrics*, **21**, 253–255.

Jackson, J.E. and Mudholkar, G.S. (1979). Control procedures for residuals associated with principal component analysis. *Technometrics*, **21**, 341–349.

James, G.M., Hastie, T.J. and Sugar, C.A. (2000). Principal component models for sparse functional data. *Biometrika*, **87**, 587–602.

Jaupi, L. and Saporta, G. (1993). Using the influence function in robust principal components analysis. In *New Directions in Statistical Data Analysis and Robustness*, eds. S. Morgenthaler, E. Ronchetti and W.A. Stahel, 147–156. Basel: Birkhäuser.

Jeffers, J.N.R. (1962). Principal component analysis of designed experiment. *Statistician*, **12**, 230–242.

Jeffers, J.N.R. (1967). Two case studies in the application of principal component analysis. *Appl. Statist.*, **16**, 225–236.

Jeffers, J.N.R. (1978). *An Introduction to Systems Analysis: With Ecological Applications.* London: Edward Arnold.

Jeffers, J.N.R. (1981). Investigation of alternative regressions: Some practical examples. *Statistician,* **30**, 79–88.

Jensen, D.R. (1986). The structure of ellipsoidal distributions II. Principal components. *Biom. J.,* **28**, 363–369.

Jensen, D.R. (1997). Conditioning and concentration of principal components. *Austral. J. Statist.,* **39**, 93–104.

Jensen, D.R. (1998). Principal predictors and efficiency in small second-order designs. *Biom. J.,* **40**, 183–203.

Jia, F., Martin, E.B. and Morris, A.J. (2000). Non-linear principal components analysis with application to process fault detection. *Int. J. Syst. Sci.,* **31**, 1473–1487.

Jmel, S. (1992). *Application des Modèles Graphiques au Choix de Variables et à l'analyse des Interactions dans une Table de Contingence Multiple.* Unpublished doctoral dissertation. Université Paul Sabatier, Toulouse.

Jolicoeur, P. (1963). The multivariate generalization of the allometry equation. *Biometrics,* **19**, 497–499.

Jolicoeur, P. (1984). Principal components, factor analysis, and multivariate allometry: A small-sample directional test. *Biometrics,* **40**, 685–690.

Jolicoeur, P. and Mosimann, J.E. (1960). Size and shape variation in the painted turtle. A principal component analysis. *Growth,* **24**, 339–354.

Jolliffe, I.T. (1970). *Redundant Variables in Multivariate Analysis.* Unpublished D. Phil. thesis. University of Sussex.

Jolliffe, I.T. (1972). Discarding variables in a principal component analysis 1: Artificial data. *Appl. Statist.,* **21**, 160–173.

Jolliffe, I.T. (1973). Discarding variables in a principal component analysis II: Real data. *Appl. Statist.,* **22**, 21–31.

Jolliffe, I.T. (1982). A note on the use of principal components in regression. *Appl. Statist.,* **31**, 300–303.

Jolliffe, I.T. (1987a). Selection of variables. *Appl. Statist.,* **36**, 373–374.

Jolliffe, I.T. (1987b). Rotation of principal components: Some comments. *J. Climatol.,* **7**, 507–510.

Jolliffe, I.T. (1989). Rotation of ill-defined principal components. *Appl. Statist.,* **38**, 139–147.

Jolliffe, I.T. (1995). Rotation of principal components: Choice of normalization constraints. *J. Appl. Statist.,* **22**, 29–35.

Jolliffe, I.T., Jones, B. and Morgan, B.J.T. (1980). Cluster analysis of the elderly at home: A case study. In *Data Analysis and Informatics,* eds. E. Diday, L. Lebart, J.P. Pagès and R. Tomassone, 745–757. Amsterdam: North-Holland.

Jolliffe, I.T., Jones, B. and Morgan, B.J.T. (1982a). An approach to assessing the needs of the elderly. *Clearing House for Local Authority, Social Services Research,* **2**, 1–102.

Jolliffe, I.T., Jones, B. and Morgan, B.J.T. (1982b). Utilising clusters: A case study involving the elderly. *J. R. Statist. Soc. A*, **145**, 224–236.

Jolliffe, I.T., Jones, B. and Morgan, B.J.T. (1986). Comparison of cluster analyses of the English personal social services authorities. *J. R. Statist. Soc. A*, **149**, 253–270.

Jolliffe, I.T., Morgan, B.J.T. and Young, P.J. (1996). A simulation study of the use of principal components in linear discriminant analysis. *J. Statist. Comput. Simul.*, **55**, 353–366.

Jolliffe I.T., Trendafilov, N.T. and Uddin, M. (2002a). A modified principal component technique based on the LASSO. Submitted for publication.

Jolliffe, I.T. and Uddin, M. (2000). The simplified component technique. An alternative to rotated principal components. *J. Computat. Graph. Statist.*, **9**, 689–710.

Jolliffe I.T., Uddin, M and Vines, S.K. (2002b). Simplified EOFs. Three alternatives to rotation. *Climate Res.*, **20**, 271–279.

Jones, M.C. and Sibson, R. (1987). What is projection pursuit? *J. R. Statist. Soc., A*, **150**, 1–38 (including discussion).

Jones, P.D., Wigley, T.M.L. and Briffa, K.R. (1983). Reconstructing surface pressure patterns using principal components regression on temperature and precipitation data. *Second International Meeting on Statistical Climatology*, Preprints volume. 4.2.1–4.2.8.

Jong, J.-C. and Kotz, S. (1999). On a relation between principal components and regression analysis. *Amer. Statistician*, **53**, 349–351.

Jordan, M.C. (1874). Mémoire sur les Formes Bilinéaires. *J. Math. Pures Appl.*, **19**, 35–54.

Jungers, W.L., Falsetti, A.B. and Wall, C.E. (1995). Shape, relative size, and size-adjustments in morphometrics. *Yearbook of Physical Anthropology*, **38**, 137–161.

Kaciak, E. and Sheahan, J.N. (1988). Market segmentation: An alternative principal components approach. In *Marketing 1998, Volume 9, Proceedings of the Annual Conference of the Administrative Sciences Association of Canada—Marketing Division*, ed. T. Barker, 139–148.

Kaigh, W.D. (1999). Total time on test function principal components. *Stat. and Prob. Lett.*, **44**, 337–341.

Kaiser, H.F. (1960). The application of electronic computers to factor analysis. *Educ. Psychol. Meas.*, **20**, 141–151.

Kambhatla, N. and Leen, T.K. (1997). Dimension reduction by local principal component analysis. *Neural Computat.*, **9**, 1493–1516.

Kaplan, A., Cane, M.A. and Kushnir, Y. (2001). Reduced space approach to the optimal analysis of historical marine observations: Accomplishments, difficulties and prospects. *WMO Guide to the Applications of Marine Climatology*. Geneva: World Meteorological Organization.

Karl, T.R., Koscielny, A.J. and Diaz, H.F. (1982). Potential errors in the application of principal component (eigenvector) analysis to geophysical data. *J. Appl. Meteorol.*, **21**, 1183–1186.

Kazi-Aoual, F., Sabatier, R. and Lebreton, J.-D. (1995). Approximation of permutation tests for multivariate inference—application to species environment relationships. In *Data Science and Its Application*, eds. Y. Escoufier, B. Fichet, E. Diday, L. Lebart, C. Hayashi, N. Ohsumi and Y. Baba, 51–62. Tokyo: Academic Press.

Kazmierczak, J.B. (1985). Analyse logarithmique deux exemples d'application. *Rev. Statistique Appliquée*, **33**, 13–24.

Kendall, D.G. (1984). Shape-manifolds, procrustean matrices and complex projective spaces. *Bull. Lond. Math. Soc.*, **16**, 81–121.

Kendall, M.G. (1957). *A Course in Multivariate Analysis*. London: Griffin.

Kendall, M.G. (1966). Discrimination and classification. In *Multivariate Analysis*, ed. P. R. Krishnaiah, 165–185. New York: Academic Press.

Kendall, M.G. and Stuart, A. (1979). *The Advanced Theory of Statistics*, *Vol. 2*, 4th edition. London: Griffin.

Kent, J.T. (1994). The complex Bingham distribution and shape analysis. *J. R. Statist. Soc. B*, **56**, 285–299.

Keramidas, E.M., Devlin, S.J. and Gnanadesikan, R. (1987). A graphical procedure for comparing the principal components of several covariance matrices. *Commun. Statist.-Simul.*, **16**, 161–191.

Kiers, H.A.L. (1993). A comparison of techniques for finding components with simple structure. In *Multivariate Analysis: Future Directions 2*, eds. C.M. Cuadras and C.R. Rao, 67–86. Amsterdam: North-Holland.

Kim, K.-Y. and Wu, Q. (1999). A comparison study of EOF techniques: analysis of nonstationary data with periodic statistics. *J. Climate*, **12**, 185–199.

King, J.R. and Jackson, D.A. (1999). Variable selection in large environmental data sets using principal components analysis. *Environmetrics*, **10**, 67–77.

Klink K. and Willmott, C.J. (1989). Principal components of the surface wind field in the United States: A comparison of analyses based upon wind velocity, direction, and speed. *Int. J. Climatol.*, **9**, 293–308.

Kloek, T. and Mennes, L.B.M. (1960). Simultaneous equations estimation based on principal components of predetermined variables. *Econometrica*, **28**, 45–61.

Kneip, A. (1994). Nonparametric estimation of common regressors for similar curve data. *Ann. Statist.*, **22**, 1386–1427.

Kneip, A. and Utikal, K.J. (2001). Inference for density families using functional principal component analysis. *J. Amer. Statist. Assoc.*, **96**, 519–542 (including discussion).

Konishi, S. and Rao, C.R. (1992). Principal component analysis for multivariate familial data. *Biometrika*, **79**, 631–641.

Kooperberg, C. and O'Sullivan, F. (1994). The use of a statistical forecast criterion to evaluate alternative empirical spatial oscillation pattern decomposition methods in climatological fields. Technical report 276, Department of Statistics, University of Washington, Seattle.

Kooperberg, C. and O'Sullivan, F. (1996). Predictive oscillation patterns: A synthesis of methods for spatial-temporal decomposition of random fields. *J. Amer. Statist. Assoc.*, **91**, 1485–1496.

Korhonen, P.J. (1984). Subjective principal component analysis. *Computat. Statist. Data Anal.*, **2**, 243–255.

Korhonen, P. and Siljamäki, A. (1998). Ordinal principal component analysis: Theory and an application. *Computat. Statist. Data Anal.*, **26**, 411–424.

Korth, B. and Tucker, L.R. (1975). The distribution of chance congruence coefficients from simulated data. *Psychometrika*, **40**, 361–372.

Kramer, M.A. (1991). Nonlinear principal component analysis using autoassociative neural networks. *AIChE J.*, **37**, 233–243.

Kroonenberg, P.M. (1983a). *Three-Mode Principal Component Analysis*. Leiden: DSWO Press.

Kroonenberg, P.M. (1983b). Annotated bibliography of three-mode factor analysis. *Brit. J. Math. Statist. Psychol.*, **36**, 81–113.

Kroonenberg, P.M., Harch, B.D., Basford, K.E. and Cruickshank, A. (1997). Combined analysis of categorical and numerical descriptors of Australian groundnut accessions using nonlinear principal component analysis. *J. Agri. Biol. Environ. Statist.*, **2**, 294–312.

Kruskal, J.B. (1964a). Multidimensional scaling by optimizing goodness of fit to a nonmetric hypothesis. *Psychometrika*, **29**, 1–27.

Kruskal, J.B. (1964b). Nonmetric multidimensional scaling: A numerical method. *Psychometrika*, **29**, 115–129.

Krzanowski, W.J. (1979a). Some exact percentage points of a statistic useful in analysis of variance and principal component analysis. *Technometrics*, **21**, 261–263.

Krzanowski, W.J. (1979b). Between-groups comparison of principal components. *J. Amer. Statist. Assoc.*, **74**, 703–707 (correction **76**, 1022).

Krzanowski, W.J. (1982). Between-group comparison of principal components—some sampling results. *J. Statist. Computat. Simul.*, **15**, 141–154.

Krzanowski, W.J. (1983). Cross-validatory choice in principal component analysis: Some sampling results. *J. Statist. Computat. Simul.*, **18**, 299–314.

Krzanowski, W.J. (1984a). Principal component analysis in the presence of group structure. *Appl. Statist.*, **33**, 164–168.

Krzanowski, W.J. (1984b). Sensitivity of principal components. *J. R. Statist. Soc. B*, **46**, 558–563.

Krzanowski, W.J. (1987a). Cross-validation in principal component analysis. *Biometrics*, **43**, 575–584.

Krzanowski, W.J. (1987b). Selection of variables to preserve multivariate data structure, using principal components. *Appl. Statist.*, **36**, 22–33.

Krzanowski, W.J. (1990). Between-group analysis with heterogeneous covariance matrices: The common principal component model. *J. Classific.*, **7**, 81–98.

Krzanowski, W.J. (2000). *Principles of Multivariate Analysis: A User's Perspective*, 2nd edition. Oxford: Oxford University Press.

Krzanowski, W.J., Jonathan, P., McCarthy, W.V. and Thomas, M.R. (1995). Discriminant analysis with singular covariance matrices: Methods and applications to spectroscopic data. *Appl. Statist.*, **44**, 101–115.

Krzanowski, W.J. and Kline, P. (1995). Cross-validation for choosing the number of important components in principal component analysis. *Mult. Behav. Res.*, **30**, 149–166.

Krzanowski, W.J. and Marriott, F.H.C. (1994). *Multivariate Analysis, Part 1. Distributions, Ordination and Inference*. London: Arnold.

Kshirsagar, A.M., Kocherlakota, S. and Kocherlakota, K. (1990). Classification procedures using principal component analysis and stepwise discriminant function. *Commun. Statist.—Theor. Meth.*, **19**, 92–109.

Kuhfeld, W.F. (1990). Comment on Dawkins (1989). *Amer. Statistician*, **44**, 58–59.

Kung, E.C. and Sharif, T.A. (1980). Multi-regression forecasting of the Indian summer monsoon with antecedent patterns of the large-scale circulation. *WMO Symposium on Probabilistic and Statistical Methods in Weather Forecasting*, 295–302.

Lafosse, R. and Hanafi, M. (1997). Concordance d'un tableau avec k tableaux: Définition de $k + 1$ uples synthétiques. *Rev. Statistique Appliquée*, **45**, 111-136.

Lane, S., Martin, E.B., Kooijmans, R. and Morris, A.J. (2001). Performance monitoring of a multi-product semi-batch process. *J. Proc. Cont.*, **11**, 1–11.

Lang, P.M., Brenchley, J.M., Nieves, R.G. and Halivas, J.H. (1998). Cyclic subspace regression. *J. Mult. Anal.*, **65**, 58–70.

Lanterman, A.D. (2000). Bayesian inference of thermodynamic state incorporating Schwarz-Rissanen complexity for infrared target recognition. *Optical Engng.*, **39**, 1282–1292.

Läuter, J. (1996). Exact t and F tests for analyzing studies with multiple endpoints. *Biometrics*, **52**, 964–970.

Lawley, D.N. (1963). On testing a set of correlation coefficients for equality. *Ann. Math. Statist.*, **34**, 149–151.

Lawley, D.N. and Maxwell, A.E. (1971). *Factor Analysis as a Statistical Method*, 2nd edition. London: Butterworth.

Lawson, C.L. and Hanson, R.J. (1974). *Solving Least Squares Problems*. Englewood Cliffs, NJ: Prentice-Hall.

Leamer, E.E. and Chamberlain, G. (1976). A Bayesian interpretation of pretesting. *J. R. Statist. Soc. B*, **38**, 85–94.

Lebart, L., Morineau, A. and Fénelon, J.-P. (1982). *Traitement des Données Statistique*. Paris: Dunod.

Lee, T.-W. (1998). *Independent Component Analysis. Theory and Applications*. Boston: Kluwer.

Lefkovitch, L.P. (1993). Concensus principal components. *Biom. J.*, **35**, 567–580.

Legates, D.R. (1991). The effect of domain shape on principal component analyses. *Int. J. Climatol.*, **11**, 135–146.

Legates, D.R. (1993). The effect of domain shape on principal component analyses: A reply. *Int. J. Climatol.*, **13**, 219–228.

Legendre, L. and Legendre, P. (1983). *Numerical Ecology*. Amsterdam: Elsevier.

Lewis-Beck, M.S. (1994). *Factor Analysis and Related Techniques*. London: Sage.

Li, G. and Chen, Z. (1985). Projection-pursuit approach to robust dispersion matrices and principal components: Primary theory and Monte Carlo. *J. Amer. Statist. Assoc.*, **80**, 759–766 (correction **80**, 1084).

Li, K.-C., Lue, H.-H. and Chen, C.-H. (2000). Interactive tree-structured regression via principal Hessian directions. *J. Amer. Statist. Assoc.*, **95**, 547–560.

Little, R.J.A. (1988). Robust estimation of the mean and covariance matrix from data with missing values. *Appl. Statist.*, **37**, 23–38.

Little, R.J.A. and Rubin, D.B. (1987). *Statistical Analysis with Missing Data*. New York: Wiley.

Locantore, N., Marron, J.S., Simpson, D.G., Tripoli, N., Zhang, J.T. and Cohen, K.L. (1999). Robust principal component analysis for functional data. *Test*, **8**, 1–73 (including discussion).

Lott, W.F. (1973). The optimal set of principal component restrictions on a least squares regression. *Commun. Statist.*, **2**, 449–464.

Lu, J., Ko, D. and Chang, T. (1997). The standardized influence matrix and its applications. *J. Amer. Statist. Assoc.*, **92**, 1572–1580.

Lynn, H.S. and McCulloch, C.E. (2000). Using principal component analysis and correspondence analysis for estimation in latent variable models. *J. Amer. Statist. Assoc.*, **95**, 561–572.

Macdonell, W.R. (1902). On criminal anthropometry and the identification of criminals. *Biometrika*, **1**, 177–227.

Mager, P.P. (1980a). Principal component regression analysis applied in structure-activity relationships 2. Flexible opioids with unusually high safety margin. *Biom. J.*, **22**, 535–543.

Mager, P.P. (1980b). Correlation between qualitatively distributed predicting variables and chemical terms in acridine derivatives using principal component analysis. *Biom. J.*, **22**, 813–825.

Mandel, J. (1971). A new analysis of variance model for non-additive data. *Technometrics*, **13**, 1–18.

Mandel, J. (1972). Principal components, analysis of variance and data structure. *Statistica Neerlandica*, **26**, 119–129.

Mandel, J. (1982). Use of the singular value decomposition in regression analysis. *Amer. Statistician*, **36**, 15–24.

Mann, M.E. and Park, J. (1999). Oscillatory spatiotemporal signal detection in climate studies: A multi-taper spectral domain approach. *Adv. Geophys.*, **41**, 1–131.

Mansfield, E.R., Webster, J.T. and Gunst, R.F. (1977). An analytic variable selection technique for principal component regression. *Appl. Statist.*, **26**, 34–40.

Mardia, K.V., Coombes, A., Kirkbride, J., Linney, A. and Bowie, J.L. (1996). On statistical problems with face identification from photographs. *J. Appl. Statist.*, **23**, 655–675.

Mardia, K.V., Kent, J.T. and Bibby, J.M. (1979). *Multivariate Analysis.* London: Academic Press.

Maronna, R.A. (1976). Robust M-estimators of multivariate location and scatter. *Ann. Statist.*, **4**, 51–67.

Maronna, R.A. and Yohai, V.J. (1998). Robust estimation of multivariate location and scatter. In *Encyclopedia of Statistical Sciences, Update 2*, eds. S. Kotz, C.B. Read and D.L. Banks, 589–596. New York: Wiley.

Marquardt, D.W. (1970). Generalized inverses, ridge regression, biased linear estimation, and nonlinear estimation. *Technometrics*, **12**, 591–612.

Martens, H. and Naes, T. (1989). *Multivariate Calibration.* New York: Wiley.

Martin, E.B. and Morris, A.J. (1996). Non-parametric confidence bounds for process performance monitoring charts. *J. Proc. Cont.*, **6**, 349–358.

Martin, E.B., Morris, A.J. and Kiparissides, C. (1999). Manufacturing performance enhancement through multivariate statistical process control. *Ann. Rev. Cont.*, **23**, 35-44.

Martin, J.-F. (1988). On probability coding. In *Component and Correspondence Analysis. Dimension Reduction by Functional Approximation*, eds. J.L.A. van Rijckevorsel and J. de Leeuw, 103–114. Chichester: Wiley.

Marx, B.D. and Smith, E.P. (1990). Principal component estimation for generalized linear regression. *Biometrika*, **77**, 23–31.

Maryon, R.H. (1979). Eigenanalysis of the Northern Hemispherical 15-day mean surface pressure field and its application to long-range forecasting. *Met O 13 Branch Memorandum No. 82* (unpublished). UK Meteorological Office, Bracknell.

Mason, R.L. and Gunst, R.F. (1985). Selecting principal components in regression. *Stat. Prob. Lett.*, **3**, 299–301.

Massy, W.F. (1965). Principal components regression in exploratory statistical research. *J. Amer. Statist. Assoc.*, **60**, 234–256.

Matthews, J.N.S. (1984). Robust methods in the assessment of multivariate normality. *Appl. Statist.*, **33**, 272–277.

Maurin, M. (1987). A propos des changements de métriques en ACP. *Fifth International Symposium: Data Analysis and Informatics*, Posters, 15–18.

Maxwell, A.E. (1977). *Multivariate Analysis in Behavioural Research.* London: Chapman and Hall.

McCabe, G.P. (1982). Principal variables. *Technical Report No. 82-3*, Department of Statistics, Purdue University.

McCabe, G.P. (1984). Principal variables. *Technometrics*, **26**, 137–144.

McCabe, G.P. (1986). Prediction of principal components by variable subsets. Unpublished Technical Report, 86-19, Department of Statistics, Purdue University.

McGinnis, D.L. (2000). Synoptic controls on upper Colorado River Basin snowfall. *Int. J. Climatol.*, **20**, 131–149.

McLachlan, G.J. (1992). *Discriminant Analysis and Statistical Pattern Recognition*. New York: Wiley.

McLachlan, G.J. and Bashford, K.E. (1988) *Mixture Models. Inference and Applications to Clustering*. New York: Marcel Dekker.

McReynolds, W.O. (1970). Characterization of some liquid phases. *J. Chromatogr. Sci.*, **8**, 685–691.

Mehrotra, D.V. (1995). Robust elementwise estimation of a dispersion matrix. *Biometrics*, **51**, 1344–1351.

Meredith, W. and Millsap, R.E. (1985). On component analysis. *Psychometrika*, **50**, 495–507.

Mertens, B.J.A. (1998). Exact principal component influence measures applied to the analysis of spectroscopic data on rice. *Appl. Statist.*, **47**, 527–542.

Mertens, B., Fearn, T. and Thompson, M. (1995). The efficient cross-validation of principal components applied to principal component regression. *Statist. Comput.*, **5**, 227–235.

Mertens, B., Thompson, M. and Fearn, T. (1994). Principal component outlier detection and SIMCA: A synthesis. *Analyst*, **119**, 2777–2784.

Mestas-Nuñez, A.M. (2000). Orthogonality properties of rotated empirical modes. *Int. J. Climatol.*, **20**, 1509–1516.

Meulman, J. (1986). *A Distance Approach to Nonlinear Multivariate Analysis*. Leiden: DSWO Press.

Michailidis, G. and de Leeuw, J. (1998). The Gifi system of descriptive multivariate analysis. *Statist. Sci.*, **13**, 307–336.

Milan, L. and Whittaker, J. (1995). Application of the parametric bootstrap to models that incorporate a singular value decomposition. *Appl. Statist.*, **44**, 31–49.

Miller, A.J. (1984). Selection of subsets of regression variables (with discussion). *J. R. Statist. Soc. A*, **147**, 389–425.

Miller, A.J. (1990). *Subset Selection in Regression*. London: Chapman and Hall.

Milliken, G.A. and Johnson, D.E. (1989). *Analysis of Messy Data Vol. 2: Nonreplicated Experiments*. New York: Van Nostrand-Reinhold.

Monahan, A.H. (2001). Nonlinear principal component analysis: Tropical Indo-Pacific sea surface temperature and sea level pressure. *J. Climate*, **14**, 219–233.

Monahan, A.H., Tangang, F.T. and Hsieh, W.W. (1999). A potential problem with extended EOF analysis of standing wave fields. *Atmos.-Ocean*, **37**, 241–254.

Mori, Y., Iizuka, M., Tarumi, T. and Tanaka, Y. (1999). Variable selection in "principal component analysis based on a subset of variables". *Bulletin of the International Statistical Institute 52nd Session Contributed Papers, Tome LVIII, Book 2*, 333–334.

Mori, Y., Iizuka, M., Tarumi, T. and Tanaka, Y. (2000). Study of variable selection criteria in data analysis. *Proc. 10th Japan and Korea Joint Conference of Statistics*, 547–554.

Mori, Y., Tanaka, Y. and Tarumi, T. (1998). Principal component analysis based on a subset of variables for qualitative data. In *Data Science, Classification, and Related Methods*, eds. C. Hayashi, N. Ohsumi, K. Yajima, Y. Tanaka, H.H. Bock and Y. Baba, 547–554. Tokyo: Springer-Verlag.

Morgan, B.J.T. (1981). Aspects of QSAR: 1. Unpublished report, CSIRO Division of Mathematics and Statistics, Melbourne.

Morrison, D.F. (1976). *Multivariate Statistical Methods*, 2nd edition. Tokyo: McGraw-Hill Kogakusha.

Moser, C.A. and Scott, W. (1961). *British Towns*. Edinburgh: Oliver and Boyd.

Mosteller, F. and Tukey, J.W. (1977). *Data Analysis and Regression: A Second Course in Statistics*. Reading, MA: Addison-Wesley.

Mote, P.W., Clark, H.L., Dunkerton, T.J., Harwood, R.S., and Pumphrey, H.C. (2000). Intraseasonal variations of water vapor in the tropical upper troposphere and tropopause region. *J. Geophys. Res.*, **105**, 17457–17470.

Muller, K.E. (1981). Relationships between redundancy analysis, canonical correlation and multivariate regression. *Psychometrika*, **46**, 139–142.

Muller, K.E. (1982). Understanding canonical correlation through the general linear model and principal components. *Amer. Statistician*, **36**, 342–354.

Naes, T. (1985). Multivariate calibration when the error covariance matrix is structured. *Technometrics*, **27**, 301–311.

Naes, T. and Helland, I.S. (1993). Relevant components in regression. *Scand. J. Statist.*, **20**, 239–250.

Naes, T., Irgens, C. and Martens, H. (1986). Comparison of linear statistical methods for calibration of NIR instruments. *Appl. Statist.*, **35**, 195–206.

Naes, T. and Isaksson, T. (1991). Splitting of calibration data by cluster analysis. *J. Chemometrics*, **5**, 49–65.

Naes, T. and Isaksson, T. (1992). Locally weighted regression in diffuse near-infrared transmittance spectroscopy. *Appl. Spectroscopy*, **46**, 34–43.

Naga, R.A. and Antille, G. (1990). Stability of robust and non-robust principal components analysis. *Computat. Statist. Data Anal.*, **10**, 169–174.

Naik, D.N. and Khattree, R. (1996). Revisiting Olympic track records: Some practical considerations in the principal component analysis. *Amer. Statistician*, **50**, 140–144.

Nash, J.C. and Lefkovitch, L.P. (1976). Principal components and regression by singular value decomposition on a small computer. *Appl. Statist.*, **25**, 210–216.

Nel, D.G. and Pienaar, I. (1998). The decomposition of the Behrens-Fisher statistic in q-dimensional common principal common submodels. *Ann. Inst. Stat. Math.*, **50**, 241–252.

Nelder, J.A. (1985). An alternative interpretation of the singular-value decomposition in regression. *Amer. Statistician*, **39**, 63–64.

Neuenschwander, B.E. and Flury, B.D. (2000). Common principal components for dependent random vectors. *J. Mult. Anal.*, **75**, 163–183.

Nomikos, P. and MacGregor, J.F. (1995). Multivariate SPC charts for monitoring batch processes. *Technometrics*, **37**, 41–59.

North, G.R., Bell, T.L., Cahalan, R.F. and Moeng, F.J. (1982). Sampling errors in the estimation of empirical orthogonal functions. *Mon. Weather Rev.*, **110**, 699–706.

North, G.R. and Wu, Q. (2001). Detecting climate signals using space-time EOFs. *J. Climate*, **14**, 1839–1863.

Obukhov, A.M. (1947). Statistically homogeneous fields on a sphere. *Usp. Mat. Nauk.*, **2**, 196–198.

Ocaña, F.A., Aguilera, A.M. and Valderrama, M.J. (1999). Functional principal components analysis by choice of norms. *J. Mult. Anal.*, **71**, 262–276.

Ogasawara, H. (2000). Some relationships between factors and components. *Psychometrika*, **65**, 167–185.

O'Hagan, A. (1984). Motivating principal components, and a stronger optimality result. *Statistician*, **33**, 313–315.

O'Hagan, A. (1994). *Kendall's Advanced Theory of Statistics. Volume 2B Bayesian Inference.* London: Arnold.

Okamoto, M. (1969). Optimality of principal components. In *Multivariate Analysis II*, ed. P. R. Krishnaiah, 673–685. New York: Academic Press.

Oman, S.D. (1978). A Bayesian comparison of some estimators used in linear regression with multicollinear data. *Commun. Statist.*, **A7**, 517–534.

Oman, S.D. (1991). Random calibration with many measurements: An application of Stein estimation. *Technometrics*, **33**, 187–195.

Osmond, C. (1985). Biplot models applied to cancer mortality rates. *Appl. Statist.*, **34**, 63–70.

Ottestad, P. (1975). Component analysis: An alternative system. *Int. Stat. Rev.*, **43**, 83–107.

Overland, J.E. and Preisendorfer, R.W. (1982). A significance test for principal components applied to a cyclone climatology. *Mon. Weather Rev.*, **110**, 1–4.

Pack, P., Jolliffe, I.T. and Morgan, B.J.T. (1988). Influential observations in principal component analysis: A case study. *J. Appl. Statist.*, **15**, 39–52.

Pearce, S.C. and Holland, D.A. (1960). Some applications of multivariate methods in botany. *Appl. Statist.*, **9**, 1–7.

Pearson, K. (1901). On lines and planes of closest fit to systems of points in space. *Phil. Mag. (6)*, **2**, 559–572.

Peña, D. and Box, G.E.P. (1987). Identifying a simplifying structure in time series. *J. Amer. Statist. Assoc.*, **82**, 836–843.

Peña, D. and Yohai, V. (1999). A fast procedure for outlier diagnostics in large regression problems. *J. Amer. Statist. Assoc.*, **94**, 434–445.

Penny, K.I. and Jolliffe, I.T (2001). A comparison of multivariate outlier detection methods for clinical laboratory safety data. *Statistician*, **50**, 295-308.

Pla, L. (1991). Determining stratum boundaries with multivariate real data. *Biometrics*, **47**, 1409–1422.

Plaut, G. and Vautard, R. (1994). Spells of low-frequency oscillations and weather regimes in the Northern Hemisphere. *J. Atmos. Sci.*, **51**, 210–236.

Preisendorfer, R.W. (1981). Principal component analysis and applications. Unpublished lecture notes.*Amer. Met. Soc. Workshop on Principal Component Analysis*, Monterey.

Preisendorfer, R.W. and Mobley, C.D. (1982). Data intercomparison theory, I-V. NOAA Tech. Memoranda ERL PMEL Nos. 38–42.

Preisendorfer, R.W. and Mobley, C.D. (1988). *Principal Component Analysis in Meteorology and Oceanography.* Amsterdam: Elsevier.

Press, S.J. (1972). *Applied Multivariate Analysis.* New York: Holt, Rinehart and Winston.

Press, W.H., Teukolsky, S.A., Vetterling, W.T. and Flannery, B.P. (1992) *Numerical Recipes in C*, 2nd edition. Cambridge: Cambridge University Press.

Priestley, M.B., Subba Rao, T. and Tong, H. (1974). Applications of principal component analysis and factor analysis in the identification of multivariable systems. *IEEE Trans. Autom. Cont.*, **AC-19**, 730–734.

Qian, G., Gabor, G. and Gupta, R.P. (1994). Principal components selection by the criterion of the minimum mean difference of complexity. *J. Multiv. Anal.*, **49**, 55–75.

Radhakrishnan, R. and Kshirsagar, A.M. (1981). Influence functions for certain parameters in multivariate analysis. *Commun. Statist.*, **A10**, 515–529.

Ramsay, J.O. (1996). Principal differential analysis: Data reduction by differential operators. *J. R. Statist. Soc. B*, **58**, 495–508.

Ramsay, J.O. (2000). Functional components of variation in handwriting. *J. Amer. Statist. Assoc.*, **95**, 9–15.

Ramsay, J.O. and Abrahamowicz, M. (1989). Binomial regression with monotone splines: A psychometric application. *J. Amer. Statist. Assoc.*, **84**, 906–915.

Ramsay, J.O. and Silverman, B.W. (1997). *Functional Data Analyis*. New York: Springer.

Ramsier, S.W. (1991). A graphical method for detection of influential observations in principal component analysis. In *Proc. Section on Statistical Graphics, Joint Statistical Meetings, American Statistical Association.*

Ranatunga, C. (1989). *Methods of Removing 'Size' from a Data Set.* Unpublished M.Sc. dissertation. University of Kent at Canterbury.

Rao, C.R. (1955). Estimation and tests of significance in factor analysis. *Psychometrika*, **20**, 93–111.

Rao, C.R. (1958). Some statistical methods for comparison of growth curves. *Biometrics*, **14**, 1–17.

Rao, C.R. (1964). The use and interpretation of principal component analysis in applied research. *Sankhya A*, **26**, 329–358.

Rao, C.R. (1973). *Linear Statistical Inference and Its Applications*, 2nd edition. New York: Wiley.

Rao, C.R. (1987). Prediction of future observations in growth curve models. *Statist. Sci.*, **2**, 434–471 (including discussion).

Rasmusson, E.M., Arkin, P.A., Chen, W.Y. and Jalickee, J.B. (1981). Biennial variations in surface temperature over the United States as revealed by singular decomposition. *Mon. Weather Rev.*, **109**, 587–598.

Ratcliffe, S.J. and Solo, V. (1998). Some issues in functional principal component analysis. In *Section on Statistical Computing, Joint Statistical Meetings, American Statistical Association*, 206–209.

Raveh, A. (1985). On the use of the inverse of the correlation matrix in multivariate data analysis. *Amer. Statistician*, **39**, 39–42.

Reddon, J.R. (1984). *The Number of Principal Components Problem: A Monte Carlo Study.* Unpublished Ph.D. thesis. University of Western Ontario.

Rencher, A.C. (1995). *Methods of Multivariate Analysis*. New York: Wiley.

Rencher, A.C. (1998). *Multivariate Statistical Inference and Applications*. New York: Wiley.

Reyment, R.A. and Jöreskog, K.G. (1993). *Applied Factor Analysis in the Natural Sciences*. Cambridge: Cambridge University Press.

Richman, M.B. (1983). Specification of complex modes of circulation with T-mode factor analysis. *Second International Meeting on Statistical Climatology*, Preprints volume, 5.1.1–5.1.8.

Richman, M.B. (1986). Rotation of principal components. *J. Climatol.*, **6**, 293–335.

Richman, M.B. (1987). Rotation of principal components: A reply. *J. Climatol.*, **7**, 511–520.

Richman, M.B. (1988). A cautionary note concerning a commonly applied eigenanalysis procedure. *Tellus*, **40B**, 50–58.

Richman M.B. (1993). Comments on 'The effect of domain shape on principal component analyses.' *Int. J. Climatol.*, **13**, 203–218.

Richman, M.B. and Gong, X. (1999). Relationships between the definition of the hyperplane width to the fidelity of principal component loading patterns. *J. Climate*, **12**, 1557–1576.

Richman, M.B. and Lamb, P.J. (1987). Pattern analysis of growing season precipitation in Southern Canada. *Atmos.—Ocean*, **25**, 137–158.

Rissanen, J. and Yu, B. (2000). Coding and compression: A happy union of theory and practice. *J. Amer. Statist. Assoc.*, **95**, 986–989.

Robert, P. and Escoufier, Y. (1976). A unifying tool for linear multivariate statistical methods: The RV coefficient. *Appl. Statist.*, **25**, 257–265.

Roes, K.C.B. and Does, R.J.M.M. (1995). Shewhart-type charts in nonstandard situations. *Technometrics*, **37**, 15–24.

Romanazzi, M. (1993). Jackknife estimation of the eigenvalues of the covariance matrix. *Comput. Statist. Data Anal.*, **15**, 179–198.

Romero, R., Ramis, C., Guijarro, J.A. and Sumner, G. (1999). Daily rainfall affinity areas in Mediterranean Spain. *Int. J. Climatol.*, **19**, 557–578.

Rousseeuw, P.J. and Leroy, A.M. *Robust Regression and Outlier Detection.* New York: Wiley.

Roweis, S. (1997). EM algorithms for PCA and SPCA. *Neural Inf. Proc. Syst.*, **10**, 626–632.

Roweis, S.T. and Saul, L.K. (2000). Nonlinear dimensionality reduction by locally linear embedding. *Science*, **290**, 2323–2326.

Rummel, R.J. (1970). *Applied Factor Analysis.* Evanston: Northwestern University Press.

Ruymgaart, F.H. (1981). A robust principal component analysis. *J. Multiv. Anal.*, **11**, 485–497.

Sabatier, R., Lebreton, J.-D. and Chessel, D. (1989). Principal component analysis with instrumental variables as a tool for modelling composition data. In *Multiway Data Analysis*, eds R. Coppi and S. Bolasco, 341–352. Amsterdam: North-Holland.

Salles, M.A., Canziani, P.O. and Compagnucci, R.H. (2001). The spatial and temporal behaviour of the lower stratospheric temperature over the Southern Hemisphere: The MSU view. Part II: Spatial behaviour. *Int. J. Climatol.*, **21**, 439–454.

Sato, M. (1990). Some remarks on principal component analysis as a substitute for factor analysis in monofactor cases. *J. Jap. Statist. Soc.*, **20**, 23–31.

Schafer, J.L. (1997). *Analysis of Incomplete Multivariate Data.* London: Chapman and Hall.

Schneeweiss, H. (1997). Factors and principal components in the near spherical case. *Mult. Behav. Res.*, **32**, 375–401.

Schneeweiss, H. and Mathes, H. (1995). Factor analysis and principal components. *J. Mult. Anal.*, **55**, 105–124.

Schneider, T. (2001). Analysis of incomplete climate data: Estimation of mean values and covariance matrices and imputation of missing values. *J. Climate*, **14**, 853–871.

Schott, J.R. (1987). An improved chi-squared test for a principal component. *Stat. Prob. Lett.*, **5**, 361–365.

Schott, J.R. (1988). Common principal component subspaces in two groups. *Biometrika*, **75**, 229–236.

Schott, J.R. (1991). Some tests for common principal component subspaces in several groups. *Biometrika*, **78**, 771–777.

Schreer, J.F., O'Hara Hines, R.J. and Kovacs, K.M. (1998). Classification of dive profiles: A comparison of statistical clustering techniques and unsupervised artificial neural networks. *J. Agri. Biol. Environ. Statist.*, **3**, 383–404.

Sclove, S.L. (1968). Improved estimators for coefficients in linear regression. *J. Amer. Statist. Assoc.*, **63**, 596–606.

Sengupta, S. and Boyle, J.S. (1998). Using common principal components for comparing GCM simulations. *J. Climate*, **11**, 816–830.

Shafii, B. and Price, W.J. (1998). Analysis of genotype-by-environment interaction using the additive main effects and multiplicative interaction model and stability estimates. *J. Agri. Biol. Environ. Statist.*, **3**, 335–345.

Shi, L. (1997). Local influence in principal components analysis. *Biometrika*, **84**, 175–186.

Shibayama, T. (1990). A linear composite method for test scores with missing values. Unpublished technical report. Department of Psychology, McGill University.

Sibson, R. (1984). Multivariate analysis. Present position and potential developments: Some personal views. *J. R. Statist. Soc. A*, **147**, 198–207.

Skinner, C.J., Holmes, D.J. and Smith, T.M.F. (1986). The effect of sample design on principal component analysis. *J. Amer. Statist. Assoc.*, **81**, 789–798.

Smith, B.T., Boyle, J.M., Dongarra, J.J., Garbow, B.S., Ikebe, Y., Klema, V.C., and Moler, C.B. (1976). *Matrix Eigensystem Routines—EISPACK guide*, 2nd edition. Berlin: Springer-Verlag.

Smyth, G.K. (2000). Employing symmetry constraints for improved frequency estimation by eigenanalysis methods. *Technometrics*, **42**, 277–289.

Snook, S.C. and Gorsuch, R.L. (1989). Component analysis versus common factor analysis: A Monte Carlo study. *Psychol. Bull.*, **106**, 148–154.

Solow, A.R. (1994). Detecting change in the composition of a multispecies community. *Biometrics*, **50**, 556–565.

Somers, K.M. (1986). Multivariate allometry and removal of size with principal components analysis. *Syst. Zool.*, **35**, 359–368.

Somers, K.M. (1989). Allometry, isometry and shape in principal component analysis. *Syst. Zool.*, **38**, 169–173.

Soofi, E.S. (1988). Principal component regression under exchangeability. *Commun. Statist.—Theor. Meth.*, **17**, 1717–1733.

Sprent, P. (1972). The mathematics of size and shape. *Biometrics*, **28**, 23–37.

Spurrell, D.J. (1963). Some metallurgical applications of principal components. *Appl. Statist.*, **12**, 180–188.

Srivastava, M.S. and Khatri, C.G. (1979). *An Introduction to Multivariate Statistics*. New York: North-Holland.

Stauffer, D.F., Garton, E.O. and Steinhorst, R.K. (1985). A comparison of principal components from real and random data. *Ecology*, **66**, 1693–1698.

Stein, C.M. (1960). Multiple regression. In *Contributions to Probability and Statistics. Essays in Honour of Harold Hotelling*, ed. I. Olkin, 424–443. Stanford: Stanford University Press.

Stewart, D. and Love, W. (1968). A general canonical correlation index. *Psychol. Bull.*, **70**, 160–163.

Stoffer, D.S. (1999). Detecting common signals in multiple time series using the spectral envelope. *J. Amer. Statist. Assoc.*, **94**, 1341–1356.

Stone, E.A. (1984). *Cluster Analysis of English Counties According to Socio-economic Factors*. Unpublished undergraduate dissertation. University of Kent at Canterbury.

Stone, J.V. and Porrill, J. (2001). Independent component analysis and projection pursuit: A tutorial introduction. Unpublished technical report. Psychology Department, University of Sheffield.

Stone, M. and Brooks, R.J. (1990). Continuum regression: Cross-validated sequentially constructed prediction embracing ordinary least squares, partial least squares and principal components regression (with discussion). *J. R. Statist. Soc. B*, **52**, 237–269.

Stone, R. (1947). On the interdependence of blocks of transactions (with discussion). *J. R. Statist. Soc. B*, **9**, 1–45.

Storvik, G. (1993). Data reduction by separation of signal and noise components for multivariate spatial images. *J. Appl. Statist.*, **20**, 127–136.

Stuart, M. (1982). A geometric approach to principal components analysis. *Amer. Statistician*, **36**, 365–367.

Sugiyama, T. and Tong, H. (1976). On a statistic useful in dimensionality reduction in multivariable linear stochastic system. *Commun. Statist.*, **A5**, 711–721.

Sullivan, J.H., Woodall, W.H. and Gardner, M.M. (1995) Discussion of 'Shewhart-type charts in nonstandard situations,' by K.C.B. Roes and R.J.M.M. Does. *Technometrics*, **37**, 31–35.

Sundberg, P. (1989). Shape and size-constrained principal components analysis. *Syst. Zool.*, **38**, 166–168.

Sylvestre, E.A., Lawton, W.H. and Maggio, M.S. (1974). Curve resolution using a postulated chemical reaction. *Technometrics*, **16**, 353–368.

Takane, Y., Kiers, H.A.L. and de Leeuw, J. (1995). Component analysis with different sets of constraints on different dimensions. *Psychometrika*, **60**, 259–280.

Takane, Y. and Shibayama, T. (1991). Principal component analysis with external information on both subjects and variables. *Psychometrika*, **56**, 97–120.

Takemura, A. (1985). A principal decomposition of Hotelling's T^2 statistic. In *Multivariate Analysis VI*, ed. P.R. Krishnaiah, 583–597. Amsterdam: Elsevier.

Tan, S. and Mavrovouniotis, M.L. (1995). Reducing data dimensionality through optimizing neural network inputs. *AIChE J.*, **41**, 1471–1480.

Tanaka, Y. (1983). Some criteria for variable selection in factor analysis. *Behaviormetrika*, **13**, 31–45.

Tanaka, Y. (1988). Sensitivity analysis in principal component analysis: Influence on the subspace spanned by principal components. *Commun. Statist.—Theor. Meth.*, **17**, 3157–3175.

Tanaka, Y. (1995). A general strategy of sensitivity analysis in multivariate methods. In *Data Science and Its Application*, eds. Y. Escoufier, B. Fichet, E. Diday, L. Lebart, C. Hayashi, N. Ohsumi and Y. Baba, 117–131. Tokyo: Academic Press.

Tanaka, Y. and Mori, Y. (1997). Principal component analysis based on a subset of variables: Variable selection and sensitivity analysis. *Amer. J. Math. Manag. Sci.*, **17**, 61–89.

Tanaka, Y. and Tarumi, T. (1985). Computational aspect of sensitivity analysis in multivariate methods. Technical report No. 12. Okayama Statisticians Group. Okayama, Japan.

Tanaka, Y. and Tarumi, T. (1986). Sensitivity analysis in multivariate methods and its application. *Proc. Second Catalan Int. Symposium on Statistics*, 335–338.

Tanaka, Y. and Tarumi, T. (1987). A numerical investigation of sensitivity analysis in multivariate methods. *Fifth International Symposium: Data Analysis and Informatics, Tome 1*, 237–247.

Tarpey, T. (1999). Self-consistency and principal component analysis. *J. Amer. Statist. Assoc.*, **94**, 456–467.

Tarpey, T. (2000). Parallel principal axes. *J. Mult. Anal.*, **75**, 295–313.

ten Berge, J.M.F. and Kiers, H.A.L. (1996). Optimality criteria for principal component analysis and generalizations. *Brit. J. Math. Statist. Psychol.*, **49**, 335–345.

ten Berge, J.M.F. and Kiers, H.A.L. (1997). Are all varieties of PCA the same? A reply to Cadima and Jolliffe. *Brit. J. Math. Statist. Psychol.*, **50**, 367–368.

ten Berge, J.M.F. and Kiers, H.A.L. (1999). Retrieving the correlation matrix from a truncated PCA solution: The inverse principal component problem. *Psychometrika*, **64**, 317–324.

Tenenbaum, J.B., de Silva, V. and Langford, J.C. (2000). A global geometric framework for nonlinear dimensionality reduction. *Science*, **290**, 2319–2323.

ter Braak, C.J.F. (1983). Principal components biplots and alpha and beta diversity. *Ecology*, **64**, 454–462.

ter Braak, C.J.F. and Looman, C.W.N. (1994). Biplots in reduced rank regression. *Biom. J.*, **36**, 983–1003.

Timmerman, M.E. and Kiers, H.A.L. (2000). Three-mode principal components analysis: Choosing the numbers of components and sensitivity to local optima. *Brit. J. Math. Stat. Psychol.*, **53**, 1–16.

Thacker, W.C. (1996). Metric-based principal components: Data uncertainties. *Tellus*, **48A**, 584–592.

Thacker, W.C. (1999). Principal predictors. *Int. J. Climatol.*, **19**, 821–834.

Thurstone, L.L. (1931). Multiple factor analysis. *Psychol. Rev.*, **38**, 406–427.

Tibshirani, R. (1996). Regression shrinkage and selection via the lasso. *J. R. Statist. Soc. B*, **58**, 267–288.

Tipping, M.E. and Bishop, C.M. (1999a). Probabilistic principal component analysis. *J. R. Statist. Soc. B*, **61**, 611–622.

Tipping, M.E. and Bishop, C.M. (1999b). Mixtures of probabilistic principal component analyzers. *Neural Computat.*, **11**, 443–482.

Titterington, D.M., Smith, A.F.M. and Makov, U.E. (1985). *Statistical Analysis of Finite Mixture Distributions*. New York: Wiley.

Townshend, J.R.G. (1984). Agricultural land-cover discrimination using thematic mapper spectral bands. *Int. J. Remote Sensing*, **5**, 681–698.

Torgerson, W.S. (1958). *Theory and Methods of Scaling*. New York: Wiley.

Tortora, R.D. (1980). The effect of a disproportionate stratified design on principal component analysis used for variable elimination. *Proceedings of the Amer. Statist. Assoc. Section on Survey Research Methods*, 746–750.

Treasure, F.P. (1986). The geometry of principal components. Unpublished essay. University of Cambridge.

Trenkler, D. and Trenkler, G. (1984). On the Euclidean distance between biased estimators. *Commun. Statist.—Theor. Meth.*, **13**, 273–284.

Trenkler, G. (1980). Generalized mean squared error comparisons of biased regression estimators. *Commun. Statist.*, **A9**, 1247–1259.

Tryon, R.C. (1939). *Cluster Analysis*. Ann Arbor: Edwards Brothers.

Tucker, L.R. (1958). An inter-battery method of factor analysis. *Psychometrika*, **23**, 111–136.

Tucker, L.R. (1966). Some mathematical notes on three-mode factor analysis. *Psychometrika*, **31**, 279–311.

Tukey, P.A. and Tukey, J.W. (1981). Graphical display of data sets in three or more dimensions. Three papers in *Interpreting Multivariate Data* (ed. V. Barnett), 189–275. Chichester: Wiley.

Turner, N.E. (1998). The effect of common variance and structure pattern on random data eigenvalues: implications for the accuracy of parallel analysis. *Educ. Psychol. Meas.*, **58**, 541–568.

Uddin, M. (1999). *Interpretation of Results from Simplified Principal Components*. Unpublished Ph.D. thesis. University of Aberdeen.

Underhill, L.G. (1990). The coefficient of variation biplot. *J. Classific.*, **7**, 241–256.

van de Geer, J.P. (1984). Linear relations among *k* sets of variables. *Psychometrika*, **49**, 79–94.

van de Geer, J.P. (1986). *Introduction to Linear Multivariate Data Analysis—Volume 2*. Leiden: DSWO Press.

van den Brink, P.J. and ter Braak, C.J.F. (1999). Principal response curves: Analysis of time-dependent multivariate responses of biological community to stress. *Environ. Toxicol. Chem.*, **18**, 138–148.

van den Dool, H.M., Saha, S. and Johansson, Å. (2000) Empirical orthogonal teleconnections. *J. Climate* **13**, 1421-1435.

van den Wollenberg, A.L. (1977). Redundancy analysis. An alternative for canonical correlation analysis. *Psychometrika*, **42**, 207–219.

van Rijckevorsel, J.L.A. (1988). Fuzzy coding and B-splines. In *Component and Correspondence Analysis. Dimension Reduction by Functional Approximation*, eds. J.L.A. van Rijckevorsel and J. de Leeuw, 33–54. Chichester: Wiley.

Vargas-Guzmán, J.A., Warrick, A.W. and Myers, D.E. (1999). Scale effect on principal component analysis for vector random functions. *Math. Geol.*, **31**, 701–722.

Vautard, R. (1995). Patterns in time: SSA and MSSA. In *Analysis of Climate Variability: Applications of Statistical Techniques*, eds. H. von Storch and A. Navarra, 259–279. Berlin: Springer.

Velicer, W.F. (1976). Determining the number of components from the matrix of partial correlations. *Psychometrika*, **41**, 321–327.

Velicer, W.F. and Jackson, D.N. (1990). Component analysis versus common factor analysis—some issues in selecting an appropriate procedure. *Mult. Behav. Res.*, **25**, 1–28.

Verboon, P. (1993). Stability of resistant principal component analysis for qualitative data. In *New Directions in Statistical Data Analysis and Robustness*, eds. S. Morgenthaler, E. Ronchetti and W. A. Stahel, 265–273. Basel: Birkhäuser.

Vermeiren, D., Tavella, D. and Horovitz, A. (2001). Extending principal component analysis to identify portfolio risk contributors. Submitted for publication.

Vigneau, E. and Qannari, E.M. (2001). Clustering of variables around latent components. Submitted for publication.

Vines, S.K. (2000). Simple principal components. *Appl. Statist.*, **49**, 441–451.

Vong, R., Geladi, P., Wold, S. and Esbensen, K. (1988). Some contributions to ambient aerosol calculated by discriminant partial least squares (PLS). *J. Chemometrics*, **2**, 281–296.

von Storch, H., Bruns, T., Fischer-Bruns, I. and Hasselmann, K. (1988). Principal oscillation pattern analysis of the 30- to 60-day oscillation in general circulation model equatorial troposphere. *J. Geophys. Res.*, **93**, 11,022–11,036.

von Storch, H. and Zweirs, F.W. (1999). *Statistical Analysis in Climate Research*. Cambridge: Cambridge University Press.

Wackernagel, H. (1995). *Multivariate Geostatistics. An Introduction with Applications*. Berlin: Springer.

Walker, M.A, (1967). Some critical comments on 'An analysis of crimes by the method of principal components,' by B. Ahamad. *Appl. Statist.*, **16**, 36–38.

Wallace, J.M., Smith, C. and Bretherton, C.S. (1992). Singular value decomposition of wintertime sea surface temperature and 500-mb height anomalies. *J. Climate*, **5**, 561–576.

Wallace, T.D. (1972). Weaker criteria and tests for linear restrictions in regression. *Econometrica*, **40**, 689–698.

Walton, J.J. and Hardy, D.M. (1978). Principal components analysis and its application to wind field pattern recognition. Lawrence Livermore Laboratory Technical Report UCRL-52488.

Wang, P.C.C. (1978). *Graphical Representation of Multivariate Data*. New York: Academic Press.

Wang, S.-G. and Liski, E.P. (1993). Effects of observations on the eigensystem of a sample covariance structure. *J. Stat. Plan. Inf.*, **36**, 215–226.

Wang, S.-G. and Nyquist, H. (1991). Effects on the eigenstructure of a data matrix when deleting an observation. *Computat. Statist. Data Anal.*, **11**, 179–188.

Wang, X.L. and Zwiers, F.W. (2001). Using redundancy analysis to improve dynamical seasonal mean 500 hPa geopotential forecasts. *Int. J. Climatol.*, **21**, 637–654.

Wang, Y.M. and Staib, L.H. (2000). Boundary finding with prior shape and smoothness models. *IEEE Trans. Patt. Recog. Machine Intell.*, **22**, 738–743.

Waternaux, C.M. (1984). Principal components in the nonnormal case: The test of equality of Q roots. *J. Multiv. Anal.*, **14**, 323–335.

Weare, B.C. (1990). Four-dimensional empirical orthogonal analysis of climate variables. *Int. J. Climatol.*, **10**, 313–319.

Weare, B.C. and Nasstrom, J.S. (1982). Examples of extended empirical orthogonal function analyses. *Mon. Weather Rev.*, **110**, 481–485.

Webber, R. and Craig, J. (1978). Socio-economic classification of local authority areas. *OPCS Studies on Medical and Population Subjects*, No. 35. London: HMSO.

Webster, J.T., Gunst, R.F. and Mason, R.L. (1974). Latent root regression analysis. *Technometrics*, **16**, 513–522.

White, D., Richman, M. and Yarnal, B. (1991). Climate regionalization and rotation of principal components. *Int. J. Climatol.*, **11**, 1–25.

White, J.W. and Gunst, R.F. (1979). Latent root regression: Large sample analysis. *Technometrics*, **21**, 481–488.

Whittaker, J. (1990). *Graphical Models in Applied Multivariate Analysis.* Chichester: Wiley.

Whittle, P. (1952). On principal components and least squares methods of factor analysis. *Skand. Actuar.*, **35**, 223–239.

Wiberg, T. (1976). Computation of principal components when data are missing. In *Compstat 1976*, eds. J. Gordesch and P. Naeve, 229–236. Wien: Physica-Verlag.

Widaman, K.F. (1993). Common factor analysis versus principal component analysis: Differential bias in representing model parameters. *Mult. Behav. Res.*, **28**, 263–311.

Wigley, T.M.L., Lough, J.M. and Jones, P.D. (1984). Spatial patterns of precipitation in England and Wales and a revised, homogeneous England and Wales precipitation series. *J. Climatol.*, **4**, 1–25.

Wikle, C.K. and Cressie, N. (1999). A dimension-reduced approach to space-time Kalman filtering. *Biometrika*, **86**, 815–829.

Wilkinson, J.H. (1965). *The Algebraic Eigenvalue Problem.* Oxford: Oxford University Press.

Wilkinson, J.H. and Reinsch, C. (1971). *Handbook for Automatic Computation, Vol. 11, Linear Algebra.* Berlin: Springer-Verlag.

Winsberg, S. (1988). Two techniques: Monotone spline transformations for dimension reduction in PCA and easy-to generate metrics for PCA of sampled functions. In *Component and Correspondence Analysis. Dimension Reduction by Functional Approximation*, eds. J.L.A. van Rijckevorsel and J. de Leeuw, 115–135. Chichester: Wiley.

Witten, I.H. and Frank, E. (2000). *Data Mining. Practical Machine Learning Tools and Techniques with Java Implementations.* San Francisco: Morgan Kaufmann.

Wold, H. (1984). Partial least squares. In *Encyclopedia of Statistical Science, Vol 6*, eds. N. L. Johnson and S. Kotz, 581–591. New York: Wiley.

Wold, S. (1976). Pattern recognition by means of disjoint principal components models. *Patt. Recog.*, **8**, 127–139.

Wold, S. (1978). Cross-validatory estimation of the number of components in factor and principal components models. *Technometrics*, **20**, 397–405.

Wold, S. (1994). Exponentially weighted moving principal components analysis and projections to latent structures. *Chemometrics Intell. Lab. Syst.*, **23**, 149–161.

Wold, S., Albano, C., Dunn, W.J., Esbensen, K., Hellberg, S., Johansson, E. and Sjöström, M. (1983). Pattern recognition: Finding and using regularities in multivariate data. In *Food Research and Data Analysis*, eds. H. Martens and H. Russwurm, 147–188. London: Applied Science Publishers.

Worton, B.J. (1984). *Statistical Aspects of Data Collected in Year 1974–1975 of the Irish Wetlands Enquiry.* Unpublished M.Sc. dissertation. University of Kent at Canterbury.

Wu, D.-H., Anderson, D.L.T. and Davey, M.K. (1994). ENSO prediction experiments using a simple ocean-atmosphere model. *Tellus, A. Dynam. Meteorol. Oceanog.*, **46**, 465–480.

Xie, Y.-L., Wang, J.-H., Liang, Y.-Z., Sun, L.-X., Song, X-H. and Yu, R-Q. (1993). Robust principal component analysis by projection pursuit. *J. Chemometrics*, **7**, 527–541.

Xu, L. and Yuille, A. (1995). Robust principal component analysis by self-organizing rules based on statistical physics approach. *IEEE Trans. Neural Networks*, **6**, 131–143.

Yanai, H. (1980). A proposition of generalized method for forward selection of variables. *Behaviormetrika*, **7**, 95–107.

Yendle, P.W. and MacFie, H.J.H. (1989). Discriminant principal components analysis. *J. Chemometrics*, **3**, 589–600.

Young, G. (1941). Maximum likelihood estimation and factor analysis. *Psychometrika*, **6**, 49–53.

Yuan, K.-H. and Bentler, P.M. (1994). Test of linear trend in eigenvalues of K covariance matrices with applications in common principal components analysis. *Commun. Stat. — Theor. Meth.*, **23**, 3141–3156.

Yule, W., Berger, M., Butler, S., Newham, V. and Tizard, J. (1969). The WPPSL. An empirical evaluation with a British sample. *Brit. J. Educ. Psychol.*, **39**, 1–13.

Zheng, X., Frederiksen, C.S. and Basher, R.E. (2002). EOFs of the potentially predictable and weather noise components of seasonal mean fields. Part 1 Methodology. Submitted for publication.

Zwick, W.R. and Velicer, W.F. (1986). Comparison of five rules for determining the number of components to retain. *Psychol. Bull.*, **99**, 432–446.

Zwiers, F.W. (1999). The detection of climate change. In *Anthropogenic Climate Change*, eds. H. von Storch and G. Flöser, 161–206. Berlin: Springer.

Index

Author Index

Springer Series in Statistics (continued from p. ii)

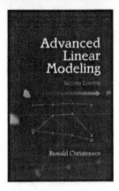

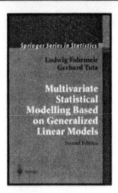

228011LV00003B/14/P

9 780387 954424